On-Site Bioreclamation

Processes for Xenobiotic and Hydrocarbon Treatment

Edited by

Robert E. Hinchee

and

Robert F. Olfenbuttel

Battelle Memorial Institute
Columbus, Ohio

BUTTERWORTH–HEINEMANN
Boston London Oxford Singapore Sydney Toronto Wellington

LC 91-073684

ISBN 0-7506-9302-9

Butterworth-Heinemann
80 Montvale Avenue
Stoneham, Massachusetts 02180

10 9 8 7 6 5 4 3 2 1

Printed in the United States of America

Contents

Technical Notes

In Situ *Bioreclamation* A Companion Volume

Contents

Articles

Technical Notes

Foreword

This book is one of two that resulted from the international symposium *"In Situ* and On-Site Bioreclamation" held in San Diego in March of 1991. The two volumes, In Situ *Bioreclamation: Applications and Investigations for Hydrocarbon and Contaminated Site Remediation* and *On-Site Bioreclamation: Processes for Xenobiotic and Hydrocarbon Treatment,* consist of selected papers submitted by symposium participants. The symposium brought together more than 700 people from 20 countries in Europe, North America, Asia, and Australia. It was apparent that the interest in the biological treatment of contaminated soil, water, and gas is tremendous. It was also obvious that the state of the art varies from proven commercially available technologies for some applications, such as reactor treatment of petroleum hydrocarbons in aqueous steams, to fundamental research in others, such as genetic engineering to develop new strains of biodegradable xenobiotics. Together, these volumes represent the most complete and up-to-date set of papers published on the topic.

In response to an invitation to submit papers, we received 100 papers. The papers were sent to two peer reviewers and, following revision, 82 were accepted for publication. Two different formats are used, articles and technical notes. Articles represent a substantial technical contribution; technical notes are brief technology descriptions or reports of preliminary or less substantial studies.

Appreciation is extended to the more than 100 people who served as technical reviewers. This relatively thankless task is essential to assure quality in publishing. The peer reviewers are listed at the end of this Foreword. Additionally, Lewis Semprini of Stanford University played an invaluable role in inviting and reviewing many of the papers on chlorinated solvent degradation; as a result of his efforts this subject is very well covered.

Numerous organizations and individuals contributed to organizing and supporting both the symposium and these resulting publications. Battelle Memorial Institute underwrote the symposium and subsequent publications and assumed all of the financial risks of this effort. The following is a list of people and organizations who either served as cosponsors or organized technical sessions contributing to this effort:

Mick Arthur, Battelle Columbus
Bruce Bauman, American Petroleum Institute
Christian Bocard, Institut Francais du Pétrole, France
Rob Booth, Environment Canada
D. B. Chan, U.S. Naval Civil Engineering Laboratory
Soon Cho, Ajou University, Korea
Giovanni Ferro, D'Appolonia, Italy
Volker Franzius, Umweltbundesamt, Germany
Giancarlo Gabetto, Castalia, Italy
Ronald Hoeppel, U.S. Naval Civil Engineering Laboratory
Hidemi Iguchi, Mitsubishi Corporation, Japan
Massimo Martinelli, ENEA, Italy
Jeff Means, Battelle Columbus
Kun Mo Lee, Ajou University, Korea
Ross N. Miller, U.S. Air Force HSD/YAQE
Donna Palmer, Battelle Columbus
Gloria Patton, U.S. Department of Energy
Chongrak Polprasert, Asian Institute of Technology, Thailand
Agusto Porta, Battelle Geneve, Switzerland
Parmely "Hap" Prichard, U.S. Environmental Protection Agency
Emilio Santucci, Castalia, Italy
Paul Schaltzberg, David Taylor Research Center
Lewis Semprini, Stanford University
John Skinner, U.S. Environmental Protection Agency
Terri Stewart, Battelle Pacific Northwest Laboratories
Klass Visscher, RIVM, The Netherlands
John Wilson, U.S. Environmental Protection Agency

Cher Paul at Battelle served diligently in the monumental task of organizing and assembling both volumes. She read and edited every paper both for consistency of style and for readability. She also supervised the logistics of actually assembling the books. Others who contributed to putting the volumes together were Karl Nehring, Tracy Melchiori, Shauna Kearney, Gina Melaragno, Jennifer Eddy, Janelle McClary, Sharon Manwering, Susan Snyder, Darlene Whyte, and Lisa Becker of Battelle and Tom Bigelow of Kenyon College. Mr. Nehring provided administrative support and handled communications with the publisher. Ms. Melchiori, Ms. Kearney, and Ms. Melaragno coordinated the communications with the authors and peer reviewers. Mr. Bigelow, an editor, supported Ms. Paul. Ms. Eddy and Ms. McClary coordinated art and layout, Ms. Manwering was the text processor, and Ms. Snyder, Ms. Whyte, and Ms. Becker proofread the work.

In the rapidly evolving field of bioremediation, much of the work presented in these volumes will become dated in a few years. A consensus of participants in the 1991 symposium called for a second symposium, which will be held in 1993. The 1993 symposium will most probably also result in publication; at this time its format is undecided, but that it will be needed is undoubtable.

Rob Hinchee
May 1993

Peer Reviewers

Dan Acton, Beak Consultants Limited, Canada
Pradeep Aggarwal, Argonne National Laboratory
Robert C. Ahlert, Rutgers University
Pedro J. J. Alvarez, University of Michigan
Mick Arthur, Battelle Columbus
Erik Arvin, Technical University of Denmark

Dr. Ballerini, Institut Français du Pétrole, France
Chris Barber, CSIRO, Australia
E. R. Barenschee, Degussa Corporation, Germany
James F. Barker, University of Waterloo, Canada
Denise M. Barnes, Ecosystems Engineering
Thomas C. Beard, Battelle Columbus Division
Philip B. Bedient, Rice University
Sanjoy K. Bhattacharya, Tulane University
Christian Bocard, Institut Français du Pétrole, France
Gary Boettcher, Geraghty & Miller, Inc.
Robert C. Borden, North Carolina State University
Edward J. Bouwer, Johns Hopkins University
Hans Breukelman, DSM Research B.V., The Netherlands
Andrew Briefer, Western Geologic Resources, Inc.
Fred Brockman, Battelle Pacific Northwest Laboratory
Thomas M. Brouns, Battelle Pacific Northwest Laboratory
Richard A. Brown, Groundwater Technology, Inc.
Gunter Brox, EIMCO Process Equipment
Gaylen R. Brubaker, Remediation Technologies, Inc.
David Burris, U.S. Air Force
Timothy E. Buscheck, Chevron Corporation

Jason A. Caplan, ESE Biosciences
Peter J. Chapman, U.S. Environmental Protection Agency
Thomas H. Christensen, Technical University of Denmark
John A. Cioffi, Ecova Corporation

Gary L. Conaway, Ebasco Services, Inc.
Robert Coutant, Battelle Columbus
Craig Criddle, Michigan State University

Greg B. Davis, CSIRO, Australia
Lois Davis, Sybron Chemicals, Inc.
Wendy J. Davis-Hoover, U.S. Environmental Protection Agency
W. P. de Bruin, Agricultural University, Wageningen, The Netherlands
Barbara A. Denovan, University of Washington
Jeffrey C. Dey, Groundwater Technology Inc.
Ludo Diels, SCK-CEN/VI TO, Belgium
Philip M. Digrazia, University of Tennessee
Ellen L. Dorwin, Computer Sciences Corporation
William Doucette, Utah State University
Rob Douglass, The Traverse Group, Inc.
Douglas C. Downey, Engineering Science, Inc.
Wayne C. Downs, U.S. Environmental Protection Agency
Jean Ducreux, Institut Français du Pétrole, France
R. Ryan Dupont, Utah State University

Elizabeth A. Edwards, Standford University
Richard Egg, Texas A&M University
Mary Pat Eisman, U.S. Naval Civil Engineering Laboratory
W. Eng, University of Tennessee

Margaret Findlay, ABB Environmental Services
Paul E. Flathman, OHM Corporation
John Flyvbjerg, Technical University of Denmark
Samuel Fogel, ABB Environmental Services
Louis B. Fournier, Groundwater Technology, Inc.
William T. Frankenberger, Jr., University of California, Riverside
James K. Fredrickson, Battelle Pacific Northwest Laboratory
David L. Freedman, University of Illinois

Dr. Gatellier, Institut Français du Pétrole, France
W. Kennedy Gauger, Institute of Gas Technology
Richard M. Gersberg, San Diego State University
James M. Gossett, Cornell University
Chris Griffen, The Traverse Group, Inc.
Mirat Gurol, Drexel University

Cathleen J. Hapeman-Somich, U.S. Department of Agriculture
Joop Harmsen, The Winand Staring Centre, The Netherlands
Kirk Hatfield, University of Florida
Edward Heyse, U.S. Air Force
Duane D. Hicks, Texas A&M University
Ronald Hoeppel, U.S. Naval Civil Engineering Laboratory
Scott G. Huling, U.S. Environmental Protection Agency

Charles Imel, Ecosystems Engineering
J. Mark Inglis, Western Geologic Resources, Inc.

Bjørn K. Jensen, Technical University of Denmark
Rex Johnson, The Traverse Group, Inc.
Mark Jones, Delta Environmental Consultants, Inc.

Paul Kemp, Woodward-Clyde Consultants
Khalique A. Khan, Ensotech Inc.
J. M. Henry King, University of Tennessee
T. Kent Kirk, U.S. Department of Agriculture

Calvin A. Kodres, U.S. Naval Civil Engineering Laboratory
Sydney S. Koegler, Battelle Pacific Northwest Laboratory
William M. Korreck, The Traverse Group, Inc.
Peter Kroopniek, Groundwater Technology, Inc.
Curtis A. Kruger, Midwest Research Institute

R. Lamar, U.S. Department of Agriculture
Robert LaPoe, U.S. Air Force
Jim League, Bogart Environmental
Kun Mo Lee, Ajou University, South Korea
Michael D. Lee, Du Pont Environmental Remediation Services
Richard F. Lee, Skidaway Institute of Oceanography
Alfred P. Leuschner, ReTec
M. Tony Lieberman, ESE Biosciences
Carol D. Litchfield, Environment America
C. Deane Little, Stanford University
John Lyngkilde, Technical University of Denmark

David Major, Beak Consultants Limited, Canada
Donn L. Marrin, InterPhase Environmental, Inc.
Michael M. Martinson, Delta Environmental Consultants, Inc.
Perry L. McCarty, Stanford University
Gloria McCleary, EA Engineering, Science and Technology
Linda McConnell, Midwest Research Institute
Mike McFarland, Utah State University
Stephen P. McGrath, AFRC Institute of Arable Crops Research, United Kingdom
Jeffrey Means, Battelle Columbus
Blaine Metting, Battelle Pacific Northwest Laboratory
Barbara J. Mickelson, Delta Environmental Consultants, Inc.
Ross N. Miller, U.S. Air Force
Ali Mohagheghi, Solar Energy Research Institute

Tom Naymik, Battelle Columbus Division
Michael J. K. Nelson, Ecova Corporation
Bruce Nielsen, U.S. Air Force
Per H. Nielsen, Technical University of Denmark
Victor K. Nowicki, WMS Associates Ltd., Canada

Joseph E. Odencrantz, Lavine-Fricke Consulting Engineers
Richard A. Ogle, Applied Earth Sciences
Say Kee Ong, Battelle Columbus
Rick L. Ornstein, Battelle Pacific Northwest Laboratory
David Ostendorf, University of Massachusetts

Donna Palmer, Battelle Columbus
Anthony V. Palumbo, Oak Ridge National Laboratory
Gene F. Parkin, University of Iowa
Christopher J. Perry, Battelle Columbus
Chuck Pettigrew, U.S. Air Force
George Philippidis, Solar Energy Research Institute
Michael Piotrowski, Woodward-Clyde Consultants
Parmely "Hap" Pritchard, U.S. Environmental Protection Agency
Barb Prosen, The Traverse Group, Inc.

John F. Quensen, III, Michigan State University

Roger D. Reeves, Massey University, New Zealand
Martin Reinhard, Stanford University
H. James Reisinger, II, Integrated Environmental Technologies, Inc.
H. S. Rifai, Rice University
Christopher J. Rivard, Solar Energy Research Institute
George Robinson, Rice University

Eric K. Schmitt, ESE Biosciences Group
Gosse Schraa, Agricultural University, Wageningen, The Netherlands
Doug Selby, Las Vegas Valley Water District
Lewis Semprini, Stanford University
Patrick Sferra, U.S. Environmental Protection Agency
Daniel R. Shelton, U.S. Department of Agriculture
Ronald Sims, Utah State University
Rodney S. Skeen, Battelle Pacific Northwest Laboratory
George J. Skladany, Envirogen
Lloyd Slezak, Dayton & Knight Ltd., Canada
Larry Smith, Battelle Columbus
Jim Spain, U.S. Air Force
Sjef Staps, Grontmij N.V. Consulting Engineers
Jan Stepek, EA Engineering
Gerald W. Strandberg, Oak Ridge National Laboratory

William C. Tacon, Battelle Columbus
James M. Tiedje, Michigan State University
Francis T. Tran, Asian Institute of Technology, Thailand
R. D. Tyagi, INRS-EAU, Canada

A. J. Valsangkar, WMS Associates Ltd., Canada
Jean-Paul Vandecastelle, Institut Français du Pétrole, France
J. van Eyk, Delft Geotechnics, The Netherlands
Al Venosa, U.S. Environmental Protection Agency
Catherine Vogel, U.S. Air Force
Timothy Vogel, University of Michigan
Evangelos A. Voudrias, Georgia Institute of Technology

Martin Werner, Washington State Department of Ecology
Godage B. Wickramanayake, Environ Corporation
Barbara H. Wilson, Rice University
Nancy Winters, Washington State Department of Ecology
Will Wright, III, California Department of Health Services
Lin Wu, University of California, Davis
Robert E. Wyza, Battelle Columbus

Lily Young, New York University Medical Center

U.S. EPA Bioremediation Research Programs

John H. Skinner[*], *Gregory G. Ondich, Thomas L. Baugh*
U.S. Environmental Protection Agency

BACKGROUND

Bioremediation is one of the most promising technologies on the horizon for more cost-effective cleanup of hazardous wastes. The basic concepts are not new. Bioremedial processes have been recognized and used for many decades. Most notable are the processes involved in wastewater and sewage treatment. Microbiologists have been familiar with biodegradation (organic breakdown) and bioconversion (action on inorganics) for nearly a century. The microbial actions involved in these processes have also been used extensively in the food processing industry—in the production of items such as beer, wine, yogurt, cheese, and sauerkraut—as well as in chemical science.

In spite of this awareness, the transfer of biological treatment technology to the environmental sector for use in soil and groundwater cleanup has been tediously slow. In the past, use of innovative remedial techniques was limited. Conventional cleanup methods, such as landfill disposal and incineration, were generally used to manage hazardous wastes. However, the increasing costs associated with traditional technologies, closure of landfills, and more stringent restrictions pertaining to air pollution provided the impetus for developing and utilizing new forms of remedial

[*] Office of Research and Development, U.S. Environmental Protection Agency RD-672, 401 M Street SW, Washington, DC 20460

methods. It was at this point that bioremediation began to receive the serious attention of the environmental community.

Over the past several years, the EPA has tested new applications of bioremediation, using it successfully in emergency responses to oil spills and cleaning up Superfund sites. Natural biodegradation of wastes occurs in most environments, but the natural rate is sometimes so slow that it is not considered an effective cleanup mechanism. Fortunately, there is much we can do to enhance the process. For example, we can enhance the efficiency of the process by adding nutrients to encourage proliferation of microorganisms or by adding cultured or acclimated microorganisms to the waste materials.

Although bioremediation has many potential advantages in treating hazardous wastes, our experience with these technologies to date has been limited. Questions remain among environmental professionals about where and how bioremediation fits into the remedial action decision tree. What factors should be considered when evaluating the appropriateness of bioremediation for a given site? How much will it cost, and how long will it take? What are the required steps in developing and implementing an effective bioremediation cleanup program? The lack of reliable information on bioremediation applications and performance is a major impediment to its use. Research is needed to improve our understanding of bioremediation and how it works. Field demonstrations are also needed to establish reliable performance data and a track record of bioremediation successes. Finally, this information must be widely disseminated to motivate site managers to consider bioremediation along with other, more conventional technologies in planning cleanups.

POTENTIAL ADVANTAGES OF BIOREMEDIATION

The potential advantages of applying biodegradation principles to the cleanup of contaminated sites have been recognized for some time. Benefits include

- *Reduced cost*. Bioremediation can be much cheaper than other technologies, perhaps one-third to one-half the cost of incineration (Lindsey 1990).

- *Reduced risk of exposure.* When used *in situ,* bioremediation reduces the risk of exposure during cleanups by avoiding the need for excavation.

- *Reduced residual contamination.* Bioremediation shows promise for further reducing the low levels of contaminants left after excavation of high-level contaminants.

- *Minimal environmental impact and liability.* Bioremediation is a natural process that has the potential of degrading toxics to harmless products—carbon dioxide, water, and fatty acids—when the process is completed.

If *in situ* application can be further developed, significant cost savings could be possible, and the risks associated with the cleanup of contaminated sites should be reduced. In addition, *in situ* bioremediation holds promise as an alternative to the sometimes costly and time-consuming practice of pumping and treating contaminated groundwater. Another potential benefit of bioremediation is the long-term prospect of developing improved microorganisms that can be tailored to degrade specific wastes or groups of wastes.

Bioremediation is not, however, a panacea for cleaning up hazardous wastes. It is not suited to all situations. For example, it does not destroy toxic metals. Also, based on our current experience, bioremediation technologies may not be capable of achieving the very high destruction rates obtainable through thermal and chemical treatment. Finally, bioremediation is a slow process, taking from days to months, depending on the wastes, the microorganism(s), and the method of application. Competing treatment processes can usually be completed in less time.

THE ROLE OF RESEARCH

Given the potential benefits of bioremediation, why is it not being used more widely to clean up hazardous wastes? In addition to the limitations noted above, one major factor is a lack of information. Our limited experience with bioremediation results in a lack of information on the actual performance and cost of bioremediation technologies. This information is essential to those responsible for selecting cleanup technologies: it enables them to evaluate bioremediation along with other methods, to assess its applicability to specific wastes, to compare the effectiveness of various methods, and to compare their costs. Information on performance is also needed to instill confidence in bioremediation technologies. The potential liabilities associated with unproven technologies are substantial. For site managers and contractors to select bioremediation instead of more proven methods, they must be confident that bioremediation will be as good as or better than other available approaches.

Research can play a major role in providing this important information. Research programs can develop treatability protocols to assist in the decision-making process. Field demonstrations can provide information on the applicability and effectiveness of specific bioremediation technologies and can begin to build a track record of successes for these technologies. Research on the basic science of bioremediation can contribute to better understanding of biodegradation processes, methods of enhancing these processes, and the byproducts of biodegradation.

The EPA is actively pursuing research in many of these areas. The goal of the EPA's research program is to develop new scientific knowledge to take bioremediation from its current stage to routine application. Some examples of the EPA's current bioremediation field activities are described in the next section. Following this is a discussion of bioremediation research questions and how EPA research is addressing these. Finally, the process that the EPA has established to identify bioremediation research needs and develop an agenda for action for the 1990s is detailed.

CURRENT FIELD RESEARCH PROJECTS

Demonstrations at Superfund Sites

The EPA has selected bioremediation as a primary cleanup technology at 25 Superfund sites to date, but few of these have progressed to the point where results can be reported. At the Traverse City Coast Guard Station in Michigan, the EPA conducted a successful *in situ* bioremediation demonstration. A groundwater plume was contaminated with benzene, toluene, and xylenes as a result of a 95,000-liter spill of aviation fuel. Hydrogen peroxide was added as an oxygen source to stimulate growth of indigenous microorganisms. As a result, the groundwater in the plume was brought within U.S. drinking water standards within six months (Wilson *et al.* 1990).

To build confidence in bioremediation among decision makers and the general public, more credible field experience is needed. The EPA's field demonstration program encompasses a dual approach. First, the EPA will identify cleanup projects that appear amenable to bioremediation and then attempt to design and implement bioremediation approaches. Some of these projects may involve modifications of already planned pump-and-treat groundwater cleanup remedies.

The second approach of the field demonstration program involves refocusing the ongoing Superfund Innovative Technology Evaluation (SITE) program to include more innovative bioremediation products or systems developed by the private sector. Currently, the program includes five reactor-based techniques and an *in situ* process using acclimated microorganisms and nutrients. Also, the SITE program may demonstrate several bioremediation systems developed in EPA laboratories. One promising organism is white rot fungus (*Phenerocheate chrysosporium*), which is being optimized at pilot scale in a rotating biological contactor. White rot fungus generates 17 enzymes that are capable of degrading a wide range of organic compounds. In a pilot-scale field test, the fungus was added through inoculated wood chips to test plots contaminated with a pentachlorophenol-based wood preservative. After 46 days, the pentachlorophenol concentration was reduced 82 to 86 percent (Glaser 1990).

The SITE program's Emerging Technologies Program has developed an artificial wetlands approach to removing metals from acid mine drainage. The mine drainage filters through the root zone of marsh plants where bacteria cause the metals to precipitate in a highly insoluble sulfide form. Zinc, cadmium, copper, and iron are readily removed; manganese is not (Haz TECH News 1990). The process will be used at a Superfund site in Colorado.

Results of this research must be made available to potential users of bioremediation technologies. Therefore, a key component of this program is information dissemination. Information from tests, demonstrations, and actual cleanups must be compiled, reviewed, and distributed. A clearinghouse is needed to provide a centralized resource for all potential users of bioremediation technologies. The EPA is planning to expand the Alternative Treatment Technology Information Center (ATTIC) to incorporate this information. ATTIC is an on-line, keyword searchable data base that includes information on innovative hazardous waste treatment technologies. It currently includes performance and cost data on technologies that have been demonstrated in the SITE program or addressed by Superfund treatability studies. ATTIC could easily be expanded to include a file on bioremediation technologies as well.

The Alaska Bioremediation Research Project

A widely publicized application of bioremediation to oil spill cleanup is the EPA's research project in Prince William Sound, Alaska. In March of 1989, the supertanker *Exxon Valdez* ran aground on Bligh Reef in Prince William Sound, flooding one of the nation's most pristine and sensitive environments in less than five hours with approximately 41.7 million liters (11 million gal) of crude oil. The spilled oil affected an estimated 1,400 kilometers (900 miles) of shoreline in the sound. The Alaska Bioremediation Research Project was initiated in the aftermath of this oil spill to evaluate the feasibility of using bioremediation to assist in cleanup operations.

The Alaska Bioremediation Research Project has been a cooperative research effort involving scientists from the EPA; Exxon Company, USA; the Alaska Department of Environmental Conservation; and the University of Alaska in Fairbanks. In 1989, the project's objective was to demonstrate the feasibility of augmenting, in an environmentally safe manner, shoreline cleanup by accelerating natural microbial oil degradation processes through the application of fertilizers. Based on the success of the 1989 and 1990 research, more than 160 kilometers (100 miles) of Alaskan coastline have been successfully treated with bioremediation by the Exxon Company, USA.

Results of field activities in 1989 (Pritchard *et al.* 1990) were

- Visual inspection of beaches treated with the fertilizer showed that within three weeks following fertilizer application, considerably less oil was observed on the rock surfaces than in the untreated control beaches. This condition became more pronounced with time and remained visually apparent through the end of the summer season (five weeks).

- No oil slicks were observed in the near-shore seawater following application of the inorganic fertilizer, indicating that oil was not released.

- Samples of the oil taken from the beach surfaces when the oil was (visually) beginning to disappear showed changes in composition that indicated extensive biodegradation.

- Visual disappearance of the oil as a result of inorganic nutrient application can be attributed only to enhanced biodegradation.

- Addition of fertilizer to oiled shorelines did not cause any increases in planktonic algae or bacteria or any measurable nutrient accumulation in adjacent areas.

- No oil was detected in tissues taken from mussels that had been placed (in flow-through plastic containers) just offshore of the fertilizer-treated beaches; the bioremediation treatment did not disperse the oil, but kept it available for biodegradation.

- Samples taken from the bioremediated beaches showed that as the oil was biodegraded, the slight mutagenicity of fresh Prudhoe Bay crude was eliminated. Thus, no mutagenic byproducts result from enhanced biodegradation.

In 1990, the project addressed two additional needs. First, the project sought to advance our understanding of the science of bioremediation as it applies to oil spill cleanup, particularly to subsurface cleanup. Second, a range of commercial products were assessed for their applicability to shoreline cleanup.

Recent results are very encouraging. A single application of fertilizer has been shown to double the rate of oil biodegradation, as measured by change in oil residue weight, over the rate of an untreated shoreline. A repetitive application of a fertilizer solution has tripled the rate of oil biodegradation when compared with untreated shoreline. This accelerated rate has been sustained for at least several weeks. We can now estimate the rate at which nutrient-enhanced bioremediation occurs. This advancement represents an important step forward in oil spill bioremediation since it will permit us to compare the results from different fertilizer treatments and different environments. Such comparisons will allow further refinement of the process, leading to more effective use of this treatment tool.

The most recent field research on risks examined the survival of mysids (shrimplike crustacea), which were maintained for 96 h in seawater taken from treated and untreated segments of shoreline. These tests found no significant differences in survival (Clark *et al.* 1990).

The EPA also plans to continue research on the application of bioremediation to oil spills, concentrating on beach bioremediation as an outgrowth of the Alaska project. The EPA has evaluated the ability of several commercial products to enhance bioremediation in Alaska. To identify candidate products, the EPA enlisted the assistance of the National Environmental Technology Applications Center (NETAC), a cooperative venture of the EPA and the University of Pittsburgh. NETAC assembled a Bioremediation Products Evaluation Panel to evaluate proposals that were submitted by product developers. The panel recommended eleven products—two nutrients, one dispersant, and eight microbial cultures—for testing. Ten vendors supplied products for testing in the EPA's laboratories. The panel then evaluated the results of these tests and recommended products for further demonstration. From this work the EPA selected the two most promising for field studies on weathered crude from the *Valdez* accident in Alaska. Both products contain naturally occurring microbes. Data on oil chemistry degradation, microbial population counts, and microbial activity are available through a series of papers. These may be obtained from the EPA's Risk Reduction Engineering Laboratory in Cincinnati, Ohio. As a follow-up over time, the EPA intends to perfect its screening protocols for routine use in evaluating the effectiveness of bioremediation products. Later, similar screening protocols will be developed for oil spill cleanup in marshes. Also, the EPA will investigate the potential for application of bioremediation for oil on open waters.

RESEARCH QUESTIONS

Research has only begun to address the various questions applicable to widespread use of bioremediation. Because so many scientific and engineering disciplines contribute to bioremediation, research needs are multidisciplinary and numerous. Critical research questions focus on a range of issues such as development of successful application techniques, understanding the ecology of biodegradation, enhancing the availability of nutrients and wastes to the microorganisms, quantitative assessment of biodegradation activity, and manipulation of the genetic material that controls degradation. Specific projects must be designed to answer each of these questions.

Application technology is the tool that takes the microorganism from the test tube to the waste. Research is needed to evaluate techniques of application as well as site preparation activities. The process by which an inoculum can be introduced *in situ* is poorly understood. To date, introduction has been to some degree haphazard, with no certainty that the inoculum will survive the transfer. We must identify the necessary methods to assure successful application and colonization.

Once application has been accomplished, the ability of the inoculum to thrive in the new environment becomes critical. Understanding the ecology of biodegradation, i.e., the interrelationship between the microorganism and its environment, is crucial to our ability to utilize bioremediation in the most optimal manner. Research that characterizes this complex interrelationship is the necessary first step. Characterization needs include identifying how biodegradation for a specific microorganism or inoculum is affected by such environmental factors as nutrient concentrations, geology, soil type, microclimate, macroclimate, and the presence or absence of oxygen. Previous exposure to organic materials is also an important characteristic, as this may facilitate acclimation to related substances. Research to identify and subsequently create the optimal environment is also important.

The effective delivery of nutrients and waste materials to the microorganism is yet another area requiring research. When bioremediation is ineffective, it is often because of the inability to establish contact between the microorganisms and either the needed nutrients or the waste substrate. Additional problems arise when the nutrients or substrate exists in a form that precludes uptake by the microorganisms.

To address these problems, an understanding of microbial cell biology and physiology is needed. Research must identify the most effective methods to interface microbes, nutrients, and wastes. To accomplish this, current methods must be tested and evaluated, and enhanced techniques must be developed. We must understand the mechanisms used to degrade pollutants. With this information, we will better understand limitations on the ability of microbes to assimilate materials of various physical forms and chemical structures.

Research to develop improved analytical methods and measurement techniques will facilitate the use of bioremediation. Currently, it is very difficult to quantify the progress of *in situ* biodegradation. In some situations, distinguishing biodegradation from volatilization, photolysis, or leaching is a problem. The lack of a controlled reaction environment increases the probability of sample contamination and variability, making analysis and comparison of data extremely difficult. Development of standard protocols for measuring the rate and extent of degradation will reduce these ambiguities and provide a framework for a better bioremediation database.

Genetic manipulation is a promising area of bioremediation research. Our ability to create microorganisms with enhanced biodegradation capabilities is limited primarily by our incomplete understanding of genetic function. Advances in this area depend upon continued research in genetics, cell physiology, and biochemistry. The application of genetically altered microorganisms to field situations may be limited to a greater extent by regulatory constraints than by technical factors.

Implicit in this discussion of needed research is the need for professionals trained in a variety of scientific and engineering disciplines. Adequate staffing is essential to the rapid advancement of bioremediation. We must ensure the removal of institutional obstacles that thwart the education and development of technical professionals. Government, industry, and academia must work cooperatively to promote the training of scientists and engineers.

CURRENT LABORATORY RESEARCH PROJECTS

The Biosystems Technology Development Program

The EPA's Biosystems Technology Development Program seeks to expand and refine biological treatment and associated engineering technologies applicable to hazardous wastes. One specific objective of the program is to advance the science of bioremediation in the areas of process characterization, process development, process engineering, environmental risk assessment, and mitigation of adverse consequences. To achieve this objective, multidisciplinary research in engineering science, microbiology and biology, measurement and analytical methods, and toxicity of organisms is ongoing at five EPA laboratories. The research program also has a demonstration component and a technology transfer/information dissemination component.

Engineering Science

The ultimate success of bioremediation depends on microorganisms staying in close physical contact with the substance to be degraded. Some methods do not provide such contact, and this has reduced the efficiency of treatment. The polluted surface may have different abilities to hold microorganisms. Pretreating the waste with some additional natural element, such as oxygen, may provide a more hospitable environment for the microorganism, assuring maximum viability and effectiveness.

An ongoing EPA research project is seeking to determine the ability of granular activated carbon to hold organic contaminants under a variety of environmental conditions. This project stems from a discovery that, in the presence of dissolved molecular oxygen, the activated carbon adsorbs greater quantities of organic compounds than in the absence of molecular oxygen. Complete knowledge of this interaction will contribute to better design of waste treatment facilities (Vidic *et al.* 1990).

Microbiology and Biology

In the area of microbiology, research is examining, at a cellular level, the process of breaking down complex substances. This requires study of the metabolism, genetics, and biochemistry of different species of microorganisms. For example, branched-chain aromatic hydrocarbons are generally more difficult to degrade than straight-chains. Also, degradation is often incomplete. In such situations, the end products are not solely carbon dioxide and water, but include a hydrocarbon that is less complex than the starting material. Are these hydrocarbons benign, or are they creating additional environmental problems? How do species vary in their completeness of degradation? Information concerning the ability of selected species to break down specific waste is being used to guide species selection for a particular site.

Recent EPA studies have resulted in the isolation of bacterial cultures capable of providing both carbon and energy for organism growth. With a more thorough understanding of these unique abilities, biological systems can be developed that are capable of destroying particularly problematic pollutants, such as complex hydrocarbons, that have not previously responded to bioremediation. For bioremediation to be recognized as a legitimate tool for hazardous waste treatment, progress must be made in degrading the more complex compounds.

Measurement and Analytical Methods

Measurement research is an important component of bioremediation investigation. To determine the best site and microorganism combination, the ecology of the polluted site must be characterized. Description of the subsurface is receiving growing attention, especially with respect to its ability to sustain microbial activity and the biodegradation reaction. Especially important will be the development of improved sampling procedures that are capable of accurately evaluating the biodegradation progress in subsurface samples.

Fundamental to the assessment of biodegradability and biotreatability is the ability to distinguish the biological from the physical processes. The EPA is planning a program to evaluate the accuracy, sensitivity, and reproducibility of commercially available instrumentation. Standard systems, whose reaction rates can be independently established, will be used to validate our ability to monitor bioremediation efforts rapidly and accurately.

Toxicity of Organisms

To assure the public regarding the safety of bioremediation, it is necessary to test the microorganisms and breakdown products for toxicity to humans and the ecosystem. A number of high-priority compounds for disposal by bioremediation are known to be carcinogenic; many are also mutagenic. Some of these compounds are procarcinogens; that is, they must first be metabolized to the carcinogenic form before they have a genotoxic effect. Determination of these toxic and carcinogenic properties requires the development of dependable testing methods.

Previous research has resulted in the development of short-term bioassays that can monitor for organic pollutants and their genotoxic byproducts within actual complex field samples. The EPA intends to take advantage of these assays to measure the degradation products of chemicals and the decrease in mutagenicity that occurs during biodegradation.

EPA'S BIOREMEDIATION ACTION COMMITTEE

The success of the Alaska Bioremediation Research Project sparked interest within the EPA, state and local governments, and others involved in cleaning up hazardous waste sites and oil spills. In response to this interest, the EPA conducted a meeting on the Environmental Applications of Biotechnology on February 22, 1990. The purpose of the meeting was to recommend a biotechnology agenda for action for the 1990s to guide the EPA, other government agencies, industry, and academia in collaborative efforts to expand the application of biotechnology.

As a follow-up to this initial meeting, the EPA established a process that will continue to provide guidance on biotechnology research and implementation issues. A joint federal agency, academic, and industry committee called the Bioremediation Action Committee was created to implement the actions from the February meeting and to guide future research. On June 20, 1990, the first meeting of this committee was held to discuss the recommendations from the February meeting. The group established four subcommittees to address different issues. Each of the subcommittees includes representatives from industry, government, and academia. These subcommittees are:

- *Research and Education Subcommittee.* This subcommittee will review federal, state, and university research on bioremediation to determine the current status of research in this area and to recommend appropriate actions to expand and coordinate this research. This subcommittee will also identify existing or needed education efforts.

- *Data Identification and Collection Subcommittee.* This subcommittee will work with states and industry to collect biotechnology field demonstration data and make them available in a centralized database source, the EPA-supported ATTIC. The National Governors' Association and Applied Biotreatment Association will coordinate collection of the data.

- *Treatability Protocol Subcommittee.* This subcommittee will provide technical input into the development of treatability protocols for testing the applicability and effectiveness of bioremediation as a cleanup technology at specific sites. The protocols currently under consideration address both aerobic and anaerobic treatment of soils and liquids containing hazardous contaminants.

- *National Bioremediation Spill Response Subcommittee.* This subcommittee will review the concept of establishing a National Bioremediation Spill Response Plan to address oil spills. The concept—advanced by the Applied Biotreatment Association—proposes to characterize a wide range of potential oil spill scenarios and the effectiveness of various microbial products for addressing them. This would provide the basis for rapid decision making by oil spill response authorities and rapid development of a bioremediation response.

CONCLUSION

Much research remains to be done if bioremediation is to reach its potential to address the multitude of hazardous waste treatment problems. The promise of favorable economics, reduced residuals generation, and minimal environmental impact give reason for optimism. We believe that sound science is the foundation upon which the future of bioremediation will be built. We are committed to conducting the necessary scientific and engineering research needed to enable bioremediation to fulfill this promise.

REFERENCES

Clark, J. R.; Lindstrom, J. E.; and Prince, R. C. "Bioremediation Monitoring Program—September 1990 Update." Prepared for presentation at the Alaska Regional Response Team Operations Meeting; September 11, 1990.

Glaser, J. A. *White Rot Fungus Applications to Hazardous Waste Treatment*; U.S. EPA Risk Reduction Engineering Laboratory: Cincinnati, OH, 1990.

Haz TECH News. "Colorado Artificial Wetlands Project Receives Engineering Excellence Award"; *Haz TECH News*; May 17, 1990; p. 74.

Lindsey, A. W. "Bioremediation—State of the Art." Presented at the Pacific Basin Conference on Hazardous Waste Management, Honolulu, HA, November 12–16, 1990.

Pritchard, P. H.; Araujo, R.; Clark, J. R; Claxton, L. D.; Coffin, R. B.; Costa, C. F.; Glaser, J. A.; Haines, J. R.; Heggem, D. T.; Kremer, F. V.; McCutcheon, S. C.; Rogers, J. E.; Venosa, A. D. "Oil Spill Bioremediation Project"; interim report to the EPA Office of Research and Development; February 28, 1990.

Vidic, R. D.; Suidan, M. T.; Sorial, G. A.; Brenner, R. C. "Effect of Oxygen on Adsorptive Capacity and Extraction Efficiency of GAC for Three Ortho-Substituted Phenols." Prepared for presentation at the 1990 Annual American Institute of Chemical Engineering Meeting; submitted to the *Journal of Hazardous Materials* (October 5, 1990).

Wilson, J.; Leach, L.; Michalowski, J.; Vandergrift, S.; Callaway, R. 1990. *In Situ* Reclamation of Spills from Underground Storage Tanks: New Approaches for Site Characterization, Project Design, and Evaluation of Performance. *Proceedings* of the Environmental Research Conference on Groundwater Quality and Waste Disposal; Washington, DC, March 1990.

In Situ Methanotrophic Bioremediation for Contaminated Groundwater at St. Joseph, Michigan

*Perry L. McCarty**, *Lewis Semprini, Mark E. Dolan, Thomas C. Harmon, Claire Tiedeman, Steven M. Gorelick*
Stanford University

INTRODUCTION

This paper summarizes the results of a 9-month study to evaluate the feasibility of *in situ* stimulation of indigenous bacteria that use methane as a source of cell carbon and energy (methanotrophs) for degrading chlorinated solvents and their degradation products in groundwater at St. Joseph MI. Wilson and Wilson (1985) first reported the transformation from soil of TCE by aerobic microbial communities that were grown on natural gas. Since the main constituent of natural gas is methane, their study suggested that TCE transformation was carried out by methanotrophs, a physiological group of aerobic microorganisms that use methane as their carbon and energy source. Methanotrophs are capable of initiating oxidation of a range of organic substrates, but obtain no benefit from such oxidations. This process is termed *cometabolism.* The characteristics of methanotrophs have been summarized in numerous reviews (Anthony 1975; Colby *et al.* 1979; Hanson 1980; Higgens *et al.* 1981; Hou 1984; Quayle 1972; Wolfe & Higgins 1979).

With methanotrophs, energy generation from methane is initiated through the activity of an enzyme, methane monooxygenase

* Department of Civil Engineering, Stanford University, Stanford CA 94305-4020

(MMO). MMO is the nonspecific enzyme that is responsible for cometabolism. Following the original report by Wilson and Wilson (1985), Fogel *et al.* (1986) showed that methane-grown enrichment cultures could degrade TCE, c-DCE, t-DCE, 1,1-dichloroethylene (1,1-DCE), and VC, but found that tetrachloroethylene (PCE) was not degraded by the methanotrophs. Henson *et al.* (1987, 1988) also reported essentially the same finding, and indicated that the less halogenated compounds were degraded more rapidly than more halogenated compounds. The Moffett field and associated laboratory studies (Roberts *et al.* 1989) confirmed these findings.

At the St. Joseph site, trichloroethylene (TCE), 1,2-*cis*-dichloroethylene (c-DCE), 1,2-*trans*-dichloroethylene (t-DCE), and vinyl chloride (VC) are present at groundwater concentrations from 0.8 to 8.0 mg/L. This site appears to be a suitable candidate for evaluating the potential for full-scale treatment using a process that has been evaluated in the laboratory and at the Moffett Naval Air Station field site in Mountain View CA, over a 4-year period (Roberts *et al.* 1989).

At the Moffett test site injection of water containing dissolved methane and oxygen into the aquifer stimulated the growth of a native population of methanotrophic bacteria. The stimulated population in turn degraded TCE, DCE, and VC at different rates. With contaminant retention times of several days at the Moffett site, percentages of biodegadation achieved were: 20 percent for TCE, 40 percent for c-DCE, 85 percent for t-DCE, and 95 percent for VC (Roberts *et al.* 1989).

The St. Joseph site appeared ideal for evaluating the potential of this process at full scale since the aquifer is relatively homogeneous and the major contaminants present downgradient from the source of contamination are c-DCE, t-DCE, and VC. These compounds were found to be most readily degraded at the Moffett test site.

Objectives

The specific objectives of this study were to

- Determine whether a native population of methanotrophic bacteria exists in the St. Joseph aquifer, and if so, to evaluate this population's ability to degrade the contaminants of concern

- Determine the degree of sorption of the contaminants to the aquifer material

- Evaluate likely designs for *in situ* treatment through use of a non-steady-state model developed for this process

To simulate results for *in situ* bioremediation, the presence of an indigenous population and its size must be known. This is the basis for the first objective. Knowledge of the sorptive properties of contaminants at a given site is also important for any bioremediation scheme as this affects how the contaminants move relative to the water. The rates of biodegradation are also functions of the amount of compound sorbed.

The third task was to evaluate possible designs for the remediation system. The model used for this purpose was developed and verified for simulation at the Moffett test site and was believed suitable for initial evaluations at the St. Joseph site. In addition, a regional flow model was developed to evaluate the important parameters affecting flow of water into and through the St. Joseph aquifer system.

PROCEDURES

Collection of Aquifer Samples

The U.S. Environmental Protection Agency's (EPA) Robert S. Kerr Environmental Research Laboratory provided their drilling crew and drilling rig for collection of aquifer samples in a manner to avoid surface contamination (Wilson & Leach 1989). Figure 1 represents data from a report by Keck (1988) and indicates the concentrations and distribution of TCE, DCE isomers, and VC at the St. Joseph site. The St. Joseph aquifer is unconfined and consists of fine sand extending from the ground surface to between 15 and 32 m below the surface (Figure 2). Samples were taken from nearly the highest concentrations of DCE and VC in the western portion of the plume that is migrating toward Lake Michigan near sampling wells OW-18 and OW-29. Here, the groundwater lies about 10 m below the surface, and the center of the contaminant plume is about 20 m below the surface.

Aseptically obtained samples of aquifer material were collected using a 15-cm-diameter hollow-stem auger through which a 10-cm pitcher barrel could be driven, as described previously (Wilson & Leach 1989). The aquifer material so collected was placed in sterile 1-L Mason® jars for transport to the Stanford laboratory, where they were stored in the dark at 4 C until used. Because of difficulties with the

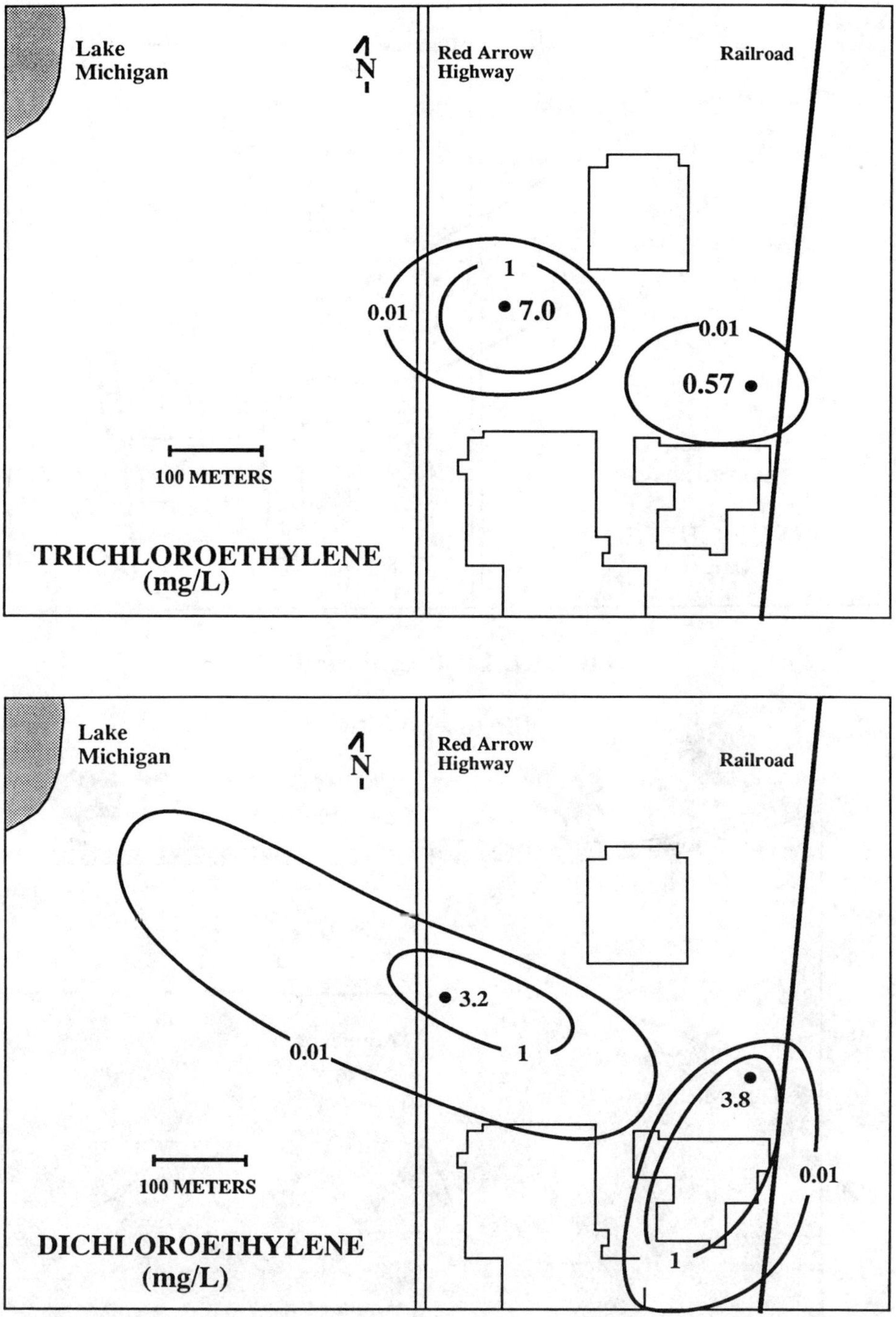

FIGURE 1. Distribution of TCE, DCE, and VC at the St. Joseph site (after Keck 1988).

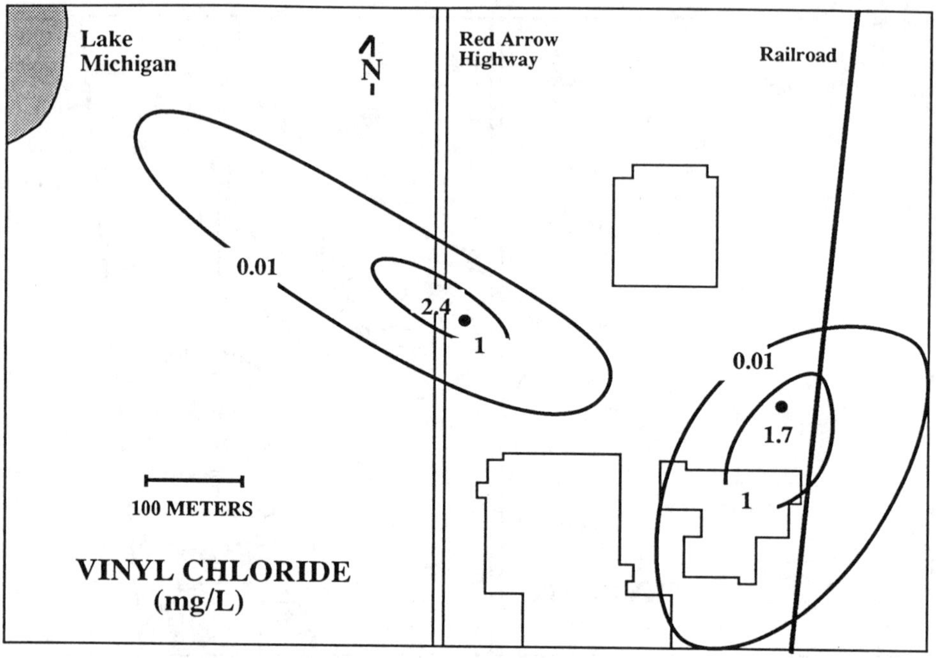

FIGURE 1. (Continued)

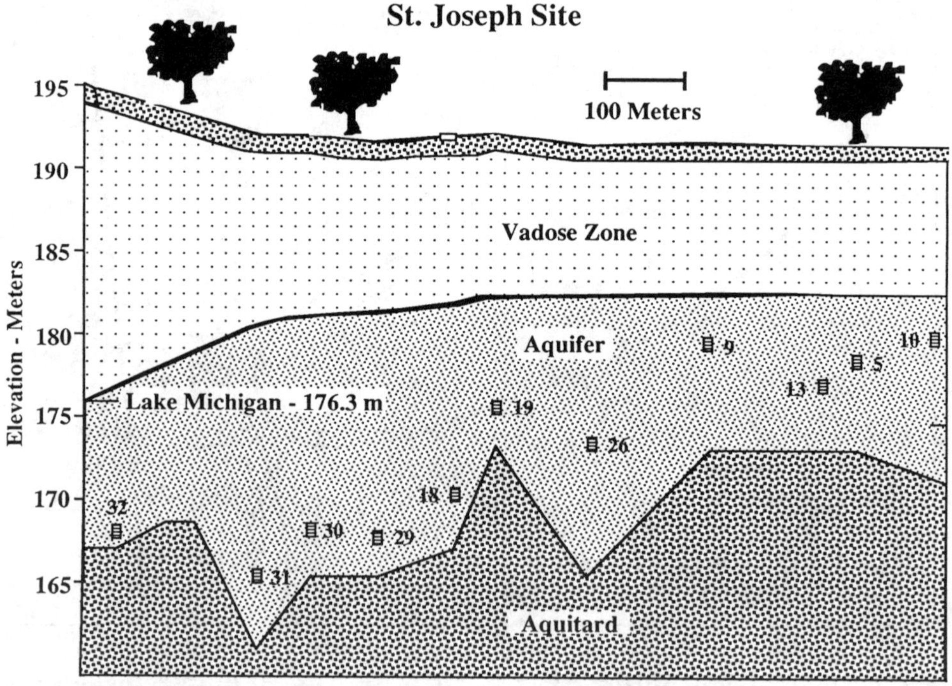

FIGURE 2. Vertical profile of subsurface at St. Joseph.

fine sandy aquifer material, aseptically collected core samples were obtained only from the upper 4 m at OW-29 and the upper 2 m of the aquifer at OW-18. However, material was obtained between 10 and 12 m below the water surface near OW-29 for physical characterization.

Column Microcosm Studies

The objectives of column microcosm studies were to determine (1) whether indigenous methanotrophic bacteria are present at the St. Joseph site, (2) the time required to increase the population of such bacteria to an adequate level for bioremediation, and (3) the capability of such a stimulated population to degade TCE, t-DCE, and VC.

The 2-cm-diameter glass column microcosms had a volume of 133 mL, contained about 200 g of aquifer material, had a void volume of 50 mL, and were constructed and operated as described previously (Lanzarone & McCarty 1990; Roberts *et al.* 1989). The microcosms were autoclaved, and aquifer material was aseptically added as was filter-sterilized groundwater from an uncontaminated location at the St. Joseph site. The columns were operated by exchange of column fluid after periods ranging from a few to 30 days. Here, replacement of column fluid was effected by downward flow of about 140 mL of amended and filter-sterilized groundwater over about a 30-min period.

Five columns (B, C, D, E, F) contained aquifer material from near well OW-18 and two (G, K) from near well OW-29. Column flow characteristics were determined by exchange of water containing a bromide tracer, the concentration of which was monitored by ion chromatography. Then, some columns were exchanged with water containing only methane and oxygen to determine whether methane-consuming methanotrophs were present. Once stimulation of methanotrophs was in evidence, VC was added to the exchange water and the possible degradation of VC was determined. Later, TCE, t-DCE, as well as VC, were added to evaluate degradation potential.

Sorption Studies

Solids Preparation. The methods for characterizing the St. Joseph aquifer solids, discussed briefly here, are presented in detail by Ball *et al.* (in press). The solids, taken from the 10.7-m depth near well OW-29, were combined in equal fractions and homogenized into a single sample. This was oven-dried at 60 C and divided into subsamples using a riffle splitter for homogeneity. Samples were examined for physical characteristics (particle size distribution, solid density, fraction of organic carbon, and mineralogy) and sorption capacity.

Solids Characterization. Particle size distribution was determined using standard dry-sieving for three samples of approximately 150 g each. To examine the possibility of variability, separate samples from 10.6-m and 11.4-m depths near OW-29 were analyzed in duplicate for organic carbon content by Desert Analytics (Tucson AZ). The solid densities of duplicate 30-g samples were determined using a water displacement method. The mineralogy of the aquifer solids was examined using X-ray diffraction (XRD).

Sorption Methods. Batch sorption studies to characterize the equilibrium behavior of TCE, c- and t-DCE, and VC with the St. Joseph aquifer solids were as previously described (Ball 1989; Curtis *et al.* 1986). Known quantities of aquifer solids and water were added to glass ampules, spiked with a known quantity of contaminant, and the ampules were flame sealed and incubated. The samples were then centrifuged, and supernatant liquid was extracted with pentane. The pentane extracts were analyzed on both a Packard 437A gas chromatograph (Packard Instrument Co., Downers Grove IL) with electron capture detector and on the HP 5890 GC system described below. Some samples were spiked with ^{14}C-labeled TCE and analyzed using liquid scintillation counting to test the accuracy of the procedure. Using control samples (containing no solids) to account for volatilization losses, the concentration of the sorbed phase was taken as the difference between the mass added and mass in the aqueous phase.

Analytical Procedures

Dissolved Oxygen. Dissolved oxygen (DO) concentration was measured with a Model 54A oxygen meter (Yellow Springs Instruments, Yellow Springs OH) that had been calibrated in air-saturated water. The coefficient of variation for DO less than 20 mg/L was 4 percent (n = 8) and for higher DO was 1.3 percent (n = 10).

Dissolved Methane. Dissolved methane concentration was determined by headspace analysis as described previously (Lanzarone & McCarty 1990). Samples (5 mL) were added to 13-mL bottles, equilibrated, and

0.5 mL of the gas phase was analyzed on a Hewlett Packard 5730A gas chromatograph (Hewlett Packard, Avondale PA) equipped with an FID detector and a 60/80 Carbosieve column (1.5 m × 4.5 cm; Packard, Avondale PA). The coefficient of variation for methane based on three sets of data was 9 percent.

Chlorinated Aliphatic Compounds. Concentrations of TCE, t-DCE, and VC were determined by headspace analysis. Approximately 5 mL of liquid sample was transferred into a 15-mL glass vial, sealed with Teflon®-faced septa and screw-top closures, and shaken for 30 min. A 300-μL sample of the headspace gas was injected into a capillary column (J&W fused silica megabore DB 624) on a HP 5890 GC fitted with a Hall electrolytic conductivity detector. Peak areas were integrated using a Nelson Analytical 3000 series chromatography data system. NBS traceable VC in nitrogen gas standards was used for preparing standard curves. Calibration for t-DCE and TCE was conducted in a similar manner.

Radioactivity Analyses. ^{14}C activity was assayed using a Tricarb Model 4530 scintillation spectrometer (Packard Instrument Co., Downers Grove IL) which made counting efficiency corrections with the external standard channels ratio method (Harrocks 1974). For each sample, three separate aliquots were counted.

Solids Characteristics

The St. Joseph aquifer solids consisted of a relatively clean, sandy material, having a size distribution dominated by fine material with 80 percent between 0.075 and 0.20 mm in diameter. For comparison, the Moffett material was much more heterogeneous, varying from coarse sands and gravels (>2 mm) to silts and clays.

The organic carbon fraction (f_{oc}) for the two St. Joseph samples differed by a factor of about 2, with duplicates from the upper sample having values of 0.018 and 0.035 percent, and from the lower sample of 0.056 and 0.067 percent. By comparison f_{oc} for the Moffett site aquifer material varied by a factor of 5 or 6, with a bulk value of about 0.11 percent. This is two to four times greater than that of the St. Joseph material.

XRD results showed that over 90 percent of the St. Joseph bulk sample consisted of quartz. Magnetic separation revealed a measurable quantity (<0.1% mass) of magnetite. The solids density was 2.65 g/cm^3, which is typical for quartz sand.

Sorption Studies

Sorption results are listed in Table 1. The distribution coefficients (K_d), developed from sorption isotherms, are based on a 10-day equilibration time, which is believed sufficient to reach true equilibrium because of the relatively fine particles involved (Harmon *et al.* 1989). Table 1 also provides a comparison between the retardation factors (R) computed for the St. Joseph and Moffett sites. R indicates how the speed of contaminant movement is retarded relative to that of the water. The inverse (1/R) indicates the fraction of contaminant that is dissolved in goundwater compared with that sorbed onto aquifer material. Table 1 values are based on a range of likely aquifer porosities (0.33 to 0.38) and bulk densities (1.65 to 1.75 g/cm^3). The 95 percent confidence intervals listed are based on the slopes of regression lines of measured K_d values over a range of contaminant concentrations (0.050 to 0.45 mg/L) using the following equation:

$$R = 1 + \frac{\rho_b K_d}{\theta}$$

where ρ_b is the aquifer bulk density, K_d (mL/g) is the distribution coefficient, and θ is the aquifer porosity.

COLUMN MICROCOSM RESULTS

Bromide Tracer. No breakthroughs of the nonreactive and nonsorptive bromide tracer were observed in preliminary column microcosm studies until the passage of about 30 mL of fluid. All effluents reached the influent concentration after about 90 mL passed through. The estimated average column pore volume is about 50 mL. The sharp bromide breakthroughs for all columns are consistent with the uniform and fine grain size of the aquifer material. These results indicated that the first 30 mL of column effluent was representative of concentrations present within the column and absent of the influence of short-circuiting by the feed solution. The results also indicated that an exchange of 90 mL of fluid during each period was required to insure that the influent materials were distributed throughout the columns.

TABLE 1. Comparison between St. Joseph and Moffett retardation factors.

Compound	St. Joseph K_d (mL/g)[(a)]	St. Joseph Retardation Factors	Moffett Retardation Factors
t-DCE	0.57 ± 0.15	2.2 – 4.6	8.4 – 18.5
c-DCE	0.46 ± 0.16	3.1 – 6.0	4.7 – 10.9
TCE	0.50 ± 0.19	2.7 – 5.2	8.8 – 13.3[(b)]
VC		1.2 – 2.9[(c)]	3.9 – 7.4

(a) Values represent mean ± standard deviation.
(b) ^{14}C-labeled experiments.
(c) Calculated from Moffett values.

Methane and Oxygen Consumption. Methane (about 3.5 mg/L) and oxygen (about 25 mg/L) were added initially to four columns (C, E, G, and K), which were then exchanged similarly at various times to determine whether methane consumption had begun. Methane consumption in Column C was complete within 31 days, and in Column E within 41 days (Figure 3). The columns contained aquifer material taken near well OW-18, but at somewhat different depths. No data reveal methane utilization before day 28, but methane was consumed in Column C by day 31. Figure 3 illustrates the averages of concentrations in two samples taken at each time period, one for the first 15 mL and the other for the second 15 mL.

The time interval for the onset of methane consumption in Columns G and K, with aquifer material from near well OW-29, were similar, although initially methane consumption was not uniform throughout the columns. Methane concentration in the initial effluent was little changed from the influent values, but in the sample from the second 15 mL, it was nearly completely removed in both columns. This suggests that the methanotrophic population may not have been evenly distributed throughout the columns. Since the activity was higher in both columns at the influent end, it is also possible that the columns might have become contaminated with methanotrophs during an exchange. Assuming this did not occur, the population levels in all the columns appeared similar. The methanotrophic populations in St. Joseph aquifer materials were much lower than in Moffett field material, where noticeable methane consumption occurred within

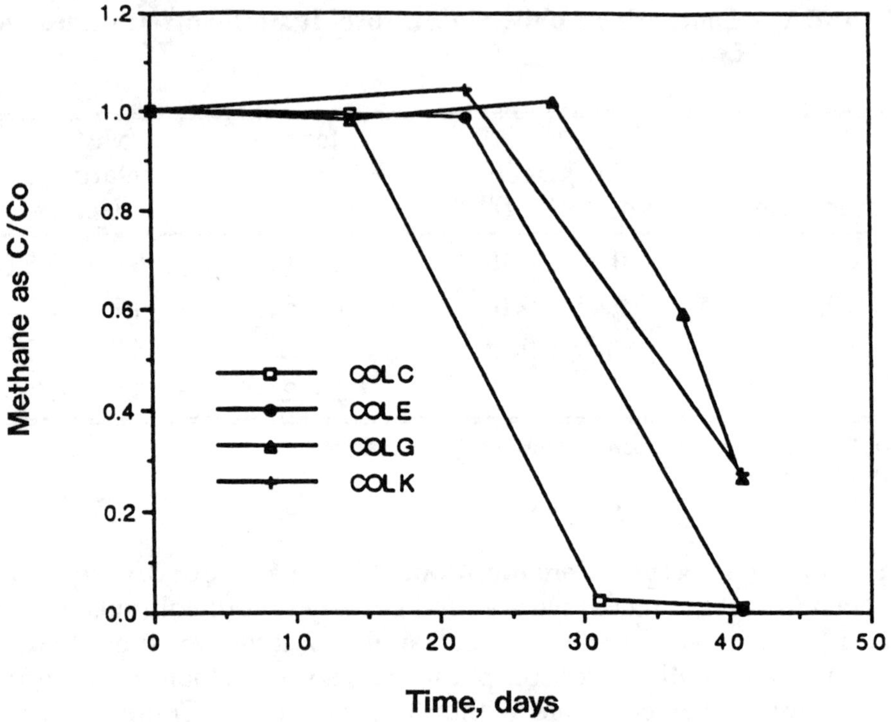

FIGURE 3. Methane removal in column microcosms.

2 weeks. Since the doubling time for methanotrophs is about 1 day, this difference indicates the St. Joseph methanotrophic population is four to five orders of magnitude smaller than the Moffett population.

Vinyl Chloride Degradation. After methane consumption occurred in Columns C and E, approximately 0.175 mg/L vinyl chloride (VC) was added to the methane- and oxygen-containing exchange fluids for these columns. The control Column B was then exchanged with fluid containing dissolved oxygen and VC, but no methane. Periodically, the initial effluents from exchanges were analyzed for dissolved oxygen, methane, and VC to determine the extent of VC degradation. The results are presented in Figure 4. Time zero represents the time of initial VC addition. At the first exchange, 1 week after VC addition, 20 percent of the VC in the control column was removed, probably from sorption to the aquifer material. Both methane-stimulated columns indicated about 75 percent VC removal, suggesting that the methane-enriched consortium was capable of VC degradation. Between this and the previous exchange, and between all further exchanges, the methane

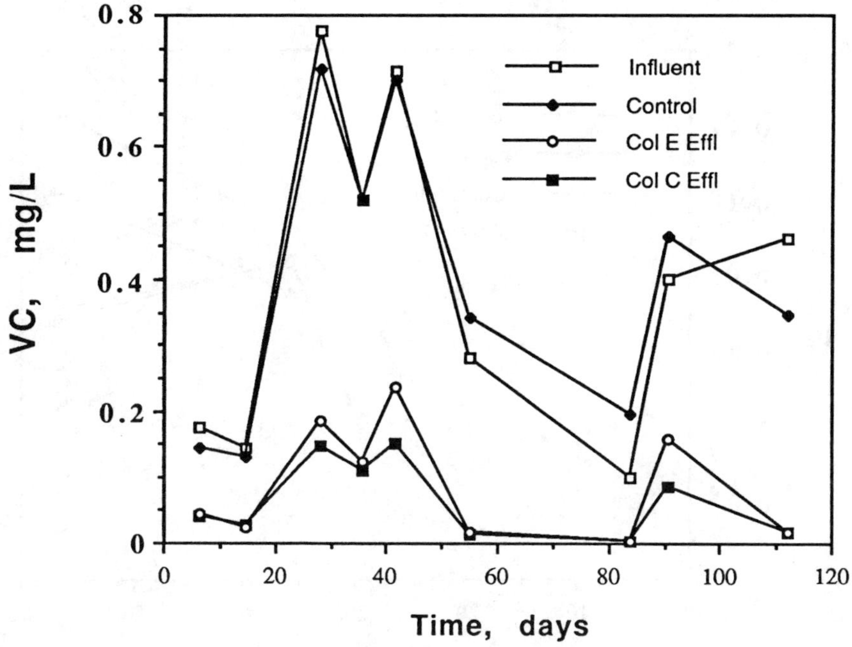

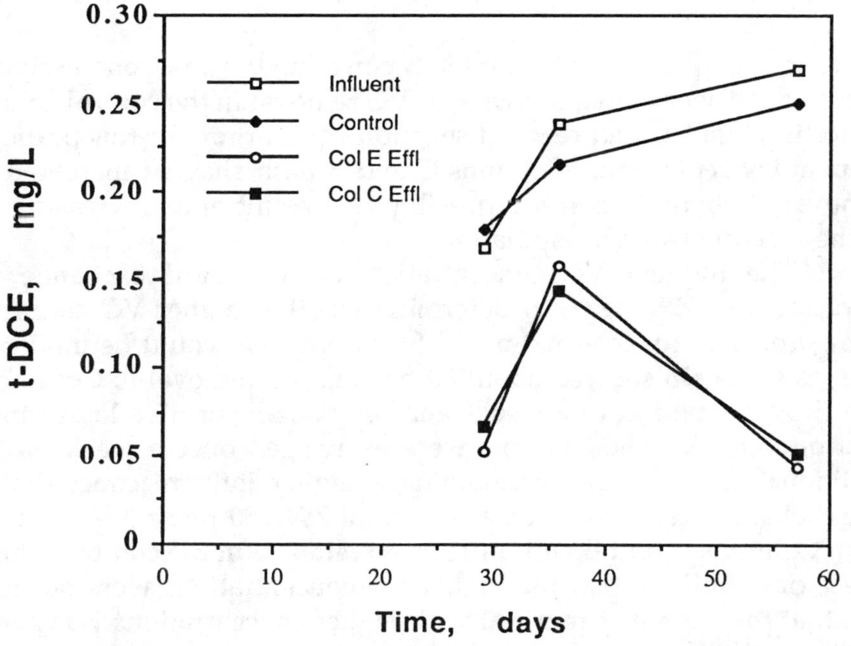

FIGURE 4. VC, t-DCE, and TCE removal in column microcosms.

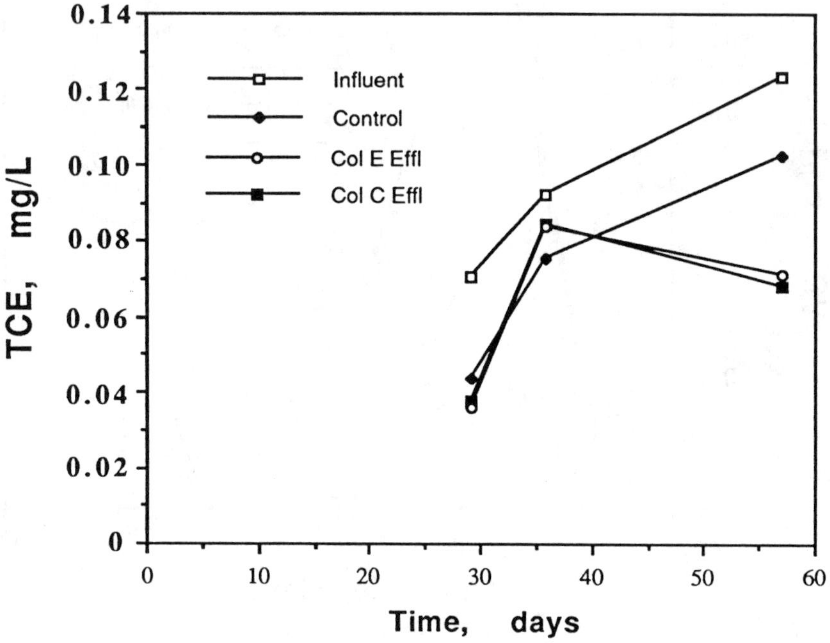

FIGURE 4. (Continued)

added to the columns was completely consumed. The second exchange, carried out 1 week later, showed no VC removal in the control column, indicating that VC had reached sorption equilibrium for this particular influent concentration. Columns C and E both showed increased VC removal of about 80 percent, possibly as a result of an increasing size of the methanotrophic population.

The influent VC concentration for the third exchange was increased to 0.775 mg/L to determine whether higher VC concentrations similar to those found at the St. Joseph site would be inhibitory. The results again showed about 80 percent VC removal in Columns C and E over a period of 2 weeks and only small sorptive losses in the control column. The columns were exchanged once a week for two additional weeks while maintaining a higher influent concentration. The exchange results showed a consistent 75 to 80 percent VC removal from Columns C and E with little or no removal in the control column. These data indicate that the higher VC concentrations were not toxic, and that the degree of removal with higher concentrations was similar to that with lower concentrations.

Following the above exchanges, the influent VC concentration was reduced to 0.28 mg/L and left in the columns for 2 weeks. After this period the control column showed VC concentrations 20 percent greater than the influent concentration resulting from desorption of VC from the aquifer solids equilibrated previously at the higher VC levels. Both Columns C and E showed percent VC removal during this period, further indicating the lack of toxic effects during the previous exchanges at higher VC concentrations.

***trans*-1,2-Dichloroethylene and Trichloroethylene.** On day 54, 0.16 mg/L t-DCE and 0.07 mg/L TCE were added together with 0.1 mg/L VC to the feed for Columns B, C, and E. The results are presented in Figure 4, with time zero for the lower two graphs representing day 54 on the upper graph when t-DCE and TCE were added. When analyzed 30 days later, VC was removed to below the detection limit in Columns C and E. However, in control Column B, it was two times greater than the influent concentration because of the desorption of VC from the aquifer solids, which had been equilibrated previously at higher VC levels. A comparison between the t-DCE concentration in the effluent from the control column and Columns C and E suggests a t-DCE removal of 60 percent by the methanotrophs. TCE levels were the same in effluents from Columns C, E, and the control.

Influent concentrations of t-DCE and TCE were then increased to 0.25 and 0. 1 mg/L, respectively, and left in the columns for 1 week. Exchange results showed removals of 60 to 75 percent for VC, 35 to 40 percent for t-DCE, and 0 percent for TCE in the active columns. The columns were fed again and left for 21 additional days, after which the active columns showed removals of 95 percent for VC, 80 percent for t-DCE, and 30 percent for TCE. The extent of VC transformation before and after the addition of t-DCE and TCE to the columns appears to be approximately the same.

Long-Term Vinyl Chloride Results. One study was conducted to simulate field conditions where methane, oxygen, and VC would be mixed together initially and allowed to incubate in the aquifer. Here, VC was added to two previously unexchanged columns and left to incubate at room temperature. Dissolved oxygen (25 mg/L) and methane (4.3 mg/L) were added with VC (0.95 mg/L) to Column D, but only oxygen and VC were added to control Column F. After 10 weeks of incubation, the columns were exchanged and the initial effluents analyzed. The control column showed a 17 percent decrease in VC, probably from sorption, whereas the methanated column showed 90 percent removal of VC. All the methane and about 15 mg/L of the dissolved oxygen were consumed in the methanated column. The theoretical oxygen demand from the methane and VC is in the range of 13 to 18 mg/L (depending on the oxygen-to-methane ratio assumed), which brackets the actual oxygen usage. In the control column about 6 mg/L-dissolved oxygen was used over the 10-week period. This test illustrates the potential effectiveness of a single aquifer exchange for removing a major portion of the VC contamination in place. VC removals in the range of 95 percent were obtained when 3 weeks rather than 1 week was allowed between exchanges. The fact that additional removal can be obtained after the depletion of methane is consistent with other column studies at Stanford (Roberts *et al.* 1989).

HYDROGEOLOGIC MODELING

Site Hydrogeology. The St. Joseph site is underlain by a glacial outwash deposit, consisting mostly of fine to medium sand with some silt (Keck 1986). This deposit forms the uppermost, unconfined aquifer, at the bottom of which is a clay unit. Lithologic logs from 35 monitoring wells indicate that the saturated thickness ranges from 6 to 23 m. These logs indicate no trends in aquifer composition in any horizontal direction, but do suggest a downward-fining trend above the clay unit in some wells. A pump test conducted in 1985 indicated values for hydraulic conductivity (K) ranging from about 1.3×10^{-4} to 2.4×10^{-4} m/sec (Keck 1986).

A groundwater divide is located at the St. Joseph site, with groundwater flowing westward toward Lake Michigan and eastward toward Hickory Creek (Figure 5). The head gradients near the divide are approximately 0.006, and increase to 0.02 near Lake Michigan and to 0.04 near Hickory Creek.

Regional Flow Modeling. To carry out groundwater remediation, a well-calibrated groundwater flow model is highly desirable, with values of recharge and hydraulic conductivity being well specified. The regional groundwater flow was simulated using the USGS Modular Finite Difference Groundwater Flow Model (MODFLOW). The system was modeled as two-dimensional, transient (September 1987 to September 1988), unconfined, and with two hydraulic conductivity zones, complex boundaries, and distributed recharge. The westem and eastern boundaries are specified head boundaries, given by the temporally varying elevation of Lake Michigan and the spatially

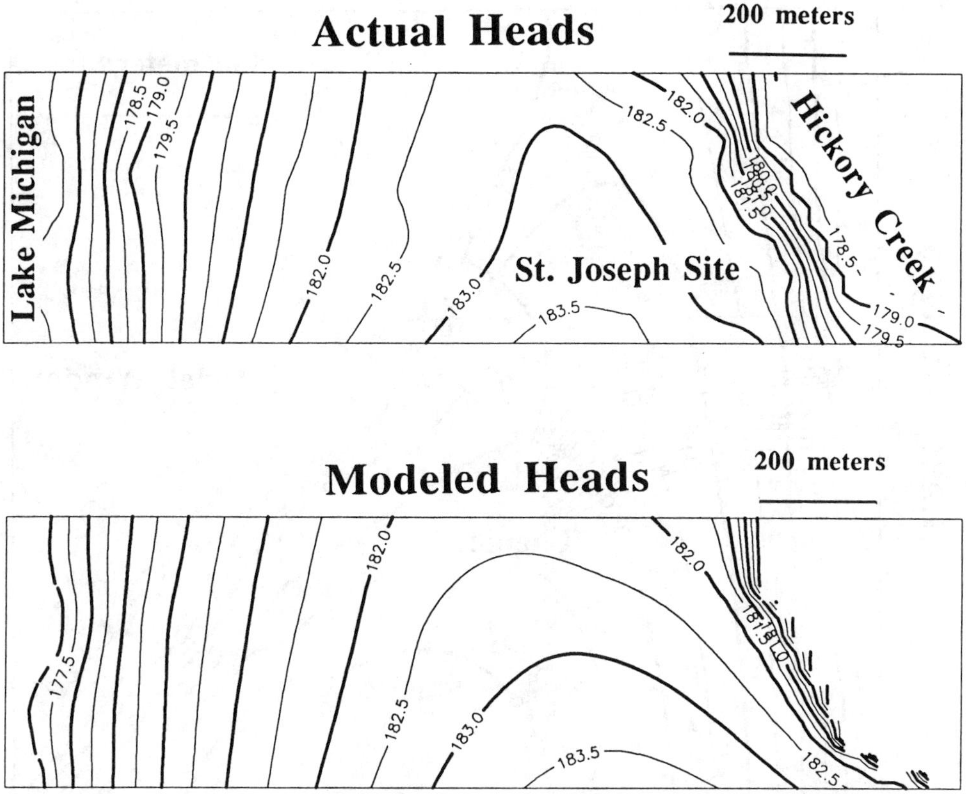

FIGURE 5. Actual and modeled heads (m), and advective flow paths for March 1988: (a) actual heads in model window, (b) modeled heads in model window, and (c) modeled heads in entire model, and advective flow paths of particles originating at possible contaminant source (lagoon).

varying elevation of Hickory Creek. As an approximation, no-flow boundaries were located several hundred meters beyond the site in the north and south directions so that they would not affect the flow near the site. The model contains 1,611 discrete active nodes. A 500 × 1,700 m model window containing the St. Joseph site and consisting of 776 nodes, 30 × 30 m, is located in the center, and nodes of larger sizes are located to the north and south.

Data used for model construction include aquifer geometry, aquifer bottom elevations, specific yield, hydraulic conductivity, and recharge. Bottom elevations were determined from borehole lithologic data, with intermediate values estimated by kriging. Specific yield was determined from the pump test. Two K zones were delineated, one

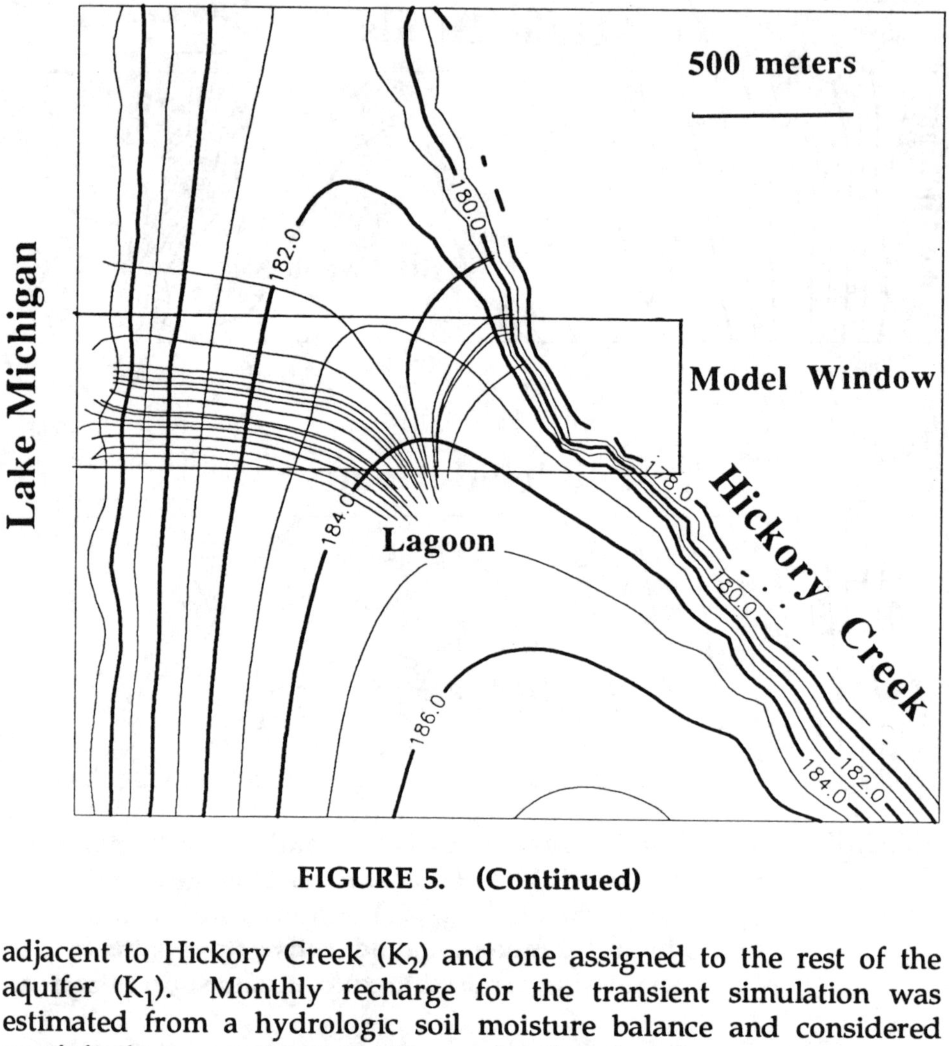

FIGURE 5. (Continued)

adjacent to Hickory Creek (K_2) and one assigned to the rest of the aquifer (K_1). Monthly recharge for the transient simulation was estimated from a hydrologic soil moisture balance and considered precipitation, evapotranspiration, and soil and plant types.

Nonlinear least squares regression linked with the groundwater simulation model was used to determine the best parameter values for K_1, K_2, and a recharge factor. Resulting values with 95 percent confidence intervals are:

$K_1 = 8.8 \times 10^{-5} \pm 3.1 \times 10^{-5}$ m/s
$K_2 = 2.4 \times 10^{-5} \pm 1.0 \times 10^{-5}$ m/s
Annual Recharge = 39 ± 17 cm/year

Contour maps of modeled and actual heads for March 1988 are shown in Figure 5. The root mean square error of all heads for March 1988 is

0.20 m and for September 1988 is 0.17 m. As an additional check, particle tracking analysis was performed with particles placed in the model at a possible contaminant source (lagoon) at the site. Results shown in Figure 5 are consistent with the observed pattern of contaminant migration both east and west of the groundwater divide. It is planned to expand the model to three dimensions so that vertical flow can be modeled and better remediation schemes can be devised.

MODEL SIMULATIONS OF *IN SITU* BIODEGRADATION

Model simulations to investigate biorestoration schemes were performed at a single location within the model window indicated above. This model used a numerical code that had been developed, verified, and calibrated for the Moffett field study (Roberts *et al.* 1989). The model includes the transport processes of advection, dispersion, and sorption. These are combined with microbial processes for the growth of methanotrophic bacteria, utilization of methane as the growth substrate and oxygen as the electron acceptor, and microbial kinetics for the cometabolic transformation of chlorinated aliphatic compounds.

The simulations used input parameters obtained from the calibration with the Moffett test results (Roberts *et al.* 1989), with additional parameters obtained from the St. Joseph laboratory study. The latter include the extent of sorption onto the aquifer solids and the estimate of initial concentration of methanotrophs in the treatment zone. Simulations were performed using the Moffett-derived rate coefficients to determine whether they were consistent with results from the St. Joseph laboratory column studies. The rate parameters in the model were then adjusted to the lower temperature of 12 C for the St. Joseph aquifer. While extrapolations from such limited results can be questioned, they can serve in a preliminary way, to indicate the order of magnitude for design factors of importance.

The modeling study did not attempt to find optimal designs for the *in situ* remediation system because remediation criteria and adequate information on site hydrogeology were not yet available. Rather, the process design developed here is one of perhaps several that might be used for comparison with alternative clean-up strategies, such as simple pump-and-treat remediation.

The remediation system assumed (Figure 6) is operated as a pump-and-treat system. However, all the treatment is carried out biologically in the aquifer itself. Here, groundwater is extracted through a

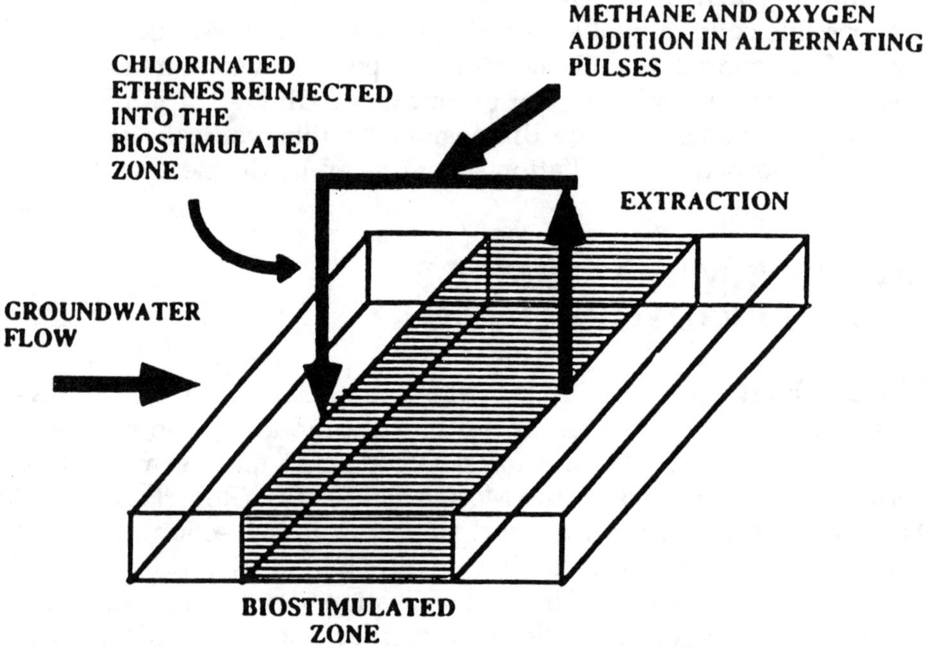

FIGURE 6. ***In situ*** **bioremediation case simulated.**

series of wells spanning the contaminant plume in a direction perpendicular to that of groundwater flow. At the surface methane and oxygen are added to the extracted groundwater, either together or in alternating pulses, depending upon the extent to which a stimulated methanotrophic population is developed. The alternating pulses help distribute the microbial growth throughout the test zone. The groundwater containing the methane and oxygen and extracted contaminants is reinjected into the treatment zone through a series of upgradient monitoring wells that are distributed parallel to the extraction wells. In the subsurface biotreatment zone both the in-place contaminants and the reinjected contaminants are biologically degraded.

The *in situ* biotreatment system was compared with a pump-and-treat system (Figure 7). In the latter case, surface treatment of some type is used to remove contaminants before the water is reinjected. Otherwise, the two systems are operated identically, and thus can be compared directly through model simulations.

For the two alternate simulations, a complete cross section through the contaminant plume was taken for treatment. The cross section assumed was 200 m wide, which is expected to span the width

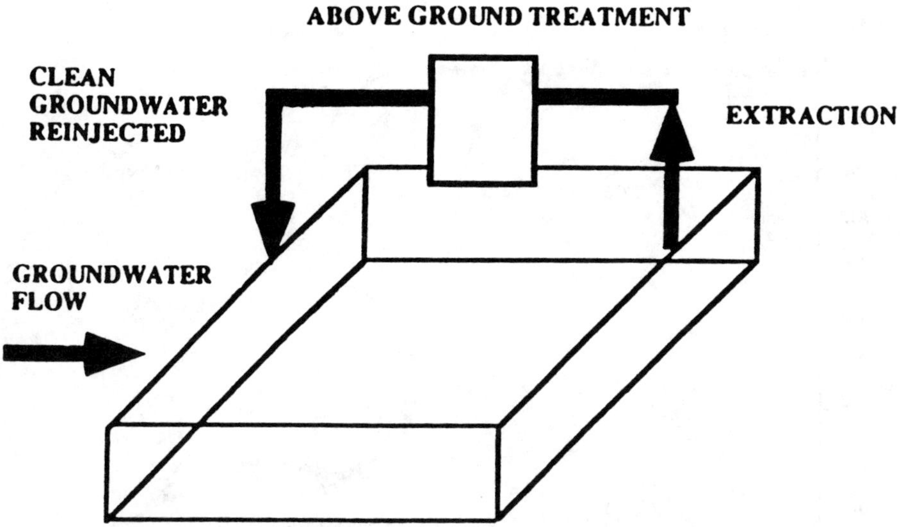

FIGURE 7. Pump-and-treat system for comparison with bioremediation.

of the plume. The length and depth of the assumed cross section are 120 and 20 m, respectively. In this cross section the contaminants VC, c-DCE, and t-DCE were assumed each to be initially present at a uniform aqueous concentration of 1 mg/L. This is within the general range of observed concentrations.

The injection and extraction flow rates used in the simulation model provide a fluid residence time in the treatment zone of 50 days. This is about one-half the time estimated for significant methanotrophic growth to occur. Thus, the chosen residence time should be sufficient to permit the mixed methane and oxygen in the injection water to permeate completely the treatment zone before observable methane and oxygen uptake occurs. After the breakthrough of methane and oxygen occurs at the extraction well(s), the induced flow (pumping) is stopped to wait for the onset of methanotrophic growth and the initial transformation of the chlorinated compounds to occur. After that, injection would again commence, but with alternating pulses of water containing methane and oxygen.

Figure 8 shows the model simulation for *in situ* biotreatment in comparison with pump-and-treat alone for VC removal. Shown are the concentrations of VC as a function of time at a node adjacent to the extraction well. After 100 days with biostimulation (50 days of no

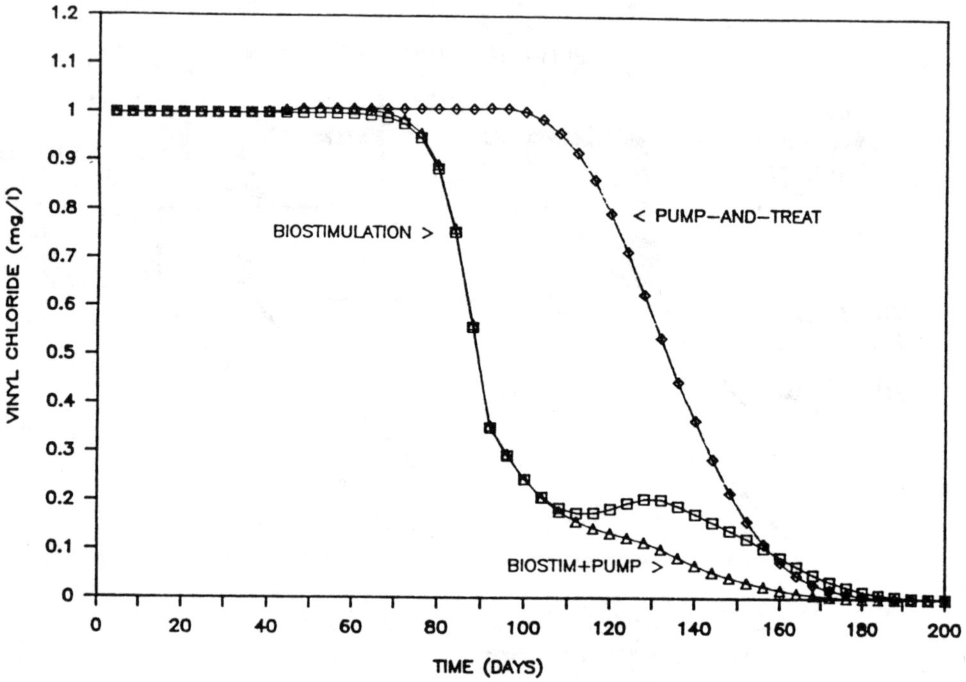

FIGURE 8. Comparison of *in situ* bioremediation of VC with pump-and-treat. Biostim plus pump is combination of surface treatment and *in situ* treatment.

pumping), all the methane should be consumed, and the model indicates that 80 percent of the VC would be biodegraded. Once methane is depleted in the aquifer, more methane and oxygen are then added to the treatment zone to accomplish additional degradation. Here, methane breakthrough is expected to occur again after 120 to 140 h of injection, and the subsequent biodegradation simulated (Figure 8) brings VC to the drinking water standard of 0.002 mg/L. The results of this model batch treatment of the test zone agree well with the results of the batch column studies presented earlier. Based upon the time required to reduce VC concentration in the treatment zone, biostimulation treatment is shown to be comparable, if not better than, the pump-and-treat method.

The simulation results are not illustrated for t-DCE as they are somewhat similar to VC. The simulation indicated that a 50 percent reduction in t-DCE concentration would occur during the initial methane injection phase. Based upon results from the Moffett study (Roberts *et al.* 1989), the reduction with t-DCE is lower than with VC

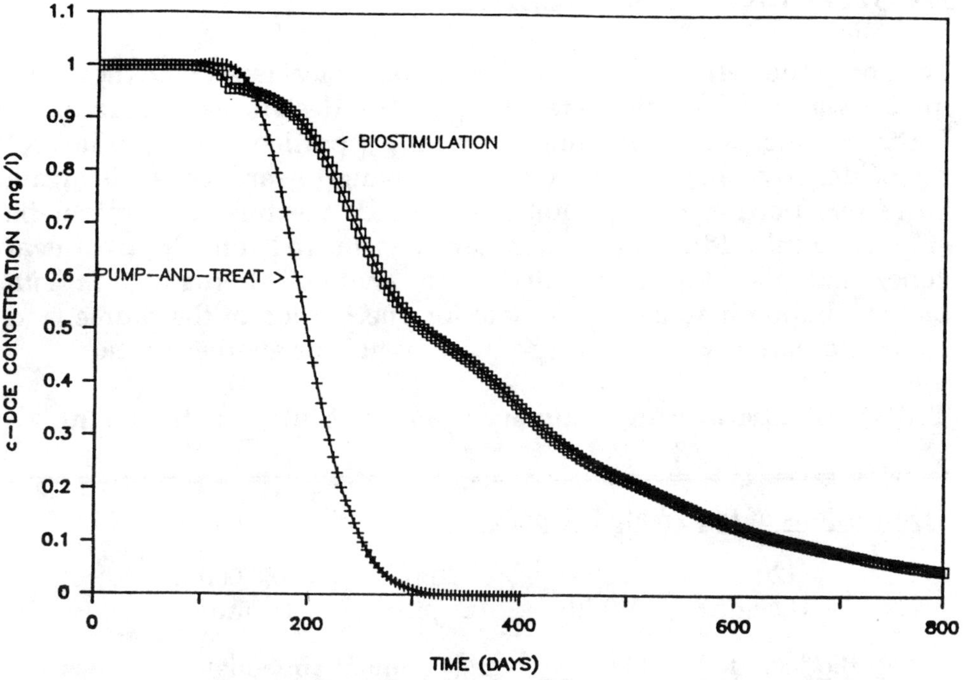

FIGURE 9. Comparison of *in situ* bioremediation of c-DCE with pump-and-treat.

primarily because of stronger sorption by the former compound. The simulation suggests that *in situ* bioremediation for t-DCE is more attractive than pump-and-treat alone.

The last simulation, which is for c-DCE, is illustrated in Figure 9. As in the Moffett field study, c-DCE was degaded at a rate an order of magnitude lower than t-DCE. With this lower rate, *in situ* bioremediation requires a longer period than simple pump and treat. However, the c-DCE that is biodegraded within the aquifer is permanently destroyed, thus *in situ* treatment has important benefits that must also be considered. Also, the proposed drinking water MCL for c-DCE is much higher than for VC. Thus, treatment that is adequate for bringing VC and t-DCE into compliance may be adequate for c-DCE as well. The simulation demonstrates that in treatment of mixtures of compounds by *in situ* bioremediation, reaction rates and degrees of sorption, as well as remediation goals, must be considered before the best strategy can be formulated. Once remediation goals are established, then a comparative evaluation of all the alternatives can be made.

IN SITU PROCESS DESIGN

To convey the pertinent aspects of *in situ* biological treatment, the above model was used to evaluate one possible design, as illustrated in Figure 6. Such a system would treat only a portion of the plume and might destroy contaminants that are moving toward Lake Michigan. Once this portion of the plume is treated, the treatment might be stopped until additional contaminant has moved into the treatment zone, and then be reactivated. In another alternative, *in situ* bioremediation may be appropriate for one portion of the plume, and some alternative scheme may be more useful for another portion.

TABLE 2. Engineering summary of *in situ* biological treatment.

Dimensions of the Treated Zone

120 m	×	200 m	×	20 m	=	480,000 m^3
(Length)		(Width)		(Depth)		(Volume)

Mass Biodegraded in 400 Days (1 mg/L initial aqueous)

	Initial Aqueous (kg)	Initial Solids (kg)	Initial Total (kg)	Treated (kg)	Percent Treated
VC	158	117	275	274	> 99
t-DCE	158	513	671	665	> 99
c-DCE	158	513	671	436	65

Methane and Oxygen Requirements

Methane	5,200 kg
Oxygen	19,200 kg

Extraction and Injection Rates

1.8 m^3/min (400 gpm)

A summary of the engineering aspects of the *in situ* bioremediation scheme as described above is contained in Table 2. Here, a total of 480,000 m^3 of aquifer would be treated in 400 days, using a total extraction rate of 1,514 L/min. A total of 1,375 kg of the original 1,617 kg of contaminants present in the treated volume would be biodegraded to nontoxic end products. This treatment would require 5,200 kg of methane and 19,200 kg of oxygen.

These initial simulations show support for the *in situ* processes for the portion of the plume considered at the St. Joseph site. The simulations, however, also show that pump-and-treat might be satisfactory for remediation as well. Here, a suitable treatment process at the surface must be used.

It is emphasized again that no attempt was made here to develop an optimized treatment system. The simulations presented are based on very limited data and serve primarily to indicate the factors involved in an *in situ* treatment scheme and the time scales that are likely. Since *in situ* bioremediation in this way has not been attempted before, there are many possibilities for improvement on the design assumed for the simulations. The simulations do help to focus on what the best possibilities may be, which should help, at least, in setting new directions for research and demonstration.

Acknowledgments

This study was supported with financial assistance from Allied-Signal, Inc., Bendix Automotive Systems–North America; and the Robert S. Kerr Environmental Research Laboratory, U.S. Environmental Protection Agency, through the U.S. EPA-sponsored Western Region Hazardous Substance Research Center, Grant No. R-815738. The content of this paper does not necessarily represent the views of these organizations. Special appreciation is extended to John T. Wilson, Lowell Leatch, and other members of the staff at the Kerr Laboratory, who provided advice on this study and all the technical assistance required for collecting samples of aquifer material from the St. Joseph site.

REFERENCES

Anthony, C. "The Biochemistry of Methylotrophic Bacteria." *Sci. Prog. (London)* **1975**, *62*, 167–206.

Ball, W. P. *Equilibrium Partitioning and Diffusion Rate Studies with Halogenated Organic Chemicals and Sandy Aquifer Material.* Ph.D. Dissertation, Stanford University, Stanford, CA, 1989.

Ball, W. P.; Buehler, C.; Harmon, T. C.; Mackay, D. M.; Roberts, P. V. "Characterization of a Sandy Aquifer Material at the Grain Scale." *J. Contam. Hydr.* in press.

Colby, J., Dalton, H. H.; Whittenbury, R. "Biological and Biochemical Aspects of Microbial Growth on Cl Compounds." *Ann. Rev. Microbiol.* 1979, *33*, 481–517.

Curtis, G. P.; Roberts, P. V.; Reinhard, M. "A Natural Gradient Experiment on Solute Transport in a Sand Aquifer: Sorption of Organic Solutes and Its Influence on Mobility." *Water Res.*, 1986, *22*, 2059–2067.

Fogel, M. M.; Taddeo, A. R.; Fogel, S. "Biodegradation of Chlorinated Ethenes by a Methane-Utilizing Mixed Culture." *Appl. Environ. Microbiol.* 1986, *51*, 720–724.

Hanson, R. S. "Ecology and Diversity of Methylotrophic Organisms." *Adv. Appl. Microbiol.* 1980, *26*, 3–39.

Harmon, T. C.; Semprini, L.; Roberts, P. V. "Sorption of Volatile Organics to Aquifer Solids at Field and Laboratory Scales." Presented at the 18th Annual Universities Forum at the Annual Conference of the American Water Works Association, Los Angeles CA, 1989.

Harrocks, D. L. *Applications of Liquid Scintillation Counting*; Academic Press: New York, 1974.

Henson, J. M.; Yates, M. V.; Cochran, J. W. "Metabolism of Chlorinated Aliphatic Hydrocarbons by a Mixed Bacterial Culture Growing on Methane." *Abstr. Q-97*, Annual Meeting American Society of Microbiology, 1987; p 298.

Henson, J. M.; Yates, M. V.; Cochran, J. W.; Shackleford, D. L. "Microbial Removal of Halogenated Methanes, Ethanes, and Ethylenes in an Aerobic Soil Exposed to Methane." *FEMS Microbiol. Ecol.* 1988, *53*, 193–201.

Higgins, I. J.; Best, D. J.; Hammond, R. C.; Scott, D. "Methane-Oxidizing Microorganisms." *Microbiol. Rev.* 1981, *45*, 556.

Hou, C. T. "Microbiology and Biochemistry of Methylotrophic Bacteria." In *Methylotrophs: Microbiology, Biochemistry, and Genetics;* Hou, C. T., Ed.; CRC Press, Inc.: Boca Raton, FL, 1984; pp 1–53.

Keck Consulting Services, Inc. "Hydrogeological Study Report." Section 10, Lincoln Township Berrien County, Michigan; Williamston, MI, June 1986.

Keck Consulting Services, Inc. "Report of September, 1988, Ground-Water Sampling." Section 10, Lincoln Township Berrien County, Michigan; Williamston, MI, November 30, 1988.

Lanzarone, N. A.; McCarty, P. L. "Column Studies on Methanotrophic Degradation of Trichloroethene and 1,2-Dichloroethane." *Ground Water* **1990**, *28*, 910–919.

Quayle, J. R. "The Metabolism of One-Carbon Compounds by Microorganisms." *Adv. Microb. Physiol.* 1972, *7*, 119–203.

Roberts, P. V.; Semprini, L.; Hopkins, G. D.; Grbić-Galić, D.; McCarty, P. L.; Reinhard, M. In Situ Aquifer Restoration of Chlorinated Aliphatics by Methanotrophic Bacteria; EPA/600/2-89/033; Center for Information Research Information: Cincinnati, OH, 1989.

Wilson, J. T.; Leach, L. E. *In Situ Bioremediation of Spills from Underground Storage Tanks: New Approaches for Site Characterization, Project Design, and Evaluation of Performance*; EPA/600/2-89/042; Center for Information Research Information: Cincinnati, OH, 1989.

Wilson, J. T.; Wilson, B. H. "Biotransformation of Trichloroethylene in Soil." *Appl. Environ. Microbiol.* 1985, *29*, 242–243.

Wolfe, R. S.; Higgins, I. J. "Microbial Biochemistry of Methane—A Study in Contrasts." *Int. Rev. Biochem.* 1979, *21*, 276–300.

In Situ Biotransformation of Carbon Tetrachloride, Freon-113, Freon-11, and 1,1,1-TCA under Anoxic Conditions

*Lewis Semprini**, *Gary D. Hopkins, Paul V. Roberts, Perry L. McCarty*
Stanford University

INTRODUCTION

Chlorinated aliphatic compounds with one or two carbon atoms are widely used as solvents, degreasing agents, and intermediates in chemical synthesis. Their widespread use and uncontrolled disposal have resulted in the contamination of groundwater supplies (Westrick *et al.* 1984). There is an urgent need to understand the behavior of these contaminants in the subsurface, to develop methods for monitoring the distribution and movement of the chemicals, and to clean up contamination once its extent is delineated. *In situ* bioremediation of contamination by halogenated aliphatics is a promising alternative for aquifer restoration, since the process may lead to complete mineralization to nontoxic end products and/or may create intermediate products that are less harmful, that are more easily removed from the aquifer, and that may be more readily treated by other subsequent processes.

This work assessed under field conditions the capacity of native organisms—i.e., bacteria indigenous to the subsurface environment—to metabolize halogenated synthetic organics when the proper conditions were provided to enhance microbial growth. Reducing conditions were promoted in the field by stimulating a consortium of denitrifying

* Department of Civil Engineering, Stanford University, Stanford CA 94305-4030

bacteria, and perhaps sulfate-reducing bacteria, through the addition of acetate as a primary substrate for growth to the aquifer that contained both nitrate and sulfate. Under biostimulated conditions the transformation of target compounds, including carbon tetrachloride (CT), 1,1,1-trichloroethane (TCA), fluorotrichloromethane (Freon-11), and 1,1,2-trichloro-1,2,2-trifluoroethane (Freon-113), was assessed by controlled addition, frequent sampling, quantitative analysis, and mass-balance comparisons.

The specific objectives of this field study were: (1) to demonstrate in a controlled field experiment the ability to biostimulate an indigenous population of denitrifying bacteria under conditions representative of groundwater environments; (2) to quantify the extent of enhanced biodegradation of CT, TCA, Freon-11, and Freon-113 in the biostimulated zone, and the formation of intermediate products; (3) to determine how to modify biostimulation conditions to achieve more complete mineralization of the halogenated aliphatics; and (4) to use mathematical models that incorporate key terms accounting for microbial and transport processes for interpreting the results of the field experiments.

Background

Studies have shown that chlorinated aliphatic compounds can be biologically transformed under a range of environmental conditions (Vogel *et al.* 1987). In reviewing chemical, biological, and enzymatic studies, Vogel *et al.* (1987) noted some clear trends between the ease of aerobic or anaerobic transformation and the oxidation state and chemical structure of the chlorinated aliphatics. Compounds that are more highly substituted with halides (more oxidized) are more likely to undergo reductive dehalogenation under anaerobic conditions, forming less halogenated intermediates. These less halogenated intermediates often are less reactive to subsequent reduction under anaerobic conditions (Vogel 1988; Vogel & McCarty 1985).

Recent studies of enhanced *in situ* transformation under aerobic conditions by methanotrophic bacteria (Semprini *et al.* 1990) have shown less chlorinated ethenes such as vinyl chloride and *trans-* and *cis*-dichloroethene (DCE) to be more rapidly transformed than trichloroethene (TCE). It was apparent from these field studies that more highly chlorinated organics such as CT and tetrachloroethene (PCE) are not amenable to aerobic transformations. Thus, enhanced anaerobic transformations for these compounds should be investigated.

The occurrence of reductive transformation of halogenated aliphatic compounds in groundwater was demonstrated by Bouwer *et al.* (1981). CT transformation has been observed in laboratory studies under reducing conditions ranging from denitrifying to methanogenic environments. Bouwer and McCarty (1983) observed the biotransformation of CT and brominated trihalomethanes in batch-fed, mixed denitrifying cultures. CT was assimilated into cell mass and mineralized to CO_2. Chloroform (CF) was detected as a transformation intermediate, indicating that reductive dehalogenation, at least to some extent, was occurring under these conditions. Bouwer and Wright (1988) studied the transformation of CT and several other chlorinated aliphatics under conditions of denitrification, sulfate reduction, and methanogenesis in anoxic biofilms. All were transformed under methanogenic conditions. Bromoform, bromodichloromethane, CT, and hexachloroethane were transformed under the less reducing conditions of denitrification. In a continuous flow column study operated under denitrifying conditions, Rittmann *et al.* (1988) found CT was nearly completely transformed, whereas bromoform, dibromethane, trichloroethylene, and PCE were transformed to lesser degees.

Little is known about the bacteria or other microorganisms that degrade CT in anaerobic environments. Egli *et al.* (1987, 1988) found pure cultures transformed CT to CF and dichloromethane (DM) quantitatively under both sulfate-reducing and methane-producing conditions. An acetogen was isolated that converted CT to nonhalogenated products. Galli and McCarty (1989) identified a *Clostridium* sp. from the effluent of an anaerobic reactor that biotransformed TCA, CF, and CT. CT transformation led to the production of CF as an intermediate, with further transformation to DM and unidentified products. Criddle *et al.* (1990a) isolated a pure denitrifying culture, *Pseudomonas* sp. strain KC, that degraded CT, 50 percent to CO_2 and about 40 percent to nonvolatile compounds. Minimal CF production was observed. Criddle *et al.* (1990b) also found CT could be transformed by *Escherichia coli* under fermenting conditions.

The extent of reported mineralization of CT to CO_2 varies ranging from 10 to 99 percent. Criddle (1989) presented parallel pathways for the degradation of CT under anaerobic conditions, leading to potential formation of CF, DM, CO_2, CO, formic acid, and hexachloroethane.

SITE CHARACTERIZATION

The Moffett Field Naval Air Station, Mountain View CA, was chosen for this demonstration; it was used earlier to study *in situ* restoration of chlorinated aliphatics by methanotrophic bacteria (Roberts *et al.* 1989). The site has been well characterized and is typical of groundwater contamination by chlorinated organics in a shallow sand-and-gravel aquifer. The shallow aquifer of the test site consists of a layer of sand and gravel, approximately 5 m below the surface and 1.2 m thick, well confined above and below by a silty clay layer of low permeability (Figure 1). The transmissivity of the test zone is high (approximately 100 m^2/day), which permits extraction of water at the design rate (approximately 10 L/min) without drawdown to the well screens of the extraction well.

The formation groundwater is also of appropriate composition for the field experiments. Dissolved oxygen concentration is below oxygen probe (0.2 mg/L) detection. Nitrate and sulfate, two potential electron acceptors, are present at concentrations of 25 mg/L (as nitrate) and 700 mg/L (as sulfate). The groundwater was contaminated with TCA (50 µg/L), Freon-113 (6 µg/L), and Freon-11 (3 µg/L). Measurement detection limits were 0.5 µg/L. The target compound, CT, was

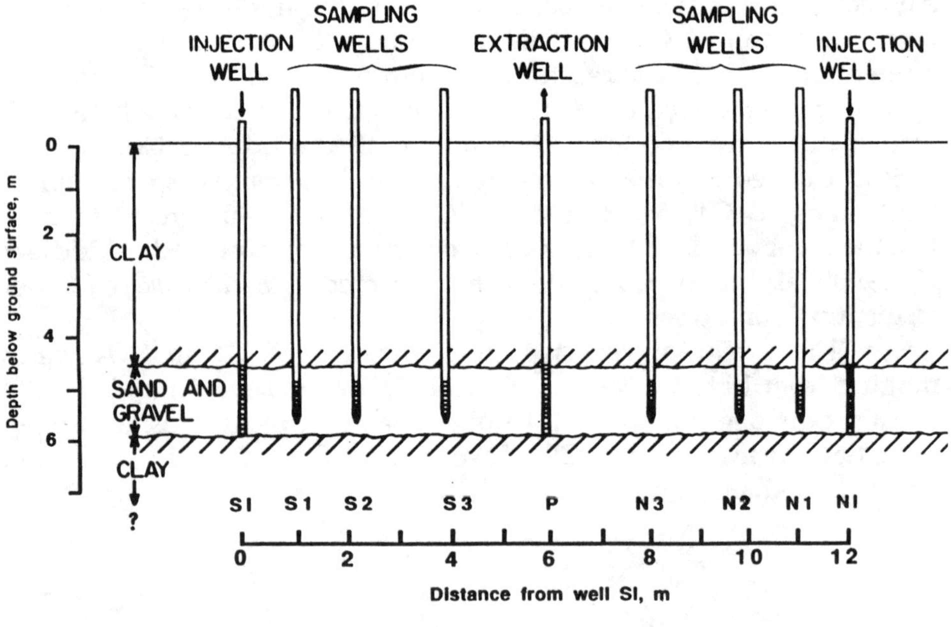

FIGURE 1. Schematic of the test zone.

not present and therefore was continuously added in a controlled manner to the injected water. The other halogenated aliphatics that were present in the extracted groundwater were reinjected along with CT into the test zone. No appreciable amounts of toxic metals were found (Roberts *et al.* 1989). Both nitrate and phosphorus, naturally present in the subsurface, served as sources of N and P so that their addition was not required during biostimulation of the test zone. Phosphate concentrations in the groundwater were low (0.1 mg/L); however, the phosphate minerals present provided a source of phosphate.

EXPERIMENTAL METHODOLOGY

A methodology was developed to evaluate objectively and quantitatively the effectiveness of the approach for stimulating anoxic microbial growth to transform the target organic compounds under natural conditions at the field site. This entailed creating a flow field dominated by pumping from an extraction well, while introducing solutes in known amounts at a nearby injection well (Figure 1). Concentrations were measured regularly at the injection, extraction, and intermediate observation points. Evidence of transformation was then assessed by quantitative examination of the concentration histories of the various solutes at the monitoring points, and comparing results under biostimulation conditions with results obtained under similar conditions in the absence of biostimulation. A specially designed automated data acquisition and control system, described in detail by Hopkins *et al.* (1988) and Roberts *et al.* (1989), provided continuous records of highly accurate data over sustained periods, which enabled mass balances to be made with relative errors of approximately ±5 percent.

The monitoring wells were located 1, 2.2, and 3.8 m from the injection well in a direct line with the extraction well located 6 m away (Figure 1). The chemicals of interest for a given experiment were added as soluble components to the injected groundwater. Groundwater was injected into well SI at a rate of 1.5 L/min and extracted at well P at a rate of 10.0 L/min.

Experimentation was conducted as a sequence of tests. The Stage 1 tests consisted of bromide tracer tests to determine which experimental leg was best suited for the field evaluation based on the transport characteristics and the ability to recover the injected tracer at the extraction well. In Stage 2 the transport of CT and bromide tracer through the selected test leg in the absence of biostimulation was studied. This stage served to determine the extent of retardation in movement of CT, evaluate whether CT was transformed either by biotic or abiotic processes before the test zone was biostimulated, and saturate the test zone with CT. Stage 3 offered the main results of the field evaluation: to determine whether a denitrifying microbial population could be easily stimulated through the addition of acetate as a growth substrate, evaluate the rates and extents of biotransformation of the chlorinated aliphatics present, and identify the intermediates transformation products formed. Stage 4 involved a transient test to determine whether enhanced biotransformation of the chlorinated aliphatics in the test zone resulted from the complete removal of nitrate from the inject fluid using a surface bioreactor fed acetate.

RESULTS OF THE FIELD EXPERIMENTS

Stage 1 Tracer Tests

The Stage 1 tracer experiments were performed along two experimental legs to determine whether the north leg (NI, N1, N2, N3, P) or the south leg (SI, S1, S2, S3, P) was best suited for the biostimulation/biotransformation experiments. Under the induced gradient conditions of injection and extraction, only 80 percent of the bromide was recovered at the extraction well when injected into the NI well, whereas over 90 percent was recovered when it was injected into the SI well. A strong regional flow of 1 to 2 m/day from south to north (Roberts *et al.* 1990) most likely caused the lower recovery along the north leg.

Because of its better recovery, the south leg was selected for the subsequent experiments. The south experimental leg had been used previously for bioremediation studies using methanotrophic bacteria, an aerobic treatment process. Thus in using the same experimental leg, it was possible to determine whether both aerobic and anoxic transformation processes could be enhanced in the same test zone.

Stage 2 Tracer Tests

The Stage 2 tracer test (Tracer 14) was performed along the south experimental leg in the absence of biostimulation. CT was continuously injected at an initial concentration of 70 µg/L using the feed method described by Roberts *et al.* (1990). Nitrate in the extracted groundwater was recycled in the injected fluid at a concentration of 25 mg/L.

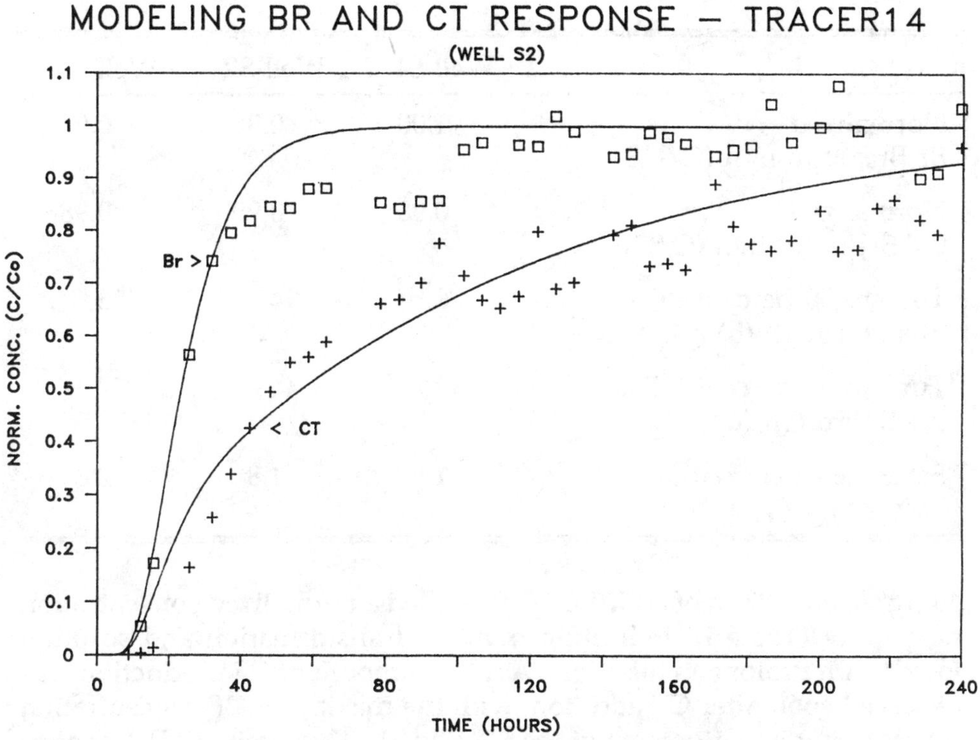

FIGURE 2. Bromide and CT breakthrough at the S2 well and the corresponding simulation using the nonequilibrium sorption model.

Figure 2 shows the normalized breakthrough of CT and bromide at the S2 observation well along with model simulations. The model simulations were performed using a 1-D advection/dispersion transport model with rate-limited sorption, as described by Semprini *et al.* (in press). The breakthrough of CT is retarded by sorption onto the aquifer solids. The hydraulic residence times (Table 1) between the injection well and the three observation wells, S1, S2, and S3, ranged from 8 to 28 h. CT residence times, longer because of sorption onto the aquifer solids, ranged from 12 to 57 h. The resulting CT retardation factors ranged from 1.5 to 2.0. CT was much less strongly sorbed than *cis-* and *trans-*DCE and TCE, for which estimated retardation factors ranged from 6 to 12 (Roberts *et al.* 1990).

The tracer test also confirmed that the injected fluid completely permeated the test zone around the S1 and S2 wells. To saturate the aquifer quickly, the CT concentration was lowered from 70 µg/L to

TABLE 1. Results of tracer 14 test.

	Well S1	Well S2	Well S3
Normalized Br Breakthrough (C/C_0)	1.00	0.98	0.94
Normalized CT Breakthrough (C/C_0)	0.98	0.99	0.98
Time to 50 percent Br Breakthrough (h)	8	24	28
Time to 50 percent CT Breakthrough (h)	12	44	57
Estimated Retardation Factor (T_{CT}/T_{Br})	1.5	1.8	2.0

40 µg/L after 400 h of addition. CT reached a normalized concentration near unity (Table 1), indicating minimal transformation and sorption losses with prolonged injection. A minor amount of CF production was observed soon after CT addition, with the maximum CF concentration representing 3 to 4 percent of the CT added. Thus, minor CT transformation was observed before biostimulation of the test zone through acetate addition.

Stage 3, Biostimulation and Biotransformation Experiments

The biostimulation and biotransformation evaluations conducted in the field were consistent in most major respects with expectations from laboratory results and theory. It was confirmed that a native bacterial community could be rapidly stimulated by introducing acetate as a growth substrate into an aquifer that contained nitrate and sulfate as potential electron acceptors, without any supplemental nutrients. In the initial biostimulation experiment, the utilization of acetate and nitrate rapidly commenced, with virtually complete nitrate utilization occurring after 100 h of acetate addition. A transitory build-up of nitrite concentration was observed within the first 60 h of addition, in response to the establishment of denitrifying conditions, which is typical (McCarty *et al.*

1969). Clogging of the injection well and borehole was effectively controlled by adding the acetate in a high concentration pulse (320 mg/L) for a period of 1 h in a 13-h pulse cycle while continuously recycling nitrate (25 mg/L) in the native groundwater. More than 80 to 90 percent of the acetate was consumed within the first meter of transport. The stoichiometric ratios of nitrate to acetate consumption were approximately 1 mg nitrate per milligram acetate, which is lower than the ratio calculated for complete respiration of nitrate to nitrogen gas, due to the incorporation of an estimated 40 percent of the acetate into cell biomass during biostimulation (McCarty *et al.* 1969). No attempt was made to measure directly the biomass concentration. Nitrogen gas formation with distinct bubbles, however, was not observed in samples at the surface, since the complete respiration of 25 mg/L of NO_3^- would produce about 11 mg/L of N_2 gas, which is a factor of two lower than its solubility in water at atmospheric pressure and aquifer temperature of 18 C.

To evaluate the transformation of CT, the target organic compound, CT was continuously injected at a concentration of 40 µg/L until the soil was saturated, as evidenced by the complete breakthrough at the monitoring wells (Table 1), in the absence of acetate addition. CT injection into the test zone was continued upon the addition of acetate.

CT transformation, as indicated by a decrease in its concentration at monitoring locations, and the formation of CF as an intermediate product significantly lagged behind the uptake of acetate and nitrate in the test zone (Figure 3). Decreases in CT concentration and increases in CF concentration were observed after approximately 400 h, with gradual decreases over the 1,250-h period (52 days) that acetate and nitrate were injected into the test zone.

Transformation of CT and the formation of CF as an intermediate product were more rapid and more complete at the S2 observation well, 2 m from the injection well (Figure 3), compared to the S1 well, 1 m from the injection well (not shown). The response indicated that the most rapid rates of transformation did not occur in the first meter of transport, where most of the acetate and nitrate were consumed, but in the zones further removed, where significantly less acetate and nitrate were consumed.

The results indicate that the main denitrifying population did not participate in the transformation process to the same extent as microbes stimulated further away. The transformation of CT by denitrifiers may have been strongly inhibited by the presence of nitrate in the test zone. We hypothesize that a secondary microbial population, living on acetate or decay products of the stimulated denitrifiers, were slowly

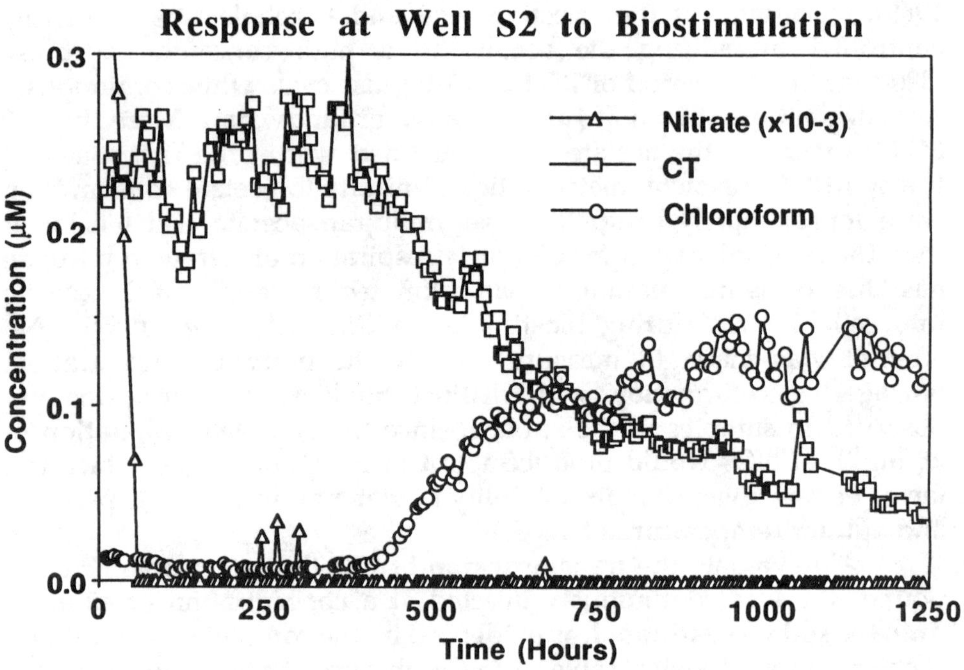

FIGURE 3. Observed CT and CF and nitrate responses at the S2 well in response to biostimulation of the test zone.

growing and were responsible for the transformation. The growth of this population and/or its transformation of CT may have been inhibited by the presence of nitrate.

Stage 4, Transient Experiment

A transient experiment was performed to study the effect of nitrate on biotransformation. Nitrate was removed from the injected fluid by a surface bioreactor that was fed acetate. The transient test was initiated after 1,260 h of acetate addition. A significant decrease in CT was observed over the 300-h period of the test, as indicated by the bar graph shown in Figure 4. CF concentration increased to a lesser extent, indicating either that less CF was being formed in a parallel transformation pathway or that CF was being degraded at higher transformation rates. Before nitrate was completely eliminated from the test zone, 55 to 67 percent of the transformed CT appeared as CF, whereas only 30 to 40 percent was observed after nitrate addition was terminated. CF was

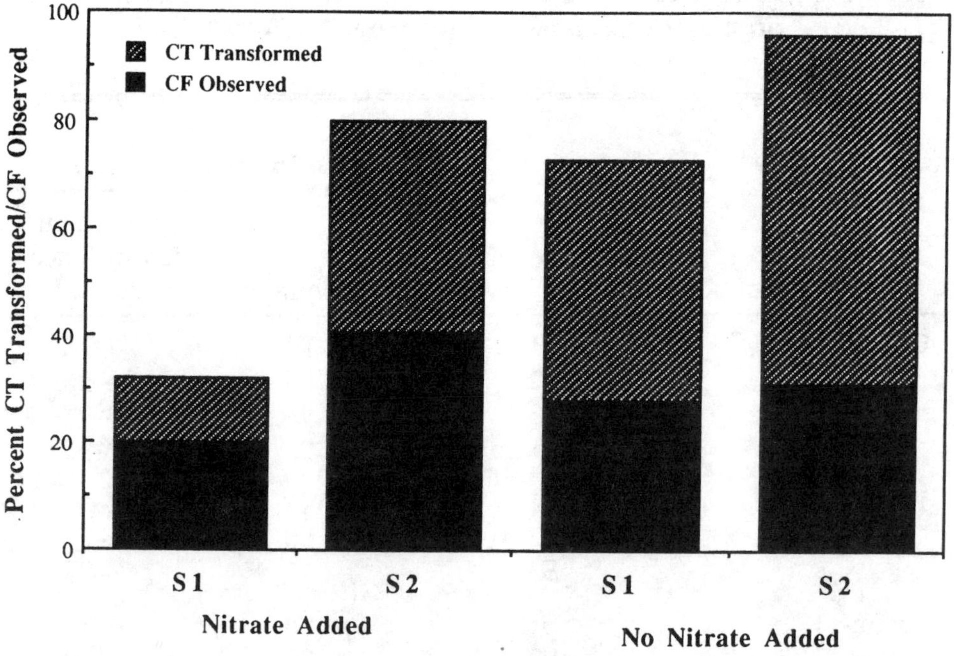

FIGURE 4. Percentage transformation of CT and fraction appearing as CF during periods with and without nitrate addition.

the main chlorinated intermediate product found. DM and chloromethane, possible intermediate products of CT transformation, were not detected at the detection limit of 1 µg/L.

No direct evidence was found for the stimulation of sulfate-reducing bacteria or methanogenic bacteria when nitrate was completely removed. Neither sulfide nor methane were detected in the groundwater extracted from the test zone. If sulfate-reducing conditions were established, however, reactions with test zone minerals may have scavenged sulfide from the groundwater. Direct measurements of sulfate uptake could not be made due to high sulfate concentration (700 mg/L) in the groundwater.

The transformation of background contaminants, including Freon-11, Freon-113, and TCA, was also observed in the biostimulated zone. The responses of the halogenated aliphatics were similar to that of CT, but with slower rates of transformation. Rates of transformation were also enhanced when nitrate was removed from the test zone. The degrees of transformation (Table 2), quantified by normalization to the bromide breakthrough, were as follows: CT, 70 to 97 percent; Freon-11,

TABLE 2. Estimates of the degree of transformation based on mean calculated values from 1,450 to 1,550 h.

		Percent Biotransformation	
Chemical	Well	Average	95 Percent Confidence Interval
CT	S1	74	70–78
	S2	95	94–96
	S3	96	95–97
	Extraction	93	89–96
Freon-11	S1	46	42–50
	S2	68	65–71
	S3	72	69–75
Freon-113	S1	8	0–16
	S2	20	10–30
	S3	18	8–27
TCA	S1	9	5–13
	S2	15	11–19
	S3	9	2–16

42 to 75 percent; Freon-113, 0 to 30 percent; and TCA, 5 to 19 percent. Of the values cited, the lower value represents the nearest observation well and the lower 95 percent confidence intervals, and the higher value represents the farther observation wells and the upper 95 percent confidence intervals. Steady-state transformation conditions had not been achieved by the end of the experiments; thus, these transformation extents are considered conservative estimates.

Overall the field results confirmed the ability of indigenous bacteria to promote the biotransformation of CT, Freon-11, Freon-113, and TCA under anoxic conditions. Denitrification was readily accomplished through the addition of acetate to the test zone. The responses indicate, however, that the main population of denitrifying bacteria was not responsible for the CT transformation, but rather that a secondary population was responsible. CT transformations of 95 percent or greater were achieved in the test zone. CF, however, was produced as an intermediate transformation product and accounted for 30 to 40 percent of the CT transformed.

MATHEMATICAL MODELING

A non-steady-state model, developed for simulating the biostimulation and biotransformation tests, proved useful in interpreting the results of the field experiments (Semprini *et al.*, in press). The model accounts for the basic phenomena of microbial growth, electron donor and electron acceptor utilization, biotransformation of the chlorinated compounds, and the formation of intermediate products. The model simulates the growth and metabolism of two microbial populations: a denitrifying population and a second assumed population that utilizes the respiration products of the denitrifiers. The approach adequately simulated the transient decreases in CT concentration from its biotransformation and the increase in CF concentration from its formation as an intermediate product. Some parameter adjustments were necessary to achieve the model fits. An heuristic fitting procedure was used. The model also fit well the observed concentration histories of other halogenated aliphatics, when different biotransformation rate coefficients were applied for different compounds.

The values for the effective first order rate coefficients (k/K_s) determined from model fits to the field observations are given in Table 3. The coefficients were in the range of those reported in the literature for microbial transformation under sulfate-reducing conditions (Bouwer & Wright 1988) and for a pure *Clostridium* culture (Galli & McCarty 1989). The rate coefficient for CT was 40 times greater than that for TCA, consistent with the results of Bouwer and Wright (1988). The simulations indicate the rate coefficient for Freon-11 transformation was less than half that for CT as a result of the substitution of one fluorine atom for a chloride atom. The substitution of a hydrogen atom for a chlorine atom has an even greater effect: CF transforms at a rate only one-fifth as great as does CT. Galli and McCarty's (1989) *Clostridium* sp. culture degraded CF at a rate 10 times slower than CT. The rate of transformation of Freon-113 ($C_2Cl_3F_3$) was estimated to be four times greater that the less-substituted TCA ($C_2Cl_3H_3$). Both absolute and relative values of rate coefficients from laboratory studies and the field and modeling study appear to be in reasonable agreement, which is encouraging.

TABLE 3. Values of k/K_s for the field study and from laboratory studies.

Compound	This Study $\frac{k}{K_s}$ $\left(\frac{L}{mg\text{-}cells \cdot d}\right)$	Galli & McCarty (1989) $\frac{k}{K_s}$ (a) $\left(\frac{L}{mg\text{-}cells \cdot d}\right)$	Bouwer & Wright (1988) $\frac{k}{K_s}$ (b) $\left(\frac{L}{mg\text{-}cells \cdot d}\right)$
CT	0.40	0.12	0.20
Freon-11	0.16	–	–
CF	0.08	0.012	–
Freon-113	0.04	–	–
TCA	0.01	0.007	0.005

(a) Reported rates on protein basis, converted to cell based on protein being 65 percent of the cell mass.
(b) Under sulfate-reducing conditions.

DISCUSSION

In this study CT was transformed to a significant extent and at a rapid rate under subsurface conditions in the absence of dissolved oxygen when a native population was biostimulated by the addition of acetate in the presence of nitrate. CF was formed as an intermediate product. Freon-11, Freon-113, and TCA, background contaminants in the test zone, were also transformed to significant extents in the field and responded like CT, but at reduced rates of transformation.

Laboratory soil column studies were also conducted under conditions similar to the field tests (Lang *et al.* 1990). The laboratory studies confirmed that a significant amount of CT was completely mineralized to CO_2, and CF was produced as an intermediate product. Laboratory soil column studies also showed a gradual increase in the extent of CT transformation, which had been delayed compared to the rapid stimulation of the denitrifying bacteria, supporting the field results and the hypothesis that a secondary microbial population, and not the main denitrifying population, was responsible for the transformation of the halogenated aliphatics.

Transformation in the field was more complete after nitrate was completely removed from the test zone and in zones that lacked the main population of denitrifiers. The response observed in the field and in the laboratory columns indicated that a secondary microbial population, and not the main denitrifying population, was responsible for the transformation. Modeling studies supported this hypothesis. The modeling results were consistent with the hypothesis that the halogenated aliphatics were transformed by the secondary population, but at slower rates than CT. The rates of transformation determined from model fits to the field response were in the range of those reported in the literature for microbial transformations under sulfate-reducing conditions and by a *Clostridium* culture. Sulfate reducers or *Clostridium,* or both, may have existed in the test zone. Sulfate reducers might have grown on the decay products of the denitrifiers under active biostimulation. Acetate, at 8 mg/L, was also consumed in the test zone after nitrate was removed from the injected fluid; sulfate reducers may have been responsible for this consumption; however, there was no direct evidence of sulfate reduction. Another possibility is that microorganisms such as *Clostridium* growing on the decay products of the stimulated denitrifiers were carrying out the transformation (Galli & McCarty 1989).

The field transformation rates were more rapid under more reducing conditions, consistent with laboratory studies of Bouwer and Wright (1988) and Criddle (1989). The question remains whether biostimulation of denitrifiers in the test zone was required to provide a source of growth substrate in the form of their respiration products for secondary bacteria, such as *Clostridium,* that promote the transformation. It is also not known whether the enhanced rates that resulted when nitrate was removed were caused by the absence of nitrate as an electron acceptor or whether nitrate inhibited growth of the transforming microorganisms.

CONCLUSIONS

This study convincingly demonstrated that CT and several other halogenated aliphatic compounds were transformed to significant extents and at rapid rates with half-lives on the order of hours to days in the absence of oxygen, when a natural bacteria population was biostimulated through the addition of acetate in the presence of nitrate and sulfate. The formation of CF as a halogenated intermediate product of CT transformation was observed, consistent with the results of laboratory studies reported in the literature. The formation of CF as a halogenated intermediate product, which is objectionable from a water quality standpoint, poses an obstacle to the deployment of this approach for enhanced aquifer remediation.

The biostimulation and biotransformation experiments demonstrated that:

1. Denitrifying organisms could be readily stimulated to remove nitrate upon the addition of acetate as a growth substrate to the subsurface.

2. Over 90 percent CT transformation was observed in the 2-m biostimulated zone. CF was an intermediate product and represented 30 to 60 percent of the CT transformed.

3. More rapid CT transformation was observed after nitrate was completely removed from the test zone and a lower fraction was observed as CF.

4. The biotransformation response and modeling exercise supported the hypothesis that a secondary population, and not the main denitrifying population, was responsible for the transformation.

5. The other halogenated aliphatics were biotransformed, but at slower rates than CT, with the following extents of transformation achieved: CT, 95 to 97 percent; Freon-11, 65 to 75 percent; Freon-113, 10 to 30 percent; and TCA, 11 to 19 percent.

Model simulations of transport, biostimulation, and biotransformation proved a useful tool in improving the understanding of the salient processes that proceeded concurrently in the complex field setting. Modeling supported the hypothesis that a secondary population was responsible for transformation of halogenated aliphatics, and that the growth of the population and/or its transformation ability was inhibited by the presence of nitrate. The modeling exercise indicated that a better understanding of the key microbial processes is required to determine how these complex transformation processes may be better represented in mathematical models.

Acknowledgment

This research was sponsored by the Kerr Environmental Research Laboratory of the U.S. Environmental Protection Agency as part of the Biosystems Program under Cooperative Agreement CR 815816. The content of this paper does not necessarily reflect the views of that organization, and no official endorsement should be inferred.

REFERENCES

Bouwer, E. J.; McCarty, P. L. "Transformation of Halogenated Organic Compounds under Denitrification Conditions." *Appl. Environ. Microbiol.* **1983,** *45,* 1295–1299.

Bouwer, E. J.; Wright, J. P. "Transformation of Trace Halogenated Aliphatics in Anoxic Biofilm Columns." *J. Contaminant Hydrol.* **1988,** 2, 155–169.

Bouwer, E. J.; Rittmann, B. E.; McCarty, P. L. "Anaerobic Degradation of Halogenated 1- and 2-Carbon Organic Compounds." *Env. Sci. Technol.* **1981,** *15,* 596–599.

Criddle, C. S. "Reductive Dehalogenation in Microbial and Electrolic Model Systems." Ph.D. Dissertation, Department of Civil Engineering, Stanford University, Stanford, CA, 1989.

Criddle, C. S; Dewitt, J. T.; Grbić-Galić, D.; McCarty, P. L. "Transformation of Carbon Tetrachloride by *Pseudomonas* sp. Strain KC Under Denitrifying Conditions." *Appl. Environ. Microbiol.* **1990a,** *56,* 3240–3246.

Criddle, C. S.; Dewitt, J. T.; McCarty, P. L. "Reductive Dehalogenation of Carbon Tetrachloride by *Escherichia coli* K-12." *Appl. Environ. Microbiol.* **1990b,** *56,* 3247–3254.

Egli, C.; Scholtz, R.; Cook, A. M.; Liesinger, T. "Anaerobic Dechlorination of Tetrachloromethane and 1,2-Dichloroethane to Degradable Products by Pure Cultures of *Desulfobacterium* sp. and *Methanobacterium* sp." *FEMS Microbiol. Lett.* **1987,** 43, 257–261.

Egli, C.; Tsuchan, T.; Scholtz, R.; Cook, A. M.; Lieslinger, T. "Transformation of Tetrachloromethane to Dichloromethane and Carbon Dioxide by *Acetobacterium Woodi.*" *Appl. Environ. Microbiol.* **1988,** *55,* 2819–2824.

Galli, R.; McCarty, P. L. "Biotransformation of 1,1,1-Trichloroethane, Trichloromethane, and Tetrachloromethane by *Clostridium* sp." *Appl. Environ. Microbiol.* **1989,** *55,* 837–844.

Hopkins, G. D.; Semprini, L.; Roberts, P. V.; Mackay, D. M. "An Automated Data Acquisition System for Assessing in-situ Biodegradation of Chlorinated Aliphatic Compounds." *Proceedings,* Second Outdoor Conference on Ground-water Monitoring and Aquifer Restoration, NWWA, Las Vegas, NV, May 23–26, 1988; Vol. 1, pp 201–203.

Lang, M.; Semprini, L.; Roberts, P. V.; McCarty, P. L.; Janssen, D. B. "Biotransformation of Carbon Tetrachloride in Batch-Exchanged Laboratory Soil Columns under Anoxic Conditions." EOS, *Trans. Amer. Geophys. Union,* **1990,** *71*(43), 1323.

McCarty, P. L.; Beck, L.; St. Amant, P. "Biological Denitrification of Wastewaters by Addition of Organic Materials." *Proceedings,* 24th Annual Industrial Waste Conference, Purdue University, 1969; pp 1271–1285.

Rittmann, B. E.; Valocchi, A. J.; Odencrantz, J. E.; Bae, W. "In-Situ Bioreclamation of Contaminated Groundwater." Illinois Hazardous Waste Research and Information Center, HWRIC RR 031.121, 1988.

Roberts, P. V.; Semprini, L.; Hopkins, G. D.; Grbić-Galić, D.; McCarty, P. L.; Reinhard, M. "In-Situ Aquifer Restoration of Chlorinated Aliphatics by Methanotrophic Bacteria." EPA/600/2-89/033, Robert S. Kerr Laboratory Research Laboratory, July 1989.

Roberts, P. V.; Hopkins, G. D.; Mackay, D. M.; Semprini, L. "A Field Evaluation of In-Situ Biodegradation of Chlorinated Ethenes: Part 1, Methodology and Field Site Characterization." *Ground Water* **1990**, *28*, 591–604.

Semprini, L.; Roberts, P. V.; Hopkins, G. D.; McCarty, P. L. "A Field Evaluation of In-Situ Biodegradation of Chlorinated Ethenes: Part 2, Results of Biostimulation and Biotransformation Experiments." *Ground Water* **1990**, *28*, 715–727.

Semprini, L.; Hopkins, G. D.; Janssen, D. B.; Lang, M.; Roberts, P. V.; McCarty, P. L. "In-Situ Biotransformation of Carbon Tetrachloride under Anoxic Conditions." Robert S. Kerr Laboratory Research Laboratory [in press].

Vogel, T. M. "Biotic and Abiotic Transformation of Halogenated Aliphatic Compounds." Ph.D. Dissertation, Department of Civil Engineering, Stanford University, Stanford, CA, 1988.

Vogel, T. M.; McCarty, P. L. "Biotransformation of Tetrachloroethylene to Trichloroethylene, Dichloroethylene, Vinyl Chloride, and Carbon Dioxide under Methanogenic Conditions." *Appl. Environ. Microbiol.* **1985**, *49*, 1080–1083.

Vogel, T. M.; Criddle, C. S.; McCarty, P. L. "Transformations of Halogenated Aliphatic Compounds." *Env. Sci. Technol.* **1987**, *21*(8), 722–736.

Westrick, J. J.; Mello, W. J.; Thomas, R. F. "The Groundwater Supply Survey." *J. Am. Water Works Assoc.* **1984**, *76*, 52–59.

The Effect of Electron Donor Concentration on the Biotransformation of Chlorinated Aliphatics

*Joseph B. Hughes**, *Gene F. Parkin*
University of Iowa

INTRODUCTION

Chlorinated aliphatic compounds are commonly used in a variety of industrial and agricultural applications. C1 chlorinated aliphatics such as chloroform (CF) and dichloromethane (DCM) and C2 chlorinated aliphatics such as 1,1,1-trichloroethane (TCA) are commonly used as cutting fluids, solvents, and dry cleaning agents. The widespread use of these compounds has resulted in their frequent detection in groundwater and landfill leachates (Vogel *et al.* 1987). While the fate and persistence of TCA, CF, and DCM in the environment are not identical, each poses long-term environmental concerns. Research concerning the ability of microorganisms to degrade these compounds has shown that microbial transformations of TCA, DCM, and CF occur, and at rates higher than abiotic transformations (Rittmann & McCarty 1980; Bouwer *et al.* 1981; LaPat-Polasko *et al.* 1984; Gossett 1985; Vogel & McCarty 1987; Galli & McCarty 1989a,b).

In general, most studies investigating the biotransformation of chlorinated aliphatics have focused on concentrations below 1 mg/L. However, TCA, CF, and DCM have high water solubilities (DCM, for example, has a reported solubility of 20,000 mg/L at 20 C [LaPat-Polasko *et al.* 1984]); therefore, concentrations of these compounds, alone or in mixtures, in excess of 1 mg/L can occur in severely

* University of Iowa, Department of Civil and Environmental Engineering, Iowa City, IA 52242

contaminated areas such as landfills and hazardous waste sites. This paper reports on research directed at the role of acetate concentration on the transformation of DCM, CF, and TCA in anaerobic filters. Information from this study is useful in evaluating the potential for bioremediation of sites contaminated with high concentrations of chlorinated aliphatics, alone and in mixtures.

POTENTIAL EFFECT OF ELECTRON DONORS

Research has been conducted in both pure and mixed cultures on the pathways of the degradation of DCM, CF, and TCA. Results from these studies may shed light on the role of the primary electron donor on the transformation of target compounds. A compounding factor in an attempt to understand this relationship in mixed cultures is that the diversity of microorganisms that may be present in the mixed culture suggests that more than one pathway may be involved in the removal of an individual chlorinated aliphatic. The rate and degree of transformations observed in the systems described herein did not allow for quantitative determination of intermediate compounds that would provide insight into the degradation pathways occurring in the systems.

The fate of TCA under methanogenic conditions was addressed by Vogel and McCarty (1987). They showed that abiotic as well as biologically mediated reactions are possible at low redox potential. The biologically mediated, reductive dechlorination of TCA to 1,1-dichloroethane (DCA) was found to be much more rapid than the abiotic elimination to 1,1-dichloroethene (DCE) or hydrolysis to acetate, and the DCA produced could be further reduced to chloroethane (CA) and eventually mineralized.

The transformation of C1 chlorinated aliphatics such as DCM and CF has been shown to have a more complicated mechanism than reductive dechlorination. Egli *et al.* (1988) concluded that the presence of organisms utilizing the acetyl CoA pathway is required for the transformation of carbon tetrachloride (CT) in anaerobic cultures. A variety of strictly anaerobic organisms possess the acetyl CoA pathway and have been classified in three groups by the manner in which acetate is utilized or produced by that organism (Brock & Madigan 1988). One of these three groups is the acetoclastic methanogens. They are responsible for the transformation of acetate to carbon dioxide and methane. The two other groups that utilize the acetyl CoA pathway are acetogenic organisms that conduct either homoacetic fermentation of sugars or the autotrophic synthesis of acetate from hydrogen and carbon dioxide. The composition of the nutrient feed (Table 1) would not select for the latter two groups, and any contribution by such organisms in comparison to the acetoclastic methanogens to the transformation of CF or DCM in this study would be minimal.

TABLE 1. Composition of nutrient media. All concentrations are listed in mg/L.

Constituent	Conc.	Constituent	Conc.
$Na_2S \bullet 4(H_2O)$	300.0	$CoCl_2 \bullet 6(H_2O)$	0.25
NH_4Cl	40.0	KI	0.25
$MgCl_2 \bullet 6(H_2O)$	40.0	$MnCl_2 \bullet 4(H_2O)$	0.05
KCl	40.0	NH_4VO_3	0.05
$(NH_4)_2HPO_4$	8.0	$ZnCl_2$	0.05
$FeCl_2 \bullet 4(H_2O)$	4.0	$Na_2MoO_4 \bullet 2(H_2O)$	0.05
$CaCl_2 \bullet 2(H_2O)$	2.5	H_3BO_3	0.05
Cysteine	1.0	$NiCl_2 \bullet 6(H_2O)$	0.05
$(NaPO_3)_6$	1.0	$NaHCO_3$	(a)

(a) Added as needed as pH buffer.

The complicating factor in the transformation of these compounds by components of the acetyl CoA pathway is that both oxidation to CO_2 and reductive dechlorination are possible transformation pathways. Both reduction and oxidation have been shown to occur in both mixed and pure cultures that possess organisms utilizing the acetyl CoA pathway (Gossett 1985; Egli *et al.* 1990).

The reductive dechlorination is a result of the reaction of the chlorinated aliphatic with corrinoids present as cofactors in the acetyl CoA utilizing organism. The rate of the reaction increases as the degree of substitution of the chlorinated aliphatic increases (CT > CF > DCM). Egli *et al.* (1990) determined that the rate of CT reduction to CF was 20 times more rapid than the reduction of CF to DCM.

The mechanism of the oxidative pathway of chlorinated aliphatics is less clear, but Egli *et al.* (1990) has presented evidence based on the heat stability of metallo-porphrin corrinoids that indicates that the corrinoids involved in the reductive dechlorination reaction are also responsible for the oxidation transformation of these pollutants. The oxidative pathway presumably becomes more evident as the degree of substitution of the chlorinated aliphatic decreases from the decrease in the rate of the competing reduction reaction.

Gossett (1985) performed batch studies on the degradation of CF in mixed methanogenic culture that demonstrated that CF was being converted to CO_2, DCM, and a nonstrippable residue. Gossett and Freedman (1988, 1989) presented evidence that an autotrophic acetogen was responsible for DCM transformation in a mixed methanogenic culture enriched with a complex media, and recently Freedman and Gossett (a,b) concluded that DCM can serve as a growth substrate for unidentified acetogenic organism(s).

Because acetate is the sole electron donor fed to the anaerobic filters used in this study, and the presence of hydrogen in the system should be minimal, it is unlikely that an autotrophic acetogen capable of growth on DCM would be present in appreciable quantities in these systems. The expected dominance of the acetoclastic methanogens in these reactors and their presumed role in the transformation of the chlorinated aliphatics is an important factor in how the primary electron donor utilization affects pollutant transformation. Acetoclastic methanogens require methanogenisis for the production of ATP and subsequent metabolic activity. Despite the fact that CF and DCM may be oxidized in significant quantities, it is unlikely that the electrons released could be coupled with a reaction that could sustain growth (i.e., result in methanogenisis). Therefore, acetate utilization will be required for the biological transformation of chlorinated aliphatics even when oxidation of the compound occurs.

Although a loss of cells from an anaerobic filter is expected after the decrease in acetate fed, the rate of loss is generally considered slow. The metabolic activity of the methanogens, and subsequently the transformation of the chlorinated aliphatics in the filter, however, should be directly proportional to the availability of the electron donor.

METHODS AND MATERIALS

Chemicals. Reagents used in this study were dichloromethane (Optima Grade), chloroform (Optima Grade), 1,1,1-trichloroethane (reagent grade), methanol (HPLC grade), acetic acid (glacial), and iso-octane (HPLC grade). All reagents were obtained from Fisher Scientific.

Anaerobic Filters. Three upflow anaerobic filters were operated during the experimental period (Figures 1 and 2) and are designated AF1, AF2, and AF3. Although the design of these reactors vary, the mode of operation was consistent for all three filters. Each filter had been in use for two or more years prior to the experiments described in this paper.

AF1 and AF2 are identical in design. They are Plexiglass cylinders 20.3 cm in length with an inside diameter of 6.8 cm. Effluent sampling ports are located 17.8 cm above the base of the filters. These filters were filled with 1.4-cm-diameter uniform glass marble support media, resulting in an empty bed void volume of 335 mL. AF3 is a Plexiglass cylinder 120 cm in length with a inside diameter of 6.8 cm. AF3 is equipped with seven sample ports spaced evenly over the length of the column at 15-cm intervals. The seventh sample port served as the effluent sample port, and samples were also taken from the first and second ports, although these data are not presented in this paper. Gravel with diameters ranging from 1.25 to 1.70 cm served as the support media, with a resulting empty bed void volume of 1.5 L.

All filters were housed in an upright incubator maintained at 35 C in the absence of light. Nutrient media and acetate were supplied with automated peristaltic pumps operating for 30 sec every 30 min. Aqueous solutions of DCM, CF, and TCA were introduced into the feed lines with syringe pumps under the same intermittent operation as the feed pumps. Initial influent acetate concentration for all filters was 1,800 mg/L (1 g COD/L-day). The acetate concentration was decreased in a stepwise manner, first to 450 mg/L, then to 100 mg/L. The hydraulic residence time (HRT) in all filters was maintained at 2 days throughout the experimental period.

Analytical Methods. Acetate concentration was measured in the effluents from AF1, AF2, and AF3 with a gas chromatograph (HP 5890A with an HP 3393A integrator and an FID). Samples were taken with disposable syringes through effluent sample ports. Samples were immediately filtered with 0.45-µ syringe tip filters (Millipore) and acidified with solid phosphoric acid. Aliquots of 1 µL were injected onto a GP 60/80 Carbopack c/0.3%, Carbowax 20M/0.1%, H3PO4 30" × 0.25" OD × 4 mm ID glass column (Supelco). Operating conditions were: inlet temperature 200 C, oven temperature 120 C, detector temperature 200 C, gas flow rate 50 mL/min He.

Chlorinated aliphatic concentrations were also determined by GC (HP 5890A with an HP 3393A integrator and an ECD). Effluent samples were taken with disposable syringes through sample ports, and influent samples were obtained by disconnecting feed lines and

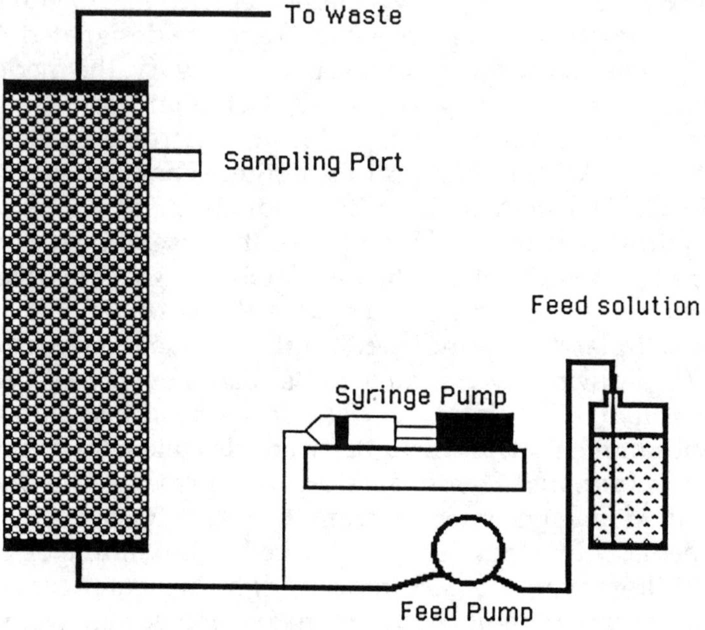

FIGURE 1. Schematic of AF1 and AF2.

collecting 10 mL directly. Filtered samples of 10 mL were extracted with 2 mL of iso-octane in 13-mL test tubes equipped with Teflon®-lined lids. After 70 sec of vigorous shaking and a short time period for phase separation, the iso-octane fraction was injected on a 30 m × 0.325 mm i.d. fused silica capillary column with DB-5 solid support at a 0.25-μ film thickness (J & W Scientific). Analyses were conducted at the following conditions: column pressure = 15 psi He, argon/methane make up gas, 15:1 split ratio, oven temperature 30 C, injection temperature 150 C, detector temperature 250 C.

Limits of detection for acetate, DCM, CF, and TCA are as follows: Acetate = 5 mg/L, DCM = 0.05 mg/L, CF = 0.025 mg/L, TCA = 0.025 mg/L. Chlorinated aliphatic standards were prepared by the method recommended by the United States Environmental Protection Agency (U.S. EPA 1987). This method involves the gravimetric dilution of DCM, CF, and TCA into methanol. The chlorinated aliphatics are allowed to dissolve completely in the methanol. Representative aqueous chlorinated aliphatic standard concentrations are prepared by a volumetric addition of the methanol solution (<20 μL) into 100 mL of deionized water. Standards are then extracted using the sample extraction method described previously.

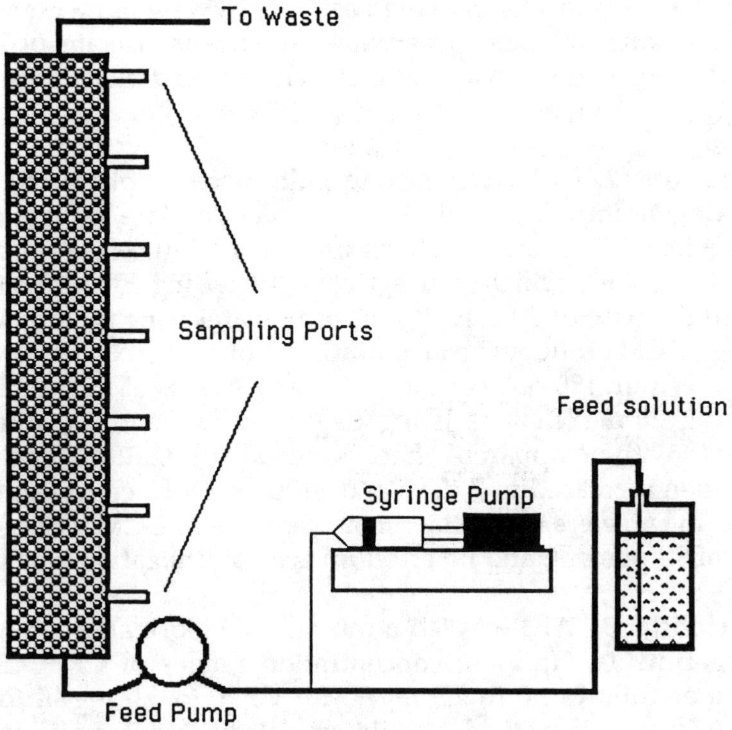

FIGURE 2. Schematic of AF3.

EXPERIMENTAL RESULTS

Previous Work

Previous work with AF1, AF2, and AF3 (Syverson 1989; Syverson *et al.* 1989; Hughes & Parkin, 1990) had developed cultures that actively degraded DCM, CF, and TCA when fed alone and in mixtures. These systems were maintained at an HRT of 2 days and influent acetate concentrations of 1,800 mg/L. A brief discussion of filter performance prior to altering electron donor concentration is presented below.

Anaerobic Filter 1. AF1 was fed CF as the sole chlorinated aliphatic. It was found to be readily degradable, in part to DCM, at concentrations not exceeding 3 to 4 mg/L. Exceeding this concentration would eventually cause filter performance to deteriorate (no CF transformation and reduced acetate utilization). Increasing the HRT from 2 to 7 days showed that at longer HRTs this concentration threshold could be exceeded. CF concentrations as high as 7 mg/L were fed with >99 percent removal and with no observable impact on acetate utilization. During this time, DCM was no longer detected as a biotransformation product.

Anaerobic Filter 2. DCM was the only chlorinated aliphatic fed to AF2. Initially, degradation was not observed, and acetate utilization was inhibited when DCM concentrations exceeded 10 mg/L. A small seed obtained from AF3, which was actively degrading DCM, was introduced into the base of AF2, and in approximately one month AF2 was degrading DCM influent concentrations of 5 mg/L to less than 0.5 mg/L. Within 120 days of the seeding of AF2, DCM concentrations approaching 20 mg/L were being degraded to nondetectable levels (<0.2 mg/L) with no apparent effect on acetate utilization. The HRT of AF2 was then increased to 7 days and influent DCM concentrations as high as 70 mg/L were fed to the filter. Removal of DCM was generally in excess of 90 percent, and no effect on acetate utilization was noticed.

Anaerobic Filter 3. AF3 was fed a mixture of chlorinated aliphatics in initial experiments. Influent concentration ranges of CF, DCM, and TCA were as follows: 0 to 2.5 mg/L for CF, 0 to 5.0 mg/L for TCA, and 0 to 6.7 mg/L for DCM. At times, chlorinated organic feed was removed from the systems due to increases in effluent acetate concentration. CF was suspected to be the cause of the decrease in acetate utilization and was eventually removed from the feed solution. For a period of 50 days, DCM and TCA were fed at 4 mg/L each with no effect on acetate utilization, confirming that CF was exerting a toxic effect on the filter. Throughout this entire experimental period, no chlorinated aliphatics were detected in the filter effluent. Eventually, TCA was removed from the feed to concentrate on DCM degradation.

Current Studies

Baseline operating conditions of three anaerobic filters degrading target chlorinated aliphatics at influent feed acetate concentrations of 1,800 mg/L had been maintained during previous experimental periods. Further experiments were designed to assess the role of the primary electron donor, acetate, on the transformation of DCM, CF, and TCA.

Anaerobic Filter 1. The initial 45 days of this experimental period were identical to previous study conditions (HRT = 2 days, feed acetate = 1,800 mg/L, CF approximately 2.0 to 2.5 mg/L). Effluent acetate data for the entire experimental period are shown in Figure 3. Percent acetate utilization ranged from 50 to 100 percent throughout the experimental period and was typically in excess of 70 percent.

CF data are also presented in Figure 3. Some variation in influent concentration was noticed in the second half of the experimental period without great effect on measured effluent concentrations. This variation was most likely the result of influent sampling problems. Slight variations in the pumping rates of the two pumps utilized in the operation of the system led to this inconsistency. Similar variations in the influent concentrations were observed in AF2 and AF3 and were attributed to variation in pumping rates. The effect of the mechanical inconsistency could be dampened by taking a larger sample, except that, under experimental temperatures, volatilization of these compounds is quite rapid. The error introduced in the longer sampling time required for a larger sample size would lead to a consistent underestimation of influent concentration.

Figure 4 is a plot of the percent of CF degradation and the influent acetate concentration throughout the experimental period. The previously noted influent CF variability introduces some scatter in this plot also, but a clear trend is observed. At the initial, higher acetate concentrations, removal consistently approached 100 percent. Upon reducing the acetate fed to the system, the percent removal decreased. At the end of the experimental period, CF degradation was approximately 40 percent.

Anaerobic Filter 2. Experimental conditions for AF2 were identical to those of AF1, except that DCM was fed instead of CF. Like AF1, previous work had developed cultures acclimated to the initial conditions of this experimental period. Figure 5 is a plot of influent and effluent acetate concentrations throughout the study. Acetate utilization in AF2 was greater than 85 to 90 percent during most of the period.

Initially, target DCM influent concentrations were set at approximately 80 mg/L. The measured influent and effluent DCM data are presented in Figure 5. On day 63, influent DCM was increased. The reason for this change was an attempt to determine if the DCM degradation was a result of an organism capable of utilizing DCM as its primary electron donor (DCM mineralization). Presumably the higher DCM concentrations would favor such an organism, assuming that no toxicity from the DCM occurs. No impact on acetate utilization was

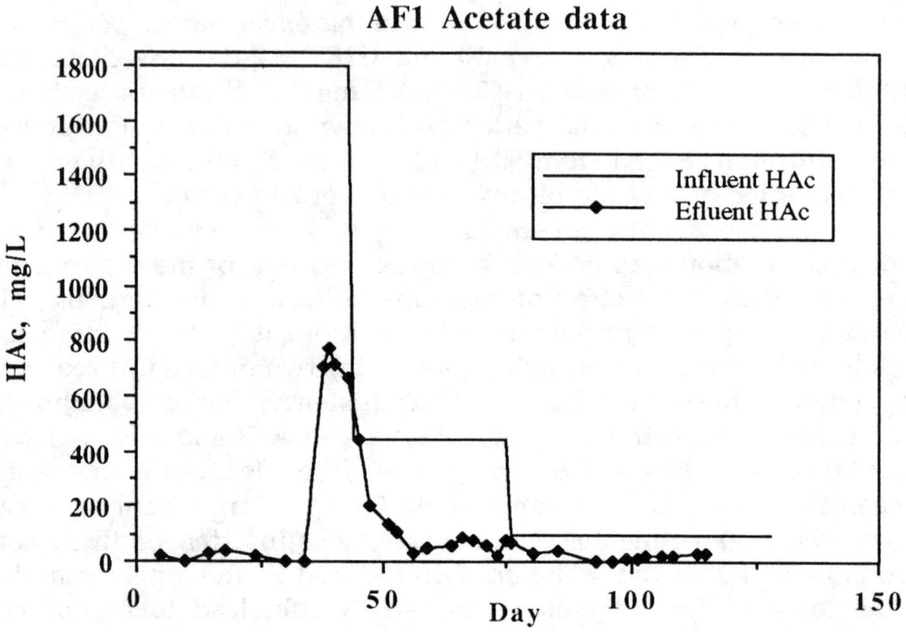

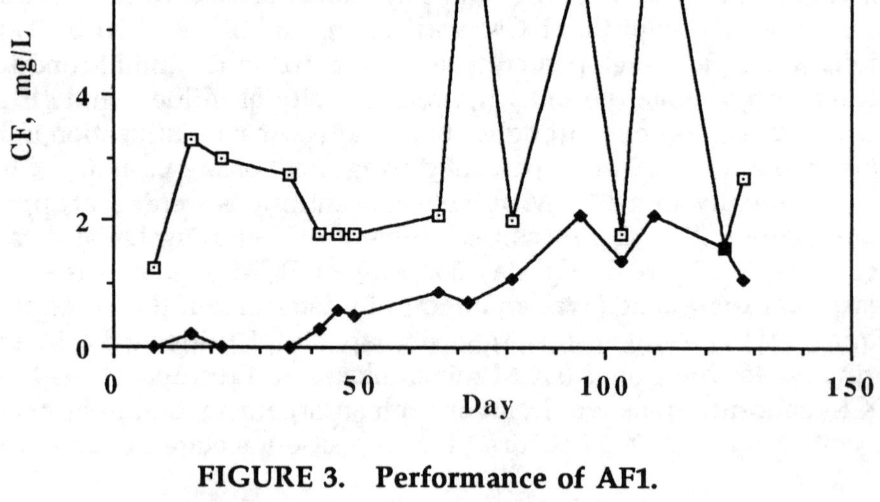

FIGURE 3. Performance of AF1.

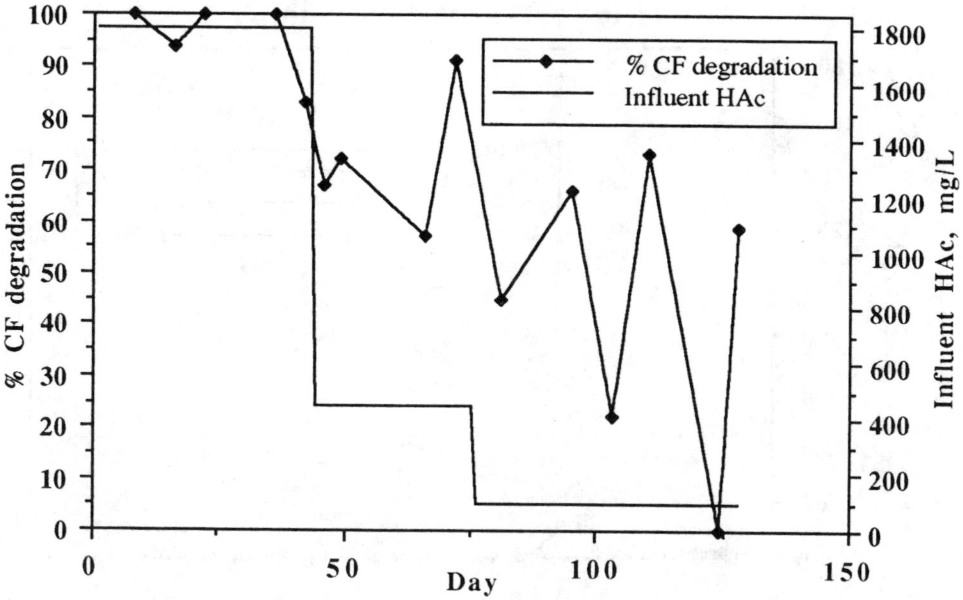

FIGURE 4. Effect of influent acetate concentration on percent CF removal in AF1.

noticed with the increased DCM concentrations (280 mg/L was the highest DCM concentration recorded); however, the percent DCM removal decreased. At the end of the experimental period, the DCM concentration was reduced to 50 mg/L in an attempt to assess whether the decrease in DCM removal was resulted from a toxic response to the high DCM levels or the result of the lower acetate loading to the filter. No increase in the percent removal of DCM was observed at lower influent DCM concentrations.

Figure 6 is a plot of the percent of DCM degradation and influent acetate concentration. At influent acetate concentrations of 1,800 and 450 mg/L, greater than 90 percent DCM removal was measured in all but two cases. However, the percent degradation of DCM dropped significantly after influent acetate concentration was lowered to 100 mg/L.

Anaerobic Filter 3. AF3 was fed a mixture of DCM and TCA. The influent and effluent acetate data are presented in Figure 7. Prior to this experimental period, AF3 was fed DCM as the sole chlorinated aliphatic. Upon the addition of TCA to AF3, effluent acetate concentrations increased from 10–20 mg/L to 200–300 mg/L. Acetate utilization

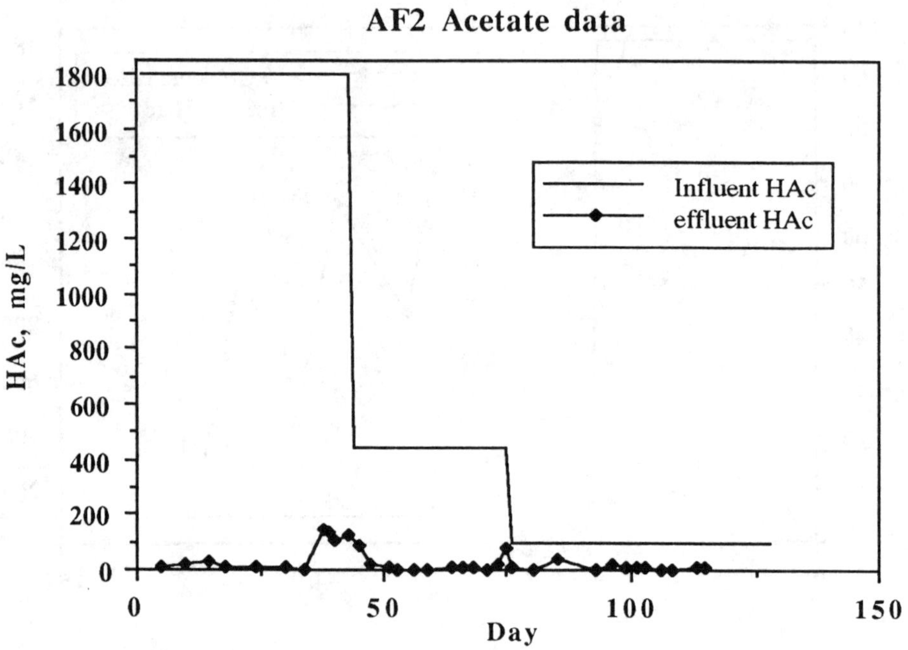

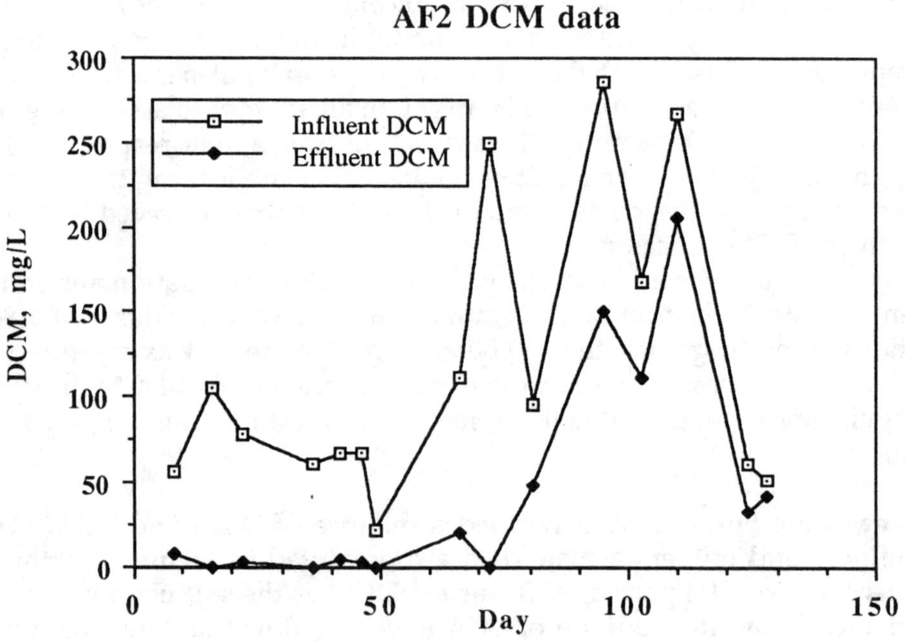

FIGURE 5. Performance of AF2.

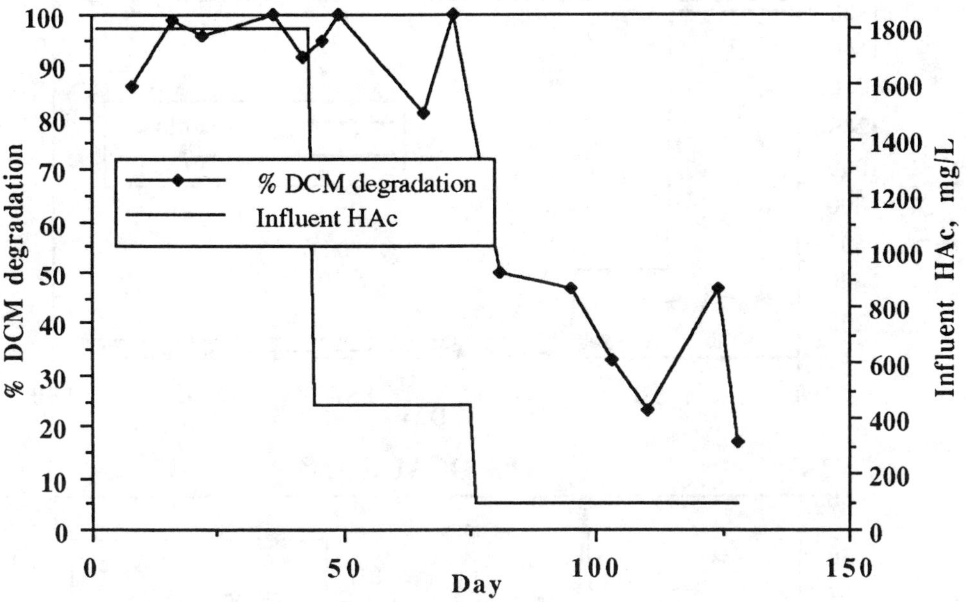

FIGURE 6. Effect of influent acetate concentration on percent DCM removal in AF2.

was typically 70 to 80 percent throughout the experimental period. This observed decrease in acetate utilization appeared to result from the presence of TCA.

Results of influent and effluent DCM and TCA analyses are presented in Figure 7. DCM and TCA concentration ranges of 10 to 20 mg/L and 2 to 4 mg/L, respectively, were typical throughout the experimental period.

The removals of TCA and DCM plotted with influent acetate data are shown in Figure 8. Removal of TCA was always in excess of 90 percent when influent acetate concentrations were 1,800 or 450 mg/L. When the influent acetate concentration was decreased to 100 mg/L, removals dropped to approximately 70 to 80 percent. A similar but more pronounced effect on the percent of DCM transformed was observed. Removal efficiencies consistently approached 100 percent until the influent acetate concentration was decreased to 100 mg/L, when DCM removal dropped to 20 to 30 percent. After over 150 days of operation with influent acetate concentration at 100 mg/L, the ranges of percent removal for DCM and TCA were 35 to 50 percent and 65 to 80 percent, respectively.

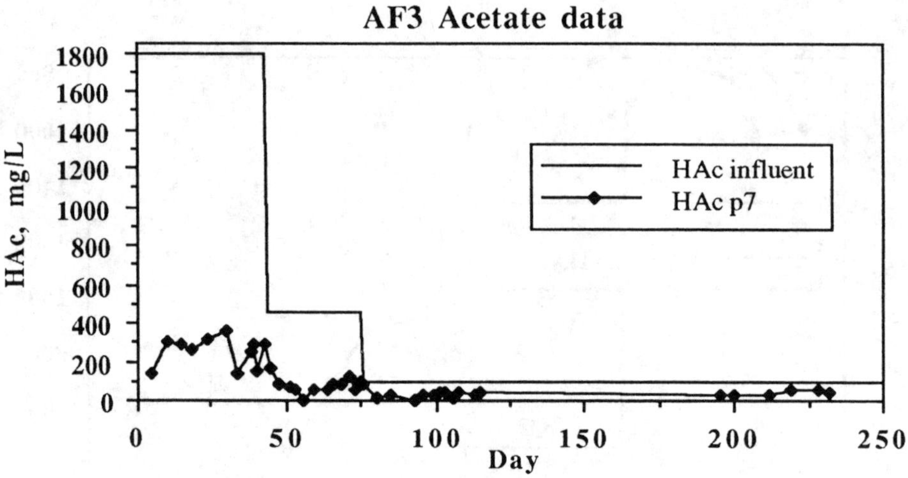

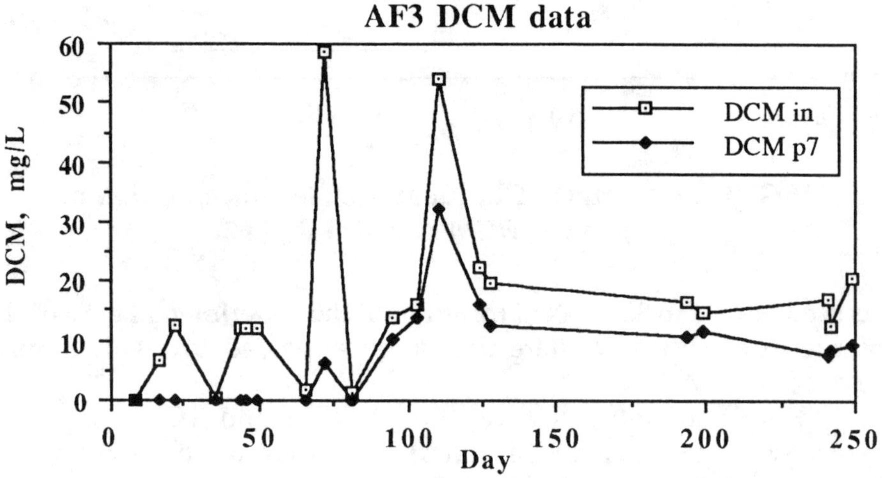

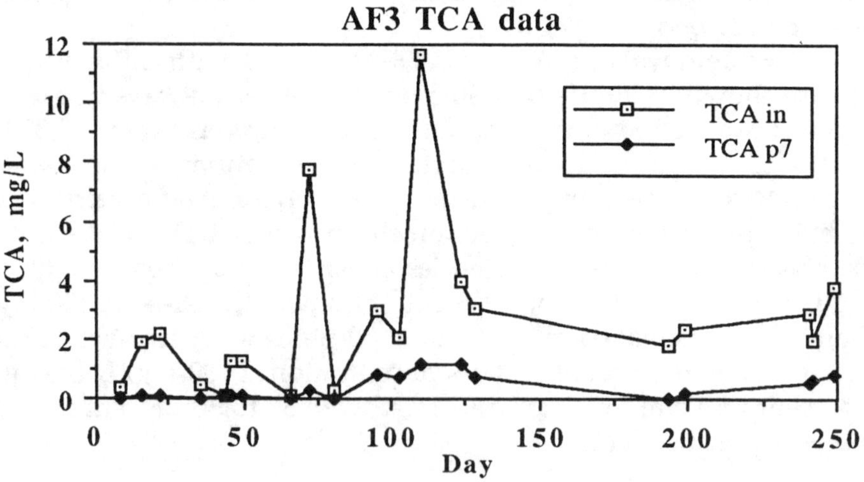

FIGURE 7. Performance of AF3.

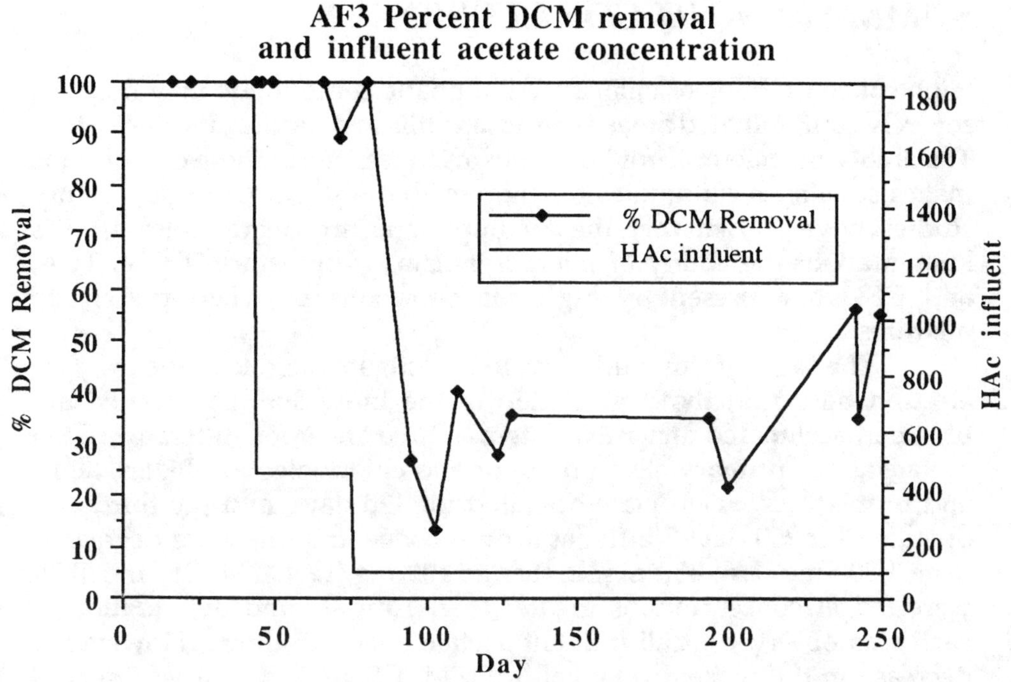

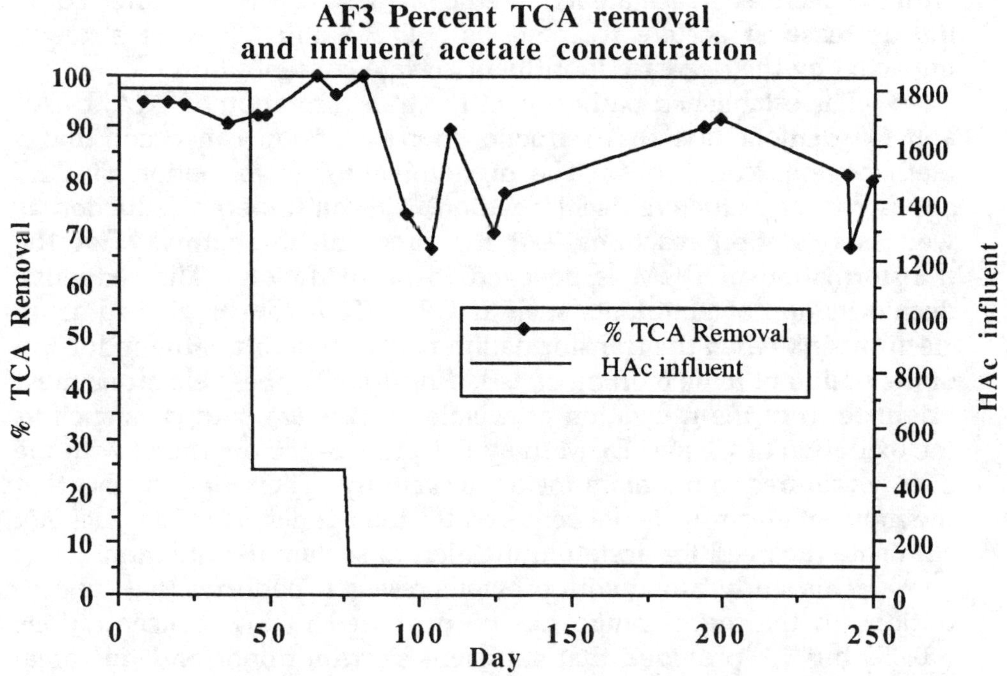

FIGURE 8. Effect of influent acetate concentration on percent DCM and percent TCA removal in AF3.

SUMMARY AND CONCLUSIONS

High concentrations of chlorinated aliphatic compounds may occur in severely contaminated areas such as landfills and hazardous waste sites. The ability of anaerobic microorganisms to transform these compounds makes such sites candidates for bioremediation. Although laboratory studies have shown that these compounds are degradable, little is known about the ability of microorganisms to transform DCM, TCA, and CF when present at high concentrations or when present in mixtures.

The focus of this study was to investigate the role of the primary electron donor on the degradation of the individual chlorinated aliphatic in acclimated anaerobic filters. Three anaerobic filters were fed acetate as the primary electron donor and chlorinated aliphatics, alone and in mixtures. Two were operated for 128 days, and the third was operated for 250 days. Influent acetate concentrations were decreased from 1,800 mg/L to 450 mg/L, then to 100 mg/L. DCM, CF, and TCA were fed at concentrations greater than 1 mg/L, and degradation of each was observed at all influent acetate concentrations. However, a decrease in the percent removal of DCM, CF, and TCA was observed with the decrease in acetate fed. Removal of TCA was least affected by the decrease in acetate fed, but both DCM and CF were strongly impacted by the decrease in influent acetate concentration.

The established pathways of the transformation of TCA, DCM, and CF demonstrate that reduction and oxidation can occur under methanogenic conditions. The predominant transformation of TCA appears to be reductive dechlorination. CF can undergo reduction as well as oxidation reactions, but the predominant pathway for the transformation of DCM is believed to be oxidation. The reductive dechlorination of pollutants such as CF or TCA can be viewed as an electron sink when the transformation pathway involves the reduction of the pollutant to be bioremediated. Presumably, these electrons must originate from the utilization of acetate. Oxidative pathways such as the oxidation of CF and DCM may not provide the organism with the energy required to maintain metabolic activity. Therefore, the fact that the input of electrons is not required for their transformation does not eliminate the need for acetate utilization to sustain degradation.

This study, along with previous research, confirms that concentrations in the mg/L range can be degraded to low concentrations (<0.050 mg/L), provided that sufficient electron donor and microbial activity is present. This information is useful in the selection and application of biological processes such as *in situ* remediation, biologically active zones, or above-ground reactors.

Results of this study show that the primary electron donor has a significant role in the transformation of TCA, CF, and DCM. The impact of the primary electron donor will be most evident in cases of severe contamination. Removals of chlorinated aliphatics at high concentrations may be underestimated when applying information from studies that have used low concentrations of chlorinated aliphatic, a medium rich in electron donor, or both. This information is of particular concern in the application of *in situ* bioremediation where the primary electron donor may be used to limit growth of organisms, and an accurate prediction of the process efficiency is required.

Acknowledgments

Funding for the research presented was provided by the Hazardous Substance Research Center, U.S. Environmental Protection Agency Regions 7 and 8.

REFERENCES

Bouwer, E.; Rittmann, B.; McCarty, P. *Environmental Science and Technology* **1981**, *15*, 596–599.

Brock, T.; Madigan, M. *Biology of Microorganisms*; Prentice-Hall: Englewood Cliffs, NJ, 1988; Chapter 19.

Egli, C.; Stromeyer, S.; Cook, A.; Leisinger, T. *FEMS Microbiology Letters* **1990**, *68*, 207–212.

Egli, C.; Tschan, T.; Scholtz, R.; Cook, A.; Leisinger, T. *Applied and Environmental Microbiology* **1988**, *54*, 2819–2824.

Freedman, D.; Gossett, J. "The Roles of Methanogens and Nonmethanogens in Biodegradation of Dichloromethane under Methanogenic Conditions," submitted to *Applied and Environmental Microbiology* (a).

Freedman, D.; Gossett, J. "Pathways for Biodegradation of Dichloromethane and Its Utilization as a Growth Substrate under Methanogenic Conditions," submitted to *Applied and Environmental Microbiology* (b).

Galli, R.; McCarty, P. *Applied and Environmental Microbiology* **1989a**, *55*, 837–844.

Galli, R.; McCarty, P. *Applied and Environmental Microbiology* **1989b**, *55*, 845–851.

Gossett, J. "Anaerobic Degradation of C1 and C2 Chlorinated Hydrocarbons"; report number ESL-TR-85-38 Engineering and Services Laboratory; U.S. Air Force Engineering and Services Center: Tyndall AFB, FL, 1985.

Gossett, J.; Freedman, D. Presented at the NATO/CCMS Pilot Study on the Demonstration of Remedial Action Technologies for Contaminated Land and Groundwater, Bilthoven, The Netherlands, 1988.

Gossett, J.; Freedman, D. Presented at the NATO/CCMS Pilot Study on the Demonstration of Remedial Action Technologies for Contaminated Land and Groundwater, Montreal, Canada, 1989.

Hughes, J. B.; Parkin, G. F. Presented at the conference on Hazardous Waste Research, Manhatten, KS, 1990.

LaPat-Polasko, L.; McCarty, P.; Zehnder, A. *Applied and Environmental Microbiology* **1984**, *47*, 825–830.

Rittmann, B.; McCarty, P. *Applied and Environmental Microbiology* **1980**, *39*, 1225–1226.

Syverson, A. Master's Thesis, The University of Iowa, 1989.

Syverson, A.; Parkin, G. F.; Loos, H.; Hughes, J. B. Presented at 10th annual SETAC Conference, Toronto, Canada, 1989.

U.S. EPA. *Title 40, Code of Federal Regulations*, 1987, Ch. 1, Pt. 141, Subpt. C, App. C, 559.

Vogel, T.; Criddle, C.; McCarty, P. *Environmental Science and Technology* **1987**, *21*, 722–736.

Vogel, T.; McCarty, P. *Environmental Science and Technology* **1987**, *21*, 1208–1213.

Effects of Diverse Organic Contaminants on Trichloroethylene Degradation by Methanotrophic Bacteria and Methane-Utilizing Consortia

*A. V. Palumbo**
Oak Ridge National Laboratory
W. Eng
University of Tennessee
P. A. Boerman, G. W. Strandberg, T. L. Donaldson, S. E. Herbes
Oak Ridge National Laboratory

INTRODUCTION

Groundwater contaminated with organic compounds, especially solvents such as benzene, trichloroethylene (TCE), perchloroethene (PCE), carbon tetrachloride, and chlorinated ethanes, is a problem at many U.S. Department of Energy (DOE) facilities, including the Oak Ridge National Laboratory (ORNL). These sites are on the National Priority List and fall under the jurisdiction of both the Comprehensive Environmental Response, Compensation, and Liability Act and DOE/Environmental Protection Agency Interagency Agreements. Regulations require consideration of alternatives for remediation of these sites. A demonstration project was initiated in the spring of 1990 that will permit evaluation of two cometabolic approaches to remediation of groundwater and may lead to remediation alternatives that prove both more effective and less costly than traditional methods. More generally,

* Environmental Sciences Division, Oak Ridge National Laboratory, P.O. Box 2008, Oak Ridge TN 37831-6036

the demonstration will provide valuable information on the applicability of bioremediation to a groundwater contamination problem at numerous DOE sites.

Bioremediation techniques for TCE removal employing two groups of microorganisms (methanotrophs and aromatic utilizers) are currently being investigated by researchers at many academic and government institutions and by private firms. The use of both groups of organisms is based on cometabolic degradation (i.e., fortuitous degradation of the target contaminant, TCE, by organisms that are utilizing another compound as a growth substrate). The research presented here focuses on methanotrophic organisms.

Cometabolic degradation of TCE and other compounds by methane-utilizing bacteria has been documented in numerous studies (*e.g.*, Fogel *et al.* 1986; Henson *et al.* 1989; Jansen *et al.* 1988; Leahy *et al.* 1989; Little *et al.* 1988; Moore *et al.* 1989; Oldenhuis 1989; Wilson & Wilson 1985). However, most of these studies were conducted using individual contaminants, and the effects of additional organics on cometabolic degradation have not been well documented. The purpose of this research is to examine the effects of contaminants commonly found in association with TCE at DOE sites and to determine the conditions required for maximizing TCE degradation rates.

Much of this research focuses on compounds found in a seep at the ORNL K-25 site, the target for this demonstration. Chemical characteristics of the K-25 seep were determined by analysis of water samples collected during May and June 1987 and in November 1990. Numerous contaminants were found (Table 1). The flow at the seep was relatively constant at several liters per minute during the spring and summer of 1990. Flow increases during and immediately following storms and reportedly has stopped completely during extended dry periods (Kucsmas, D., Department of Energy Y-12 Production Plant, Oak Ridge, TN, personal communication).

The research presented here details initial experiments on TCE degradation by methanotrophs conducted in the presence of a synthetic medium, TCE, and one or more contaminants found at the K-25 site. Formate has been reported to increase the rate of TCE degradation by pure cultures but had not been tested with mixed cultures. As part of the effort to maximize TCE degradation rates, we examined the effect of formate on degradation by a mixed culture. Results presented here are from batch culture tests although the effect of PCE was also tested in a bench-scale trickle-filter bioreactor. Ultimately, the K-25 seep water will be used in treatability studies to determine the effects on TCE degradation of the mixed organic contaminants present at the site.

TABLE 1. Concentrations (μg/L) of selected contaminants detected in storm drain SD-180-04 at the K-25 site, May–June 1987 and 1990.

Compound	1987		1990	
	Max.	Min.	Max.	Min.
Volatiles				
Benzene	250	230	1,300	1,200
Chloroethane	160	<10		
1,1-Dichloroethane	3,800	2,300	1,000	980
1,1-Dichloroethane	530	250	640	510
1,2-Dichloroethane	31	<5	810	580
Methylene chloride	8,000	1,900	460	160
Tetrachloroethene	50	<5	65	<5
Toluene	650	440	3,100	2,700
trans-1,2-Dichloroethene	950	670	810	580
1,1,1-Trichloroethane	4,600	3,000	6,800	4,900
1,1,2-Trichloroethane	24	<5	33	<5
Trichloroethene	250	140	430	330
Vinyl chloride	140	<10		
Xylene (total)				
Semivolatiles				
2-Methylnaphthalene	44	(a)	92	76
Acenaphthene	2(b)	<5	3	2
Anthracene	3	2		
Ethylbenzene	250	190	430	310
Fluorene	2	2	4	3
Naphthalene	48	14	130	93
Phenanthrene	2(b)	<5	5	4

Source: Miller, D., Oak Ridge National Laboratory, personal communication to S. E. Herbes, October 1990. Excerpted from "Site Characterization Summary: K-1070-C/O Classified Burial Ground," draft report K/ER-4/01; Environmental Restoration Division, K-25 Environmental Program; U.S. DOE: Oak Ridge, TN, March 1990; App. C, Surface Water Sampling Data.

(a) Only one analysis performed.

(b) Estimated.

MATERIALS AND METHODS

Culture Maintenance and Growth

Both pure and mixed cultures were used in these studies. The only pure culture used was *Methylosinus trichosporium* strain OB3b, a Type II obligate methanotroph. The mixed cultures were obtained by enrichments of water samples obtained from various contaminated sites, including one culture (KC) from a DOE Kansas City site (Garland *et al.* 1989), one culture (JS) from a contaminated area in Oak Ridge (Little *et al.* 1988), and two cultures (DT1 and DT2) from other contaminated areas in Oak Ridge (Tyndall *et al.*, in press).

Cultures were maintained on mineral salts agar, which is a modification (Little *et al.* 1988) of NATE medium (Whittenbury *et al.* 1970). The cultures were kept in desiccator jars under a 30 percent (v/v) methane-air atmosphere. These jars had ambient concentrations of oxygen minus the volume displaced by the methane to produce a final oxygen concentration of about 15 percent (v/v). To ensure purity of *M. trichosporium*, bacteria were plated on NATE agar, supplemented with yeast extract (0.2 g/L), and grown in the absence of methane. After 2 days the absence of growth indicated pure culture.

Starter cultures were inoculated by transfer of cells from a stock plate to 100 mL of mineral salts medium (NATE without agar) in 250-mL bottles with Teflon®-lined septum caps. A 30-mL volume of methane was injected through a sterile filter apparatus to yield a 10 percent (v/v) methane headspace, and the cultures were incubated on a shaker table (75 rpm) at 22 C. At an OD_{600} of 0.3 to 0.4, or during late exponential phase, the starter cultures were used as inoculum to grow cells for degradation assays. Either 1 or 2 mL of culture was transferred into each of a set of 250-mL bottles containing NATE medium and grown as described in the preceding paragraph.

Degradation Assays

In preparation for experiments, cultures were transferred into 250-mL centrifuge bottles and centrifuged at 3,000 Gs for 1 h at 10 C. The pellet was suspended in a volume of NATE medium, which increased the cell concentration twofold or more. Cell suspension aliquots of 12 mL were transferred to 40-mL vials and sealed with Teflon-lined septum caps (Supelco, Bellefonte PA). Chlorinated compounds were injected through the septa with Hamilton syringes. Chlorinated ethylene (CE) concentrations were calculated by assuming that all of the CE dissolved in the aqueous phase. The actual concentration of CE present in the media may be calculated from Henry's Law (Lyman 1982). The DCEs were injected in the form of a hexane-DCE mixture. TCE, PCE, toluene, and methylene chloride were added as water-saturated solutions. Triplicate vials were inoculated for each treatment. A 5-ml volume of methane was injected into each vial after CE addition, and vials were inverted and incubated under the same conditions as those described for cell growth. Experiments were terminated by injection of hexane (1.0 to 4.0 mL) followed by agitation, which consistently extracted greater than 99 percent of the remaining CE. Sterile controls contained distilled water or heat-killed cells and the highest concentration of CE tested.

A series of batch experiments was run to determine the effect of *trans*-1,2-DCE, *cis*-1,2-DCE, and 1,1-DCE on TCE degradation by *M. trichosporium* strain OB3b. These experiments compared TCE degradation in the presence and absence of DCEs. Additional experiments examined the effect of 1,1-DCE and PCE on *trans*-1,2-DCE degradation by *M. trichosporium* strain OB3b. Each of these experiments was conducted with several concentrations of *trans*-1,2-DCE, and degradation rates were calculated over a 4- to 5-h period. Only data from the highest DCE concentration tested (0.8 mM) are presented because degradation was most evident at this level.

The effects of formate, toluene, and methylene chloride on TCE degradation by a mixed methane-utilizing culture were examined separately in batch experiments. The effect of formate on TCE degradation was determined by incubating the KC culture with methane (5 mL), or sodium formate (20 mM), or methane plus formate. Experiments examining the effect of 200 µg/L toluene, the approximate toluene concentration found at the K-25 site in 1987 and 1990, on TCE degradation were conducted using the JS and DT2 mixed cultures. The effect of methylene chloride on TCE degradation was examined using the KC and DT2 mixed cultures. Methylene chloride was added at concentrations of 0.023 and 0.095 mM, matching those present at the site in 1987.

Analytical Procedures

Protein concentrations were determined using a Bio-Rad protein determination kit (Bio-Rad, Richmond CA) with an ultrafast centrifuge analyzer (COBAS-FARA model, Roche Diagnostic Systems, Nutley NJ) (Palumbo *et al.*, in press; Uchiyama *et al.* 1989).

Chlorinated ethenes were quantified by gas chromatography. After a minimum 24-h extraction period, 1 µL of hexane extract was analyzed using a Perkin Elmer Sigma 2000 (Norwalk CT). The gas chromatograph was equipped with a 2-ft by 1/8-in. i.d. glass column packed with 60/80 Carbopack B/1% SP-1000 (Supelco, Bellefonte PA) and an electron capture detector and used nitrogen (30 mL/min) as the carrier gas. The detector temperature was 300 or 350 C, and the oven temperature was varied. The oven temperature was held at 75 C to detect DCEs and their intermediates. The injector temperature was set at 250 C or 50 C higher than the oven temperature. A temperature program was designed to resolve mixtures of *trans*-1,2-DCE and PCE. For the first 5 min, the oven remained at 75 C, which resolved *trans*-1,2-DCE. After 5 min, the oven temperature was raised 20 C/min to 150 C. The oven was held at 150 C for the next 20 min to resolve PCE.

Data Analysis

Standard curves were used to convert gas chromatography data (area units) to CE concentrations. These concentrations were then used to calculate degradation rates by linear regression analysis of concentration over time. Lotus 1-2-3 software was used for these calculations (Lotus Development Corporation, 1-2-3, Cambridge MA). Degradation rates were expressed in terms of micromoles per hour per milligram protein. Additional data analysis (*e.g.*, analysis of variance and additional regression analysis) was done on a CompuAdd 386 using SAS software (SAS Institute Inc., Version 6, Raleigh NC), and all tests of significance were made at the 95 percent confidence level.

RESULTS

Degradation of Mixed CEs in Batch Experiments

Patterns of TCE degradation at initial concentrations of 0.019 mM TCE and various DCE concentrations are presented in Figures 1a, 1c, and 2a. At the highest DCE concentrations (6.52 or 3.09 mM), TCE degradation is strongly inhibited (Figures 1a and 1c). At both DCE concentrations, however, TCE degradation was greater in the presence of *trans*-1,2-DCE than either *cis*-1,2- or 1,1-DCE. At 0.309 mM DCE (Figure 2a), TCE was completely degraded in the presence of *trans*-1,2-DCE, although at a somewhat different rate than in the absence of DCE. Again, *cis*-1,2- and

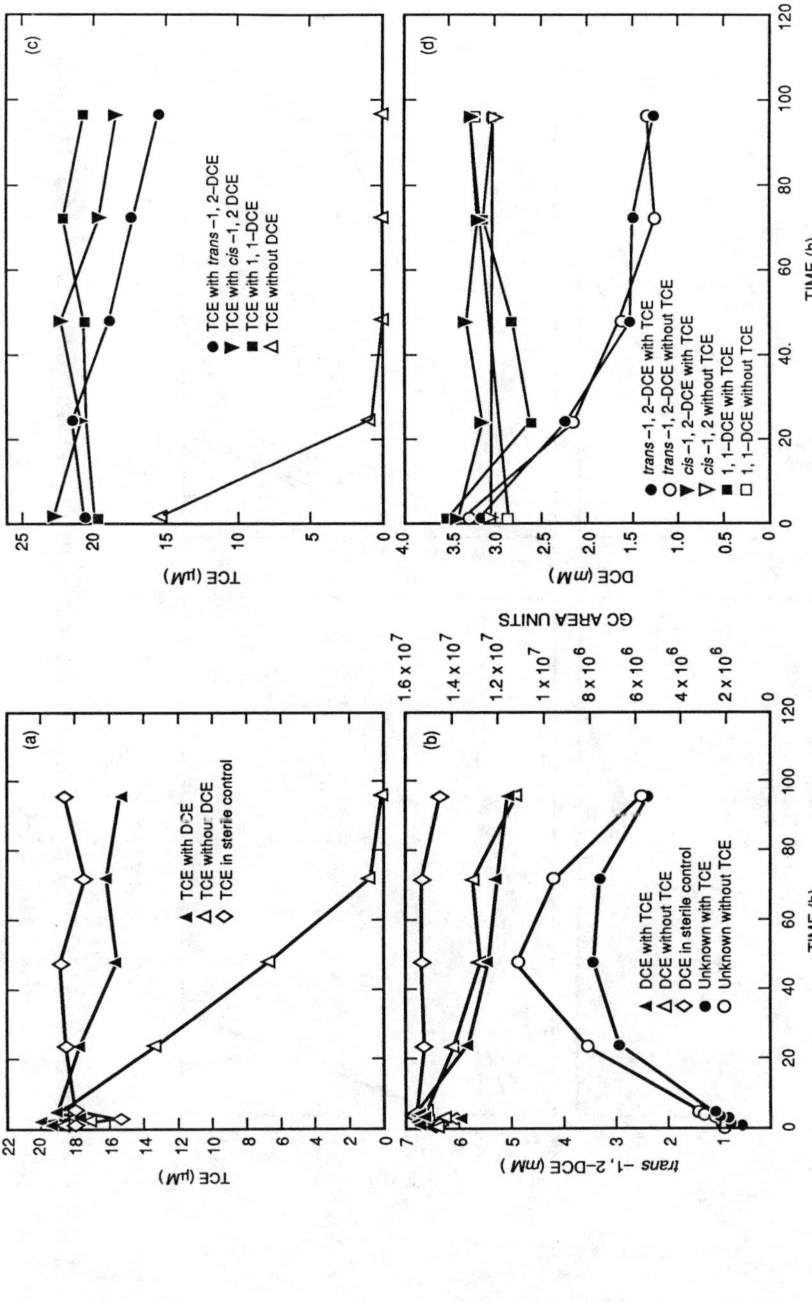

FIGURE 1. Degradation of TCE (a) and DCE (b) in mixtures containing 0.019 mM TCE and 6.52 mM *trans*-1,2-DCE or degradation of TCE (c) and DCE (d) in mixtures containing 0.019 mM TCE and 3.09 mM DCEs (*trans*-1,2- *cis*-1,2-, and 1,1-).

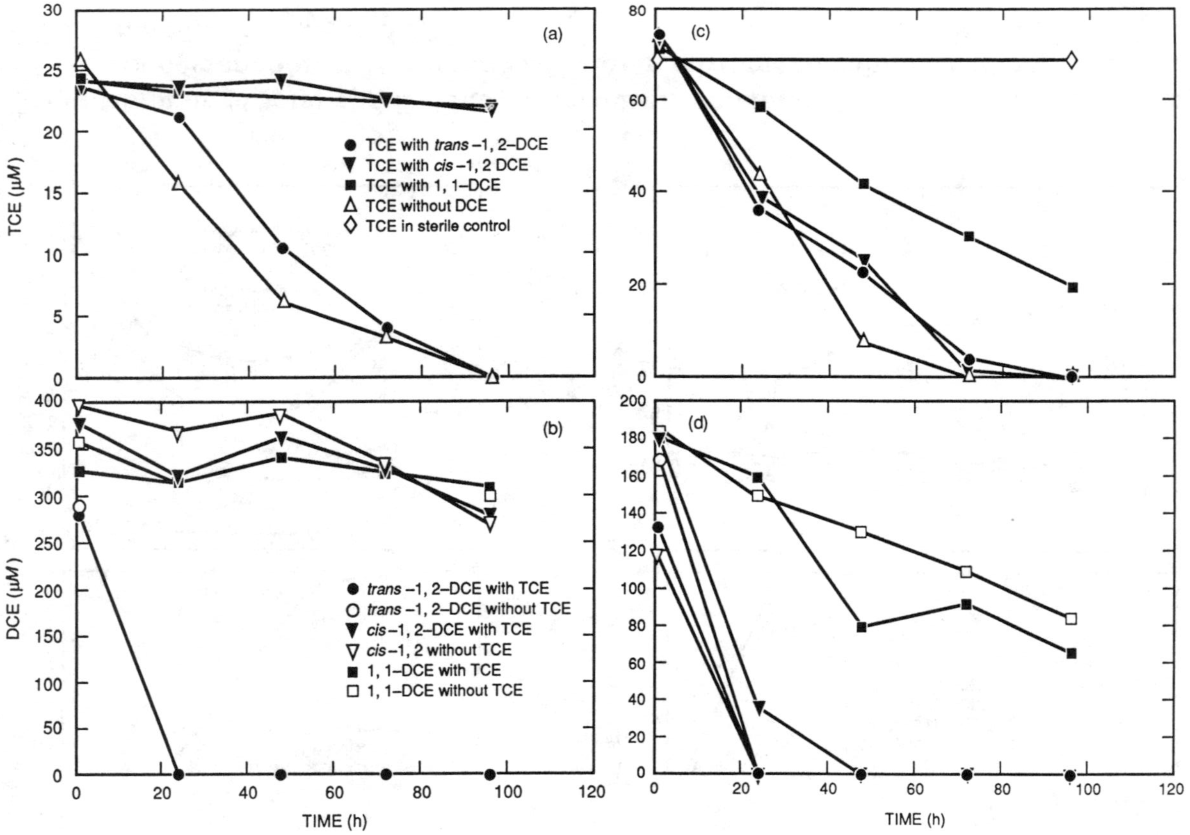

FIGURE 2. Degradation of TCE (a) and DCE (b) in mixtures containing 0.019 mM TCE and 0.309 mM DCEs (*trans*-1,2-, *cis*-1,2-, and 1,1-).

1,1-DCE prevented significant degradation of TCE over the course of the experiment.

The presence of 1,1-DCE was the most inhibitory to TCE degradation, whereas *trans*-1,2-DCE or *cis*-1,2-DCE did not affect the degree of TCE degradation in experiments when the TCE concentration was increased to 0.068 mM and the DCE level decreased to 0.206 mM (Figure 2c). At these concentrations (Figure 2c), TCE degradation was either more rapid (*trans*-1,2- or *cis*-1,2-DCE) or occurred to a greater extent (1,1-DCE) than at lower TCE and higher DCE levels (Figure 2a).

Data are also presented comparing DCE degradation rates from the same experiments discussed in the preceding paragraph. At 0.019 mM TCE and DCE concentrations of 6.52 mM, 3.09 mM, or 0.309 mM, *trans*-1,2-DCE was the only DCE to undergo significant degradation (Figures 1b, 1d and 2b). It appears that TCE had no effect on DCE degradation because results were the same whether or not TCE was present (Figures 1b, 1d, and 2b). At 0.068 mM TCE and 0.206 mM DCE (Figure 2b), degradation of all DCEs occurred and was not affected by the presence of TCE.

At 0.019 mM TCE and 6.52 mM *trans*-1,2-DCE (Figure 1b), formation of an unknown compound, believed to be DCE epoxide, was observed in conjunction with DCE degradation. Comparison of Figures 2a and 2b or 2c and 2d most clearly demonstrate that TCE degradation was occurring simultaneously with DCE degradation.

Velocities were calculated over the course of 4- to 6-h incubations for *trans*-1,2-DCE alone and *trans*-1,2-DCE in the presence of either 0.006 mM PCE, 0.006 mM 1,1-DCE, or 0.063 mM TCE. With *trans*-1,2-DCE only, V was equal to 8.2 ± 2.5 µM/h/mg. When PCE was added, V equaled 4.4 ± 1.5 µM/h/mg, and in the presence of 1,1-DCE the value was 1.8 ± 0.5 µM/h/mg. It appears that the addition of PCE or 1,1-DCE reduced the rate of *trans*-1,2-DCE but TCE did not.

Effects of Other Organics in Batch Experiments

Formate had a significant effect on the rate of TCE degradation by the KC culture (Figure 3). The degradation rate was twice as high in the presence of formate (0.027 ± 0.005 µm/L/h, r = 0.91, N = 9) as in the presence of methane alone (0.013 ± 0.003 µm/L/h, r = 0.82, N = 9) or methane plus formate (0.013 ± 0.004 µm/L/h, r = 0.80, N = 8). There was no significant loss of TCE in the sterile controls during the experiment (r = 0.04, N = 6).

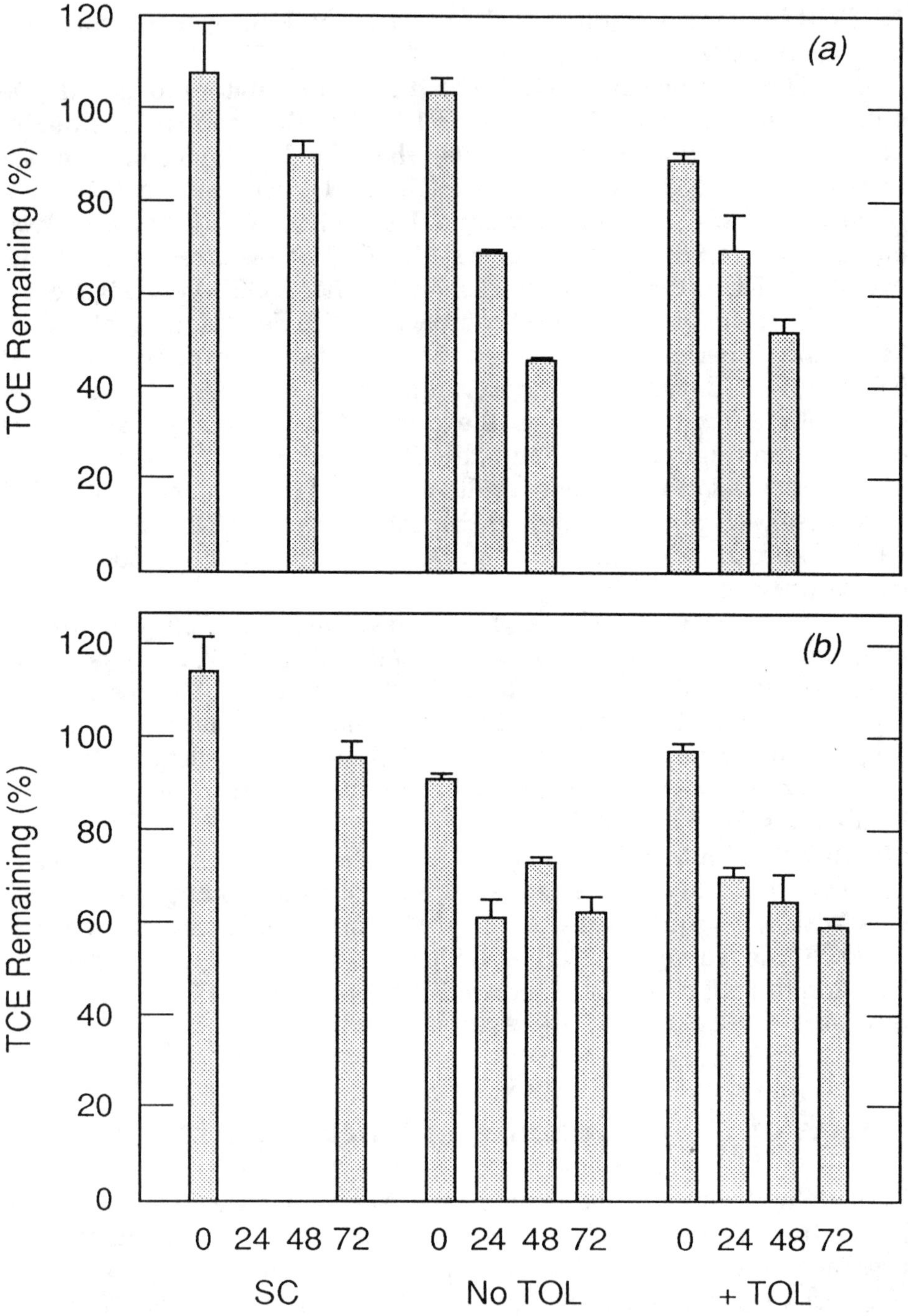

FIGURE 3. Degradation of TCE (a) and DCE (b) in mixtures containing 0.068 mM TCE and 0.206 mM DCE (*trans*-1,2-, *cis*-1,2-, and 1,1-).

Toluene did not have a significant effect on the extent of TCE degradation by either the JS or DT2 mixed cultures (Figure 4). There were no significant differences at any sampling time between treatments with and without toluene. With the DT2 culture, TCE loss occurred throughout the 48-h incubation period. In the presence of the JS culture, most degradation occurred during the first 24 h, and TCE concentrations did not fall significantly over the next 48 h.

Methylene chloride had a negative effect on TCE degradation (data not shown). An analysis of variance indicated that TCE degradation was significantly greater than the sterile control only in the absence of methylene chloride. At both 0.023 and 0.095 mM, methylene chloride inhibited TCE degradation by both the KC and the DT2 mixed cultures. In neither of these treatments was TCE degradation significantly different from the sterile controls.

DISCUSSION

Inhibitory Effects of CEs and Effects of Other Organics

Based on results obtained in batch cultures, the extent of TCE degradation in the presence of DCE appeared to be a function of DCE degradation, which in turn depended on the DCE isomer and concentration. PCE appeared inhibitory to degradation of other CEs by methanotrophs in both the batch and bioreactor experiments. At low concentrations (0.006 mM) in batch cultures, PCE appeared to inhibit degradation of *trans*-1,2-DCE. At higher concentrations (0.6 mg/L) in trickle-filter bioreactors, inhibition of *trans*-1,2-DCE degradation was not evident, but TCE degradation was completely inhibited (unpubl. data).

The significant effect of formate on TCE degradation evident in our experiments has also been reported in pure culture studies and may provide a mechanism for increasing TCE degradation in pilot-scale bioreactors. The highest TCE degradation rates were observed in the presence of formate without methane. This is consistent with the hypothesis that the TCE degradation is inhibited by methane. Although formate provides an alternative carbon source that does not inhibit TCE degradation, it is not clear that TCE degradation can be maintained over long periods of time on formate. Methane monooxygenase (MMO) production may fall if methane is not present as an inducer. Also, in mixed cultures, bacteria that do not produce MMO, and thus do not degrade TCE, may outcompete the methanotrophs when formate is

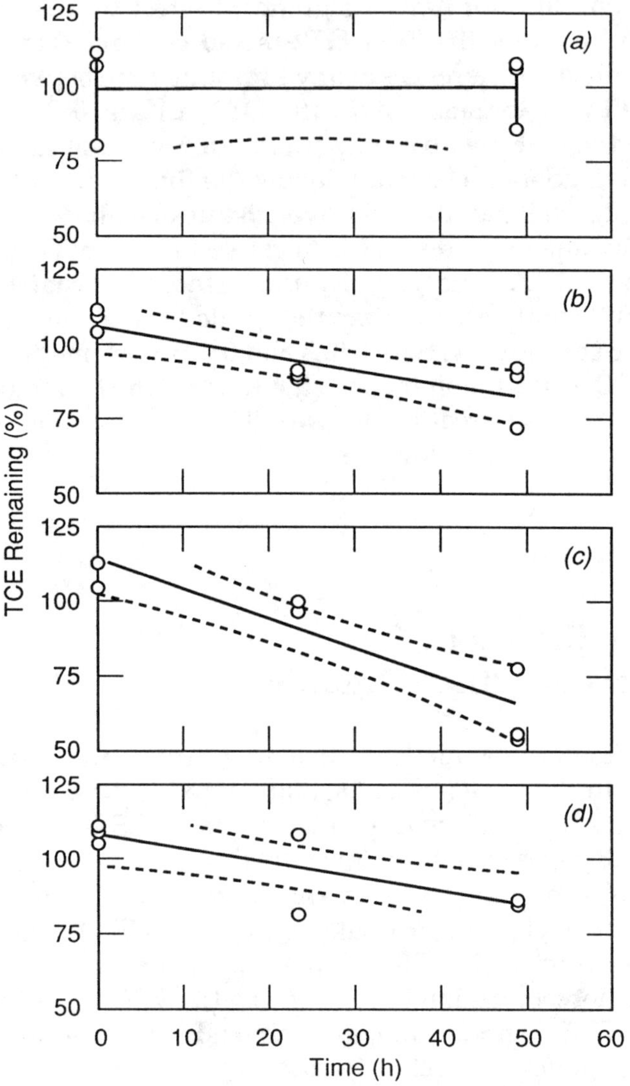

FIGURE 4. Percentage of TCE remaining (mean and standard error) in (a) sterile controls, (b) cells incubated with 5 mL of methane, (c) cells incubated with 20 mM sodium formate, and (d) cells incubated with sodium formate and methane. Open circles are data points, the solid line represents predicted values, and the dotted lines are 95 percent confidence intervals for the regression of TCE and time.

present as the carbon source. As an alternative to methane, pulsing of formate and methane may be the best way to promote continued MMO production and consequent TCE degradation.

Addition of toluene at the levels present in the site water appeared to affect neither the extent nor the rate of the TCE degradation by the JS and DT2 mixed cultures. Future experiments will be conducted to see if the combined concentrations of toluene, benzene, and xylene found on site have an effect on TCE degradation by methane-utilizing consortia.

Results using the highest concentrations of methylene chloride indicate that this compound could have a significant effect on remediation of TCE-contaminated groundwater, although the effect may be concentration dependent. Results from batch cultures will be supplemented by further data from bench-scale bioreactors. Past results indicate that bench-scale bioreactors are less sensitive (Palumbo *et al.*, in press) to some types of perturbation (although the effect of PCE was evident in the bioreactors), probably because of the continuous feed of methane, which facilitates continuous synthesis of the MMO and regeneration of biomass.

Implications for Full-Scale Bioremediation

The data indicate the potential for PCE and 1,1-DCE to limit degradation of TCE by methanotrophs and for the addition of formate to stimulate TCE degradation. Toluene at the levels tested apparently will not interfere with the methanotrophs' ability to degrade the TCE. The effect of benzene and xylene in combination with toluene remains to be determined. Clearly, further work in the bench-scale bioreactors is needed to evaluate the magnitude and importance of both the inhibitory and stimulatory effects.

The presence of PCE presents two problems. The first, resistance to degradation of PCE by methanotrophs has been generally recognized, but the second, inhibition of TCE degradation, may be more important if it is a factor at low PCE levels. Concentrations of PCE and 1,1-DCE present at potential sites must be evaluated prior to initiation of remediation using methanotrophic organisms. If concentrations of PCE are above 0.006 mM, it may be necessary to pretreat the water in an anaerobic bioreactor (Vogel & McCarty 1985) prior to the use of methane-oxidizing cultures. In addition, conditions need to be such that 1,1-DCE is not generated because it can also inhibit TCE degradation. The results presented here indicate that the type and concentration of CE dictate whether TCE biodegradation by methanotrophs can occur and at what rate. The success of bioremediation in bioreactors or *in situ* may be determined by how well inhibition of TCE degradation can be overcome.

Acknowledgments

We thank B. Faison, R. Burlage, and other members of the staff of ORNL for providing critical reviews of the manuscript. This research was supported by the U.S. Department of Energy's Office of Environmental Restoration and Waste Management, Technology Development Program. Oak Ridge National Laboratory is operated by Martin Marietta Energy Systems, Inc., under contract DE-AC05-84OR21400 with the U.S. Department of Energy.

REFERENCES

Eng, W. M.S. Thesis, University of Tennessee, Dec. 1990.

Fogel, M. M.; Taddeo, A. R.; Fogel, S. "Biodegradation of Chlorinated Ethenes by a Methane-utilizing Mixed Culture." *Appl. Environ. Microbiol.* **1986**, *51*, 720–724.

Garland, S. B., II; Palumbo, A. V.; Eng, W.; Little, C. D.; Strandberg, G. W.; Donaldson, T. L.; Bolla, L. L. "The Use of Methanotrophic Bacteria for the Treatment of Groundwater Contaminated with Trichloroethylene at the Bendix Kansas City Plant"; ORNL/TM-11084; Oak Ridge National Laboratory: Oak Ridge, TN, 1989.

Henson, J. M.; Yates, M. Y.; Cochran, J. W. "Metabolism of Chlorinated Ethanes, Ethenes and Ethylenes by a Mixed Bacterial Culture Growing on Methane." *J. Ind. Microbiol.* **1989**, *4*, 29–35.

Jansen, D. B.; Grobben, G.; Witholt, B. "Toxicity of Aliphatic Hydrocarbons and Degradation by Methanotrophic Cultures." In *Proceedings*, 4th European Congress on Biotechnology; Neijssel, O. M.; Van der Meer, R. R.; Luyben, K.C.A.M., Eds.; Elsiver Science Publishers: Amsterdam, 1988; Vol. 3, pp 515–518.

Leahy, M. C.; Findlay, M.; Fogel, S. "Biodegradation of Chlorinated Aliphatics by a Methanotrophic Consortium in a Biological Reactor." In *Biotreatment: The Use of Microorganisms in the Treatment of Hazardous Materials and Hazardous Wastes*, Proceedings of the 2nd National Conference; HMCRI: Silver Spring, MD, 1989.

Little, C. D.; Palumbo, A. V.; Herbes, S. E.; Lindstrom, M. E.; Tyndall, R. L.; Gilmer, P. J. "Trichloroethylene Biodegradation by Pure Cultures of a Methane-oxidizing Bacterium." *Appl. Environ. Microbiol.* **1988**, *54*, 951–956.

Lyman, W. J.; Reehl, W. F.; Rosenblatt, D. H. *Handbook of Chemical Property Methods*; McGraw Hill: New York, 1982; Table 15-4.

Moore, A. T.; Vira, A.; Fogel, S. "Biodegradation of *trans*-1,2-Dichloroethylene by Methane-utilizing Bacteria in an Aquifer Simulator." *Environ. Sci. Technol.* **1989**, *23*, 403–406.

Oldenhuis, R.; Vink, R.L.J.M.; Janssen, D. B.; Witholt, B. "Degradation of Chlorinated Aliphatic Hydrocarbons by *Methylosinus trichosporium* OB3b Expressing Soluble Methane Monooxygenase." *Appl. Environ. Microbiol.* **1989**, *55*, 2819–2826.

Palumbo, A. V.; Eng, W.; Strandberg, G. W. "The Effects of Groundwater Chemistry on Co-metabolism of Chlorinated Solvents by Methanotrophic Bacteria." In *Organic Substances and Sediments;* Baker, R., Ed.; Lewis Publishers: Chelsea, MI, in press.

Tyndall, R. L.; Ironside, K. S.; Little, C. D.; Katz, D. S.; Kennedy, J. R. "Free-living Amoeba Used to Isolate Consortia Capable of Degrading TCE." *Applied Biochemistry and Biotechnology*, in press.

Uchiyama, H.; Nakajima, T.; Yagi, O. "Aerobic Degradation of Trichloroethylene at High Concentration by a Methane-utilizing Mixed Culture." *Agric. Biol. Chem.* **1989**, *53*, 1019–1024.

Vogel, T. M.; McCarty, P. "Biotransformation of Tetrachloroethylene to Trichloroethylene, Dichloroethylene, Vinyl Chloride, and Carbon Dioxide under Methanogenic Conditions." *Appl. Environ. Microbiol.* **1985**, *49*, 1080–1083.

Whittenbury, R.; Phillps, K. C.; Wilkinson, J. F. "Enrichment Isolation and Some Properties of Methane-utilizing Bacteria." *J. Gen. Microbiol.* **1970**, *61*, 205–218.

Wilson, J. T.; Wilson, B. H. "Biotransformation of Trichloroethylene in Soil." *Appl. Environ. Microbiol.* **1985**, *49*, 242–243.

Biochemistry and Kinetics of Aerobic Degradation of Chlorinated Aliphatic Hydrocarbons

Dick B. Janssen[*], *Arjan J. van den Wijngaard, Jaap J. van der Waarde, Roelof Oldenhuis*
University of Groningen

INTRODUCTION

Halogenated aliphatic hydrocarbons have found extensive application as degreasing agents, solvents, intermediates in chemical synthesis, and pesticides. Numerous emissions and spills into the environment have been reported, which has raised concern about the behavior and effects of these chemicals. The environmental fate of chlorinated aliphatics is mainly determined by their resistance to chemical decomposition, their relatively high water solubility and volatility, and the low numbers of microorganisms that are able to degrade these chemicals (Vogel *et al.* 1987). As a result, chlorinated compounds may persist in the subsurface environment for decades and spread in aquifers.

The cause of the low biodegradation rates is the absence of microorganisms that produce enzymes that catalyze transformations resulting in the cleavage or labilization of carbon-halogen bonds. Such bonds are scarce in naturally occurring compounds, and evolution has apparently not kept up with the rate at which human activities have introduced synthetic organochlorine compounds into the environment. As a result, only a limited number of halogenated aliphatic compounds

[*] Department of Biochemistry, University of Groningen, Nijenborgh 16, 9747 AG Groningen, The Netherlands

have been found to be able to function as a growth substrate for microbial cultures (Janssen *et al.* 1989 a, b; Vogel *et al.* 1987). However, several cometabolic conversions have been demonstrated, including trichloroethylene (TCE) oxidation by methane- (Fogel *et al.* 1986), toluene- (Nelson *et al.* 1987), and alkane-oxidizing organisms (Wackett *et al.* 1989).

The potential use of bioremediation for aquifers or soil contaminated with chlorinated compounds is largely determined by the ability to develop a technology that can be used for stimulating microorganisms with special degradative capabilities, either under field conditions or in a suitable reactor system. Degradation rates will depend on conditions such as oxygen and nutrient availability, moisture content, and pH. At least as important, however, are microbiological factors, such as growth requirements and competitive strength of the organisms involved, as well as the kinetics of pollutant degradation. Only a limited number of cultures can use these compounds for growth, and the growth rates of the organisms and their substrate affinity will influence degradation rates. Data, however, on the magnitude of these kinetic constants and how they are determined by properties of the compounds or characteristics of the microorganisms are still lacking.

With organisms that cometabolically degrade chlorinated compounds, degradation rates will be influenced by additional factors. Aerobic cometabolic degradation of chlorinated ethylenes also requires molecular oxygen, both for growth of the organisms and for the oxygenation reactions. Growth substrate(s) for the organisms, and a reducing cosubstrate when a monooxygenase is involved, are also required. The reducing cosubstrate may or may not be identical to the growth substrate, which usually is also the natural substrate for the monooxygenase (*e.g.*, methane or toluene). If the natural substrate for the enzyme is used, competitive inhibition may be expected. The accumulating products may inhibit the degradation process (Wackett & Householder 1989), causing a loss of degradation activity.

The development of aerobic treatment technologies for locally polluted environments and waste streams will require an understanding of the microbial potential and the ecophysiology of the most suitable organisms. Therefore, we have studied physiological pathways and some kinetic aspects of the biotransformation of halogenated aliphatics. Special emphasis was given to the identity and kinetics of dehalogenation reactions since this is the step where toxicity is lost. The possibility of using methanotrophs for cometabolic conversion of compounds that do not function as a growth substrate also was investigated. The results suggest significant differences in the rates at which halogenated aliphatics can be removed and indicate that cometabolic conversions mediated by methanotrophic bacteria may be suitable for low concentrations of several compounds that do not serve as a growth substrate.

USE OF CHLOROALIPHATICS AS A CARBON SOURCE

We have conducted enrichment experiments using soil or sediment obtained from polluted environments or activated sludge as an inoculum and halogenated aliphatic compounds as the sole carbon source. Enrichment contains a mineral medium (Janssen *et al.* 1985) supplemented with 10 mg/L of yeast extract and 2 mM of the xenobiotic compound to be tested. For most haloaliphatics, no growth occurred with any inoculum tested (Table 1). Only for several of the lesser chlorinated compounds could actively growing enrichment cultures be obtained. From most of the positive enrichments, a pure culture that utilizes the carbon compound as a growth substrate could be obtained. Although clear chemical structure-degradability relationships are difficult to derive from this, a few general conclusions may be drawn.

All 1-monohalo-*n*-alkanes tested were found to be degradable. Strains GJ10 (Janssen *et al.* 1985), AD20, and GJ70 (Janssen *et al.* 1988b), as well as organisms obtained in other laboratories (Scholtz *et al.* 1987b; Yokota *et al.* 1987) could utilize a broad range of such compounds for growth. Some α,ω-dichloroalkanes were also degradable, including dichloromethane, 1,2-dichloroethane and 1,3-dichloropropane. 1-Bromo-*n*-alkanes were also degraded by pure cultures. 1,2-Dibromoethane, however, was repeatedly found not to serve as a growth substrate in enrichments. The carbon-bromine bond is weaker than the carbon-chlorine bond, but brominated compounds are apparently not always easier to degrade. Other recalcitrant chloroalkanes of considerable environmental relevance included chloroform, carbon tetrachloride, 1,2-dichloropropane, 1,1,1- and 1,1,2-trichloroethane, and 1,1-dichloroethane.

Another observation was that organisms that utilize halogenated carboxylic acids, such as 2-chloropropionic acid, 3-chloroacrylic acid, and mono- or dihaloacetic acids, are more easily obtained than organisms that grow with chlorinated hydrocarbons. The presence of a hydroxy group in some cases also allowed degradation. Several organisms that degrade 2-chloroethanol, epichlorohydrin, chloropropanols, and similar compounds have been isolated (Table 1).

TABLE 1. Pure bacterial isolates that degrade chlorinated compounds.[a]

Strain no.[b]	Identity	Isolated On	Degrades Also
GJ1, GJ3	*Pseudomonas*	2-chloroethanol	chloroacetic acid
GJ10–12	*Xanthobacter*	1,2-dichloroethane	toluene, methanol 1-propanol, acetone C_4 chloro- and bromoalkanes
AD20	*Ancylobacter*	1,2-dichloroethane	2-chloroethanol, methanol
GH20–22	*Hyphomicrobium*	methylene chloride	formaldehyde
GJ70	*Arthrobacter*	1,6-dichlorohexane 1,9-dichlorononane	1-chloroalkanes
AD1–3	*Pseudomonas*	epichlorohydrin	*vic*-chlorohydrins 1,3-dichloro-2-propanol
AD2	*Arthrobacter*	1-chloro-2,3-propanediol	
AD3	coryneform bact.	epichlorohydrin	
CFZ–7	*Ancylobacter*	1-chloro-2-propanol	2,3-dichloro-1-propanol
AD18 + AD19[c]	n.d.[d]	1-chloro-2-propanol	2-chloro-1-propanol
GJ84	*Corynebacterium*	*trans*-3-chloroacrylic acid	*cis*-3-chloroacrylic acid
AD25, AD27	*Ancylobacter*	chloroethylvinylether	2-chloroethanol 1,2-dichloroethane

(a) Inocula were obtained from various soils contaminated with chlorinated compounds, sediment from the Rhine River, the Eems Channel, and activated sludge.

(b) All enrichments were negative with chloroform, 1,1-dichloroethane, 1,1,1-trichloroethane, 1,1-dichloroethylene, *cis*-1,2- and *trans*-1,2-dichloroethylene, trichloroethylene, perchloroethylene, 1,2-dichloropropane, and hexachlorobutadiene.

(c) Mixed culture of two organisms that could not utilize 1-chloro-2-propanol without each other.

(d) n.d. = not determined.

Growth with di- and trichloroethylenes as sole carbon source has not been described. Of the chloroethylenes, only a vinyl chloride degrading organism has been described (Hartmans *et al.* 1985).

We have observed that the outcome of an enrichment experiment was strongly influenced by the source of the inoculum and the identity of the compounds. All soil and sediment samples tested were positive when tested for chloroacetic acid degradation, but only a limited number of these samples gave rise to dichloromethane-utilizing enrichments, and 1,2-dichloroethane degradation was even more seldom observed. Recalcitrance of 1,2-dichloroethane in samples of aquifer material that had been contaminated with this compound for several years has been observed, although the other contaminants present, toluene and *p*-cymene, were rapidly degraded (Keuning, S., & Jager, D., personal communication).

The negative outcome of many enrichment experiments suggests that specific biochemical factors make the utilization of these chlorinated alkanes difficult. These factors may include

- Compounds that are inert, i.e., not converted by microbial enzymes
- Products of initial transformation that are not further converted
- Degradation that does not yield energy or carbon for growth
- Compounds that are toxic
- Compounds that are converted to toxic metabolites

The above factors may be caused by the absence of the correct combination of microbial enzymes that rapidly dehalogenate or otherwise metabolize chlorinated organics and the intermediates produced. To understand the causes of recalcitrance and to be able to take specific measures, i.e., the use of special adaptation or selection techniques, a detailed insight in the ecophysiology of the organisms is required. Therefore, effort has been directed at identifying the dehalogenation enzymes and studying their characteristics.

Biochemistry of Dehalogenation Reactions

The conversion of halogenated compounds to mineral products and cell components requires the activity of efficient dehalogenation mechanisms. With aliphatic compounds, only a limited number of dehalogenation reactions have been described; Figure 1 illustrates four of them. In Figure 1A, nucleophilic displacement is catalyzed by glutathione-*s*-transferase (GST) in methylene chloride-utilizing bacteria. Dehalogenation of 1,2-dichloroethane proceeds in two steps: a haloalkane dehalogenase (HAD) cleaves off a first chlorine, whereas the second chlorine is removed after oxidation to chloroacetic acid by halocarboxylic acid dehalogenase (HCD) (Figure 1B). Epichlorohydrin is dehalogenated in *Pseudomonas* AD1 after hydrolysis of the epoxide ring by an epoxidase (Figure 1C). The dehalogenating enzyme is a halohydrin dehalogenase (HHD) that produces an epoxide; in Figure 1D hydration is mediated by chloroaerylic acid hydratase (CAH) in a coryneform bacterium that uses this compound for growth.

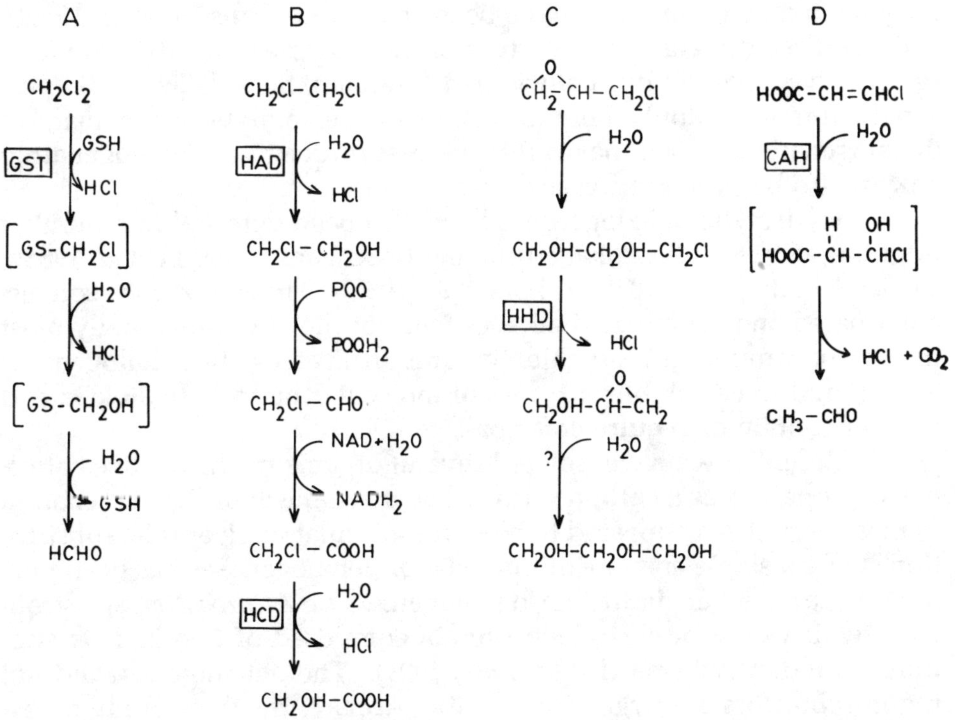

FIGURE 1. Proposed dehalogenation pathways in bacteria that utilize chlorinated aliphatics. (A) Nucleophilic displacement catalyzed by glutathione-*S*-transferase (GST). (B) Dehalogenation of 1,2-dichloroethane. (C) Epichlorohydrin dehalogenation in *Pseudomonas* AD1. (D) Hydration mediated by chloroacrylic acid hydratase.

A broad range of dehalogenases have been detected in a variety of bacteria that use halogenated short-chain carboxylic acids for growth (Motosugi & Soda 1983). The dehalogenases have been divided in two classes: enzymes that convert their substrate with inversion of configuration and enzymes that dehalogenate 2-chloropropionic acid with retention of stereochemical configuration (Weightman *et al.* 1982).

Hydrolysis of carbon-chlorine bonds in chlorinated hydrocarbons that carry no other functional groups was first identified in a strain of *Xanthobacter autotrophicus* that utilizes 1,2-dichloroethane (Janssen *et al.* 1985; Keuning *et al.* 1985). The haloalkane dehalogenase of this organism had a very broad substrate range. More than 22 different substrates have been identified, among which are several environmentally important compounds (Table 2). Since the gene encoding the wild type enzyme and some mutants have been sequenced (Janssen *et al.* 1989b) and the three-dimensional structure of the enzyme that carries out this reaction has recently been elucidated (Franken, S., & Dijkstra, B., personal communication), there is insight in the catalytic mechanism of dehalogenation, which makes this the best understood dehalogenating enzyme (to be published elsewhere).

Hydrolytic dehalogenases have also been detected in a number of gram-positive chloroalkane-utilizing organisms (Janssen *et al.* 1988b, 1989a; Scholtz *et al.* 1987a; Yokota *et al.* 1987). These enzymes convert chlorinated and brominated alkanes, but not the environmentally most important compound 1,2-dichloroethane. In all cases, the dehalogenases were found to be soluble enzymes of molecular mass 28 to 36 kDa and did not contain or require cofactors.

Recently, we have isolated several organisms that utilize chlorinated propanols as a carbon source. The mechanism of dehalogenation in these organisms appeared to be intramolecular nucleophilic substitution. Only a single enzyme of this class of dehalogenases has been purified so far, i.e., an inducible dehalogenase of *Arthrobacter* sp. strain AD2, which was found to be a protein composed of two 29 kDa subunits (van den Wijngaard *et al.* 1989, 1991). The dehalogenase did not require cofactors or oxygen for activity. Apparently, there are two possible routes for the direct dehalogenation of vicinal haloalcohols; hydrolysis to produce glycols or intramolecular substitution to produce epoxides (Figure 1).

A third dehalogenation mechanism that is currently studied in our laboratory is the hydration of 3-chloroacrylic acids. Novel types of inducible dehalogenating enzymes appeared to be present in the organisms that utilize *cis*- and *trans*-3-chloroacrylic acid for growth.

TABLE 2. Substrates of haloalkane dehalogenases.[a]

Compound	Activity (U/mg) GJ10	Activity (U/mg) GJ70	Compound	Activity (U/mg) GH10	Activity (U/mg) GH70
methylchloride	1.68	–	2-bromoethanol	0.42	1.66
methylbromide	0.84	4.29	3-bromopropanol	–	3.70
methyliodide	0.84	2.24	1-chloro-6-hexanol	0.42	1.80
dibromomethane	–	–	1-bromo-6-hexanol	–	2.04
bromochloromethane	–	0.15			
ethylchloride	1.44	–	bis(2-chloroethyl)ether	–	0.90
ethylbromide	1.44	4.29	chloroethylvinylether	–	0.39
ethyliodide	–	2.79			
1,2-dichloroethane	6.00	0.39	1-phenyl-2-bromopropane	–	0.54
1,2-dibromoethane	5.64	5.16			
1-chloropropane	3.06	0.45	1-chloropentane	<0.3	1.95
1-bromopropane	1.74	3.00	1-bromopentane	1.92	1.80
2-bromopropane	–	2.91	2-bromopentane	–	1.14
1,3-dichloropropane	4.80	3.06	1-chlorohexane	0.18	2.55
3-chloropropene	2.70	4.17	1,6-dichlorohexane	0.24	2.01
1,3-dichloropropene	–	3.99	1,9-dichlorononane	0.24	0.78
1,2-dibromopropane	7.14	4.44	1,2-dichloropropane	0.036	–
1-chlorobutane	1.86	1.98	epichlorohydrin	0.84	–
1-bromobutane	1.62	2.70	epibromohydrin	7.74	–
2-bromobutane	–	1.80			

(a) Activities were determined at 30 C with purified dehalogenase of *X. autotrophicus* GJ10 (Keuning *et al.* 1985) and *Arthrobacter* GJ70 (Janssen *et al.* 1988a). Substrate concentrations were 5 mM.

Another mechanism of dehalogenation has been discovered in bacteria that use the important compound dichloromethane for growth (Kohler-Staub *et al.* 1986; Stucki *et al.* 1981). The dichloromethane dehalogenase, which in fact is a glutathione transferase, converted several dihalomethanes. The reaction and the amino acid sequences indicate a similarity to eukaryotic detoxification enzymes (La Roche & Leisinger 1990).

Reactions in which carbon-halogen bond fission occurs by spontaneous decomposition of enzymatically formed unstable products have also been proposed, and although these have not unequivocally been proven to be relevant in organisms that use haloaliphatics for growth, their role in vinyl chloride utilization is very probable (Hartmans *et al.* 1985).

Growth Rates and Substrate Affinities

It may be expected that microorganisms capable of utilizing chlorinated hydrocarbons have evolved only since the introduction of these compounds in the environment started some 50 years ago. Nevertheless, efficient dehalogenation enzymes have developed and degradation rates are not necessarily low. Generation times in a number of cultures that we have tested were generally in the range of 3 h (epichlorohydrin) to 11 h (dichloromethane).

The data in Table 3 indicate that there is some relationship between the K_s of the organism and the K_m of the first enzyme, but not with the lipophility or solubility of the substrate. K_s values were significantly lower than K_m values of the haloalkane dehalogenase, indicating that rapid growth can be obtained at lower substrate concentrations than suggested by the rather low affinity of the dehalogenase for chlorinated alkanes. Active transport of chlorinated hydrocarbons has not yet been investigated.

OXIDATIVE COMETABOLISM OF CHLOROALIPHATICS

Since 1985 (Wilson & Wilson 1985), the possibility to convert chlorinated ethylenes by cometabolic reactions using methane oxidizing bacteria has received increasing attention (Fogel *et al.* 1986; Janssen *et al.* 1988a; Little *et al.* 1988; Oldenhuis *et al.* 1989). The methane monooxygenases (MMO) involved have a broad substrate range (Burrows *et al.*

TABLE 3. Kinetic constants of bacteria that use halogenated compounds for growth.

Compound	μ_{max}	K_s[(a)] (µM)	K_m[(a)] (µM)
1,2-dichloroethane	0.102	150 ± 50	670[(d)]
2-cloroethanol	0.158	6 ± 3	70[(c)]
1-chlorobutane	0.088	243 ± 30	1330[(d)]
methanol	0.126	n.d.[(b)]	n.d.

(a) K_s = Monod constant, K_m = Michaelis-Menten constant of first enzyme (dehalogenase, d, or chloroethanol dehydrogenase, c). Values were determined at steady state in continuous cultures.
(b) n.d. = not determined.

1984; Colby *et al.* 1977) and can convert alkanes to alcohols, alkenes to epoxides, and alkylbenzenes to phenols or alcohols.

Our initial experiments with methanotrophic bacteria showed that *trans*-1,2-dichloroethylene was more easily degraded than trichloroethylene (Janssen *et al.* 1988a). This work was carried out with mixed cultures, containing predominantly cells of the genus *Methylomonas,* and with pure cultures of *Methylomonas methanica* NCIB11130 and GJ6 and *Methylosinus trichosporium* OB3b grown in batch culture. Much more rapid degradation of trichloroethylene and several other chloroalkanes was obtained with *M. trichosporium* cells grown under copper starvation (Oldenhuis *et al.* 1989; Tsien *et al.* 1989). Copper limitation is known to result in derepression of a soluble type of MMO that has a broader substrate range than the particulate enzyme from cells grown under copper sufficiency (Stanley *et al.* 1983). At copper/biomass ratios exceeding 2.5 µmol/g cells (dry weight), soluble MMO was fully repressed. The biochemical characteristics of soluble MMOs have been described by others (Fox *et al.* 1989, 1990; Green & Dalton 1989).

Summarizing the results (Table 4), we conclude that the perchloro compounds carbon tetrachloride and tetrachloroethylene are converted neither by soluble nor by particulate monooxygenase; that trichloroethylene, 1,1-dichloroethylene, 1,1,1-trichloroethane, 1,1-dichloroethane, and *trans*-1,3-dichloropropylene are only or much faster converted by the soluble enzyme than by particulate monooxygenase; and that chloroform, 1,2-dichloroethane, and *trans*-1,2-dichloroethylene can

TABLE 4. Degradation of halogenated compounds by soluble (sMMO) and particulate (pMMO) methane monoxygenase.

Compound	Concentration (μM) Remaining[a]				Products Identified[c]
	sterile	pMMO[b]	sMMO[c]	pMMO[c]	
dichloromethane	167	<1	<1	24	chloride
chloroform	124	12	<0.01	6	chloride
carbon tetrachloride	46	48	45	40	
1,1-dichloroethane	49	58	<0.1	45	chloride
1,2-dichloroethane	117	<0.1	<0.1	<0.1	chloride
1,1,1-trichloroethane	65	70	25	70	trichloroethanol
1,1-dichloroethylene	39	38	18	32	
trans-1,2-dichloroethylene	143	0.5	<0.1	<0.1	chloride, *trans*-epoxide
cis-1,2-dichloroethylene	110	111	<0.1	45	chloride, *cis*-epoxide
trichloroethylene	70	59	<0.1	50	chloride, trichloroethanol
tetrachloroethylene	69	71	72	73	
1,2-dichloropropane	170	158	<0.1	83	1,2-dichloro-3-propanol
trans-1,3-dichloropropylene	118	52	19	60	chloride

(a) Cells (0.1 mg/mL) incubated aerobically in the presence of formate (20 mM) for 48 h at 30 C.
(b) Degradation with cells of *Methylomonas* GJ6.
(c) Degradation with cells of *M. trichosporium* OB3b expressing soluble (sMMO) or particulate (pMMO) methane monooxygenase (Oldenhuis *et al.* 1989).

be converted by either of the two enzymes (Oldenhuis *et al.* 1989). In several cases, (partial) dehalogenation was demonstrated by the accumulation of inorganic chloride, sometimes via the formation of chemically unstable intermediates.

A consequence of the requirement for expression of soluble MMO for efficient TCE degradation is that, for practical application, specific stimulation of growth of organisms that can form this enzyme will be necessary. It is assumed that *Methylomonas* strains, which we found usually to be the predominant organisms in enrichment cultures carried out in the presence of copper in the growth medium, are not able to produce such a soluble MMO, in agreement with the observation that these organisms do not grow to high cell densities in the absence of added copper.

The rates that we have found for cometabolic oxidation of TCE were rather attractive: specific degradation activities of up to 200 nmol/min • mg of cells were obtained under appropriate conditions, i.e., the use of formate as a cosubstrate at temperature 30 C and fresh cells from a copper-limited chemostat (Oldenhuis *et al.* 1989, 1991). The kinetics of TCE degradation by methanotrophs thus compares favorably with that of the toluene oxidizing organism strain. V_{max} values were 290 nmol/min • mg (cell dry weight) for *M. trichosporium* OB3b, compared to only 4 nmol/min • mg for *Pseudomonas cepacia* G4 (Folsom *et al.* 1990). Although the methanotroph has a lower affinity for TCE, the first order rate constant (K_1), describing rates at very low substrate concentrations ([S] << K_m), is still higher for the methanotroph: 2.0 mL/min • mg of cells, compared with 1.3 for the *Pseudomonas*.

K_1 values for *M. trichosporium* OB3b (Table 5) were in the order of chloroform > dichloromethane > methane, and for the ethylenes *cis*-1,2-dichloroethylene > trichloroethylene = *trans*-1,2-dichloroethylene > 1,1-dichloroethylene. No direct correlation was found between toxicity and conversion rate. Thus, 1,1,1-trichloroethane and chloroform, which were a very poor and a good substrate, respectively, showed low toxicity when compared with *trans*-1,2-dichloroethylene, which was a moderately good substrate. The identity of the degradation products will be very important. Alcohols are known to inhibit methanotrophs, whereas epoxides could either directly, or after decomposition to even more reactive compounds like acylchlorides, react with cellular proteins.

The toxicity of TCE (Oldenhuis *et al.* 1989) appeared to be caused by the transformation of TCE (Fox *et al.* 1990), and the degree of inactivation was proportional to the amount of TCE converted. Figure 2A illustrates the amount of TCE degraded after repeated addition of 45 µm TCE to cells of *M. trichosporium* OB36. A depletion curve

TABLE 5. Kinetic parameters of chlorinated hydrocarbon degradation by *M. trichosporium* OB3b.[(a)]

Compound	V_{max}[(b)] (nmol/mg cells/min)	K_{mb} (μM)	K_1[(b)] (mL/min/mg cells)	C_{tox}[(c)] (μM)
methane	92 ± 30	363 ± 60	3.9	n.d.[(d)]
dichloromethane	4 ± 1	33 ± 1	8	>90
chloroform	34 ± 6	548 ± 55	16	195 ± 35
1,2-dichloroethane	77 ± 9	65 ± 6	1	60 ± 10
1,1,1-trichloroethane	214 ± 18	24 ± 1	0.1	>700
1,1-dichloroethylene	5 ± 2	6 ± 1	1	28 ± 6
trans-1,2-dichloroethylene	148 ± 61	331 ± 89	2	98 ± 22
cis-1,2-dichloroethylene	30 ± 10	182 ± 29	6	257 ± 73
trichloroethylene	145 ± 61	290 ± 99	2	65 ± 12

(a) Values were obtained by following substrate disappearance with cells grown in continuous culture in the absence of copper (data from Oldenhuis *et al.* 1991).
(b) Michaelis-Menten constants derived from degradation rates determined at different initial concentrations of substrate. K_m = apparent K_m value; K_1 = first order rate constant V_{max}/K_m.
(c) Concentration above which an increase of substrate concentrations led to lower transformation rates.
(d) n.d. = not determine.

after each addition (number), as well as TCE depletion in a control incubation that received TCE for the first time (*c*), are shown. The effect of the amount of TCE converted on the TCE degradation rate (Figure 2B) and measured and calculated TCE depletion data (Figure 2C) are also shown. Lines indicate fitting with the first order rate constant (*FO*, k, = 2 mL/min/mg of cells), with the Michaelis-Menten constant (*MM*, apparent K_m = 145 μM and V_{max} = 290 mol/min/mg of cells), and with the Michaelis-Menten kinetics corrected for the inactivation rate shown in Figure 2B (*IN*). From the proportional relationship between the degree of inactivation and the amount of TCE converted, we calculated that only 2.1 μmol (0.27 mg) of TCE could be converted per mg of *M. trichosporium* OB3b cells (dry weight) (Oldenhuis *et al.* 1991). The amount of TCE that could be degraded per amount of cells was also determined in the presence of activated carbon to prevent acute toxicity of TCE. After conversion of 2.6 ± 0.5 μmol of TCE per mg of cells, TCE degradation stopped, which is in reasonable agreement with the above data on the rate at which cells became inactivated upon transformation of TCE. This rate of inactivation is too low to have much effect on the shape of a substrate depletion curve. It will,

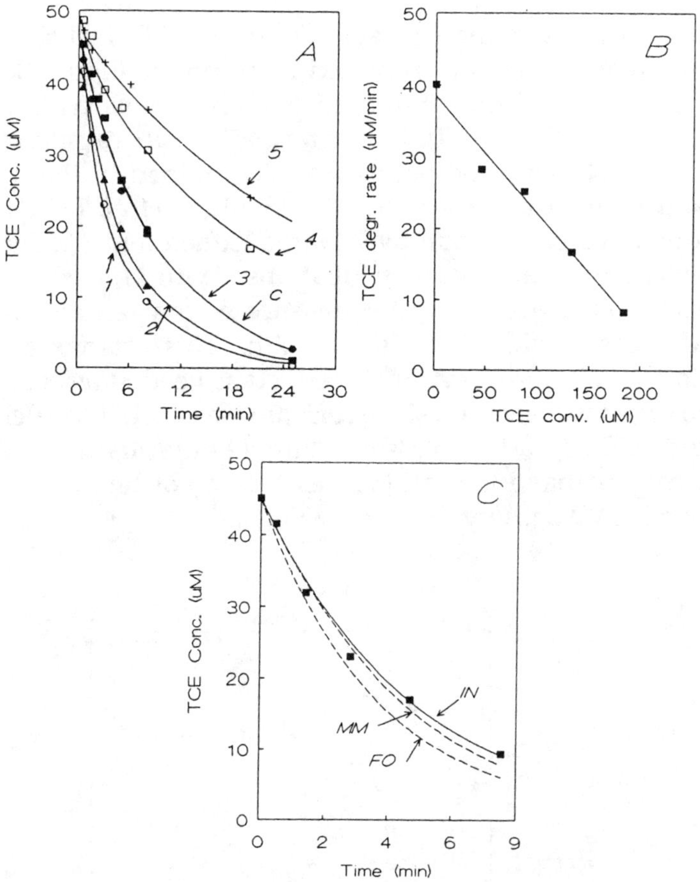

FIGURE 2. Inactivation of TCE transformation activity by TCE conversion. (A) TCE degradation after repeated addition of 45 µM TCE to cells of *M. trichosporium* OB3b. (B) Effect of the amount of TCE converted on the TCE degradation rate. (C) Measured and calculated TCE depletion data.

however, limit the amount of TCE that can be converted in a continuous system.

Another important factor is the stability of the organism under conditions of limited substrate supply. Since methane inhibits TCE degradation (Oldenhuis *et al.* 1991), it could be advantageous for continuous treatment systems to separate growth on methane and degradation of TCE. Therefore, we have cultivated *M. trichosporium* OB3b in continuous culture and studied the effect of methane and oxygen limitation on the TCE transformation capacity. When methane was replaced by nitrogen in the incoming gas (at 0, Figure 3), cells lost activity much faster than when oxygen was replaced by nitrogen; Figure 3 illustrates this in continuous culture under copper limitations as described in Oldenhuis *et al.* (1989). The TCE transformation rate of cells taken from the culture at different time points was determined in the presence of formate and oxygen as described in Oldenhuis *et al.* (1991). Similar observations have been made by Alvarez-Cohen and McCarty (1991), using mixed cultures and batch incubations. With *M. trichosporium*, the TCE degradation capacity in the presence of oxygen and absence of methane decreased with a half life of 2.7 h. This was not caused by a decrease in the cellular reserve of a reductant needed as cosubstrate in the monooxygenase reaction, since formate was added as such a cosubstrate to the TCE degradation assays. Possibly the absence of an energy supply during methane starvation caused decay of cellular components essential for MMO activity.

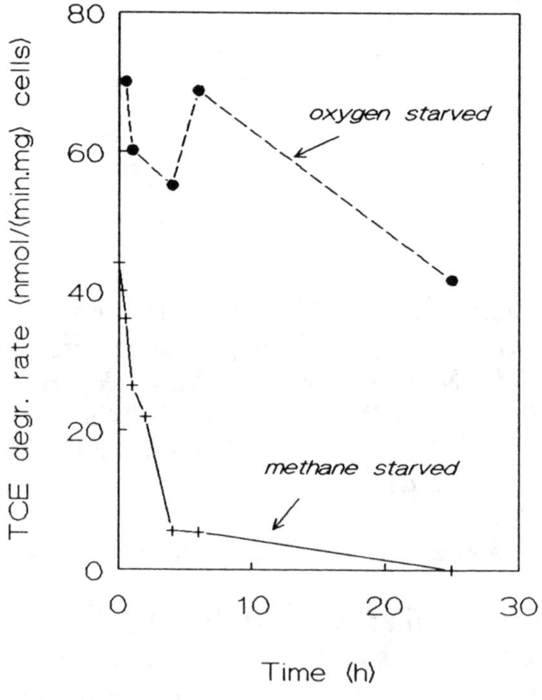

FIGURE 3. Effect of methane and oxygen starvation on TCE transformation capacity of *Methylosinus trichosporium* OB3b growing in continuous culture under copper limitation.

DEGRADATION IN SOIL

Degradation of Chlorinated Compounds in Soil Slurries

We have tested whether addition of specific cultures to slurries of contaminated soil can decrease adaption periods or increase degradation rates. It was found that dichloromethane removal occurred faster when dichloromethane degrading organisms (*Hyphomicrobium* GJ21 or *Methylobacterium* DM4) were added to contaminated soil. Without inoculation, no significant degradation took place within 100 h. In Figure 4, the slurries contained 50 percent sandy soil and were supplemented with phosphate and ammonium. Cells (5 mg/L) of *Hyphomicrobium* GJ21 or *Methylobacterium* DM4 that were precultivated with dichloromethane were added to some incubations. During incubation at 30 C, dichloromethane addition was followed by mixing in a carousel.

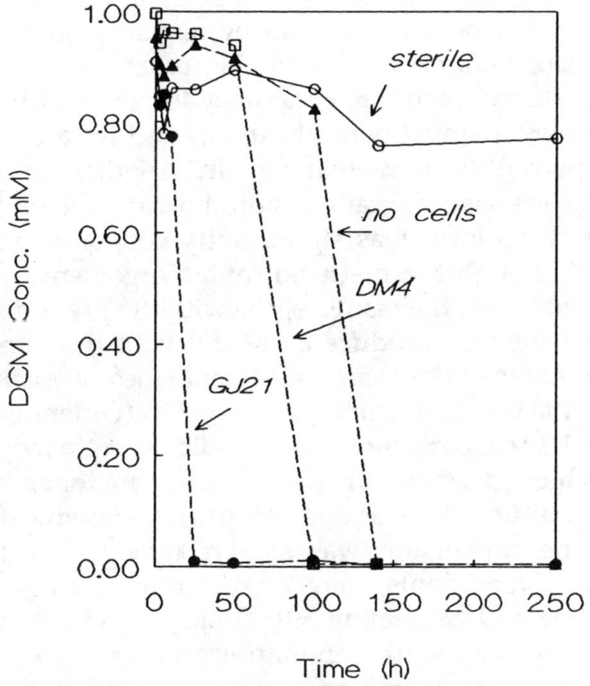

FIGURE 4. Effect of inoculation on the degradation of dichloromethane in soil slurries.

Similar results have been obtained with the degradation of 1,2-dichloroethane in slurries prepared from subsurface soils contaminated with 1,2-dichloroethane, toluene, and *p*-cymene (Keuning, S. & Jager, D., personal communication). Even after addition of nutrients and control of pH and temperature, no degradation of 1,2-dichloroethane was found.

These data show that, although some chlorinated aliphatics can serve as growth substrates for bacterial cultures, organisms that have this capability are not always present in numbers high enough to allow a rapid onset of degradation in polluted soil. Inoculation with organisms that have the required catabolic trait may thus reduce adaptation periods, which could make contaminant removal possible if the indigenous microflora does not adapt rapidly enough for accelerated degradation.

Cometabolic Degradation in Soil

Experiments with the addition of methane to stimulate cometabolic transformation have been carried out by us using sandy soil obtained from a dump site contaminated with chloroform, TCE, and perchloroethylene. The slurry (Figure 5) contained 45 percent (dry w/v) sandy soil and was supplemented with phosphate and nitrate. The gas phase contained 20 percent (v/v) methane in air. Another contaminant present *n*-butanol, was also degraded (not shown). Of these three compounds, only chloroform was significantly converted when methane was added. Although the methanotrophic organisms present in this soil were not analyzed, the results agree with the predominant involvement of organisms that produce an MMO with the narrow substrate range characteristic of the (particulate) enzymes of *Methylomonas* GJ6 (Janssen *et al.* 1988a) or *M. trichosporium* OB3b (Oldenhuis *et al.* 1989).

In slurries that contained *trans*-1,2-dichloroethylene, rapid degradation was achieved when either methane or methane plus cells of a *Methylomonas* culture were added (Figure 6), showing that *trans*-1,2-dichloroethylene conversion was also relatively easy to achieve in polluted soil. Experiments were carried out in bottles closed with viton-lined septa and containing 50 mL slurry and 240 mL gas phase. Slurries (60% dry w/v soil) were supplemented with phosphate and nitrate. The gas phase contained 20 percent (v/v) methane in air.

Field experiments have also indicated a more rapid conversion of *trans*-1,2-dichloroethylene than of TCE (McCarty *et al.* 1989). These data could be explained by methanotrophs that express particulate

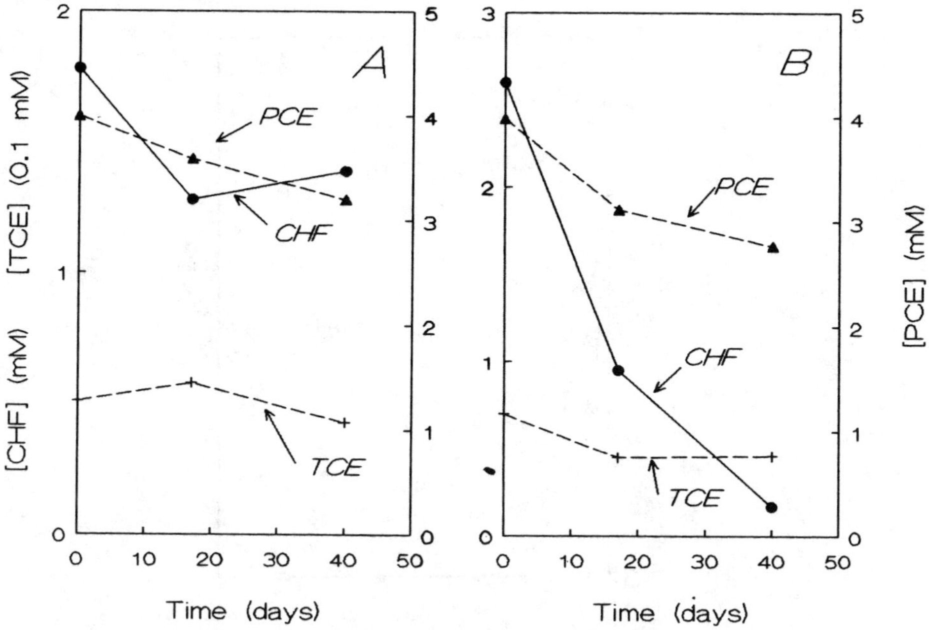

FIGURE 5. Degradation of chloroform (CHF), trichloroethylene (TCE), and perchloroethylene (PCE) in a soil slurry exposed to methane.

MMO as the predominant organisms present. Similar observations were made with water treatment (Strandberg *et al.* 1989). More rapid TCE degradation, therefore, will require the specific stimulation of cells that produce soluble MMO. However, apart from copper availability, which will be difficult to manipulate in a natural environment, it is not yet known how this can be achieved.

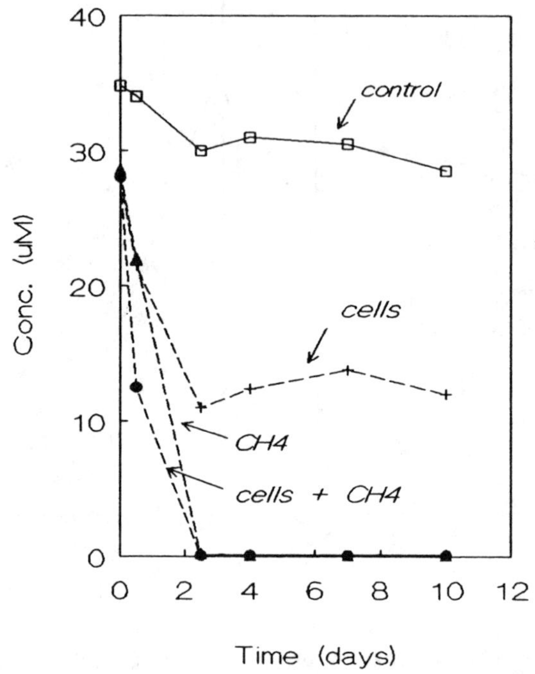

FIGURE 6. Degradation of *trans*-1,2-dichloroethylene in soil slurry.

REFERENCES

Alvarez-Cohen, L.; McCarty, P. L. "TCE Transformation by a Mixed Methanotrophic Culture: Effects of Toxicity, Aeration and Reductant Supply." *Appl. Environ. Microbiol.* **1991**, *57*, 228–235.

Burrows, K. J.; Cornish, A.; Scott, D.; Higgins, I. J. "Substrate Specificities of the Soluble and Particulate Methane Monooxygenases of *Methylosinus trichosporium* OB3b." *J. Gen. Microbiol.* **1984**, *130*, 3327–3333.

Colby, J.; Stirling, D. I.; Dalton, H. "The Soluble Methane Monooxygenase of *Methylococcus capsulatus* (Bath): Its Ability to Oxygenate *n*-Alkanes, *n*-Alkenes, Ethers, and Alicyclic, Aromatic and Heterocyclic Compounds." Biochem. J. **1977**, *165*, 395–402.

Fogel, M. M.; Taddeo, A. R.; Fogel, S. "Biodegradation of Chlorinated Ethenes by a Methane-Utilizing Mixed Culture." *Appl. Environ. Microbiol.* **1986**, *51*, 720–724.

Folsom, B. R.; Chapman, P. J.; Pritchard, P. H. "Phenol and Trichloroethylene Degradation by *Pseudomonas cepacia* G4: Kinetics and Interaction Between Substrates." *Appl. Environ. Microbiol.* **1990**, *56*, 1279–1285.

Fox, B. G.; Froland, W. A.; Dege, J. E.; Lipscomb, D. "Methane Monooxygenase from *Methylosinus trichosporium* OB3b: Purification and Properties of a Three-Component System with High Specific Activity from a Type II Methanotroph." *J. Biol. Chem.* **1989**, *264*, 10023–10033.

Fox, B. G., Borneman, J. G.; Wackett, L. P.; Lipscomb, J. D. "Haloalkene Oxidation by the Soluble Methane Monooxygenase from *Methylosinus trichosporium* OB3b: Mechanistic and Environmental Implications." *Biochemistry* **1990**, *29*, 6419–6427.

Green, J.; Dalton, H. "Substrate Specificity of Soluble Methane Monooxygenase: Mechanistic Implications." *J. Biol. Chem.* **1989**, *264*, 17698–17703.

Hartmans, S.; de Bont, J.A.M.; Tramper, J.; Luyben, K. Ch. A. M. "Bacterial Degradation of Vinylchloride." *Biotechnol. Lett.* **1985**, *7*, 383–388.

Janssen, D. B.; Scheper, A.; Dijkhuizen, L.; Witholt, B. "Degradation of Halogenated Aliphatic Compounds by *Xanthobacter autotrophicus* GJ10." *Appl. Environ. Microbiol.* **1985**, *163*, 635–639.

Janssen, D. B.; Grobben, G.; Hoekstra, R.; Oldenhuis, R.; Witholt, B. "Degradation of *trans*-1,2-Dichloroethene by Mixed and Pure Cultures of Methanotrophic Bacteria." *Appl. Microbiol. Biotechnol.* **1988a**, *29*, 392–399.

Janssen, D. B.; Gerritse, J.; Brackman, J.; Kalk, C.; Jager, D.; Witholt, B. "Purification and Characterization of a Bacterial Dehalogenase with Activity Toward Halogenated Alkanes, Alcohols, and Ethers." *Eur. J. Biochem.* **1988b**, *171*, 67–72.

Janssen, D. B.; Pries, F.; van der Ploeg, J.; Kazemier, B.; Terpstra, P.; Witholt, B. "Cloning of 1,2-Dichloroethane Degradation Genes of *Xanthobacter autotrophicus* GJ10 and Expression and Sequencing of the *dhlA* Gene." *J. Bacteriol.* **1989a**, *171*, 6791–6799.

Janssen, D. B.; Oldenhuis, R.; van den Wijngaard, A. J. "Hydrolytic and Oxidative Degradation of Halogenated Aliphatic Compounds by Aerobic Microorganisms." In *Biotechnology and Biodegradation*; Kamely, D.; Chakrabarty, A. M.; and Omenn, G., Eds.; Gulf Publishing Co.: Houston, TX; pp 105–125.

Keuning, S.; Janssen, D. B.; Witholt, B. "Purification and Characterization of Hydrolytic Haloalkane Dehalogenase from *Xanthobacter autotrophicus* GJ10." *J. Bacteriol.* **1985**, *163*, 635–639.

Kohler-Staub, D.; Hartmans, S.; Gälli, R.; Suter, F.; Leisinger, T. "Evidence for Identical Dichloromethane Dehalogenases in Different Methylotrophic Bacteria." *J. Gen. Microbiol.* **1986**, *132*, 2837–2843.

La Roche, S. D.; Leisinger, T. "Sequence Analysis and Expression of the Bacterial Dichloromethane Dehalogenase Structural Gene, a Member of the Glutathione *S*-transferase Supergene Family." *J. Bacteriol.* **1990**, *172*, 164–171.

Little, C. D.; Palumbo, A. V.; Herbes, S. E.; Lidstrom, M. E.; Tyndall, R. L.; Gilmer, P. J. "Trichloroethylene Biodegradation by a Methane-Oxidizing Bacterium." *Appl. Environ. Microbiol.* **1988**, *54*, 951–956.

McCarty, P. L.; Semprini, L.; Roberts, P. V. "Methodologies for Evaluating the Feasibility of *In-Situ* Biodegradation of Halogenated Aliphatic Groundwater Contaminants by Methanotrophs." *Proceedings*, AWMA/EPA International Symposium on Biosystems for Pollution Control, Cincinnati, OH, 1989.

Motosugi, K.; Soda, K. "Microbial Degradation of Synthetic Organochlorine Compounds." *Experientia* **1983**, *39*, 1214–1220.

Nelson, M. J. K.; Montgomery, S. O.; Mahaffey, W. R.; Pritchard, P. H. "Biodegradation of Trichloroethene and Involvement of an Aromatic Biodegradative Pathway." *Appl. Environ. Microbiol.* **1987**, *53*, 949–954.

Oldenhuis, R.; Vink, R.L.J.M.; Janssen, D. B.; Witholt, B. "Degradation of Chlorinated Aliphatic Hydrocarbons by *Methylosinus trichosporium* OB3b Expressing Soluble Methane Monooxygenase. *Appl. Environ. Microbiol.* **1989**, *55*, 2819–2826.

Oldenhuis, R.; Oedzes, J. Y.; van der Waarde, J. J.; Janssen, D. B. "Kinetics of Chlorinated Hydrocarbon Degradation by *Methylosinus trichosporium* OB3b and Toxicity of Trichloroethylene." *Appl. Environ. Microbiol.* **1991**, *57*, 7–14.

Scholtz, R.; Leisinger, T.; Suter, F.; Cook, A. M. "Characterization of 1-Chlorohexane Halidohydrolase, a Dehalogenase of Wide Substrate Range from an *Arthrobacter* sp." *J. Bacteriol.* **1987a**, *169*, 5016–5021.

Scholtz, R.; Schmuckle, A.; Cook, A. M.; Leisinger, T. "Degradation of Eighteen 1-Monohaloalkanes by *Arthrobacter* sp. strain HA1." *J. Gen. Microbiol.* **1987b**, *133*, 267–274.

Stanley, S. H.; Prior, S. D.; Leak, D. J.; Dalton, H. "Copper Stress Underlies the Fundamental Change in Intracellular Location of Methane Monooxygenase in Methane-Oxidizing Organisms: Studies in Batch and Continuous Cultures." *Biotechnol. Lett.* **1983**, *5*, 487–492.

Strandberg, G. W.; Donaldson, T. L.; Farr, L. L. "Degradation of Trichloroethylene and *trans*-1,2-Dichloroethylene by a Methanotrophic Consortium in a Fixed-Film, Packed-Bed Bioreactor." *Environ. Sci. Technol.* **1989**, *23*, 1422–1425.

Stucki, G. R.; Gälli, R.; Ebersold, H.; Leisinger, T. "Dehalogenation of Dichloromethane by Cell Extracts of *Hyphomicrobium* DM2." *Arch. Microbiol.* **1981**, *130*, 366–371.

Tsien, H. C.; Brusseau, G. A.; Hanson, R. S.; Wackett, L. P. "Biodegradation of Trichloroethylene by *Methylosinus trichosporium* OB3b." *Appl. Environ. Microbiol.* **1989**, *55*, 3155–3161.

van den Wijngaard, A.; Janssen, D. B.; Witholt, B. "Degradation of Epichlorohydrin and Halohydrins by Three Bacterial Cultures Isolated from Freshwater Sediment." *J. Gen. Microbiol.* **1989**, *135*, 2199–2208.

van den Wijngaard, A. J.; Reuvekamp, P.; Janssen, D. B. "Purification and Characterization of Halohydrin Dehalogenase from *Arthrobacter* sp. Strain AD2." *J. Bacteriol.* **1991**, *173*, 124–129.

Vogel, T. M.; Criddle, C. S.; McCarty, P. L. "Transformations of Halogenated Aliphatic Compounds." *Environ. Sci. Technol.* **1987**, *21*, 722–736.

Wackett, L. P.; Householder, S. R. "Toxicity of Trichloroethylene to *Pseudomonas putida* F1 Is Mediated by Toluene Dioxygenase." *Appl. Environ. Microbiol.* **1989**, *55*, 2723–2725.

Wackett, L. P.; Brusseau, G. A.; Householder, S. R.; Hanson, R. S. "Survey of Microbial Oxygenases: Trichloroethylene Degradation by Propane-Oxidizing Bacteria." *Appl. Environ. Microbiol.* **1989**, *55*, 2960–2964.

Weightman, A. J.; Weightman, A. L.; Slater, J. H. "Stereospecificity of 2-Monochloropropionate Dehalogenation by the Two Dehalogenases of *Pseudomonas putida* PP3: Evidence for Two Different Dehalogenation Mechanisms." *J. Gen. Microbiol.* **1982**, *128*, 1755–1762.

Wilson, J. T.; Wilson, B. H. "Biotransformation of Trichloroethylene in Soil." *Appl. Environ. Microbiol.* **1985**, *49*, 242–243.

Yokota, T.; Omori, T.; Kodama, T. "Purification and Properties of Haloalkane Dehalogenase from *Corynebacterium* sp. Strain m15-3." *J. Bacteriol.* **1987**, *169*, 4049–4054.

Biodegradation of Dichloromethane in a Fixed-Film Reactor under Methanogenic Conditions

*David L. Freedman**,
University of Illinois
James M. Gossett
Cornell University

INTRODUCTION

Since 1972, U.S. production of dichloromethane (DCM) has ranged from 212 to 287 million kilograms per year (U.S. ITC 1970–1985). Principal applications include paint removal and stripping, use in aerosols (*e.g.*, for spray paints and lubricants), use in manufacture of polycarbonate resin for the production of thermoplastics, and metal cleaning and degreasing. A consequence of this widespread use has been the appearance of DCM in groundwater systems. Concern over the presence of DCM in drinking water supplies is based on relatively recent studies which indicate the compound causes lung and liver cancer in mice (*Chemical Week* 1988). Although not the most widely occurring groundwater contaminant (trichloroethylene is), DCM is among the more difficult chlorinated aliphatics to remove by conventional remediation techniques (such as air stripping) because of its comparatively low Henry's constant (Gossett 1987). Interest in biological processes for remediation has grown recently because these offer

* University of Illinois, Department of Civil Engineering, Newmark Lab, Room 3230, MC-250, 205 North Mathews Avenue, Urbana IL 61801

the prospect of converting DCM to environmentally acceptable products.

Work we recently conducted with DCM, using suspended growth cultures, focused on the fundamental factors affecting its biodegradation under methanogenic conditions. Enrichment cultures were developed (in 160-mL serum bottles, maintained at 35 C) that used DCM as a sole organic carbon and electron donor source. A model for DCM biodegradation was proposed, indicating that the two principal modes of transformation were oxidation to CO_2 and fermentation to acetic acid (HAc). Thus, DCM degradation was a disproportionation, since oxidation of DCM made reducing equivalents available for reduction of an equal amount of DCM. The organism mediating DCM biodegradation was most likely an acetogen. Methanogenic bacteria in the mixed culture then used the products of acetogenic DCM degradation to form methane. CO_2-reducing methanogens used reducing equivalents gained from DCM oxidation to form methane; at least a portion of the reducing equivalents were available as hydrogen. Acetoclastic methanogens produced methane from the HAc formed by acetogenic fermentation of DCM. Assuming acetogenic bacteria did mediate DCM degradation, it is likely that the recently elucidated acetyl-CoA pathway was involved (Ljungdahl 1986). Since acetyl-CoA is a key intermediate in this pathway, it becomes apparent how DCM can serve as a growth substrate (Freedman & Gossett 1991).

Environmental conditions in DCM-contaminated groundwater are quite different from those employed in laboratory studies. First, attached growth conditions—rather than suspended growth conditions—will prevail, whether the mediating organisms are employed *in situ* or in an engineered, on-site, fixed-film reactor. Second, the operating temperature will be considerably lower than the optimum mesophilic one of 35 C. Pilot studies with a fixed-film reactor system were therefore undertaken to assess the applicability of laboratory results to conditions more like those expected during actual remediation.

The specific objectives of this study were: (1) to determine if DCM degradation can be sustained under attached growth conditions, at hydraulic residence times (HRT) ranging from 0.25 to 2 days; (2) to determine if DCM degradation can be sustained at 20 C, when DCM serves as the major organic carbon and electron donor source; (3) to examine the level of micronutrients—provided as yeast extract (YE)—needed in the system feed; (4) to determine the highest influent concentration of DCM that can be degraded at a 2-day hydraulic retention time (HRT); and (5) to analyze the fate of DCM during biodegradation in the fixed-film reactor system, using [^{14}C]DCM.

MATERIALS AND METHODS

System Description

The treatment system, located in a constant temperature room, consisted of two columns in series (Figure 1). A precolumn was used solely to reduce the redox potential of the feed to the treatment column, thereby ensuring that methanogenic conditions prevailed in the treatment column. Both columns were constructed from conical glass pipe (51 mm i.d. × 60 mm o.d. × 914 mm; Corning), filled to capacity with 6 mm glass beads. A peristaltic pump (Cole-Parmer) delivered refrigerated influent through the system. All components in the treatment column consisted of either glass, stainless steel, or Teflon®. A valve was located in the tubing coming from the precolumn, making it possible to seal off the treatment column (as was necessary when changing sampling port septa; see below). DCM-saturated water was added via a syringe pump (Sage Instruments); Pressure-Lok® gas syringes (Supelco, Inc.) were used. A valve in the DCM addition tubing made it possible to shut off the syringe when refilling it. The influent sampling port consisted of a tee modified to hold an 11-mm Teflon-faced rubber septum (Supelco, Inc.) in an ULTRA-TORR® Adapter (Cajon Co.).

Five sampling ports were located in the treatment column, spaced every 152 mm from the base of the column. Each port was constructed from a threaded glass connector tube (Ace Glass); inside the connector was a custom-fabricated stainless steel sampling tube that extended into the center of the treatment column to ensure that wall effects were minimized during sampling. The outside end of the sampling tube was machined to accept a Teflon-faced rubber septum (11 mm) and aluminum crimp cap (Supelco, Inc.). A Teflon bushing and O-ring secured the sampling tube inside the connector, and made a leak-tight seal.

The top of the column consisted of a glass cap that contained the final effluent sampling port, constructed as above, and located 200 mm above the fifth port on the treatment column. A glass outlet tube on the cap was fitted with a stopcock so that the effluent end of the treatment column could be sealed off. Tubing from the stopcock was connected to a U-tube. The discharge side of the U-tube consisted of

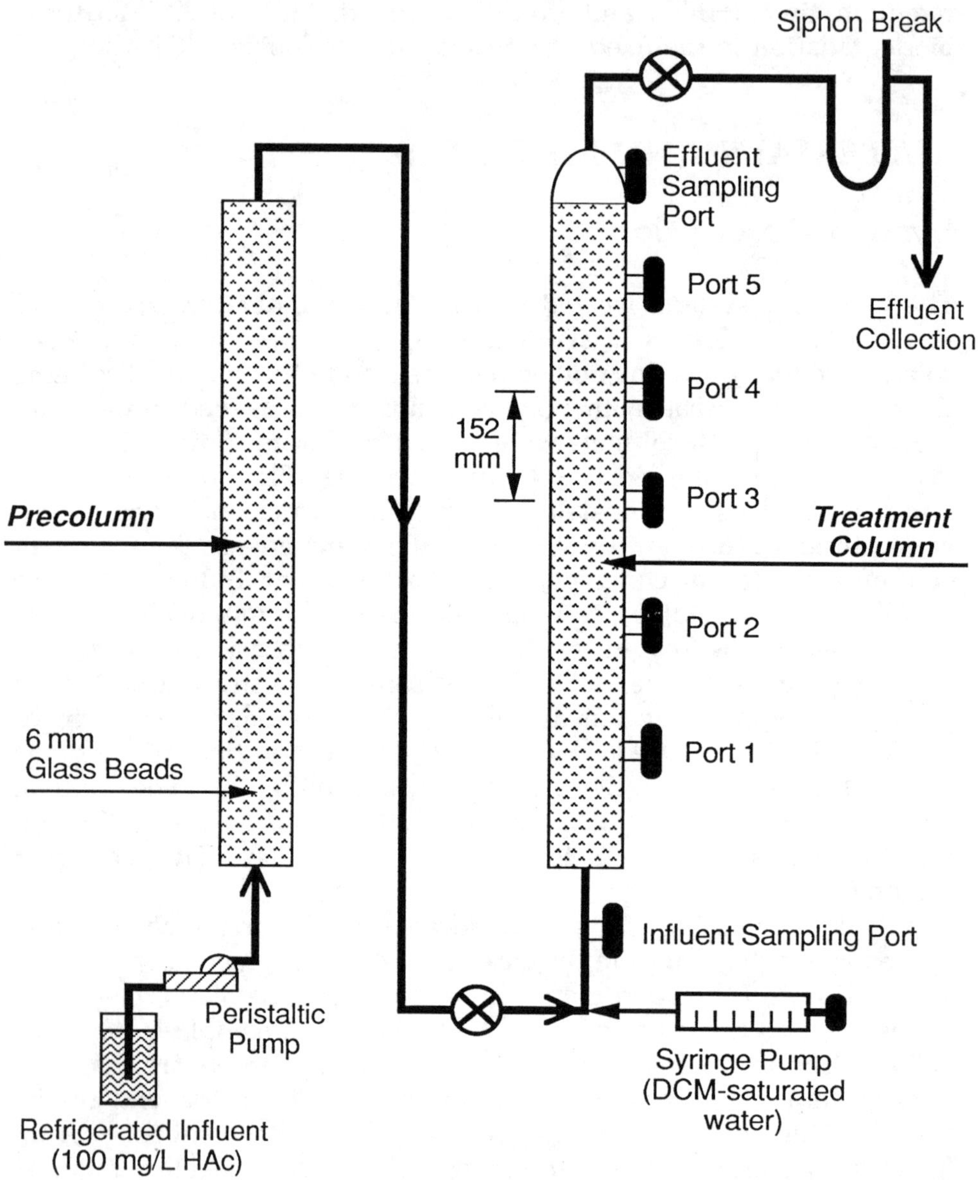

FIGURE 1. Schematic diagram of the fixed-film reactor system. Yeast extract was added along with HAc to the system feed (to provide micronutrients), in amounts shown in Table 2.

a 250-mL aspirator bottle that provided a reservoir of liquid to flow back into the column when samples were removed, thereby preventing the entry of air. A glass tee at the top of the aspirator bottle was open to the atmosphere and therefore served as a siphon break. The void volumes of the precolumn and treatment column were 660 mL and 730 mL, respectively.

When the sampling port septa were punctured several times (usually 5 to 10) they were replaced to prevent leakage and the introduction of air. The treatment column was placed on its side and, as a stream of N_2/CO_2 was placed close to a septum, the crimp cap was removed, and a new septum and cap were quickly slid into position as the old septum was pushed away.

TABLE 1. Basal medium used for the system feed.

Component	Quantity Per Liter of DDIW[(a)]
NH_4Cl	200 mg
$K_2HPO_4 \bullet 3H_2O$	101 mg
KH_2PO_4	55 mg
$MgCl_2 \bullet 6H_2O$	200 mg
Trace metal solution[(b)]	10 mL
$NaHCO_3$	760 mg
Resazurin	1 mg

(a) Distilled, deionized water.
(b) 100 mg/L $MnCl_2 \bullet 4H_2O$; 170 mg/L $CoCl_2 \bullet 6H_2O$; 100 mg/L $ZnCl_2$; 200 mg/L $CaCl_2$; 19 mg/L H_3BO_4; 50 mg/L $NiCl_2 \bullet 6H_2O$; 20 mg/L $Na_2MoO_4 \bullet 2H_2O$; and 360 mg/L $FeCl_2 \bullet 4H_2O$.

Table 1 lists the basal salts medium used in the system feed. It was prepared in 15-L batches and refrigerated. The feed was not augmented with HAc and YE until just prior to its use, to reduce the possibility of growth in the feed reservoir. HAc (present at 100 mg/L) provided the electron donor needed to consume oxygen and lower the redox potential in the precolumn; YE provided micronutrients. No reductant (Na_2S) was used, since the precolumn accomplished this task.

Analytical Methods

The concentrations of DCM and methane in the treatment column were determined by withdrawing 4-mL samples from each port, beginning with the effluent end and progressing sequentially to the influent port. A 5-mL Pressure-Lok syringe (Series C Supelco, Inc.) and specially fabricated 102-mm side port needle (Dynatech Precision Sampling Corp.) were used. The syringe and needle were flushed with N_2/CO_2, in a 70:30 ratio, prior to puncturing the sampling port septum. Immediately after removing 4 mL, it was injected into a 15-mL serum bottle stoppered with a slotted grey-butyl septum and aluminum crimp cap. The seven samples were placed in a 35 C reciprocating water bath, where DCM and methane in the serum bottles equilibrated between the aqueous and gaseous phases. A 0.5-mL headspace sample was then injected into to a Perkin-Elmer (Model 8500) gas chromatograph (GC), equipped with a flame ionization detector and a 3.2 mm × 2.44 m stainless steel column packed with 1 percent SP-1000 on 60/80 Carbopack-B (Supelco, Inc.), as previously described (Freedman & Gossett 1989).

The GC response factor for DCM was determined by adding 4 mL of water to replicate 15-mL serum bottles. A known amount of DCM (about 2.3 μM) was added to the bottles. Once equilibrated at 35 C, a 0.5-mL headspace sample was injected into the GC, as above. DCM additions (determined gravimetrically) were made from a stock solution containing DCM dissolved in methanol; the final concentration of methanol in the serum bottles was about 0.5 percent (v/v), well below the level that affects DCM's Henry's constant (Gossett 1987). The detection limit for DCM was approximately 0.63 μM (0.053 mg/L). For CH_4, 0.15 mL of pure gas was injected into replicate serum bottles containing 4 mL of water. The bottles were equilibrated at 35 C, and then a 0.5-mL headspace sample was injected into the GC. The coefficients of variation for the GC response factors were 1.89 percent for DCM, 1.21 percent for CH_4.

The concentration of DCM in the DCM-saturated water added to the treatment column was determined by injecting 50 μL into a 160-mL serum bottle filled with 100-mL of distilled, deionized water. Once the serum bottle was equilibrated at 35 C, a 0.5-mL headspace sample was analyzed by GC, as described previously (Freedman & Gossett 1989). Based on 41 samples, the DCM-saturated water contained an average concentration of 0.225 μM DCM/μL (coefficient of variation = 1.04%). Loss of DCM from the Pressure-Lok delivery syringe while it infused DCM-saturated water to the treatment column influent flow was minimal. The concentration of DCM in the

DCM-saturated water after it had been in the syringe for at least 2 days (the average time between filling) was 0.218 µM DCM/µL (n = 41; coefficient of variation = 5.08%), or a decrease of only 3.1 percent.

[^{14}C]DCM Studies

The final operational phase of the study involved addition of [^{14}C]DCM along with DCM-saturated water at a 0.25-day HRT. A solution was prepared by combining approximately 11 mL of DCM-saturated water and 1.2 mL of [^{14}C]DCM stock (containing approximately 2×10^7 dpm/mL), resulting in 1970 dpm/µL of solution. Analysis of the [^{14}C]DCM stock (used in the 35 C suspended growth studies) indicated 97 percent of the label consisted of DCM (Freedman & Gossett 1991). Based on an average delivery syringe flow rate of 50.8 µL/h and an average treatment column flow rate of 118 mL/h, the expected level of radioactivity in the treatment column was 848 dpm/mL. The DCM/[^{14}C]DCM solution contained 0.145 µM DCM/µL; this was less than the routine saturated solution because the [^{14}C]DCM stock used in the mixture was less than saturated.

When the concentration of total ^{14}C activity in the treatment column effluent was within 10 percent of the expected level in three consecutive tests, 4-mL samples were withdrawn from each of the treatment column ports and addition of [^{14}C]DCM was stopped. Total DCM and CH_4 levels were determined by headspace GC analysis (0.5 mL sample), as described above. Then the distribution of ^{14}C was determined; the first step was to raise the sample pH above 10 (to force all CO_2 into the aqueous phase of the serum bottles) by adding 5 µL of 8 M NaOH. Levels of dpm in each serum bottle were then measured by adding the following subsamples to 15 mL of liquid scintillation cocktail: (A) 0.5-mL headspace sample from the serum bottle, injected into a liquid scintillation vial modified to hold a septum in the cap, as previously described (Freedman & Gossett 1989); (B) 1 mL of basified sample; (C) 1 mL of basified sample that had been purged for at least 10 min with N_2 (>50 mL/min); and (D) 1 mL of acidified sample (approximately 10 µL of HCl added to the basified sample) that had been purged for at least 10 min with N_2 (>50 mL/min).

Total dpm in the serum bottle was the sum of the headspace (A) plus basified sample counts (B); however, only the headspace sample from the influent port was significantly different from a blank. The fraction purgeable at high pH was A + B – C; this was presumably either DCM or CH_4. The fraction purgeable only at low pH was C – D;

this was presumably CO_2. Compounds not purgeable at low pH (D) were identified as nonstrippable residue (NSR).

^{14}C activity was assayed with a model 9800 liquid scintillation counter (Beckman Instruments), following addition of 15 mL of liquid scintillation cocktail (Fisher Scientific) to each 1-mL sample. Corrections for counting efficiency were made according to a quench curve (sample H# versus efficiency).

RESULTS

Initial Testing with Water and DCM Only

Before starting the DCM degradation studies, the treatment column was operated at a 2-day HRT with only distilled, deionized water and DCM to evaluate abiotic losses. Results are shown in Figure 2. For the first 10 days, the DCM level in the column was well below the targeted concentration of 100 µM. On day 10, a length of Viton® tubing used to connect the DCM-saturated water syringe needle to the treatment column influent tubing was replaced with stainless steel tubing (1.59 mm), and all Viton O-rings were replaced with Teflon ones. Shortly thereafter, the concentration of DCM in the influent and effluent reasonably approached the target level, and remained so for an extended period of time. Although Viton is chemically resistant to DCM, it appears that it is also quite permeable with respect to DCM. The results from Figure 2 indicated that disappearance of DCM from the treatment column, once inoculated, was a result of biodegradation, not sorptive or diffusive losses.

As shown in Figure 2, the observed influent concentration of DCM varied considerably more than the effluent level. A likely reason was the poor degree of mixing between the DCM-saturated water and treatment column bulk flow. Even in the 3.18-mm tubing prior to the influent sampling port, where the velocity was highest, laminar flow conditions prevailed (Reynolds number = 28, based on $Q = 5.47 \times 10^{-8}\ m^3/s$, i. d. = 2.10×10^{-3} m, and $\upsilon = 7.27 \times 10^{-7}\ m^2/s$). For this reason, the influent DCM levels shown in all subsequent figures have been calculated, using measured values for the treatment column flow rate, the syringe pump flow rate, and the concentration of DCM in the DCM-saturated water.

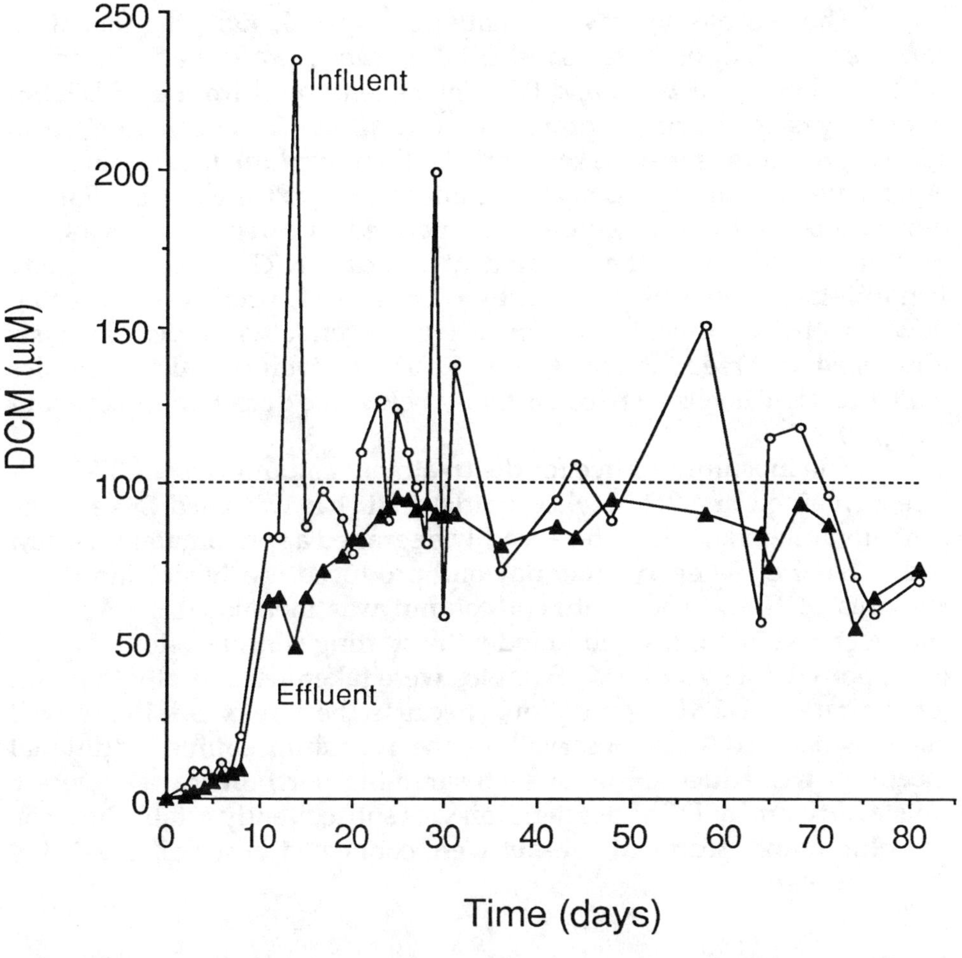

FIGURE 2. DCM levels in the treatment column when only DCM and distilled, deionized water were added. 100 µM was the DCM influent target concentration. Abiotic losses were minimal after replacing Viton tubing and O-rings on day 10.

Inoculation of the Precolumn and a Treatment Column

Before operating the two columns in series as a single system, each was inoculated and operated independently to demonstrate proper performance of its designated task, i.e., for the precolumn to reduce the redox potential of the system feed and for the treatment column to degrade DCM. All of these initial operations were conducted at 35 C.

The precolumn was inoculated on April 6, 1989; the inoculum consisted of 2 L of basal salts (in the same proportions listed in Table 1), 336 mg of HAc, and 0.5 L of mixed liquid from a 15-L laboratory digester, described previously (Freedman & Gossett 1989). No special precautions were taken to make the inoculum anoxic, since the organisms present created such conditions once in the column. For the first five days, the precolumn was operated in a recirculation mode; continuous flow was then started at a 4-day HRT. Approximately 1 month later, this was reduced to a 2-day HRT, prior to introducing flow to the treatment column. HPLC/refractive index analysis (described in Freedman & Gossett 1991) of several effluent samples indicated HAc levels were consistently below the detection limit (about 5 mg/L).

The inoculum source for the treatment column was a 1.37-L culture, contained in a 2.20-L glass bottle; a stir bar was used to keep the contents agitated. This culture readily degraded approximately 120-µM additions of DCM every other day and produced nearly stoichiometric amounts of CH_4. The treatment column was inoculated on April 25 and operated in recirculation mode; the starting concentration of DCM was approximately 120 µM. Samples were taken periodically to monitor the rate of DCM degradation. Because the rate was initially well below what had been observed in the inoculum source, additional inoculum was added (5 mL at each sampling port) on May 3. Once a satisfactory rate of DCM degradation was subsequently established, the precolumn and treatment column were connected in series, at a 2-day HRT, on May 17.

DCM Degradation Profiles and Continuous Flow Operation

Once the precolumn and treatment column were connected in series on May 17, continuous flow operation was started using the system feed described in Table 1, plus 100 mg/L HAc and 50 mg/L of YE. Table 2 lists the changes in operational parameters over time, including temperature, YE concentration in the system feed, DCM influent concentration, and HRT.

Operation at a 2-Day HRT. During the first phase of operation (from May 17 to June 8), the columns were operated at 35 C with 50 mg/L of YE in the system feed, a DCM influent concentration of approximately 100 µM, and a 2-day HRT. Samples taken on three occasions indicated

TABLE 2. Summary of DCM fixed-film system operating parameters.

Dates of Operation From	To	HRT (days)	Temperature (°C)	Yeast Extract (mg/L)	Influent DCM (μM)
5/17/89	6/8/89	2	35	50	100
6/8/89	6/22/89	2	35 ——> 20	50	100
6/22/89	7/3/89	2	20	50	100
7/3/89	7/19/89	2	20	5.0	100
7/19/89	10/4/89	2	20	0.5	100
10/4/89	10/11/89	2	20	0.5	150
10/11/89	10/16/89	2	20	0.5	250
10/16/89	10/25/89	2	20	0.5	300
10/25/89	11/2/89	2	20	0.5	400
11/2/89	11/8/90	2	20	0.5	530
11/8/89	1/5/90	2	20	0.5	780
1/5/90	1/30/90	2	20	0.5	315
1/30/90	3/21/90	2	20	0.5	460
3/21/90	3/27/90	2 —> 1	20	0.5	–
3/27/90	4/20/90	1	20	0.5	230
4/20/90	4/30/90	1 —> 0.5	20	0.5	–
4/30/90	6/17/90	0.5	20	0.5	190
6/17/90	6/23/90	0.5 —> 0.25	20	0.5	–
6/23/90	6/28/90	0.25	20	0.5	130
6/28/90	7/21/90	0.25	20	0.5	90.6
7/21/90	7/27/90	0.25	20	0.5	60.3[(a)]

(a) Includes [^{14}C]DCM.

nearly 99 percent of the DCM was degraded by port 1, with no detectable amount present by port 2. A sample profile (for June 2) is given in Figure 3. Methane output rose in accordance with the disappearance of DCM, but the amount produced was well above what could be explained by stoichiometric conversion of DCM (i.e., 0.5 M CH_4/M DCM degraded). The large amount of YE in the system feed most likely contributed to this higher than expected methane yield from the treatment column.

Having demonstrated DCM degradation at 35 C, the temperature was gradually lowered to 20 C over a 14-day period, holding all other conditions the same. The lower temperature did not affect the ability of the treatment system to biodegrade DCM, although complete conversion was not observed until port 3. Total methane output was similar, but occurred more gradually along the length of the column.

The next phase of operation (July 3 to 19) involved reducing the amount of YE in the system feed from 50 to 5 mg/L. Two sets of column samples taken during this interval indicated no affect on the ability of the treatment column to degrade DCM. Methane output gradually declined to about 290 μM.

YE in the system feed was then lowered to the target level of 0.5 mg/L and was kept there for all subsequent studies. Between July 19 and October 4, the system was operated at 20 C with a DCM influent concentration of 100 μM, 0.5 mg/L YE, and a 2-day HRT. Figure 4 presents average results for this period. The DCM degradation profile appeared similar to what was observed at 35 C and a higher level of YE; i.e., essentially all of the DCM was degraded in the first 152 mm of the column. However, methane output was lower, and significant amounts were not observed until the fifth port and the effluent port. One reason for this may have been the increased sensitivity of methanogens to DCM at 20 C compared to 35 C.

The next phases of system operation involved step increases in the DCM influent concentration; the aim was to determine how high a level could be reached before breakthrough and/or failure occurred. The system maintained its ability to degrade DCM when the influent concentration was raised to approximately 150 μM, then 250 μM, then 300 μM, then 400 μM, and then 530 μM. Breakthrough did not occur until the average DCM influent concentration was stepped up to approximately 780 μM. Three sets of samples were taken at this level, and the effluent DCM concentration increased with each analysis. Concurrent with the appearance of DCM in the effluent was a sharp reduction in total methane output, to only 85 μM on January 5, 1990 (Figure 5); breakthrough of DCM inhibited methanogenesis at all levels of the treatment column.

In an effort to reestablish complete DCM degradation, the influent concentration was reduced to about 315 μM beginning on January 5. The profile obtained was very similar to what had been observed when the same level of DCM was added as when the DCM influent concentration was being stepped up. However, methane

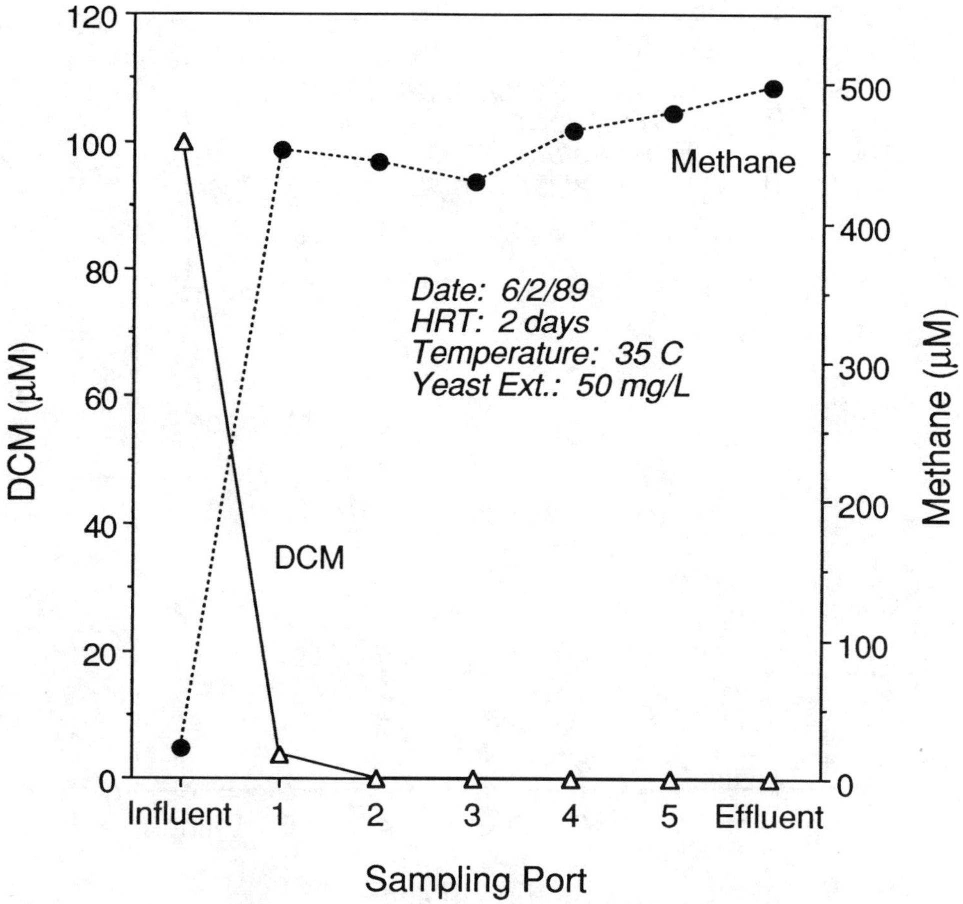

FIGURE 3. Performance of the treatment column shortly after starting continuous flow operation.

output remained quite low (approximately 100 µM), even though DCM no longer appeared in the upper reaches of the column.

The influent concentration of DCM was boosted again to approximately 460 µM. DCM disappeared below the detection limit even sooner than at a similar influent DCM level used previously, but methane output (about 100 µM) was nearly seven times less. Thus, even though nearly three months had passed since the influent DCM was lowered below the breakthrough concentration (on January 5; see Table 2), methanogenesis did not recover to the level observed prior to breakthrough of DCM.

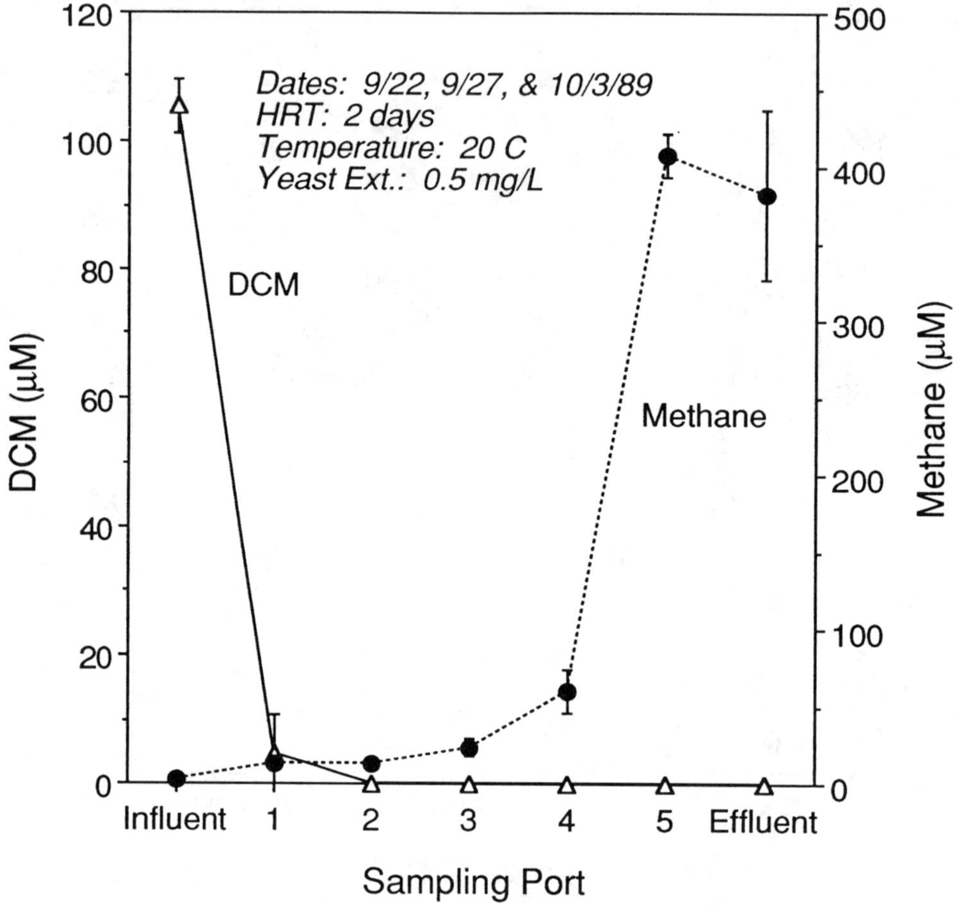

FIGURE 4. Average performance of the treatment column when operated at a 2-day HRT. Error bars represent one standard deviation.

Operation at <2-Day HRT. Having established the highest DCM concentration that could be degraded (approximately 530 µM)—at a 2-day HRT, 20 C, and 0.5 mg/L of YE in the system feed—before reaching breakthrough, the next phase of operation involved lowering the HRT to one day. This was done over a 6-day interval, beginning on March 21, 1990. Three sets of column samples were taken once stable operation at this HRT was demonstrated. At an influent DCM level of 230 µM, most of the DCM was consumed by port 1, and all of it was degraded below the detection limit by port 2, even at the lower HRT. Average methane output at a one-day HRT was only 29 µM.

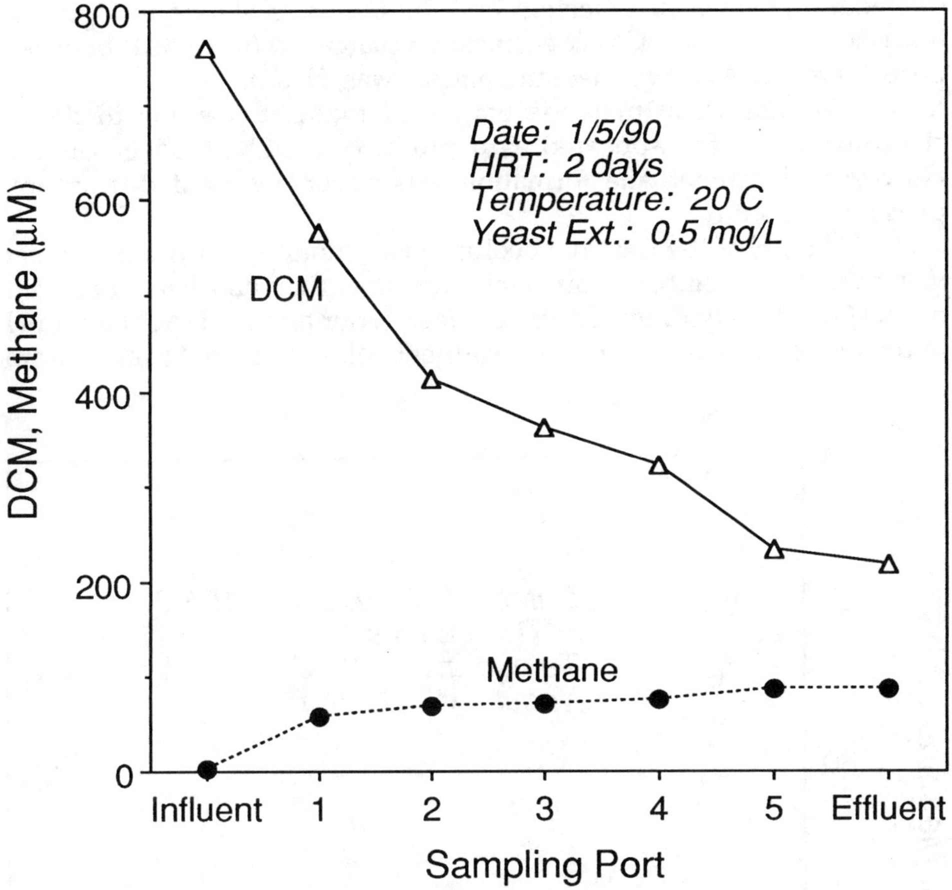

FIGURE 5. Breakthrough of DCM from the treatment column when added at an influent concentration of 780 µM, equivalent to a DCM loading rate of 11.9 µmole/h.

Over a 10-day interval, the treatment column HRT was lowered from one-day to 0.5-day, beginning on April 20. The column successfully degraded an influent DCM concentration of 190 µM, but complete degradation was not achieved until port 5. Methane output (average = 18 µM) continued to fall in accordance with the lower influent DCM concentration and decrease in HRT.

The final phase of column operation involved lowering the HRT to 0.25-day, over a 6-day interval. At 130 µM DCM and a 0.25-day HRT, DCM broke through the column, exiting at a concentration of 26 µM (CH_4 = 7 µM). When the DCM influent concentration was lowered to 90.6 µM (on June 28), triplicate sampling events indicated degradation below the detection limit by port 4, as shown in Figure 6. It is unclear why the DCM concentration plateaued (at 10 µM) between ports 1 and 2. Average methane output was 11 µM.

The analytical methods employed made it possible to detect chloromethane, if it appeared as a product of DCM biodegradation. However, chloromethane formation was never observed during any phase of the study.

The pH of the treatment column was monitored from the outset using the 4-mL samples from each port after they had been analyzed on the GC. Newly prepared system feed (after adding HAc) had a pH in the range of 6.8 to 7.4. In the treatment column, the pH ranged from

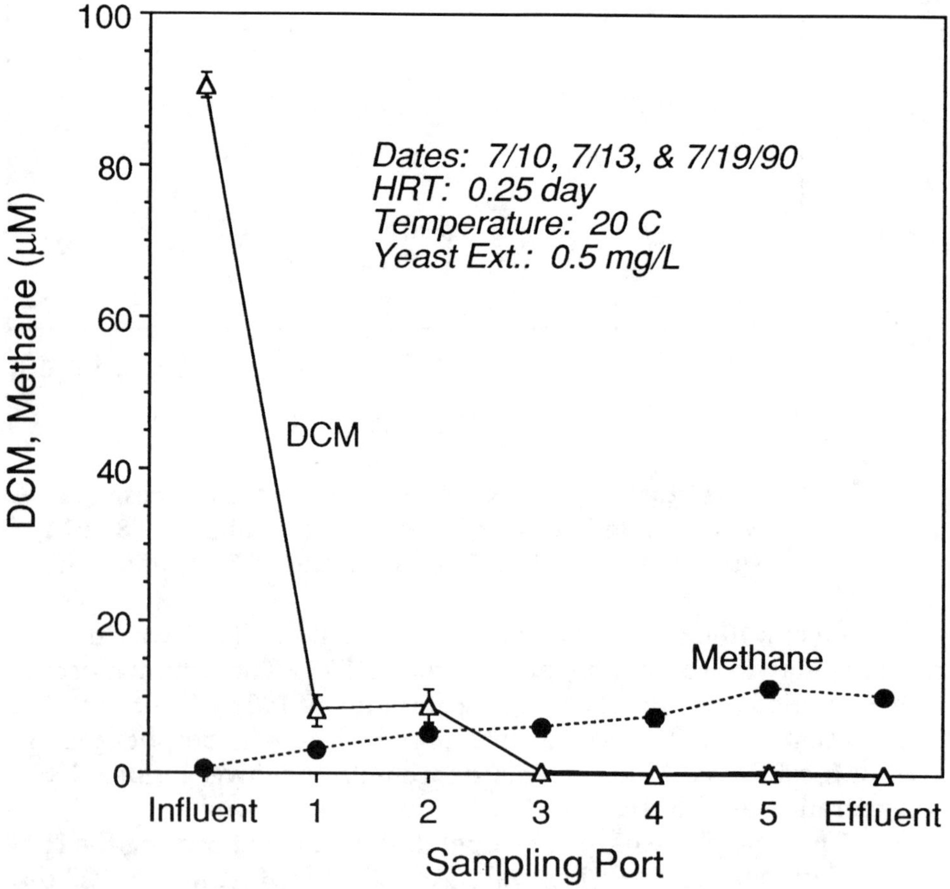

FIGURE 6. Average performance of the treatment column when operated at a 0.25-day HRT. Error bars represent one standard deviation.

7.1 to 7.7. There tended to be a slight drop in pH from the influent port to all subsequent ports, indicating that most of the DCM was degraded in the first 152 mm of the column, thereby releasing the most amount of HCl.

Fate of [^{14}C]DCM

Having demonstrated stable system operation at a 0.25-day HRT, an experiment was performed to determine the fate of [^{14}C]DCM. The HRT was maintained at 0.25-day, but the influent DCM concentration (beginning on July 21) fell from 90.6 µM to 60.3 µM since the DCM/[^{14}C]DCM syringe solution contained a lower level of DCM. Results from the [^{14}C]DCM study are summarized in Figure 7. The fraction purgeable at high pH (A + B – C) was presumed to be all DCM. Although it could have included methane, this possibility was unlikely because methane production in the treatment column was considerably inhibited: at the effluent port the methane level was 9.5 µM, nearly 70 percent less than what could have been produced based on stoichiometric conversion of DCM (0.5 M CH_4/M DCM). Under similar conditions observed during serum bottle studies, none of the methane formed was labeled (Freedman & Gossett 1991).

Disappearance of [^{14}C]DCM coincided with total DCM, based on GC analysis of sample bottle headspaces—i.e., essentially complete removal by port 4. Corresponding to the decrease in [^{14}C]DCM along the length of the treatment column was an increase in ^{14}C-labeled NSR and the fraction purgeable at low pH (C – D), presumed to be $^{14}CO_2$. At the effluent port, approximately 68 percent of the ^{14}C activity was $^{14}CO_2$ and 30 percent was ^{14}C-labeled NSR.

The distribution of ^{14}C activity in the ^{14}C-labeled NSR fraction from the effluent port was evaluated by HPLC analysis, as previously described (Freedman & Gossett 1991). Approximately 85 percent of the ^{14}C-labeled NSR was recovered as [^{14}C]HAc.

Recovery of ^{14}C activity was determined by comparing the measured dpm (A + B) at each port along the treatment column to the expected dpm (848 dpm/mL); on average, 85.1 percent was accounted for.

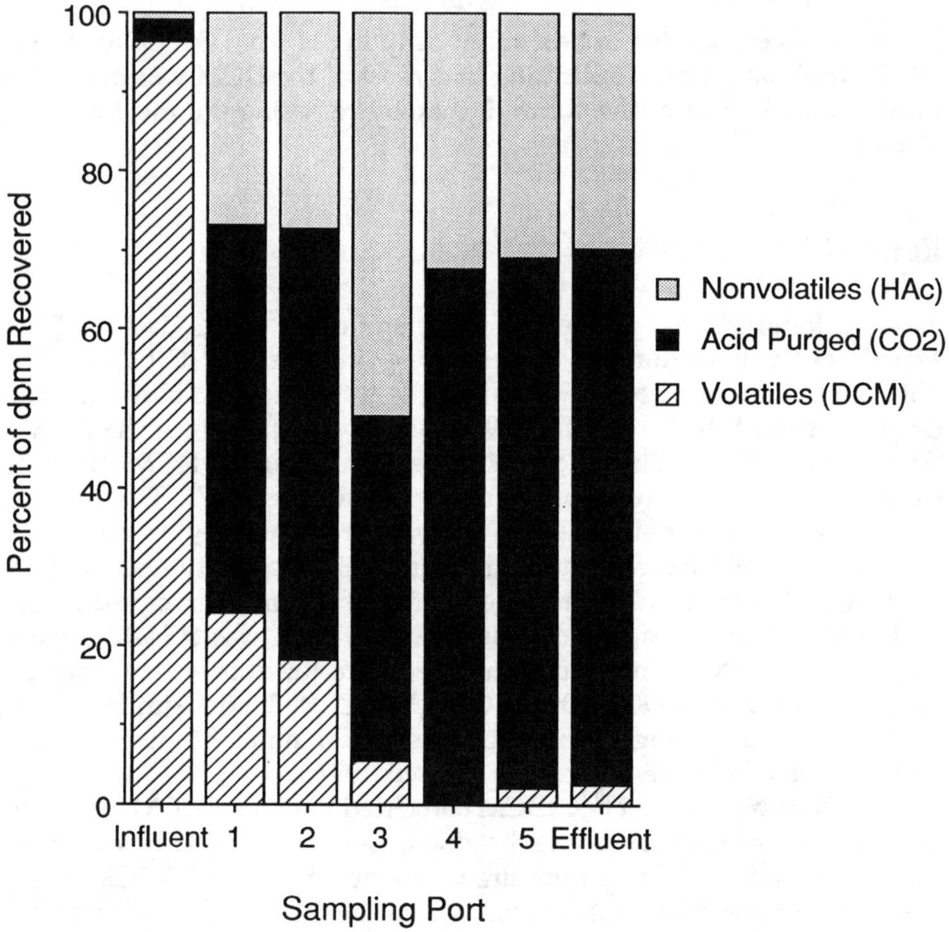

FIGURE 7. Distribution of ^{14}C following addition of [^{14}C]DCM to the treatment column, operated at a 0.25-day HRT, 20 C, and 60.3 µM DCM in the influent. Fraction A + B – C was presumably DCM; fraction C – D was presumably CO_2; fraction D was nonstrippable residue, most of which was HAc.

DISCUSSION

Based on over 14 months of operating data, the results of this study indicate that DCM degradation can be sustained under methanogenic, fixed-film conditions, at a residence time as low as 0.25-day, DCM influent levels as high as 530 µM (2-day HRT), 20 C, and a relatively low level of micronutrients (0.5 mg/L YE). These findings appear to be the first demonstration of a system capable of degrading DCM in a fixed-film reactor, under methanogenic conditions. Numerous other studies (Bouwer & McCarty 1982, 1983, 1985; Bouwer & Wright 1988; McCarty *et al.* 1984; Vogel & McCarty 1985, 1987) have previously demonstrated the capability of attached film reactors to degrade a variety of halogenated aliphatics—including tetrachloroethylene, trichloroethylene, carbon tetrachloride, and chloroform—under methanogenic conditions, but none of these evaluated biodegradation of DCM. The only attached-film research performed prior to this study with DCM was conducted under aerobic conditions (Galli & Leisinger 1985; LaPat-Polasko *et al.* 1984; Rittmann & McCarty 1980).

An important finding from the serum bottle studies was the ability of the DCM-degrading organisms to use this compound as a sole organic carbon and electron donor source (Freedman & Gossett 1991). Based on this result, only DCM was added directly to the treatment column. It was possible, however, that some biodegradable organic matter other than DCM entered the treatment column via the precolumn effluent. A direct measure of organic carbon leaving the precolumn was not made. During the initial phase of operation there was evidence of carry over of organics from the precolumn (as a consequence of the large amount of YE initially present in the system feed), since more methane was produced than could be attributed to DCM degradation. However, the system operated successfully for more than 12 months with only 0.5 mg/L YE in the precolumn feed, after which methane output in the treatment column fell well below stoichiometric levels. A YE level lower than 0.5 mg/L may also be acceptable. Of the 100 mg/L of HAc added continuously to the precolumn, less than the detectable limit (about 5 mg/L) was observed in the treatment column influent. Methane was also at very low levels (typically below 3 µM) in samples taken from the treatment column influent port. This suggests (but does not prove) that the HAc was consumed principally by organisms in the precolumn reducing the redox potential of the system, leaving little carry over of organics to the treatment column. A dense growth of biomass was observed in the first 100 mm of the precolumn. If the need for an auxiliary substrate to sustain DCM biodegradation is minimal or nonexistent, as suggested, considerable savings in operating costs will be realized during bioremediation (assuming methanogenic conditions already exist).

Another encouraging aspect of the fixed-film work was the ability of the treatment reactor to sustain DCM degradation at 20 C, a temperature approximating field conditions, rather than the 35 C environment used for the serum bottle experiments (Freedman &

Gossett 1991). Whether or not even lower temperatures can be tolerated is yet to be determined.

At a 2-day HRT (Q = 365 mL/day), a DCM influent concentration as high as 530 μM (45 mg/L) was degraded below the detection limit; this amounted to a DCM loading rate of 8.1 μM/h. Because the syringe pump operated at discrete rates, the next DCM influent level tried was 780 μM (66 mg/L), at which point breakthrough occurred (DCM loading rate = 11.9 μM/h). Breakthrough was observed on one other occasion: at a 0.25-day HRT and a DCM influent concentration of 130 μM, or 11.0 mg/L (DCM loading rate = 15.8 μM/h). Once the influent concentration was lowered to 90.6 μM (7.7 mg/L), complete degradation was achieved; this amounted to a loading rate of 11.0 μM DCM/h. Thus, the maximum DCM degradation rate at 20 C appeared to be in the range of 8 to 11 μM/h. Successful biodegradation of DCM in the treatment column at a 0.25-day HRT was encouraging because operation at or below this level is essential in order to keep the size of an on-site, engineered reactor within economical limits.

The DCM-degrading organism(s) exhibited good resilience with regard to DCM overloading. As soon as the loading rate was reduced, the column regained its ability to completely degrade DCM. Methanogenesis was far less resilient; the breakthrough of DCM at an influent concentration of 780 μM caused a reduction in methane output well below the level predicted based on stoichiometric conversion. This inhibition persisted long after the influent DCM concentration was lowered below the breakthrough value. The observed inhibition of methanogenesis by DCM was consistent with our work that employed suspended growth cultures (Freedman & Gossett 1991).

Less than stoichiometric formation of methane raised the question of what happened to the biodegraded DCM. Results from the [^{14}C]DCM experiment indicated that the principal products were (presumably) ^{14}CO2 and [^{14}C]HAc, both of which are environmentally acceptable. Our findings, using DCM-degrading enrichment cultures in serum bottles (Freedman & Gossett 1991), showed that when methanogenesis was partially inhibited, the products of [^{14}C]DCM degradation were approximately two-thirds $^{14}CO_2$ and one-third $^{14}CH_3COOH$. Although $^{14}CH_4$ formation was demonstrated when methanogenesis was uninhibited, none was produced when methanogenesis was partially inhibited. No attempt was made to measure $^{14}CH_4$ during the [^{14}C]DCM fixed-film reactor experiment; however, it seems likely that the same phenomenon occurred in the DCM treatment column when methanogenesis was partially inhibited.

Results from this study reinforce a conclusion of our work with suspended growth cultures (Freedman & Gossett 1991): to the extent that widespread use of halogenated aliphatic compounds continues, only those posing the least significant environmental hazard should be employed. Because DCM degrades to environmentally acceptable products under a variety of environmental conditions (including aerobic), it is one such compound that meets this criterion.

Acknowledgments

This research was supported by the U.S. Air Force Engineering and Services Center (AFESC), Tyndall AFB, FL, under contract no. F08635-86-C-0161, and by a post-doctoral fellowship from the Air Force Office of Scientific Research, administered by the American College of Toxicology.

REFERENCES

Bouwer, E. J.; McCarty, P. L. *Environmental Science and Technology* **1982**, *16*, 836–843.

Bouwer, E. J.; McCarty, P. L. *Applied and Environmental Microbiology* **1983**, *45*, 1286–1294.

Bouwer, E. J.; McCarty, P. L. *Biotechnology and Bioengineering* **1985**, *XXVII*, 1564–1571.

Bouwer, E. J.; Wright, J. P. *Journal of Contaminant Hydrology* **1988**, 2, 155–169.

Chemical Week. **1988**, *143*, 11–12.

Freedman, D. L.; Gossett, J. M. *Applied and Environmental Microbiology* **1989**, *55*, 2144–2151.

Freedman, D. L.; Gossett, J. M. "Pathways for Biodegradation of Dichloromethane and its Utilization as a Growth Substrate under Methanogenic Conditions." Submitted to *Applied and Environmental Microbiology*, 1991.

Galli, R.; Leisinger, T. *Conservation & Recycling* **1985**, *8*, 91–100.

Gossett, J. M. *Environ. Sci. Technol.* **1987**, *21*, 202–208.

LaPat-Polasko, L. T.; McCarty, P. L.; Zehnder, A.J.B. *Applied and Environmental Microbiology* **1984**, *47*, 825–830.

Ljungdahl, L. G. *Annual Reviews of Microbiology* **1986**, *40*, 415–450.

McCarty, P. L.; Rittmann, B. E.; Bouwer, E. J. In *Groundwater Pollution Microbiology*; Bitton, G.; Gerba, C. P., Eds.; John Wiley & Sons: New York, 1984; pp 89–116.

Rittmann, B. E.; McCarty, P. L. *Applied and Environmental Microbiology* **1980**, *39*, 1225–1226.

U.S. ITC (International Trade Commission). *Synthetic Organic Chemicals, U.S. Production and Sales.* U.S. ITC: Washington, DC, 1970–1985.

Vogel, T. M.; McCarty, P. L. *Applied and Environmental Microbiology* **1985**, *49*, 1080–1083.

Vogel, T. M.; McCarty, P. L. *Environmental Science and Technology* **1987**, *21*, 1208–1213.

Reductive Dechlorination of Tri- and Tetrachloroethylene by Nonmethanogenic Enrichment Cultures

Matthias Kästner[*]
Technical University of Hamburg-Harburg

INTRODUCTION

Trichloroethylene (TCE) and tetrachloroethylene (PCE) are the most frequently found chlorinated contaminants of groundwater in the industrialized nations (U.S. EPA 1985). These compounds are widely used as solvents or degreasing agents and are required in various technical processes. Contaminations of soil and groundwater were observed at nearly all sites at which TCE and PCE were handled to a great extent. Since the compounds are suspected to be carcinogens (Infante & Tsongas 1982), many contaminated sites have to be remediated. Biological treatment techniques may solve part of the remediation problems; however, extensive understanding of the fate of the chlorinated compounds in soil and groundwater environments is necessary to apply these techniques successfully.

In 1982, TCE and PCE were considered to be persistent in groundwater environments (Infante & Tsongas 1982). However, recent studies show aerobic biodegradation of TCE by different types of bacteria (Oldenhuis *et al.* 1989; Wilson & Wilson 1985; Nelson *et al.* 1988). Aerobic degradation of PCE was not reported. Under strictly anaerobic conditions, TCE and PCE are subject to reductive dechlorination.

[*] Department of Biotechnology II, Technical University of Hamburg-Harburg, Postfach 901403, 2100 Hamburg 90, Federal Republic of Germany

Biotransformation was observed with mixed cultures under methanogenic conditons (Vogel & McCarty 1985). The reaction proceeds by sequential reductive dechlorination via dichloroethylenes and vinyl chloride to CO_2 or ethylene (Vogel & McCarty 1985; Freedman & Gossett 1989). Vinyl chloride, which is carcinogenic, accumulates during the reaction sequence (Vogel & McCarty 1985). The accumulation explains the appearance of vinyl chloride in methanogenic aquifers contaminated with TCE and PCE.

High concentrations of cDCE were often observed in contaminated groundwaters in Germany (Brauch *et al.* 1987). Frequently, cDCE was the main component of the actual contamination and only traces of PCE or TCE were still found (Kästner *et al.* 1988). At these sites, cDCE was not an original contaminant; therefore, the component may have to be generated by transformation of PCE or TCE. The measured concentrations of cDCE exceeded those found in laboratory experiments under methanogenic conditions by orders of magnitude. Concentrations of vinyl chloride were independent of concentrations of cDCE. Often only traces of vinyl chloride were found together with high amounts of cDCE. High concentrations of vinyl chloride were observed with increased concentrations of methane, ethane, ethene, and detectable amounts of trans-1,2-dichloroethylene (Brauch *et al.* 1987). Reductive dechlorination under methanogenic conditions with the formation of vinyl chloride depended on high concentrations of degradable organic compunds (Freedman & Gossett 1989), on the formation of methane (Fathepure & Boyd 1988), and on low concentrations of sulfate. Sulfate inhibits methanogenesis and reductive dechlorination as shown for chloroaromatic compounds (Gibson & Suflita 1986; Kohring *et al.* 1989), but high concentrations of sulfate were frequently found in contaminated groundwaters.

This paper will present conditions of reductive dechlorination that led to the accumulation of cDCE without formation of vinyl chloride. The reaction may be useful in bioremediations.

METHODS

Batch experiments were carried out in closed 250-mL serum bottles with 200 mL of medium and 50 mL of air. The medium contained the following compounds per liter (pH 6.9): $Na_2HPO_4 \times 2\ H_2O$ 2.4 g, KH_2PO_4 1.5 g, $(NH_4)_2SO_4$ 0.5 g, $MgSO_4 \times 7\ H_2O$ 0.2 g, 3 mL vitamin solution, 10 mL trace element solution, and 500 mL autoclaved groundwater (well 5, Table 1). Added to the serum bottles with the medium was 2 µL of the chlorinated compounds. The partition ratios of the compounds between the gas and liquid phases were determined. The measured concentrations were corrected to total amounts to determine the molar ratio of the transformation.

TABLE 1. Analysis of groundwater from a contaminated site.

well #		FR*	4	240	5	6	7	8
Eh	mV	457	-259	-184	-342	-342	8	-94
COD	mg O_2/l	5	6100	2760	19050	1990	2440	5150
TOC	mg/l	–	–	751	4269	–	–	–
Cl^-	"	15	372	269	372	177	248	141
NO_3^-	"	34	–	0.17	0.5	–	–	–
SO_4^{2-}	"	330	–	260	100	–	–	–
CH_4	"	n.d.	0.9	0.3	1.3	0.5	0.7	0.3
Vinylchloride	mg/l	n.d.	0.96	0.79	1.67	0.71	1.02	0.27
Vinylidenechloride	"	n.d.	0.11	0.19	0.11	0.17	0.17	0.08
Dichloromethane	"	n.d.	28.90	18.47	43.29	7.72	12.14	13.59
1,1-Dichloroethane	"	n.d.	1.40	0.96	1.58	1.17	1.84	0.75
cis-1,2-Dichloro-ethylene	"	n.d.	103.35	124.18	73.79	172.28	156.39	87.13
Chloroform	"	0.002	0.71	0.77	0.53	0.78	0.67	0.85
1,1,1-Trichloroethane	"	0.001	0.38	0.33	0.21	0.37	0.24	0.21
1,2-Dichloroethane	"	n.d.	1.69	1.38	0.78	n.d.	n.d.	0.36
Carbon tetrachlor.	"	n.d.	n.d.	n.d.	n.d.	n.d.	n.d.	n.d.
Trichloroethylene	"	0.042	13.92	10.21	34.62	3.01	1.85	5.01
Perchloroethylene	"	0.005	2.83	1.80	3.50	1.29	1.28	1.20
Benzene	mg/l	n.d.	0.57	0.93	2.0	0.56	1.32	0.6
Toluene	"	n.d.	22.6	29.3	45.0	37.9	48.0	48.0
Xylene	"	n.d.	18.2	39.8	36.1	28.4	28.7	36.4

* Uncontaminated well of the same aquifer.

Mineral nutrient solution (2,050 mL $(NH_4)_2PO_4$) was fed to the groundwater treatment plant (Figure 1) in a volume ratio of 1:1 with the contaminated groundwater (well 5). The feeding rate amounted to 864 mL/day. This corresponds to a hydraulical retention time of 7.5 days in 6.5-mL liquid volume in the reactor (Hoppenheidt *et al.* 1989).

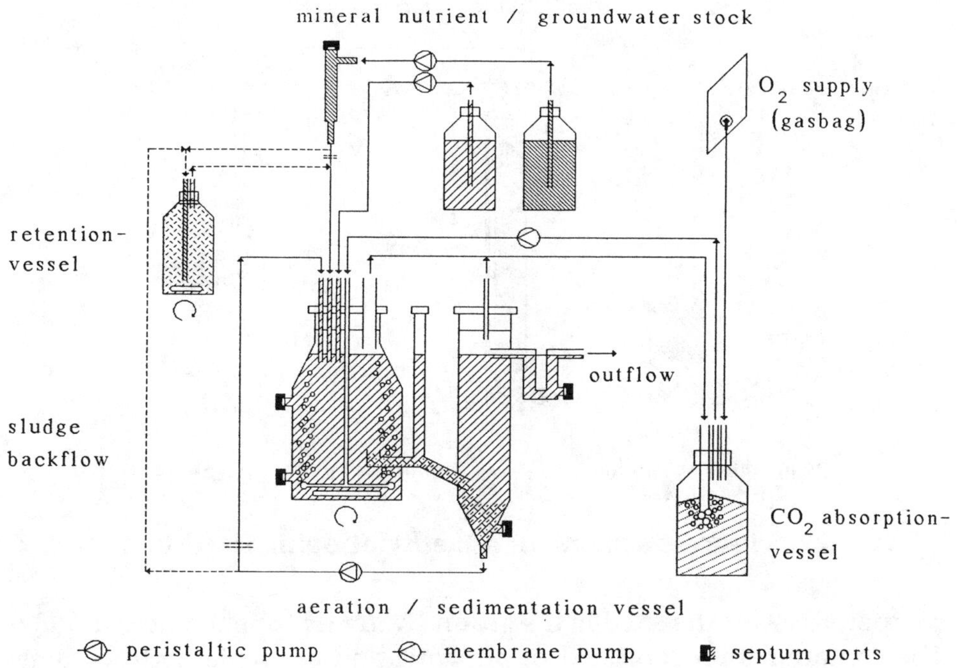

FIGURE 1. Process scheme of the continuous-flow groundwater treatment plant (full lines = aerobic process; dotted lines = introduction of reductive conditions by insertion of a sludge retention vessel).

The mineral medium fed to the methylotrophic fixed-bed reactor (Figure 2) contained the following compounds per liter (pH 6.9): Na_2HPO_4 0.24 g, KH_2PO_4 0.15 g, $(NH_4)_2SO_4$ 0.2 g, $MgSO_4$ 0.2 g, $CaSO_4$ 0.2 g, 2.5 mL trace element solution SL 4, and 7.5 mg TCE. The feeding rate was 1,653 mL/day (retention time 0.71 days).

Chlorinated hydrocarbons were measured using a gas chromatograph with an electron capture detector («0.5 mg/L) and a flame ionization detector (»500 mg/L). The number of bacteria in batch cultures was estimated by microscopical counting of the cells.

RESULTS AND DISCUSSION

Table 1 shows analyses of the groundwaters from different wells of a contaminated site in Germany. The contaminated area was the former site of a solvent recycling factory. More than 100 single organic

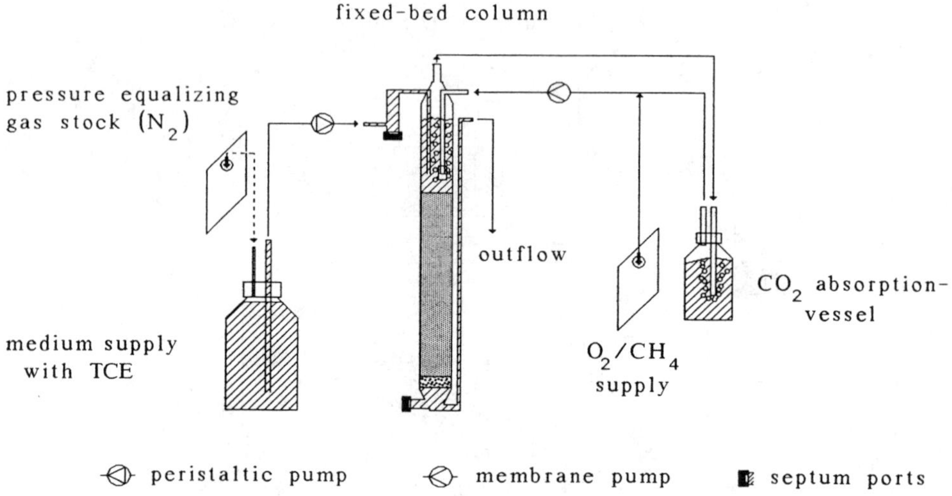

FIGURE 2. Process scheme of a methylotrophic fixed-bed reactor.

compounds were detected in the groundwater (Hoppenheidt *et al.* 1989). The contamination consisted of aliphatic and aromatic hydrocarbons, alcohols, ketones, and chlorinated aliphatic compounds. The highest contamination was found in well 5. The content of organic compounds amounted to 19,000 mg/L chemical oxygen demand (COD) measured as a summarizing parameter. cDCE found in concentrations of about 170 mg/L was the main chlorinated contaminant. The highest contents of organic compounds were found together with the highest PCE and TCE residual concentrations and with the lowest cDCE concentrations. No trans-1,2-dichloroethylene was detected. Vinyl chloride concentrations were measured up to only 3 percent of the cDCE concentration, and methane was detected in traces. Higher concentrations of vinyl chloride and methane should have been found if only methanogenic bacteria had transformed PCE and TCE. As indicated by the redox potential and the nitrate concentrations of the uncontaminated well (FR), the aquifer was originally aerobic. These data suggested that the accumulation of cDCE without extensive formation of vinyl chloride in contaminated aquifers is a process which is different from that under methanogenic conditions. The suggestion is supported by the redox potentials of the pairs PCE/TCE and TCE/cDCE which range between those of O_2/H_2O and NO_3^-/NO_2^- (Vogel *et al.* 1987). Therefore, TCE and PCE can be reduced by most biochemical coenzymes and electron carrier systems. This implies that aerobic or denitrifying bacteria may be involved in the first two dechlorination steps of PCE.

Reductive Dechlorination Experiments

Batch Cultures. Aerobically enriched bacteria (from well 5, Table 1) were able to degrade the organic contaminants in the groundwater medium. In these cultures, reductive dechlorination of PCE and TCE to cDCE occurred if the oxygen supply was limited. The transformation was highly stereoselective. Only traces of 1,1-dichloroethylene and no trans-1,2-dichloroethylene were found. Further dechlorination of the dichloroethylenes was not observed even after 4 more months of incubation. These findings are different from those described for methanogenic conditions.

Stable, dechlorinating mixed cultures could be established with three species of bacteria, but the transformation could not yet be related to axenic cultures. The three species cultures consisted of a facultative anaerobic sporogenic bacterium and a sulfidogenic sporogenic bacterium together with an aerobic bacterium. Of seven aerobic strains, each was tested in co-culture with the two sporogenic bacteria. Only three of these strains exhibited reductive dechlorination of PCE and TCE, whereas all cultures exhibited sulfidogenic conditions. A time-course experiment with one of these cultures is shown in Figures 3 and 4 to represent the kinetics of the reaction in batch cultures. Growth of the bacteria was found during the first 6 days of incubation. At the end of growth (day 7, arrow in Figure 3), sulfidogenic conditions were observed in the medium by a change in color from yellow to gray. Reductive dechlorination began at this time. Within 4 days, 16.2 mg/L = 97.7 µg M/L PCE were transformed to 9.46 mg/L = 97.5 µg M/L cDCE. (The partition ratio of PCE between the gas and liquid phases was determined to 0.79. Therefore 12.8 mg/L, initially measured in the liquid phase, represented 16.2 mg/L PCE in gas and liquid phases. The partition ratio of cDCE, corrected to the volume reduction from sampling until day 10, was determined to be 0.77; the measured concentration of 7.3 mg/L represented 9.46 mg/L overall.) This indicated an equimolar transformation of PCE to cDCE. TCE concentrations increased during the dechlorination, but, at the end, TCE was not detected. At day 12, traces of 1,1-dichloroethylene appeared. The intermediary formation of TCE is evidence of a sequential dechlorination mechanism. The concentration of vinyl chloride, which was initially found in the groundwater medium, did not increase during the transformation (Figure 3).

Reductive dechlorinations occurred with the transition from aerobic to anaerobic conditions and were restricted to this change of conditions. No further transformation of PCE and TCE was found after

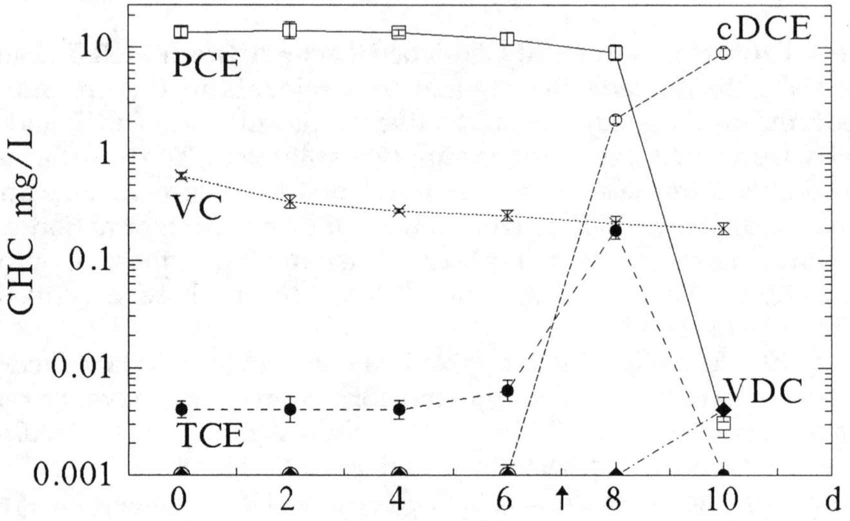

FIGURE 3. Kinetics of the reductive dechlorination of PCE and concentrations of chlorinated hydrocarbons (CHC), vinyl chloride (VC), and 1,1-dichloroethylene (VDC).

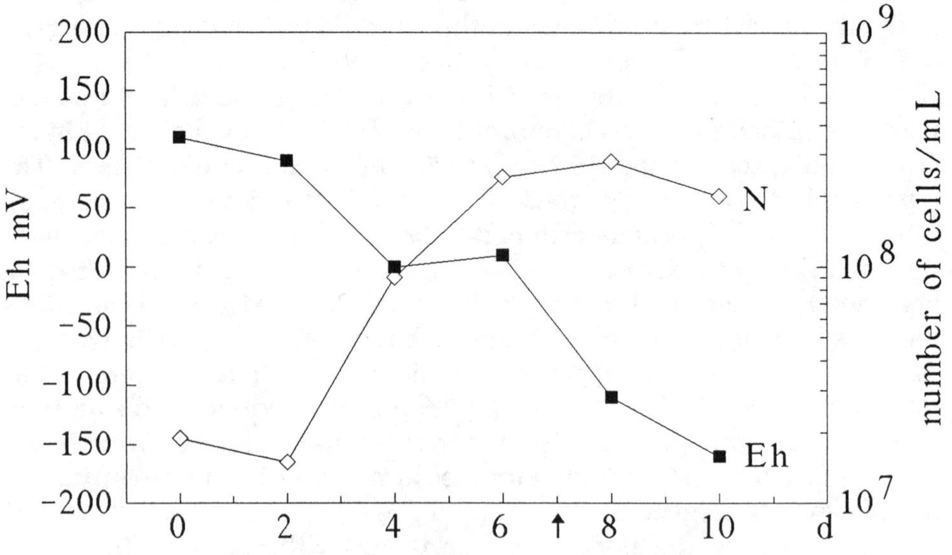

FIGURE 4. Kinetics of the reductive dechlorination of PCE, redox potential (Eh), and number of cells (N).

the change, even after repeated addition of the compounds. Anaerobic cultures as well as aerobic cultures with unlimited oxygen supply showed no dechlorination. In addition to the change of conditions, a decrease of redox potential (–50 mV to –150 mV) caused by sulfide was necessary for the reductive dechlorination in batch cultures. Sulfide could be derived either from the degradation of organic sulfur compounds or from the successive growth of sulfidogenic bacteria. In addition, sulfide release resulted in the abiotic formation of traces of 1,1-dichloroethylene from PCE and TCE (Kästner, in press).

The data led to the conclusion that aerobic or facultative anaerobic bacteria might be involved in the reductive dechlorination of PCE and TCE to cDCE. If anaerobic bacteria were involved, the reaction should not have been restricted to the transition conditions. But anaerobic bacteria able to produce cDCE were not found in these cultures. To prove this conclusion, further investigations with axenic cultures are necessary.

Dechlorination of TCE to cDCE and traces of 1,1-dichloroethylene has been described for aerobic TCE-degrading cultures with limited oxygen supply (Fliermans *et al.* 1988). In addition, batch cultures under oxygen-limited conditions with PCE-contaminated groundwater from different aquifers yielded formation of cDCE after supplementation with ethanol, glycerol, acetate, or lactate (Hoppenheidt & Hanert 1989). The described conditions of dechlorination are evidence that different physiological types of bacteria take part in the sequential reductive dechlorination of PCE and TCE to cDCE. Therefore, high concentrations of cDCE, found together with low concentrations of vinyl chloride in contaminated groundwaters, can be explained by different reactions. Vinyl chloride is a dechlorination product that depends on methanogenic conditions. Low concentrations of vinyl chloride, independent of those of cDCE, can be observed if the activity of methanogenic bacteria in an aquifer is small.

Continous-Flow Reactors. Further experiments with aerobic continuous-flow reactors were suggested by the conclusion that aerobic or facultative anaerobic bacteria may participate in the reductive dechlorination of PCE and TCE under oxygen-limited conditions. The first of these experiments was carried out with a bench-scale sewage treatment plant constructed to degrade groundwater contaminants (well 5, Table 1).

The plant consisted of a sedimentation vessel (2 L) and an aeration vessel (7 L) with a closed gas-circulating system. The reactor was aerated with pure oxygen, and the produced CO_2 was absorbed by

NaOH. Oxygen concentrations of about 9 mg/L were maintained in the aeration vessel. The process scheme of the plant is shown in Figure 1.

The average space loading of the xenobiotic contaminants amounted to 780 ± 45 mg/L day measured as chemical oxygen demand (COD). The efficiency of degradation was determined to be 730 ± 41 mg/L day COD. The space loading of the chlorinated compounds amounted to 3.6 ± 0.33 mg/L • day with an efficiency of degradation of 3.4 ± 0.38 mg/L • day. The degradation of the chlorinated compounds was determined by the release of chlorine ions (Hoppenheidt *et al.* 1989). The aerobic removal of cDCE and TCE was calculated to be 94.5 percent and 95.8 percent, respectively (inflow concentration: cDCE 21.34 ± 2.8 mg/L, TCE 5.47 ± 1.3 mg/L; outflow concentration: cDCE 1.17 ± 0.095 mg/L, TCE 0.227 ± 0.04 mg/L). Reductive dechlorination experiments were conducted by inserting a retention vessel of the aerated sludge (Figure 1, dotted lines). The inflow of the goundwater was connected into this vessel. Increased retention times of the sludge and the groundwater generated oxygen-limited conditions in this vessel. The oxygen limitation led to decreased TCE and increased cDCE outflow concentrations from the plant, as shown in Figure 5. Vinyl chloride and 1,1-dichloroethylene concentrations did not increase in this experiment. Initial values of the outflow concentrations were found after connecting the groundwater inflow back to the aeration vessel (day 36). The data indicated the dechlorination of TCE to cDCE in the sludge retention vessel. Therefore, conditions of reductive dechlorination could be introduced in aerobic suspended-sludge reactors by altering the process scheme.

The second experiment was conducted with a fixed-bed reactor. The reactor was inoculated with a methylotrophic, TCE-degrading culture enriched from an aquifer solely contaminated with TCE. Methylotrophic TCE degradation proceeds by oxygen incorporation into the molecule (Little *et al.* 1988). The reactor consisted of a glass column (2 L) filled with quartz gravel. The process scheme is shown in Figure 2. The top of the reactor was aerated with a mixture of oxygen and methane in a volume ratio of 7:3. Oxygen was supplied in excess to ensure aerobic conditions in the fixed bed. Oxidative TCE degradation was determined to be 2.2 ± 0.18 mg/day, calculated from control experiments with bacteria poisoned with Na-Azide. A change of the O_2/CH_4 ratio to 3:7 (CH_4 in excess) yielded decreased TCE outflow concentrations and the appearance of cDCE (Table 2).

After three exchanges of the reactor liquid volume, oxygen-limited conditions were established in the fixed bed and the formation

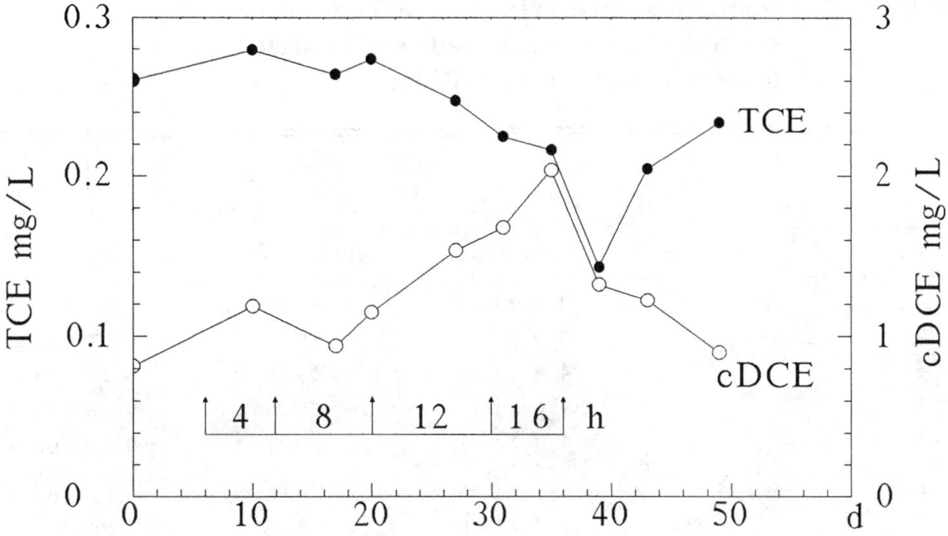

FIGURE 5. TCE and cDCE outflow concentrations of the continuous-flow groundwater treatment plant at different retention times of the activated sludge.

of cDCE was initiated. No other dechlorination product was detected besides cDCE. After another change back to unlimited conditions of the gas mixture, the reactor exhibited original methylotrophic TCE degradation rates. The experiment shows that this aerobically enriched methylotrophic mixed culture from a contaminated aquifer is also able to dechlorinate TCE to cDCE under oxygen-limited conditions. Therefore, the conditions of reductive dechlorination could be initiated in the fixed-bed reactor only by limitating the oxygen supply.

The experiments demonstrated that reductive dechlorination found with aerobic enrichment cultures could be transferred to continuous-flow reactors. Immediately after introducing oxygen-limited conditions, dechlorination products were detected. This indicated the involvement of bacteria which were already grown under unlimited aerobic conditions. Contrary to previous reports, it was shown that reductive dechlorination of PCE and TCE also occurred under non-methanogenic, not strictly anaerobic conditions. The described type of dechlorination represents an equimolar transformation of PCE to cDCE. No carcinogenic vinyl chloride is generated in the process. Therefore, the reaction can be used in bioremediation techniques. Compounds like PCE, which are resistant to oxidative microbial degradation, can be transformed to bio-oxidizable reduced products by these reaction

TABLE 2. Reductive dechlorination of TCE in the methylotrophic fixed-bed reactor after change to oxygen-limited conditions.

time	vol. ratio	TCE		cDCE [a]
	O_2/CH_4	mg/L		
days		inflow	outflow	outflow
1	3/7	4.9 ± 0.51	2.3 ± 0.19	<0.002
2	"	3.9 ± 0.44	2.7 ± 0.09	0.17 ± 0.012
3	"	4.8 ± 0.62	1.6 ± 0.11	0.33 ± 0.015
4	"	4.6 ± 0.20	1.6 ± 0.06	0.46 ± 0.021
5	"	4.3 ± 0.41	1.2 ± 0.07	0.49 ± 0.020
6	7/3 [b]	4.1 ± 0.48	2.2 ± 0.11	0.12 ± 0.011
7	"	4.3 ± 0.26	2.4 ± 0.09	<0.002

a cDCE inflow concentration < 0.002.
b Unlimited conditions of methylotrophic TCE degradation.

conditions. The aerobic degradation of cDCE by different bacteria has already been reported (Hoppenheidt *et al.* 1989; Wackett & Gibson 1989).

CONCLUSIONS

In this paper, the reductive dechlorination under oxygen-limited conditions in batch cultures and continuous-flow reactors was demonstrated. The reaction can be applied in technical plants, but further examinations for scaling up are necessary. Mineralization of TCE and PCE can be achieved when the compounds are dechlorinated reductively to cDCE with the subsequent oxidation of this molecule.

Two types of processes can be used for bioremediation techniques. PCE- and TCE-contaminated groundwaters with high contents of degradable organic compounds can be treated with a suspended sludge process as described in the experiments. Or groundwaters solely contaminated with the chlorinated compounds may be treated by a two-stage process in fixed-bed reactors. The first reactor has to be supplied with a nonequimolar ratio of oxygen and methane to establish oxygen-limited conditions for the reductive dechlorination. The second reactor has to be supplied with oxygen in excess to degrade cDCE cometabolically by methylotrophic bacteria. Additional experiments are necessary to estimate whether the transformation rates can be increased by adding other carbon sources.

The reaction sequence with methylotrophic bacteria might also be applied *in situ*. In situ bioremediation of a TCE-contaminated groundwater by supplementation of oxygen and methane was demonstrated in a pilot study (Roberts *et al.* 1989). By alternating the supplementation of oxygen and methane, changes from aerobic to anaerobic conditions could be initiated in the soil and the aquifer. With this treatment scheme, even PCE might be susceptible to microbial attack.

Acknowledgments

This work was conducted at the Department of Microbiology, Biocenter, Technical University of Braunschweig, Federal Republic of Germany. The author is very grateful to Dr. H. H. Hanert for supervising these investigations. The work was supported by a grant from the Federal Ministry of Research and Technology, Federal Republic of Germany (Grant No. 02 WT 4564, April 1985 to April 1988).

REFERENCES

Brauch, H. J.; Kühn, W.; Werner, P. *Vom Wasser* **1987**, *68*, 23–32.

Fathepure, B. Z.; Boyd, S. A. *Appl. Environ. Microbiol.* **1988**, *54*, 2976–2980.

Fliermans, C. B.; Phelps, T. J.; Ringelberg, D.; Mikell, A. T.; White, D. C. *Appl. Environ. Microbiol.* **1988**, *54*, 1709–1714.

Freedman, D. L.; Gossett, J. M. *Appl. Environ. Microbiol.* **1989**, *55*, 2144–2151.

Gibson, S.; Suflita, J. M. *Appl. Environ. Microbiol.* **1986**, *52*, 681–688.

Hoppenheidt, K.; Hanert, H. H. *Das Gas- und Wasserfach, Wasser/Abwasser* **1989**, *130*, 706–711.

Hoppenheidt, K.; Kästner, M.; Hanert, H. H. *Das Gas- und Wasserfach, Wasser/Abwasser.* **1989**, *130*, 697–705.

Infante, P. F.; Tsongas, T. A. *Environ. Sci. Res.* **1982**, *25*, 301–327.

Kästner, M. Submitted to *Appl. Environ. Microbiol.* (In press).

Kästner, M; Hoppenheidt, K.; Hanert, H. H. In *Contaminated Soil, 1988 (Second International TNO/BMFT Conference)*; Wolf, K.; van den Brink, W. J.; Colon, F. J., Eds.; Kluwer Academic Publishers: Dordrecht/Boston/London, 1988; 1263–1265.

Kohring, G. W.; Zhang, X.; Wiegel, J. *Appl. Environ. Microbiol.* **1989**, *55*, 2735–2737.

Little, C. D.; Palumbo, A. V.; Herbes, S. E.; Lidstrom, M. E.; Tyndall, R. L.; Gilmer, P. J. *Appl. Environ. Microbiol.* **1988**, *54*, 951–956.

Nelson, M.J.K.; Montgomery, S. O.; Pritchard, P. M. *Appl. Environ. Microbiol.* **1988**, *54*, 604–606.

Oldenhuis, R.; Vink, R.L.J.M.; Janssen, D. B.; Witholt, B. *Appl. Environ. Microbiol.* **1989**, *55*, 2819–2826.

Roberts, P. V.; Semprini, L.; Hopkins, G. D.; Grbić-Galić, D.; McCarty, P. L.; Reinhard, M. "*In-situ* Aquifer Restoration of Chlorinated Aliphatics by Methanotrophic Bacteria"; final report, CR-812220, Robert S. Kerr Environmental Protection Agency; Ada, OK, 1989.

U.S. EPA (Environmental Protection Agency). *Substances Found at Proposed and Final NPL Sites through Update Number Three*; U.S. Environmental Protection Agency: Washington, DC, 1985; Document NPL-U3-6-3.

Vogel, T. M.; McCarty, P. L. *Appl. Environ. Microbiol.* **1985**, *49*, 1080–1083.

Vogel, T. M., Criddle, C. S., McCarty, P. L. *Environ. Sci. Technol.* **1987**, *21*, 722–736.

Wackett, L. P.; Gibson, D. T. *Appl. Environ. Microbiol.* **1989**, *55*, 2960–2964.

Wilson, J. T.; Wilson, B. H. *Appl. Environ. Microbiol.* **1985**, *49*, 242–243.

Field and Laboratory Evidence of *In Situ* Biotransformation of Tetrachloroethene to Ethene and Ethane at a Chemical Transfer Facility in North Toronto

David W. Major[*], *Eric W. Hodgins*
Beak Consultants Limited
Barbara J. Butler
University of Waterloo

INTRODUCTION

A potential real estate transaction initiated a groundwater investigation at a 4.5-acre chemical transfer facility located in North Toronto. The fenced facility is comprised of a small office, boiler and weigh scale building, road with turnabout, railway spur line, and above- and below-ground transfer lines (Figure 1). Organic solvents are delivered to the site by rail and transferred by below- and above-ground pipes to the above-ground tank farm and eventually to tanker trucks. Both the previous and current owners of the facility stored and transferred similar organic solvents (methanol, methyl ethyl ketone, vinyl and ethyl acetate, butyl acrylate) with one notable exception. The previous owner had stored tetrachloroethene (PCE) at the site approximately 10 years ago. Preliminary groundwater investigations revealed both free and dissolved PCE, and the dechlorination intermediates of PCE had contaminated the groundwater below the facility. The absence of PCE, or its intermediates, between the former PCE storage tank and the spur line suggests that small amounts of free-phase PCE were released

[*] Beak Consultants Limited, 595 Woolwich Street, Guelph, Ontario N1H 3Y5, Canada

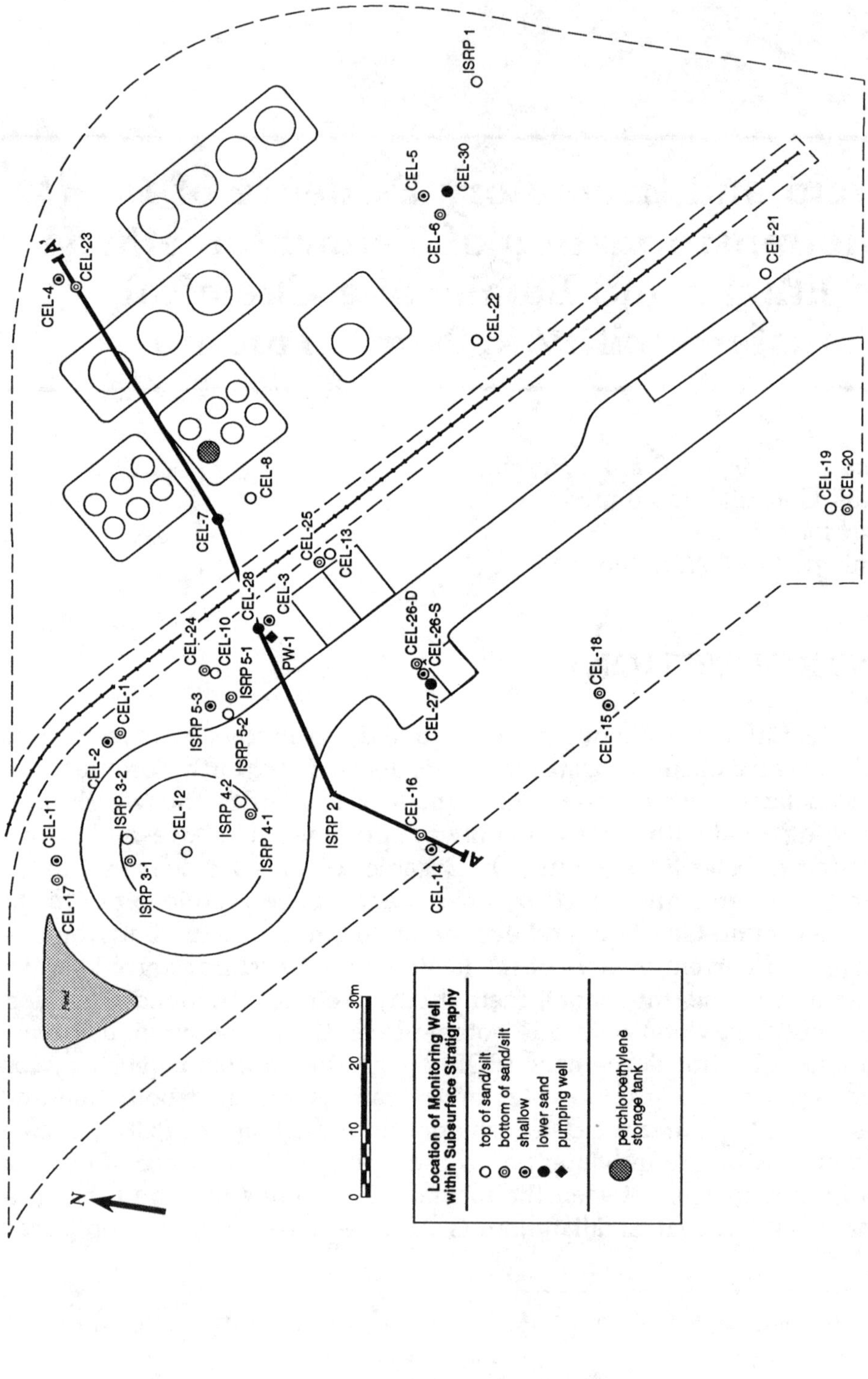

FIGURE 1. Plan map showing location of monitoring wells and cross section A-A′.

during the coupling and uncoupling of railway tanker cars, and not from the above-ground storage tank.

Pumping test data indicated that pumping could effectively contain the dissolved PCE on site, but the hydraulic conductivities of the aquifer were too low for the feasible application of pump and treat strategies to reduce source area PCE concentrations. However, the presence of dechlorination products suggested that anaerobic biotransformation of PCE might be an attenuating mechanism, one that could be exploited in remediating the site if vinyl chloride (VC) was not accumulating. As a result, we undertook a field and laboratory investigation to determine if VC was being mineralized or biotransformed to innocuous end products. If VC was being biotransformed, we wished to understand the physical, geochemical, and microbial processes that promote the conversion of PCE past VC. This paper presents our initial field and laboratory data that provide evidence of *in situ* biotransformation of PCE to ethene.

MATERIALS AND METHODS

Geology

The site geology was assessed with continuous coring techniques utilizing hollow stem augers and a 1.5-m coring tool advanced with the lead auger, and driller's logs. Downhole geophysical surveys (EM-39 and Gamma) were performed in completed monitoring wells. Core material collected during continuous coring operations was subsampled for grain size analysis and f_{oc} determinations. Hydrogeological data were obtained from the site by monthly measurements of the piezometric level in monitoring wells, well response tests, and drawdown responses during pumping.

Sample Collection

Groundwater. For use in microcosm studies, groundwater was collected using a Waterra pump. Sterile 4-L glass bottles were filled to overflowing, capped, and transported to the laboratory. Before use, the collected groundwater was purged with sterile N_2 for a minimum of 8 h to remove dissolved halocarbons and oxygen. The water had a pH of 6.4 and a temperature of 10 C. Groundwater samples collected for halocarbon, volatile fatty acid (VFA), and hydrogen sulphide analysis were dispensed without headspace into 250-mL amber glass vials, which were sealed with Teflon®-lined caps. Halocarbon samples were preserved with sodium thiosulphate and $HgCl_2$ (4 percent), and VFA samples with sodium thiosulphate only. Hydrogen sulphide samples were filtered through 0.45-µm acrylic copolymer filters and preserved with 2 drops of a zinc acetate/NaOH solution (33 g of zinc acetate per 150 mL plus 32 mL of a 50% NaOH solution). Groundwater samples for metal analysis were collected in 250-mL polyethene bottles, filtered, and preserved with HNO_3. One-litre samples were collected in amber glass bottles for methane analysis. Groundwater samples for anion analysis were collected in 400-mL polyethene bottles. Quality control and quality assurance samples included replicate groundwater samples from selected wells, field, and trip blanks.

Aquifer Sediments. The Waterloo Saturated Sand Sampler (WSSS) was used to secure and retrieve aquifer sediments from an area known to have concentrations of PCE (ISRP 5-1). The WSSS is a 1.5-m stainless steel core barrel with an inside diameter of 5 cm. The 5-cm aluminum core liner was swabbed with alcohol and steam-cleaned on site prior to its insertion into the core barrel. A piston, attached to a wireline, was sterilized with alcohol and inserted into the shoe end of the liner, and a shoe was attached to the barrel. A hollow stem auger was used to drill to an approximate depth of 8 m below ground surface (mbgs). The WSSS was then attached to the drill rods and hammered 1.5 m in front of the lead auger. The piston remained stationary as the WSSS was pushed into the sediments, resulting in the piston being drawn up the liner; this caused a slight vacuum to develop, which helped to hold the sediments within the liner. The drill rods were then decoupled downhole and the WSSS brought to the surface by wireline retrieval. During core collection and retrieval, a stream of N_2 gas was blown downhole to minimize exposure of the core to atmospheric oxygen. Once the WSSS was brought to the surface, the core liner was removed and quickly cut into 60- to 70-cm lengths, capped, and the liner cut lengthwise with a circular saw. One core was not split but was capped, wrapped, placed on dry ice, and express-shipped to the University of Tennessee for phospholipid fatty acid analysis. Split cores were immediately placed into an ethanol-swabbed, N_2-flushed glovebox for further manipulation. The split core liners were pried apart, and sterile tools were used to pare off the outermost portions of the core. The inner, aseptic core material was collected and distributed into sterile Whirl-Pak bags. When full, the bags were placed in 1-L sterile canning jars that had been flushed and filled with argon before placement inside the glovebox. The jars were sealed, then removed and stored on ice for transport from the site to the University of Waterloo.

Enumeration of Microbial Populations

Sulphate-reducing and methanogenic bacteria were enumerated using a three-tube, most probable number (MPN) procedure. All culture tubes were incubated at room temperature for up to 3 months. Under anaerobic conditions, sediments were diluted appropriately in a sulphate-free mineral salts medium (Zehnder & Wuhrmann 1977). Sulphate reducers were enumerated in Postgate's (1984) medium B modified by the addition of Pfennig's vitamins and minerals (McInerney *et al.* 1979), the reduction of yeast extract to 0.5 mg/L, and the addition of 10 g/L of purified agar (L28, Oxoid). Acetate or lactate (10 mM/L) was provided as a potential electron donor. Development of black iron sulphide precipitate in a culture tube was taken to be indicative of sulphate reduction.

Methanogens were enumerated in a medium modified from Zehnder and Wuhrmann (1979), containing 10 mM/L acetate or methanol, or H_2-CO_2 (80:20 v/v, in the headspace of the tube) as electron donors. Modifications to the medium were the addition of 0.5 mg/L yeast extract, the replacement of Zehnder and Wuhrmann's trace mineral solution with that of Zeikus *et al.* (1979), except calcium chloride and sodium chloride were omitted, and the use of NaS_2 (0.25 g/L), cysteine-HCl (0.5 g/L), but not titanium (III) as reductants. Evolution of methane gas in a culture tube was scored positive for methanogens. Methane in the headspace of culture tubes was detected with a Gow-Mac 69-750 gas chromatograph equipped with a 6 ft by 1/8 inch PoraPak N (80-100 mesh) column and a flame ionization detector, using nitrogen as the carrier gas.

Total phospholipids were extracted from frozen subsurface sediments by a modified single-phase chloroform-methanol method as previously described (Bobbie & White 1980). Methyl esters of the phospholipids (phospholipid fatty acids, PLFA) were analyzed by capillary GC as described by Bobbie and White (1980) and Guckert *et al.* (1986). Analysis of PLFA has been demonstrated to provide reproducible, quantitative, and qualitative information of the total biomass, community structure, and nutritional status of subsurface microorganisms (Balkwill *et al.* 1988).

Microcosm Studies

Sediment and water samples were stored at 4 C until microcosm construction was initiated. Microcosm construction was started approximately 18 h after on-site sample collection and was conducted in an anaerobic glovebox. The microcosms consisted of 20 g (dwt) sediment plus site water, with or without amendments, in sterile 60-mL hypovials. Any required amendments were added to 1 L of the purged groundwater, which was then stirred in a sealed, N_2-filled flask with minimal headspace for 1.5 to 2 h before it was placed into the glovebox. A sample of the purged groundwater was taken for halocarbon analysis as described below. After the addition of sediment, each hypovial was filled to capacity with water, sealed with a sterile, Teflon-lined Tuf-Bond septum (Pierce) and aluminum crimp seal. Filling and sealing a microcosm treatment set (*e.g.*, PCE plus methanol) took approximately 1 h. Upon removal from the glovebox, the completed microcosms were inverted and placed into argon-filled paint cans equipped with GasPak Plus H_2/CO_2 generator envelopes (BBL). The cans were sealed and incubated at 10 C. Periodically, a set of microcosms was removed from the incubator, warmed to room temperature, and sacrificed for analysis. A vent needle was inserted into the microcosm and a 15-mL aliquot of water was withdrawn into a glass, gas-tight syringe for methane, ethene, and ethane analysis as described below. The microcosms were shaken and left for 1 h to ensure partitioning of the halocarbons between the liquid and newly created headspace had reached equilibrium. The headspace gas was analyzed for halocarbons as described below.

In addition to sediment and site water, some microcosms were amended to contain the following: PCE only, PCE plus methanol, PCE plus acetate, or PCE plus acetate and methanol. The amendments were added to site water at concentrations of 5 mg/L for PCE and 50 mg/L for acetate or methanol; when both primary substrates were present, they were added at a concentration of 25 mg/L each. Control microcosms were similarly constructed as treatment microcosms and consisted of: PCE-H_2O controls containing only PCE in sterile, distilled water; autoclaved/poisoned controls containing PCE, autoclaved sediment (1 on each of 3 successive days), autoclaved site water, and 0.5 mL of a 4% $HgCl_2$ solution; and dissolved oxygen controls containing PCE, sediment, and site water.

Each treatment sample point is the mean result from three replicate microcosms. However, dissolved oxygen controls consisted of two replicates, and only a single PCE-H_2O control was prepared for each sample time. In addition, some autoclaved/poisoned control values (days 83 and 145) represent the mean of two instead of three replicates. The number of controls was minimized because of limitations in paint can volume and sediment availability.

Geochemical Analysis

Methane, Ethene, and Ethane Analysis. A 20-mL aliquot was withdrawn from the bottom of the sample into a 30-mL gas-tight glass syringe, and 10 mL of prepurified helium was added. The aliquot was shaken and allowed to equilibrated for 2 h. A 2-mL aliquot of the gas phase was injected into a Hewlett Packard 5840A GC equipped with an FID and 300-µL sample loop. These gases were separated on a 30-m GSQ megabore column. The GC was calibrated (ESTD method) using an analyzed gas mixture. Dissolved concentrations were determined using the Ideal Gas Law, Henry's Law, and solubility coefficients. The detection limit for each gas was estimated to be 0.01 mg/L.

Oxygenate Analysis. Methanol, ethyl acetate, vinyl acetate, butanol, and butyl acrylate were detected by direct aqueous split injection on a Hewlett Packard 5890 GC equipped with FID and 60-m megabore DB-WAX capillary column. The detection limit for these compounds was estimated to be 0.1 mg/L.

Volatile Fatty Acid Analysis. Samples of acetate, butanoate, propanoate, and pentanoate were acidified with ultrapure formic acid, and the VFAs were detected by direct aqueous split injection on a Hewlett Packard 5890 GC equipped with FID and a 15-m large-bore DB-FFAP WAX capillary column. The detection limit for these compounds was estimated to be 0.1 mg/L.

Halocarbon Analysis. Two methods were used to quantify two groups of halocarbons. Vinyl chloride (VC), 1,1-dichloroethene (1,1-DCE), and *cis*-dichloroethene and *trans*-dichloroethene (cDCE and tDCE) were determined by headspace injection onto a Photovac Model 10S55 GC equipped with a 4 ft by 1/8 inch SE-30 (packed column) and detected by a PID. Trichloroethene (TCE) and PCE were quantified by pentane microextraction and injection onto a Hewlett Packard GC equipped with ECD and a 60-m DB-5 large-bore capillary column. The detection limits for the halocarbons was estimated to be 0.001 mg/L.

Inorganic Ions and Hydrogen Gas Analysis. Anions were analyzed by ion chromatography following Standard Methods for the Examination of Water and Wastewater, method No. 429. Hydrogen gas was determined by injecting 0.5 mL headspace gas onto a 3.2 mm × 3.2 m stainless steel column packed with 100/120 Carbosieve SG (Supelco). This column was flushed with helium for 1.5 min (to remove halogenated compounds), and then the flow was directed to a reduction gas detector (Trace Analytical RGD 2).

RESULTS

Field

The geologic cross section (Figure 2) along transect A-A′ (Figure 1) intersects the area of PCE contamination and parallels the NE to SW direction of groundwater flow. The geology is composed of clay till (upper clay) underlain in order by a sand-silt unit (upper sand/silt), clay unit (lower clay), layered sand/silt unit (lower sand/silt), and bedrock. The upper clay unit is weathered and fractured and contains numerous sand and silt lenses and stringers of uncertain lateral extent.

Dissolved chlorinated organics have been detected in groundwater samples collected from monitoring wells installed in the upper sand/silt and upper clay units. Free-phase PCE was obtained from the monitoring well CEL-3 installed in the upper clay. The lower clay unit is very stiff and appears to be a competent aquitard. The lower sand/silt unit appeared to be uncontaminated. The hydraulic conductivity of the upper clay unit was estimated by grain size analysis to be in the range of 10^{-8} to 10^{-9} cm/s. By contrast, the lower sand/silt layer had hydraulic conductivities of 10^{-6} to 10^{-7} cm/s in the northern portion of this unit, and 10^{-5} cm/s in the central area as determined by response tests. Differences in the hydraulic conductivities affect the ratio of vertical to lateral groundwater flow and velocity (Figure 2). Generally, groundwater flow was from NE to SE with a horizontal velocity in the central area of 0.02 to 2 m/y.

Monitoring wells were located within the upper sand/silt and upper clay units at approximate depths of 3 and 7 mbgs. The geochemical results presented focus on samples collected at 7 mbgs monitoring wells because there were more sampling points and greater hydraulic conductivity (and thus groundwater velocity) at this depth. The distribution of PCE, TCE, cDCE, and VC (Figure 3A, B, C, D, respectively) show similar lateral extent. cDCE was found to be the

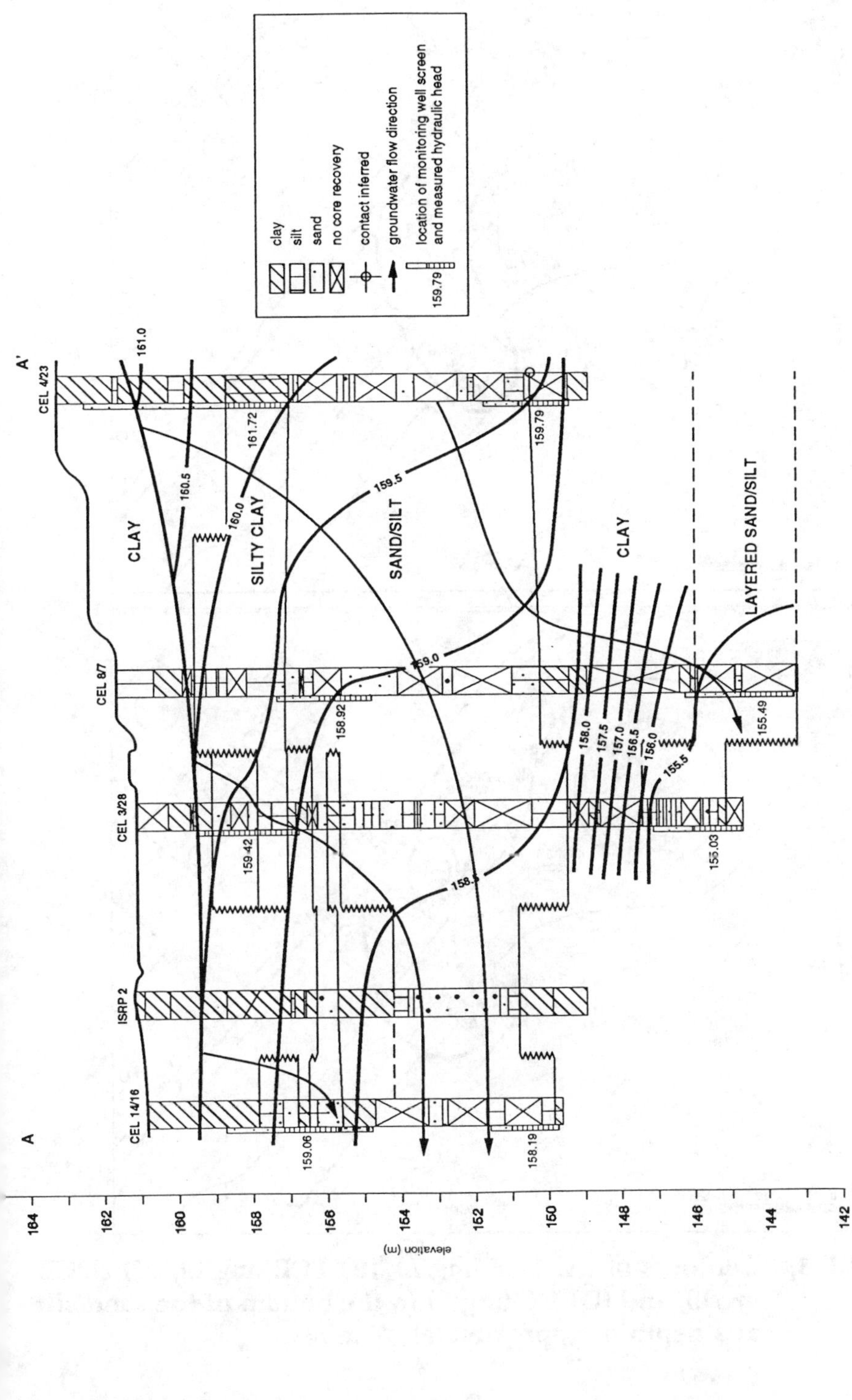

FIGURE 2. Hydrostratigraphic cross section A-A′ showing subsurface geology, hydraulic head equipotentials, and direction of groundwater flow.

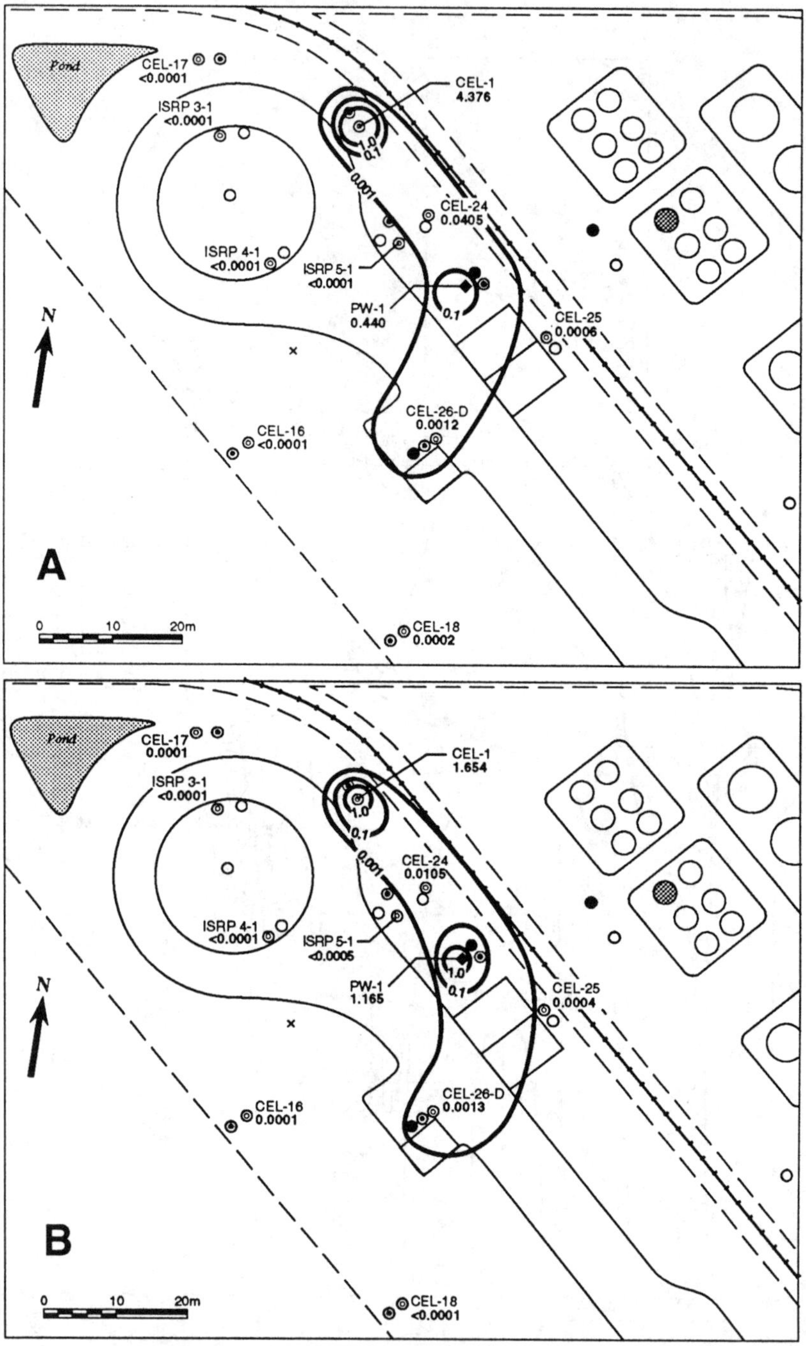

FIGURE 3. **Contours of (A) PCE (mg/L), (B) TCE (mg/L), (C) cDCE mg/L), and (D) VC (mg/L) in the bottom of the sand/silt at a depth of approximately 7 mbgs.**

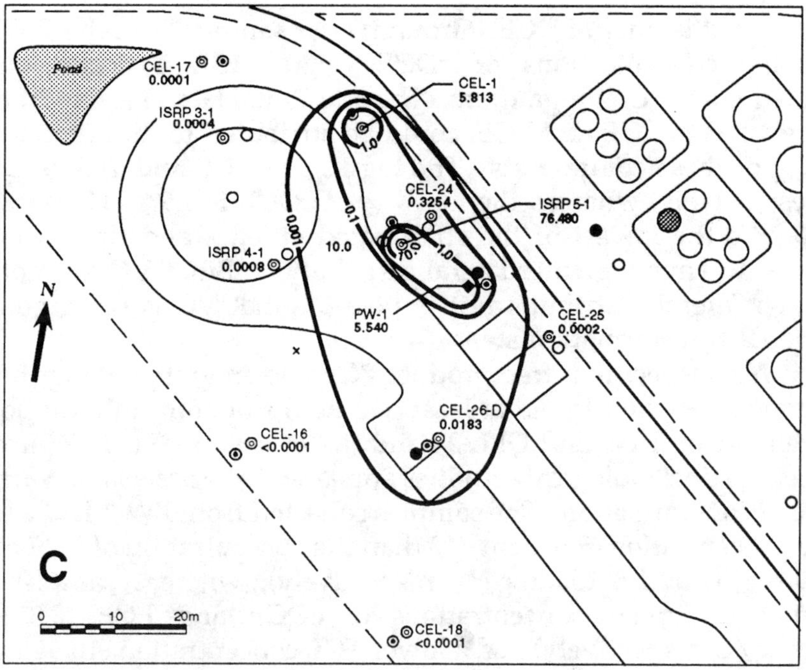

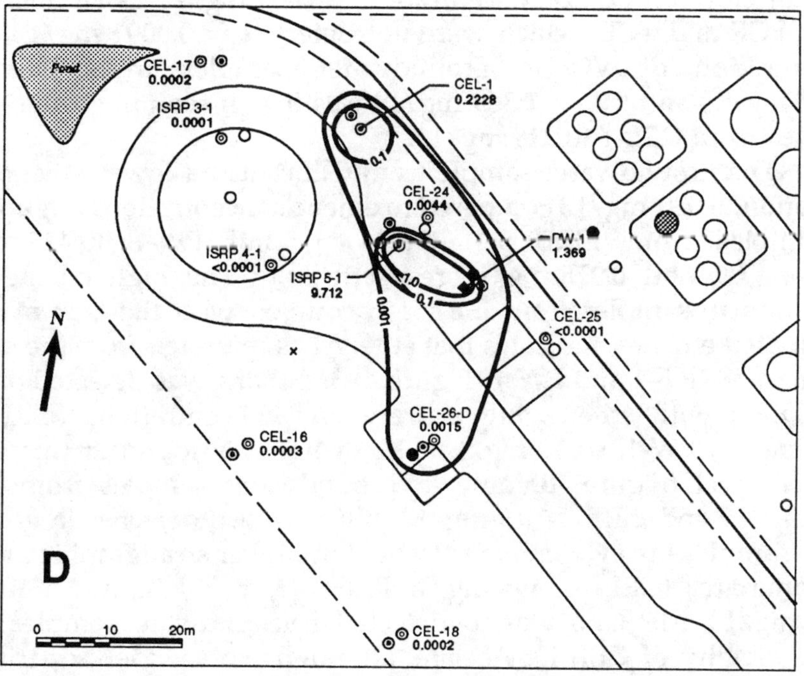

FIGURE 3. (Continued)

predominant isomer of DCE. Groundwater samples from ISRP 4-1 had detectable concentrations of cDCE (Figure 3C), indicating it has migrated farther down-gradient. The additional lateral extent of cDCE, as compared to PCE and TCE, could be attributed to differences in the log K_{ow} of these compounds. The log K_{ow} of PCE and TCE is 2.6 and 2.5, respectively, whereas the log K_{ow} of cDCE is 1.9. The log K_{ow} of VC (0.9) is the lowest of the chlorinated ethenes and thus would be expected to have a greater lateral distribution. Since VC did not have a greater lateral distribution, it is possible that VC is undergoing an additional transformation step.

We suspect that free-product PCE had migrated along the spur line gravel pack and found at least two points of entry into the aquifer: one centred on PW- and CEL-24 and the other on CEL-1 (Figure 3A). Further, distinct biological activities appeared to be associated with each centre of contamination. The samples collected from PW-1 had a higher cDCE concentration (5.540 mg/L) than the concentration of PCE or TCE (0.440 mg/L and 1.165 mg/L, respectively), whereas samples from CEL-1 had similar concentrations of cDCE and PCE (5.813 and 4.376 mg/L, respectively) and a lower TCE concentration (1.654 mg/L). The samples from ISRP 5-1 exhibited a trend similar to that of PW-1, with an elevated cDCE concentration (76.480 mg/L) as compared to either PCE and TCE, which were not detected (< 0.0001 mg/L). The concentration of VC in groundwater samples from ISRP 5-1 (9.712 mg/L) and PW-1 (1.369 mg/L) was higher than in groundwater samples from CEL-1 (0.228 mg/L).

The groundwater sample from CEL-1 had a lower ethene concentration (0.020 mg/L) compared to ethene concentrations (Figure 4A) in samples from ISRP 5-1, ISRP 4-1, and PW-1 (0.423 mg/L, 0.152 mg/L, and 0.278 mg/L, respectively). The high ethene concentration in samples from ISRP 4-1, coupled with the lack of other chlorinated ethenes, indicates that ethene has migrated from the source area near ISRP 5-1 and PW-1 (Figure 4A). Ethane was detected only in association with ethene, but at very low concentrations (ISRP 5-1, 0.010 mg/L; PW-1, 0.010 mg/L). Methanol was found at this stratigraphic level (Figure 4B) only in groundwater samples from PW-1 (810 mg/L) and ISRP 5-1 (5 mg/L), but was also detected in groundwater samples taken from wells located in higher stratigraphic units in the same area (CEL-10, 498 mg/L; ISRP 5-2, 40 mg/L; and ISRP 5-3, 1,770 mg/L). Methanol was not detected in groundwater samples taken in the vicinity of CEL-1. Acetate, although more widely distributed than methanol, still appears to originate from around PW-1 and ISRP 5-1 (Figure 4C). It was detected in groundwater samples taken from

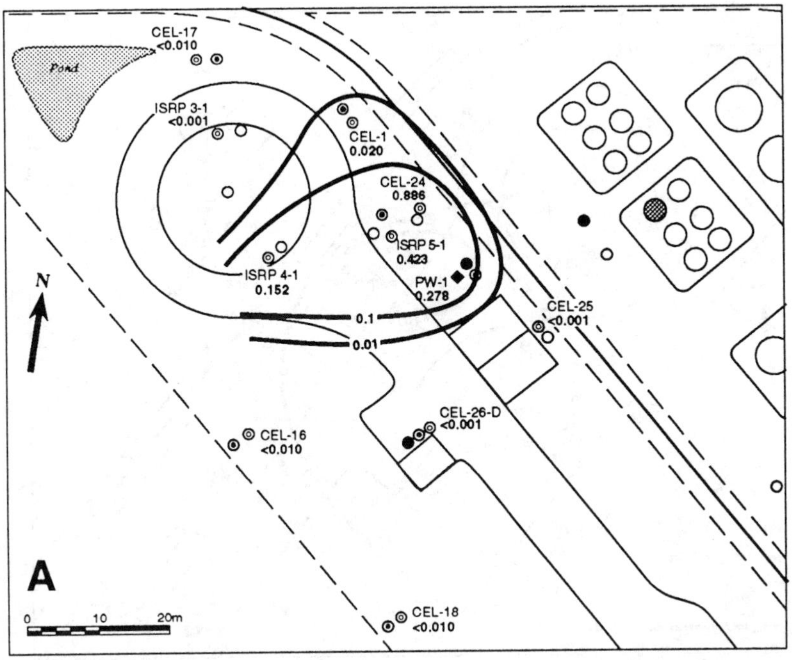

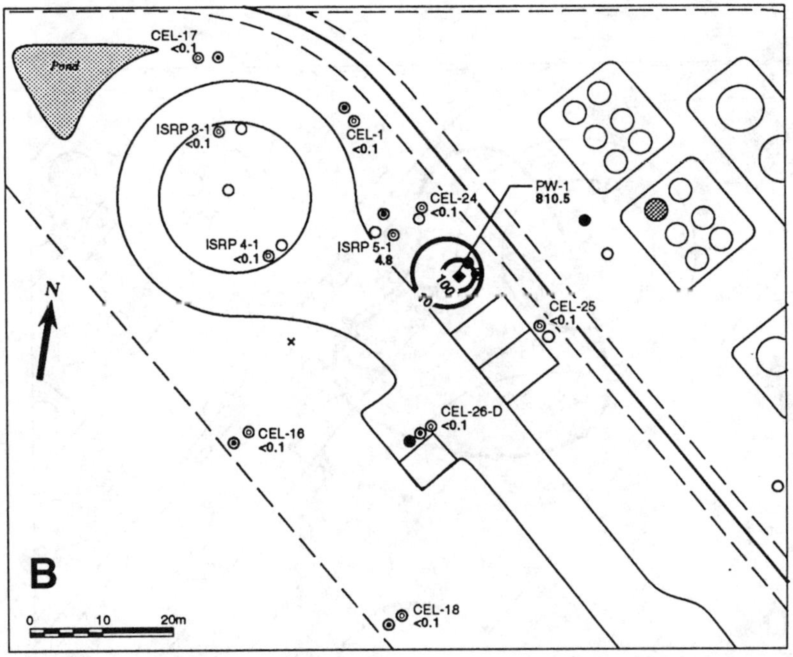

FIGURE 4. **Contours of (A) ethene (mg/L), (B) methanol (mg/L), (C) acetic acid (mg/L), and (D) chloride (mg/L) in the bottom of the sand/silt at a depth of approximately 7 mbgs.**

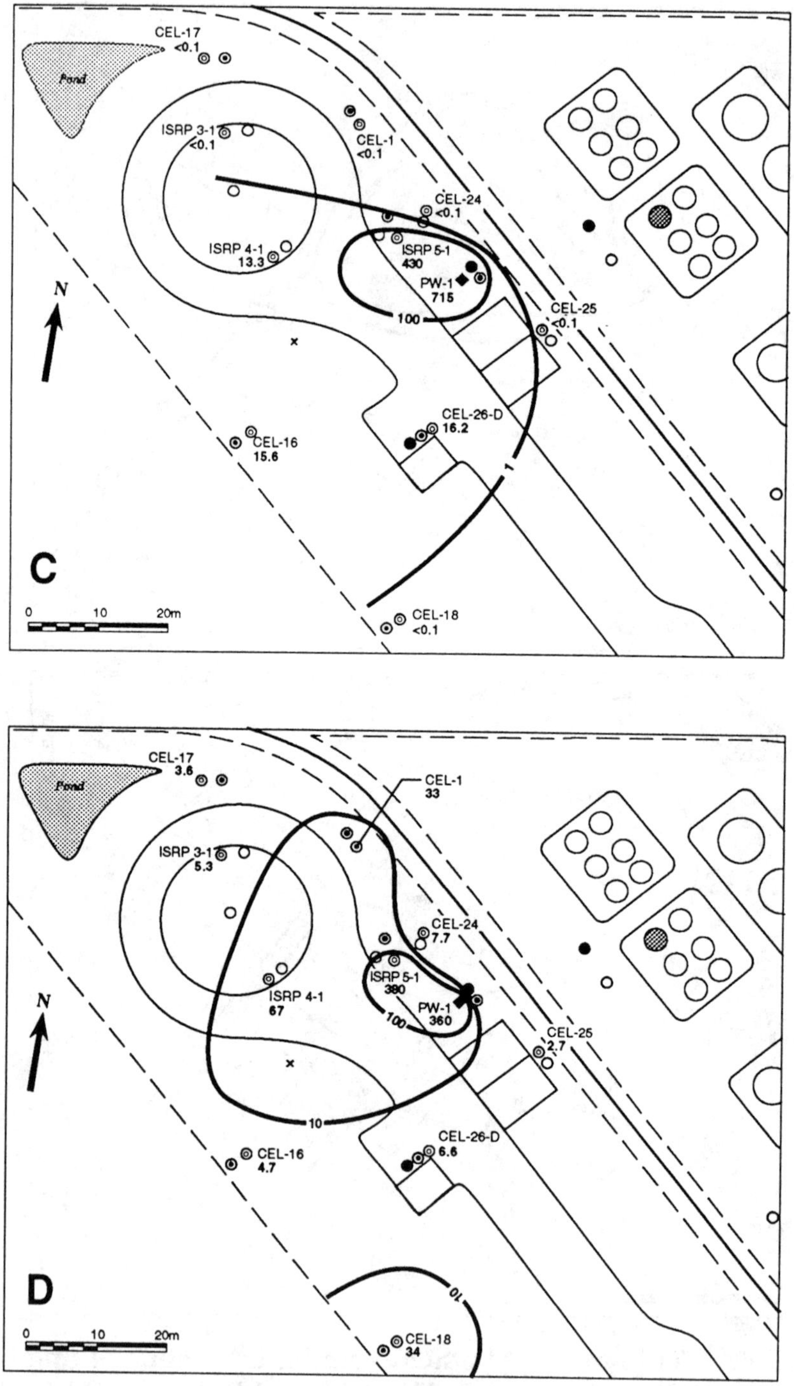

FIGURE 4. (Continued)

ISRP 5-1 (430 mg/L) and PW-1 (715 mg/L), but not CEL-1. Methane (data not shown) was detected throughout the site and did not appear to be correlated with previously presented parameters. Sulphate concentrations generally decrease in the central area of chlorinated organic contamination at the 3 mbgs level plan map (data not shown). At 7 mbgs, sulphate concentrations are low in samples from PW-1 (< 10 mg/L), but are significantly greater in samples from ISRP 5-1 (230 mg/L). Chloride concentrations were elevated in groundwater samples taken from monitoring wells located in the contaminated zones. Elevated chloride levels in groundwater samples could result from dechlorination of PCE and its daughter products. The chloride plume overlaps the chlorinated organic contours and is distributed farther down-gradient (Figure 4D). The highest chloride concentrations were observed in groundwater samples taken from PW-1 and ISRP 5-1 (360 and 380 mg/L, respectively). Chloride concentrations outside the contaminated area were typically in the 10 mg/L range.

Laboratory

Enumeration of Microbial Population. Viable sulphate-reducing and methanogenic populations were detected in the collected subsurface sediment based upon both culturing and phospholipid analysis. Methanogens were estimated to be less than 40 MPN per gdwt with the various substrates (acetate, 4 MPN; methanol, 21 MPN; H_2-CO_2, 39 MPN). Sulphate reducers are present at approximately 1,000 to 3,000 MPN per g (dwt) soil (acetate, 1,400 MPN; lactate 3,600 MPN).

PLFA analysis (Table 1) of the core material indicated that the microbial community is sparse (1 to 3 × 10^5 bacteria/g [dwt] based on 10 pmole of PLFA = 10^5 bacteria), anaerobic as indicated by the dominance of anaerobic desaturase phospholipids (moenoic ω7,16 and 18's), predominantly Gram negative (ratio of PLFA to LPS lipid A hydroxy fatty acids), and stressed (cy17:0). Evidence of sulphate-reducing bacteria (SRB) was the presence of short-chain saturated PLFA. The detection of i17:1 and i17:0 PLFA was indicative of *Desulfovibrio*, and 10Me16:0 (and no detection of 10Me18:0) PLFA indicates the presence of *Desulfobacter*. The biomass of SRB as estimated by pmole of SRB-PLFA (assuming 10 pmole is approximately 10^5 bacterial cells) was equivalent to SRB estimates by MPN. Isoprenoid ethers, which would indicate methanogens, were not detected at this low density of cells. These data suggest that neither sulphate reducers nor methanogens

TABLE 1. PFLA core sediment analysis.[a]

PFLA	Sample Location in Core: Top	Bottom
12:0	1.25	0.58
i14:0	0.05	0.54
14:0	5.81	3.75
i15:0	5.87	5.33
a15:0	2.76	2.46
15:0	3.54	3.32
i16:0	0.68	0.40
16:1w7c	16.21	26.16
16:1w5c	1.05	1.72
16:0	30.73	18.26
i17:1	0.00	0.79
10Me16:0	0.88	0.96
i17:0	0.31	0.44
a17:0	0.08	0.38
cy17:0	9.01	15.49
17:0	0.21	0.83
18:3w6	1.31	0.00
18:2w6	0.00	0.00
i18:/3w3	0.00	0.00
18:1w9c	7.47	9.21
18:1w7c	5.98	2.41
18:0	6.79	6.23
10Me18:0	0.00	0.00
br19:1	0.00	0.75
cy19:0	0.00	0.00
20:1w9c	0.00	0.00
Total	100.00	100.00
pmole/gdws	29.9	11.7
LPS Lipid-A OH-FA Analysis		
3OH12:0	0.00	0.00
2OH12:0	100.00	0.00
2OH14:0	0.00	0.00
Total	100.00	100.00
pmole/gdws	0.13	0.00

(a) Nomenclature: Total number of carbon atoms: number of double bonds with position of double bond closest to the ω end with geometry *c* for *cis*, *t* for *trans*; *i*, *a*, and *br* refer to iso-, anteiso-, and methyl-branching, respectively. Methyl-branching from the Δ end is indicated by its position followed by Me and the total number of carbon atoms; cyclopropyl fatty acids are designated as *cy*.

comprise a significant proportion of the microbial biomass in the sediment.

Microcosm Studies. Purging site groundwater to remove residual halocarbons was not entirely successful. Table 2 lists the halocarbon content of the five treatment flasks of site groundwater used to prepare the various microcosm treatments. All but the unamended water received 5 mg/L added PCE, but no other halogenated compound. However, residual halocarbon, principally cDCE and excess PCE, remained. Unamended water, for example, was purged for approximately 24 h, but contained 24 mg/L cDCE. Low levels of VC were detected in the purged groundwater containing cDCE. Since VC is a highly volatile compound, we can only explain its presence as resulting from abiotic reactions of residual halocarbons with iron complexes (Vogel *et al.* 1987) or from microbial activity that may have occurred in the unsterilized groundwater after sampling but prior to analysis. The difference between the halocarbon concentrations in the treatment flasks (Table 2) and day 0 of the microcosms (Table 3) is attributed to volatilization from the flask during dispensing of the amended groundwater. However, the concentration of PCE delivered to the microcosms appears to be constant over the microcosm construction period. This is evident in the sterile PCE-H_2O and PCE-soil control microcosms (Figure 5). The construction time difference between the day 0 and day 145 microcosms was approximately 1 h; however, the concentration of PCE in these microcosms is relatively stable over the 145-day incubation period. The aquifer sediments did not significantly contribute to the day 0 halocarbon concentrations in the microcosms because sediment samples collected from the central area did not have detectable concentrations of sorbed halocarbons (data not shown).

Microbiologically active microcosms were successfully constructed from the collected subsurface material, as methane evolution and changes in the halocarbon species and concentrations were evident in active microcosm treatments, but not in the PCE-H_2O or the autoclaved/poisoned sterile controls (Figure 5). The decrease in PCE concentration in the autoclaved/poisoned controls may reflect irreversible sorption of PCE to the sediment.

Figure 6 demonstrates biologically mediated transformation of PCE in microcosms amended by PCE plus acetate/methanol. The same general transformation pattern was observed in all active (i.e., nonsterile) microcosms (Table 3). Over time, PCE was dechlorinated to TCE, then cDCE, VC, and finally, in most instances, to ethene. The

TABLE 2. Halocarbon concentrations in purged site water after addition of PCE and substrate amendments prior to distribution to microcosms (mg/L).

Amendment	PCE	TCE	cDCE	tDCE	1,1-DCE	VC
unamended	0.36	0	24.14	0	0	0.26
PCE	2.53	0.08	3.47	0	0	0.08
PCE + methanol	7.95	0	9.46	0.002	0.001	0.42
PCE + acetate	13.92	0	0	0	0	0
PCE + methanol/acetate	11.87	0	0.67	0	0	0.003

PCE added: 5 mg/L (3.1 µL per L), except to unamended.

transformation pattern appears strictly sequential. Small amounts of tDCE and 1,1-DCE were detected in some active microcosms up to day 46, but these were minor in comparison to the cDCE levels encountered. The prevalence of cDCE is in agreement with others (Parsons & Lage 1985; Scholz-Muramatsu *et al.* 1990). In microcosms that contained initial concentrations of cDCE, the concentration remained unchanged while PCE decreased and TCE increased in concentration. Only when TCE was being depleted did the cDCE concentration rise (Table 3). This pattern held for each successive dechlorination daughter product. The rate of product formation and transformation also slowed as the less chlorinated products appeared.

Microcosms amended with PCE plus methanol and acetate showed more evolved ethene (0.5 mg/L at day 145) than any other amendment (Table 3). By day 145, VC was the sole halogenated compound remaining and was decreasing in concentration. Dechlorination was slowest in the PCE plus methanol treatment with vinyl chloride concentrations still increasing by day 145, whereas in all other treatments (except unamended), VC concentrations were decreasing (Table 3). The unamended treatment was probably exceptional in this regard because of the very large cDCE concentration in those microcosms. It is also notable that methane evolution was delayed in the PCE plus methanol treatment (Table 3) and no ethene was ever detected. This is in contrast to other treatments, in which ethene was present and increasing in concentration from day 83 to day 145.

TABLE 3. Biotransformation products (mg/L) of PCE by treatment, standard deviation in parentheses.

Treatment	Day	PCE	TCE	cDCE	VC	Ethene	Methane
PCE	0	0.372 (0.104)	0.145 (0.077)	2.350 (0.272)	0.034 (0.003)	0 (0)	0.027 (0.005)
	12	0.041 (0.059)	0.339 (0.052)	2.122 (0.321)	0.185 (0.027)	0 (0)	0.190 (0.027)
	25	0.004 (0.002)	0.029 (0.004)	2.436 (0.479)	0.279 (0.039)	0 (0)	0.491 (0.017)
	46	0.006 (0.011)	0.002 (0.004)	1.469 (0.144)	1.167 (0.071)	0 (0)	0.800 (0.123)
	83	0 (0)	0 (0)	0 (0)	1.689 (0.057)	0.092 (0.005)	4.258 (0.715)
	145	n.d	0 (0)	0 (0)	7.290 (0.321)	0.061 (0.017)	2.624 (0.543)
PCE + Methanol	0	1.965 (0.800)	0.004 (0.006)	5.116 (0.337)	0.021 (0.007)	0 (0)	0.005 (0.005)
	12	1.019 (1.015)	0.265 (0.277)	6.015 (0.925)	0.060 (0.001)	0 (0)	0.010 (0.003)
	25	0.050 (0.006)	0.012 (0.003)	5.845 (0.954)	0.060 (0.011)	0 (0)	0.094 (0.045)
	46	0.040 (0 015)	0.010 (0.010)	4.772 (0.229)	0.177 (0.22)	0 (0)	0.185 (0.123)
	83	0 (0)	0 (0)	4.380 (0.832)	1.074 (0.033)	0 (0)	1.751 (0.519)
	145	n.d	0 (0)	1.250 (0.581)	2.434 (0.147)	0 (0)	11.003(1.708)
PCE + Acetate	0	0.672 (0.076)	0.003 (0.003)	0.676 (0.283)	0.002 (0.003)	0 (0)	0.008 (0.001)
	12	0.076 (0.025)	0.237 (0.037)	0.650 (0.069)	0.001 (0.002)	0 (0)	0.063 (0.005)
	25	0.024 (0.004)	0.023 (0.006)	1.095 (0.156)	0.031 (0.005)	0 (0)	0.756 (0.050)
	46	0.020 (0.015)	0.007 (0.002)	0.863 (0.058)	0.201 (0.036)	0 (0)	3.145 (0.190)
	83	0 (0)	0 (0)	0 (0)	1.055 (0.038)	0.035 (0.006)	23.818(7.919)
	145	n.d	0 (0)	0 (0)	0.813 (0.051)	0.109 (0.025)	40.916(8.633)
PCE + Methanol and Acetate	0	1.641 (1.054)	0.085 (0.057)	0.627 (0.014)	0.007 (0.005)	0 (0)	0.008 (0.001)
	12	0.012 (0.002)	0.186 (0.026)	1.752 (0.162)	0.098 (0.005)	0 (0)	0.076 (0.014)
	25	0.004 (0.001)	0.013 (0.002)	1.685 (0.040)	0.229 (0.003)	0 (0)	0.655 (0.067)
	46	0 (0)	0.002 (0.004)	0.559 (0.073)	1.093 (0.026)	0 (0)	1.617 (0.101)
	83	0 (0)	0 (0)	0 (0)	1.188 (0.038)	0.238 (0.133)	4.074 (1.983)
	145	n.d	0 (0)	0 (0)	0.175 (0.153)	0.512 (0.082)	6.033 (0.187)

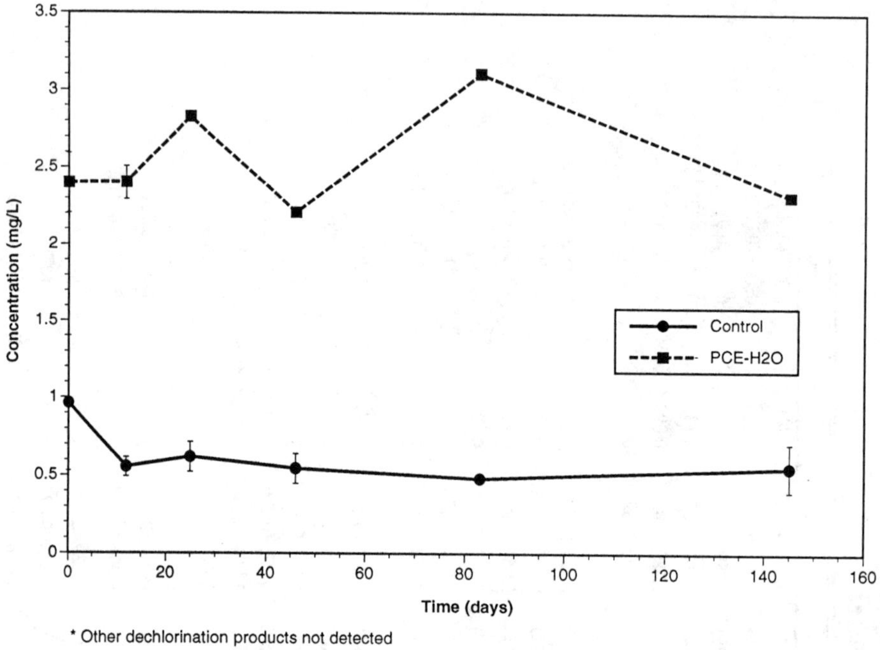

FIGURE 5. PCE controls (other dechlorination products not detected).

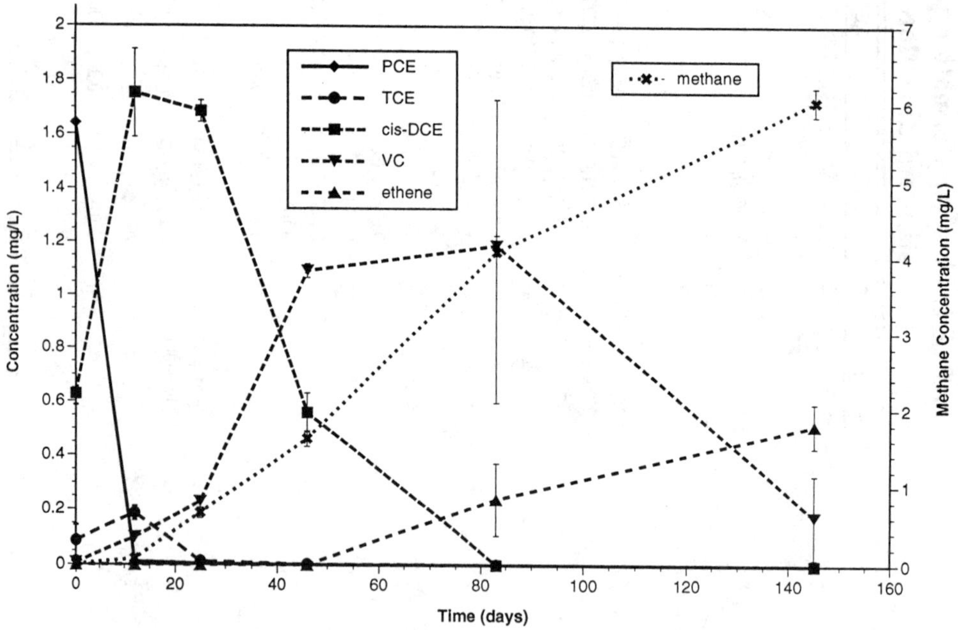

FIGURE 6. PCE + methanol + acetate amendment.

Indeed, in unamended microcosms ethene was detected (at the minimum detection levels) as early as day 12. Further reduction of ethene to ethane was not observed in any of the microcosms (data not shown).

Dechlorination and methane production in both unamended and amended microcosms with only PCE (Table 3) indicate that an electron donor was available in the microcosms without additional methanol or acetate. Chemical analysis of groundwater samples from ISRP 5-1 (obtained near the area from which the core material used in this study was taken) confirmed that both methanol and acetate were present in the area at concentrations from 5 to more than 810 mg/L.

Although methane was evolved in all active microcosms, no correlation between methane production and dechlorination was perceived. The methane concentration reached 41 mg/L in PCE plus acetate amended microcosms at day 145, but dechlorination was more complete in PCE plus acetate and methanol amended microcosms where methane concentration reached 6 mg/L. Dechlorination (and methane evolution) were relatively delayed in PCE plus methanol amended microcosms, but methane concentrations in that treatment were 11 mg/L at day 145.

DISCUSSION

Chlorinated aliphatics have been demonstrated to undergo anaerobic dechlorination in continuous-flow fixed-film reactors (Bouwer & McCarty 1983; Bouwer & Wright 1988; Vogel & McCarty 1985, 1987) in soil (Kloepfer *et al.* 1985), sediment (Barrio-Lage *et al.* 1986, 1987; Parsons *et al.* 1985; Parsons & Lage 1985), and aquifer microcosms (Parsons *et al.* 1984; Wilson *et al.* 1983, 1986). Although few studies have been conducted on the biotransformation of PCE in the field, many sites contaminated with PCE also have concentrations of dechlorination products in sampled groundwater and aquifer solids. Unfortunately, VC tends to persist in anaerobic environments and is more toxic than the parent compounds.

For anaerobic bioremediation to be a useful remedial method, PCE and TCE must be completely dechlorinated to nonchlorinated products. Few studies have shown the complete dechlorination of chlorinated alkene. Vogel and McCarty (1985, 1987) have shown the conversion of PCE and VC to CO_2. Freedman and Gossett (1989) studied the conversion of [^{14}C]PCE to [^{14}C]ethene. Their experiments showed insignificant amounts of [^{14}C]CO_2 and [^{14}C]CH_4. They also observed that the rate limiting step was the conversion of VC to ethene and that the presence of an electron donor was necessary. Methanol was the most effective electron donor. They found that 2-bromoethane-sulfonate inhibited the conversion of PCE to ethane, which suggests that methanogens are playing a role in the conversion.

If the dechlorination products of PCE were persisting at our study site, then greater down-gradient transport of these products would be expected to occur. This is because of two reasons. First, the log K_{ow} value decreases with the removal of a chlorine atom. This results in the dechlorinated products having higher mobility than their more chlorinated predecessor. Second, dechlorination products will form at the leading edge of their predecessor's plume. The good degree of distribution overlap of PCE, TCE, cDCE, and VC at the study site strongly indicates the dechlorination of these compounds.

The predominance of the cDCE (as opposed to the *trans-* or 1,1-DCE isomers) is in agreement with our and other laboratory biotransformation studies of PCE (Barrio-Lage *et al.* 1986; Parsons *et al.* 1984, 1985; Parsons & Lage 1985). Methanol and acetate appears to be promoting PCE biotransformation to ethene. The evidence supporting this conclusion is the strong relationship between PCE, methanol, ethene, and acetate contours originating from the PW-1 and ISRP 5-1 monitoring well locations. In contrast, PCE originating from the CEL-1 does not overlap with the methanol, acetate, and ethene plumes.

Contamination of the microcosms with residual halogenated compounds created three complications in the study: (1) the initial PCE levels were not the same in all PCE-containing microcosms; (2) the total concentration of halogenated compounds in the different microcosm treatments was highly varied, although within a set of microcosms (*e.g.*, the "PCE plus methanol" treatment) the initial halocarbon concentration and speciation were constant; and (3) the unamended microcosms, which ideally would be devoid of halogenated compounds, contained very large amounts of cDCE. These complications limit the amount of information gained from the study, but a number of useful results and conclusions were still acquired.

The presence of cDCE in unamended microcosms prevents the unequivocal conclusion that ethene production was the consequence of vinyl chloride dechlorination because we had no active microcosms devoid of ethene. Ethene was certainly a metabolic product because it was not formed in sterile controls, but ethene is synthesized naturally by some soil microorganisms (Alexander 1977). However, it is likely that ethene was the final product of the PCE transformation sequence in the sediment, as Freedman and Gossett (1989) found in their enrichment cultures. If ethene is assumed to be the daughter product of VC, then the early appearance of this nonchlorinated daughter product in unamended microcosms was probably related to the absence of PCE and TCE. In those microcosms, the reaction sequence started at cDCE, and although the bulk of the reductive activity was directed toward VC production, some dechlorination of VC to ethene also occurred. Despite its early appearance, ethene did not accumulate in the unamended microcosms (0.06 mg/L by day 145) nearly to the extent that it did in other ethene-positive treatments (PCE only, 0.34 mg/L; PCE plus acetate, 0.11 mg/L; PCE plus acetate and methanol, 0.51 mg/L; all after day 145), again suggesting that throughout the entire period of monitoring, most dechlorinating activity in the unamended microcosms was directed toward the large cDCE mass. This evidence, and the observation of strict sequential dechlorination of the halogenated compounds in the microcosms, indicate that reduction order, although biologically driven, was dictated by the reduction potential of the halocarbon parent-daughter product pair in question.

The microcosm experiments corroborate field evidence that PCE dechlorination is biologically driven, and ethene is probably the final, nonchlorinated product of the sequence. Although methane was produced in all active microcosms, there was no correlation between methane evolution and dechlorination. The treatments could be ranked as follows, in terms of methane production: PCE plus acetate > PCE plus methanol > PCE only > PCE plus methanol and acetate > unamended. However, if ranked in terms of dechlorination over 145 days (indicated by final ethene concentration: initial halocarbon concentration), the ranking was quite different: PCE plus methanol and acetate > PCE only > PCE plus acetate > unamended > PCE plus methanol.

The mechanism or the microorganisms responsible for dechlorination of PCE to ethene is not clear. Few studies have examined which anaerobic microorganisms are responsible for reductive dechlorination. Methanogenic enrichment cultures (utilizing acetate) have been shown to dechlorinate aliphatics (Bouwer & McCarty 1983). Galli and McCarty (1989) have shown that a *Clostridium* sp. was able to biotransform 1,1,1-trichloroethane forming volatile acids (*e.g.*, acetate). Egli *et al.* (1987) have demonstrated that chlorinated ethanes can be dechlorinated by *Desulfobacter autotrophicum*. Bagley and Gossett (1990) found that PCE was dechlorinated to TCE and cDCE in sulphate-reducing enrichment cultures, but that this activity decreased over time. In contrast Fathepure *et al.* (1987) found that *Desulfovibrio desulfuricans* and two species of *Clostridium* did not produce significant quantities of TCE from PCE. The authors also tested the ability of four acetate-utilizing methanogens to dechlorinate PCE. *Methanosarcina acetivorans* and *Methanothrix* did not produce significant quantities of TCE from PCE, but the dechlorination rate of *Methanosarcina* sp. and *M. mazie* cultures paralleled methane production, and maximum yields of TCE were obtained with methanol as opposed to acetate.

In our study, the lack of a relationship between methane production and dechlorination, and the marked differences in microcosm behaviour depending upon treatment (for example, the methanol- versus acetate-amendments) suggest that: (a) activity of methanogenic bacteria was not necessarily mediating the observed dechlorination reactions; and (b) the provision of the different electron donors enriched different segments of the microbial population native to the subsurface sediment.

These points and the low population densities of methanogens and SRB in the subsurface sediments suggest that other anaerobes may be involved in the dechlorination reactions observed *in situ*. The detection of acetate in groundwater samples containing concentrations of methanol hints that acetogens may play an important role in the dechlorination of PCE to ethene. We hypothesize that methanol is serving as a substrate for acetogenic microorganisms that may be responsible for dechlorination of PCE at the study site.

Acknowledgments

We thank Tom Phelps for assisting us in collecting authentic core material, David White for lipid analysis, and Anne Flemming for preparing our figures and editing. We also thank Jim Gossett, Jim Barker, Colin Mayfield, Stan Feenstra, and Ron Nicholson for their excellent input on various aspects of this study. We gratefully acknowledge support for this work through grants from General Electric Company, AT&T, Celanese Canada, ICI Canada, and the Ontario Ministry of the Environment.

REFERENCES

Alexander, M.; *Introduction to Soil Microbiology*; 2nd Edition; John Wiley and Sons: New York, 1977.

Bagley, D. M.; Gossett, J. M. "Tetrachloroethene transformation to trichloroethene and *cis*-dichloroethene by sulfate-reducing enrichment cultures." *Appl. Environ. Microbiol.* **1990**, *56*, 2511–2516.

Balkwill, D. L.; Leach, F. R.; Wilson, J. T.; McNabb, J. F.; White, D. C. "Equivalence of microbial biomass measures based on membrane lipid and cell wall components, adenosine triphosphate, and direct counts in subsurface aquifer sediments." *Microb. Ecol.* **1988,** *16,* 73–84.

Barrio-Lage, G.; Parsons, F. Z.; Nassar, R. S.; Lorenzo, P. A. "Sequential dehalogenation of chlorinated ethenes." *Environ. Sci. Technol.* **1986,** *20,* 96–99.

Barrio-Lage, G.; Parsons, F. Z.; Nassar, R. S.; Lorenzo, P. A. "Biotransformation of trichloroethene in a variety of subsurface materials." *Environ. Toxicol. Chem.* **1987,** *6,* 571–578.

Bobbie, R. J.; White, D. C. "Characterization of benthic microbial community structure by high resolution gas chromatography of fatty acid methyl esters." *Appl. Environ. Microbiol.* **1980,** *39,* 1212–1222.

Bouwer, E. J.; McCarty, P. L. "Transformations of 1- and 2-carbon halogenated aliphatic organic compounds under methanogenic conditions." *Appl. Environ. Microbiol.* **1983** , 45, 1286–1294.

Bouwer, E. J.; Wright, J. P. "Transformations of trace halogenated aliphatics in anoxic biofilm columns." *J. Contam. Hydrol.* **1988,** 2, 155–169.

Egli, C.; Scholtz, R.; Cook, A. M.; Leisinger, T. "Anaerobic dechlorination of tetrachloromethane and 1,2-dichloroethane to degradable products by pure cultures of *Desulfobacterium* sp. and *Methanobacterium* sp." *FEMS Microbiol. Lett.* **1987,** *43,* 257–261.

Fathepure, B. Z.; Nengu, J. P.; Boyd, S. A. "Anaerobic bacteria that dechlorinate perchloroethene." *Appl. Environ. Microbiol.* **1987,** *53,* 2671–2674.

Freedman, D. L.; Gossett, J. M. "Biological reductive dechlorination of tetrachloroethylene and trichloroethylene to ethylene under methanogenic conditions." *Appl. Environ. Microbiol.* **1989,** *55,* 2144–2151.

Guckert, J. B.; Hood, M. A.; White, D. C. "Phospholipid ester-linked fatty acid profile changes during nutrient deprivation of *Vibrio cholerae*: increase in the *trans/cis* ratio and proportions of cyclopropyl fatty acid." *Appl. Environ. Microbiol.* **1986,** *52,* 794–801.

Galli, R.; McCarty, P. L. "Biotransformation of 1,1,1-trichloroethane, trichloromethane, and tetrachloromethane by *Clostridium* sp." *Appl. Environ. Microbiol.* **1989,** *55,* 837–844.

Kloepfer, R. D.; Easley, D. M.; Haas, B. B.; Deihl, T. G.; Jackson, D. E.; Wurrey, C. J. "Anaerobic degradation of trichloroethylene in soil." *Environ. Sci. Technol.* **1985,** *19,* 277–280.

McInerney, M. J.; Bryant, M. P.; Pfennig, N. "Anaerobic bacterium that degrades fatty acids in syntrophic association with methanogens." *Arch. Microbiol.* **1976,** *122,* 129–135.

Parsons, F.; Barrio-Lage, G.; Rice, R. "Biotransformation of chlorinated organic solvents in static microcosms." *Environ. Toxicol. Chem.* **1985,** *4,* 739–742.

Parsons, F.; Lage, G. B. "Chlorinated organics in simulated groundwater environments." *J. Am. Water Works Assoc.* **1985,** *77,* 52–59.

Parsons, F.; Wood, P. R.; DeMarco, J. "Transformations of tetrachloroethene and trichloroethene in microcosms and groundwater." *J. Am. Water Works Assoc.* **1984,** *76,* 56–59.

Postgate, J. R.; *The Sulphate Reducing Bacteria*; 2nd edition; Cambridge University Press: Cambridge, 1984.

A Comparison of Soil-Phase and Slurry-Phase Bioremediation of PNA-Containing Soils

*Bruce S. Yare**
Monsanto Chemical Company

BACKGROUND

The Brio Refining Superfund Site is a 23-ha abandoned petrochemical reprocessing facility and jet fuel refinery located approximately 24 km south of downtown Houston TX. From 1957 to 1978, major reprocessing operations included regeneration of acrilonitrile synthesis catalysts, reprocessing of styrene still bottoms to recover ethylbenzene, and attempts to recover 1,2-dichloroethane from vinyl chloride process residues. Raw feed was stored in a series of unlined impoundments, which were closed between 1976 and 1979 under a Texas Air Control Board order. Pit closure was accomplished by removing pumpable liquids; stabilizing nonpumpable liquids with sand, clay, or calcined clay; filling the impoundment with clay; and covering the clay backfill with clean soil. This closure method produced four distinct layers in the backfilled impoundments: Layer I—clean soil cover, Layer II—clay backfill, Layer III—organic sludge and liquid, and Layer IV—pit subsoil. Table 1 presents analyses of the organic sludge and liquid found in the backfilled impoundments that were used as sources of material for the bioremediation demonstrations discussed in this paper.

Two pilot-scale bioremediation demonstrations were undertaken at the Brio Refining site. The first demonstration, done as part of the

* Monsanto Chemical Company, 800 North Lindbergh Boulevard, St. Louis MO 63167

TABLE 1. Maximum constituent concentrations in organic sludge and liquid from backfilled impoundments used as source material for solid-phase and slurry-phase bioremediation demonstrations, Brio Refining Superfund Site, Houston TX.

	Pit B	Pit E	Pit J	Pit O	Pit Q	Pit R	Pit H/V
Volatile Organics (mg/kg)							
Benzene	ND	3	242	ND	ND	<1	ND
Chlorobenzene	ND	15	1,100	ND	1,150	37	ND
Chloroform	4,690	43	3,880	ND	ND	111	629
1,1-Dichloroethane	5,480	433	3,600	ND	ND	126	725
1,2-Dichloroethane	245,000	10,800	179,000	ND	8,140	5,100	32,000
1,1-Dichloroethylene	ND	ND	326	ND	ND	98	1,570
1,2-Dichloroethylene	4,620	29	3,380	ND	ND	ND	ND
1,2-*trans*-Dichloroethylene	10,100	717	8,130	ND	14	203	1,810
Ethylbenzene	ND	286	2,190	340	557	588	3,370
Methylene Chloride	ND	133	1,000	ND	1,000	106	1,000
1,1,2,2-Tetrachloroethane	ND	ND	ND	ND	ND	5	ND
Tetrachloroethylene	ND	11	1,180	ND	ND	51	ND
Toluene	ND	57	ND	23	ND	186	59
1,1,2-Trichloroethane	166,000	12,500	89,100	5	65,700	2,270	13,000
Trichloroethylene	3,650	15	1,790	ND	128	83	40
Vinyl Chloride	ND	1,100	22,700	ND	ND	75	5,820
TOTAL	439,540	26,142	317,618	368	76,689	9,040	60,028
Semivolatiles (mg/kg)							
Anthracene	ND	11	ND	<1	921	651	43
Benzo(a)anthracene	ND	ND	ND	ND	18	ND	ND
bis(2-Chloroethyl)ether	3,040	15	1,410	ND	1,810	ND	ND
1,2-Dichlorobenzene	ND	ND	ND	ND	115	ND	ND
1,3-Dichlorobenzene	546	ND	238	ND	430	ND	ND
1,4-Dichlorobenzene	ND	ND	ND	ND	194	ND	ND
Hexachlorobenzene	ND	ND	ND	ND	150	ND	ND
Hexachlorobutadiene	ND	ND	ND	ND	13	ND	ND
Di-*n*-octyl Phthalate	ND	ND	ND	12	ND	ND	ND
Fluoranthene	199	8	ND	ND	94	86	2
Fluorene	184	14	ND	ND	ND	50	ND
Napthalene	891	168	653	7	400	401	101
Phenanthrene	3,440	838	2,910	29	599	758	ND
Pyrene	ND	3	ND	ND	5	ND	762
TOTAL	8,300	1,057	5,211	49	4,749	1,946	908

Remedial Investigation/Feasibility Study (RI/FS) in 1987, successfully used solid-phase bioremediation to establish the feasibility of biologically treating the organic compounds found at the site. This demonstration allowed incorporation of bioremediation into the Feasibility Study as an alternative to incineration. After toxicity screening of affected material and soil indicated that two of the backfilled impoundments were amenable to biological treatment without dilution, the pit with the lowest concentration of volatile organics (Pit O) was selected for bench-scale testing. Since the site is located immediately adjacent to a 600-home subdivision, minimizing the potential for release of volatile organics was an important consideration in selecting feed material for this demonstration. Bench-scale testing indicated that the organics present in the pit could be removed by air stripping or biological degradation. Based on this testing, a pilot-scale, solid-phase bioremediation demonstration was completed using feed material from Pit O. After 21 days of operation, 99 percent of the volatile organics were removed by air stripping. Of the semivolatile organics present, 89 percent were biologically degraded after 94 days of operation, demonstrating the feasibility of biologically treating the nonvolatile portion of the organics found at the site.

A second demonstration, completed after issuance of the Record of Decision (ROD) in 1988, established that slurry-phase bioremediation was capable of providing the same level of environmental protection as incineration for almost all of the affected material and soil found at the site. This demonstration was required by the ROD before bioremediation could be used instead of incineration to treat affected material and soil from the site. Treatment levels required by the Consent Decree, based on Land Ban Best Demonstrated Available Technology (BDAT) standards for organic process residues similar to those found at the site, are as follows:

Individual Chlorinated Solvents	5,960 μg/kg
bis(2-Chloroethyl)ether	5,440 μg/kg
Individual Carcinogenic Polynuclear Aromatic	3,400 μg/kg
Total Noncarcinogenic Polynuclear Aromatics	63,840 μg/kg

Carcinogenic polynuclear aromatics (PNAs) found at the site included benzo(a)anthracene, benzo(b)fluoranthene, benzo(k)fluoranthene, benzo(a)pyrene, dibenzo(a,h)anthracene, and indeno(1,2,3,c,d)pyrene. Noncarcinogenic PNAs consisted of acenaphthene, anthracene, chrysene, fluoranthene, fluorene, naphthalene, phenanthrene, and pyrene. Twelve batches of slurry, two batches from each of six pits

(Pits B, E, J, Q, R, and H/V) with residual organic material and soil concentrations above acceptable risk levels, were treated by air stripping and biological degradation to determine if the required treatment levels could be achieved. All but one batch achieved these levels of treatment, demonstrating that bioremediation provided the same level of environmental protection as incineration for almost all of the affected material and soil encountered at the Brio Refining site.

A Consent Decree requiring on-site incineration or biological treatment of a minimum of 62,000 m^3 of residual organic material and soil from 12 of the backfilled impoundments was lodged in 1990 but has not yet been entered.

This paper compares the results obtained with two different biological treatment methods using essentially the same feed material.

SOLID-PHASE BIOREMEDIATION DEMONSTRATION

Treatment Facility Construction. A solid-phase biological treatment facility was constructed at the Brio site using a lined treatment bed enclosed by a plastic film greenhouse. After ground surface preparation, two sets of greenhouse frames were erected side by side, creating an enclosure 22 m wide and 43 m long. To provide bottom containment, an 80-mil HDPE flexible membrane liner with heat-welded seams was fastened to the greenhouse frame uprights, and a sand layer was placed on top of the liner to provide drainage. Two lateral drains, running along the outside edge of the long side of the treatment bed, were routed through the liner to a 530-L sump. Leachate collected in the sump was transferred to the fermenter used to prepare nutrient solutions and inoculum for treatment. An overhead spray system was used to deliver inoculum and nutrients evenly over the treatment bed. A slight negative pressure was maintained within the greenhouse using three 85 cm^3/min fans. Volatile emissions and odors were contained by the plastic film greenhouse and routed to three carbon adsorbers for treatment.

Pit Sampling and Treatment Bed Preparation. Approximately 150 m^3 of soil were excavated from Pit O using a backhoe and transferred to the treatment bed with a crawler-mounted front-end loader. Excavated material was placed directly on top of the sand underdrain and allowed to dry for several days. During this drying period, the front-end loader worked the wet clay soil to break up large blocks, promote drainage of excess water, and distribute material uniformly over the entire treatment bed. Three days of physical manipulation were necessary before the soil was dry enough to allow use of a small, tractor-mounted rototiller for aeration, and a total of 6 days elapsed before the soil was dry enough to allow addition of nutrients.

Operation and Sampling. The treatment bed was tilled daily to increase exposed soil surface area, provide oxygen to the microorganisms, and promote air stripping of the volatile organics. The rototiller was pressure washed after each use at a nearby decon pad. Wash water was routed to the site wastewater handling facility. Daily measurements of soil pH, moisture, and temperature and addition of water, nutrients, and inocula as required allowed maintenance of the treatment bed at conditions conducive to microbial activity. Oxygen demand was not measured. On January 30, 100 to 110 kg of commercial ammonium phosphate was added to three of the four treatment lanes to achieve a carbon:nitrogen:phosphorous ratio of 50:2:1. Nutrients were added to the treatment bed by dissolving the ammonium phosphate in well water and applying the solution to the three treatment lanes through the overhead spray system. Every sampling period, each soil sample was analyzed for soluble ammonium, phosphate, and nitrate.

All personnel working in the treatment facility wore respiratory protection. For several days during soil excavation, movement, and preparation, individual organic vapor monitoring badges were worn by workers. The time weighted averaged (TWA) concentrations of volatile organics detected on these badges were two orders of magnitude lower than the appropriate American Conference of Governmental Industrial Hygenists (ACGIH) threshold limit values (TLV). Two independent methods were used to determine the amount of volatile organics stripped from the soil during the treatment period. Analysis of volatile organic compound (VOC) concentrations in the emission control carbon beds indicated 137 kg of VOCs were air stripped during treatment. An estimate based on organic vapor analyzer data collected throughout the treatment period indicated 159 kg of VOCs were air stripped. The amount of VOCs removed during treatment bed preparation was not measured separately; however, the emission control system was in operation during this step.

The treatment bed was divided into four 5.5 by 43 m lanes running the length of the treatment facility. Each lane was subdivided into four 5.5 by 10.7 m quadrants, creating a total of 16 sampling quadrants in the treatment bed. Three 3 × 3 m sampling cells were located within each quadrant. A quadrant sample consisted of a composite of three individual, randomly selected soil subsamples collected from each of the sampling cells. Sixteen quadrant samples were collected every sampling period. Soil samples were collected on January 26 (Day 0), February 16 (Day 21), March 25 (Day 58), and April 30 (Day 94), a total treatment period of 94 days. Samples were analyzed for volatile organics, semivolatile organics, total organic carbon, ammonium, nitrate, phosphate, and plate count.

SLURRY-PHASE BIOREMEDIATION DEMONSTRATION

Pit Sampling and Slurry Preparation. Two, and sometimes three, 208-L drums of sample were collected from each layer found in Pits B, E, J, Q, R, and H/V to provide raw feed for the slurry-phase bioremediation demonstration. Two batches of slurry were prepared for each pit; one batch contained all of the layers found in the pit, and the other batch included all of the layers except for the organic sludge and liquid layer found in the bottom of the pit. The former were called high-carbon batches, the latter low-carbon batches. An agitated, 380-L tank was used to mix the slurry. This tank, equipped with a 2-hp, marine propellor agitator, was filled with 151 L of water. With the agitator running, 227 kg of soil were added slowly to the tank to produce a slurry with a 40 to 50 percent solids (dry weight) content. Residual organic material and soil were added to the slurry mixer in proportion to the thickness of the layer they came from. Prepared slurry was transferred to the bioreactors, where water and inoculum were added to produce a 30 percent solids slurry. Inoculum was prepared on site in 19-L carboys using indigenous microorganisms.

Bioreactor Configuration. Two baffled, flat-bottom, 380-L tanks equipped with center-mounted, 1.74-hp, variable-speed agitators were used as continuously stirred bioreactors. Each agitator had a six-bladed, radial-flow impeller mounted just above the bottom of the tank and a three-bladed, axial-flow impeller mounted in the middle of the tank. Up to 85 L/min of compressed air were introduced into each bioreactor through a perforated 2.5-cm-diameter black iron pipe running across the bottom of the tank. Both bioreactors were covered so that emissions could be collected and routed to a carbon bed.

Operation and Sampling. High-carbon and low-carbon slurries from each pit were run for a total of 10 days, with the first batches starting on June 22, 1988, and the last batches finishing on September 2, 1988. Prior bench-scale treatability testing, conducted on samples collected from Pit J, indicated a 10-day treatment period would be long enough to remove most, if not all, of the volatile and semivolatile organic compounds found in residual organics and soil at the site. The bioreactors, operated under ambient conditions, did not require heating, but did require cooling to keep temperatures less than 41 C. Dissolved oxygen was maintained at a minimum of 2.0 mg/L, pH was kept between 5.5 and 8.0, and water was added to the bioreactor to make up losses. Temperature, pH, dissolved oxygen, conductivity, adenosine triphosphate (ATP), oxygen uptake rate (OUR), total solids, density, and viscosity were measured daily. Oxygen demand was monitored by measuring OUR in each bioreactor at least once a day, and air flow was adjusted as appropriate. After the first few days of the slurry-phase bioremediation demonstration, air flow was kept at 90 percent of full flow for all batches to make system operation and monitoring easier.

Nutrients were added at the start of each batch to obtain a carbon:nitrogen:phosphorous ratio of 100:5:1. Total organic carbon concentrations of 6 and 3 percent were used in calculating nutrient dosages for the high-carbon and low-carbon slurry batches, respectively. These organic carbon concentrations were based on elemental analyses of pit contents samples. Ammonium nitrate and ammonium phosphate were used as nutrients in the first six slurry batches; Miracle Gro™ fertilizer was used for the last six batches. At least one additional dose of nutrients, at half the initial concentration, was added to each batch when ATP measurements indicated a decrease in microbial activity.

Slurry samples were collected from each of the bioreactors on Days 0, 1, 2, 3, 4, 6, 8, and 10 for determination of total extractable organic compounds (TEOC) by gravimetric analysis to provide a measure of all methylene chloride extractable organics present in the samples. A single sample was collected from each bioreactor on Days 0 and 4 for volatile organic compound analyses and duplicate samples were collected on Day 10. Duplicate slurry samples were collected from both of the bioreactors on Days 0, 4, and 10 for determination of semivolatile organic compound concentrations. Contrary to expectations, no large differences in volatile organic concentrations were observed in high-carbon and low-carbon slurries from Pits E, Q, and H/V and semivolatile organic constituent concentrations in slurries made from residual organics and soils found in Pits J, Q, and R. Consequently, no distinction was made between the high-carbon and low-carbon slurry batches when interpreting the results discussed below.

RESULTS

Organic Compound Removal

Volatile Organics. Volatile organics were removed by air stripping in both bioremediation demonstrations. No attempt was made to recycle and biodegrade the volatiles released by tilling the treatment bed in the solid-phase bioreactor or by agitation and aeration in the slurry-phase reactors. In both cases, volatiles were contained and routed to carbon adsorbers. Initial average volatile organic compound concentrations were higher in the solid-phase treatment bed (1,064,455 μg/kg) than in the slurry bioreactors (498,209 μg/kg). Some of this difference in starting concentration results from the variability of the raw feed material, but most of the difference results from the loss of volatiles during slurry preparation. Both processes achieved greater than 99 percent volatile organic compound removal (Table 2) and produced final total volatile organic compound concentrations lower than the 5,960 μg/kg concentration required for individual volatile constituents. Although both treatment methods removed volatiles, the slurry-phase process removed them approximately six times faster than the solid-phase process, based on half-lives of 1.21 days and 7.57 days, respectively.

Semivolatile Organics. Total PNAs and phenanthrene were effectively removed in both bioremediation demonstrations. Final concentrations of total PNAs did not exceed the 63.84 mg/kg required treatment concentration, although this was not a very high hurdle since starting concentrations of 54,573 μg/kg (solid-phase) and 56,618 μg/kg (slurry-phase) were less than this limit. In the solid-phase demonstration, total PNA concentrations were reduced from 54,573 μg/kg to 5,031 μg/kg, a 91 percent reduction (Table 2). Phenanthrene concentrations decreased from 36,303 μg/kg to 4,866 μg/kg over the 94-day treatment period, a reduction of 87 percent. Total PNA and phenanthrene half-lives were 27.33 and 32.41 days, respectively. Removals were lower in the slurry-phase demonstration than in the solid-phase demonstration because the treatment period was shorter (10 days versus 94 days). Total PNA concentrations were reduced 62 percent from a starting concentration of 56,618 μg/kg to a final concentration of 21,252 μg/kg (Table 2). Phenanthrene concentrations decreased from 46,837 μg/kg

TABLE 2. Solid-phase and slurry-phase bioremediation demonstration results, Brio Refining Superfund Site, Houston TX.

Solid-Phase (µg/kg)		Day 0	Day 21	Day 58	Day 94	Removal (percent)	Half-Life (days)
Total Volatiles	Mean	1,064,455	253	126	196	99.98	7.57
	Std. Dev.	1,676,533	311	69	137		
Total PNAs	Mean	54,573	14,269	11,908	5,031	90.78	27.33
	Std. Dev.	68,267	6,475	4,536	2,312		
Phenanthrene	Mean	36,303	13,931	11,828	4,866	86.60	32.41
	Std. Dev.	43,018	6,248	4,524	2,227		
Total Organic Carbon	Mean	21,888,000	19,875,000	20,375,000	17,750,000	18.91	310.86
	Std. Dev.	10,365,000	3,052,000	4,303,000	1,438,000		
Slurry-Phase (µg/kg)		**Day 0**	**Day 4**	**Day 10**		**Removal (percent)**	**Half-Life (days)**
Total Volatiles	Mean	498,209	31,778	1,624		99.67	1.21
	Std. Dev.	471,530	94,573	3,111			
Total PNAs	Mean	56,518	26,806	21,252		62.46	7.07
	Std. Dev.	42,576	27,241	26,578			
Phenanthrene	Mean	46,337	23,824	19,683		57.98	7.99
	Std. Dev.	37,214	24,009	25,365			
Total Organic Carbon	Mean	15,804,000	14,721,000	11,477,000		27.38	21.66
	Std. Dev.	10,338,000	10,357,000	8,145,000			

Notes:
(1) Reported concentrations are the average of all samples collected.
(2) Half-life = $0.693[(\ln C_o/C)/t]$, where: C_o = initial concentration, C = final concentration, and t = treatment period.
(3) Slurry-phase results represent high-carbon slurry and low-carbon slurry data combined.

to 19,683 μg/kg, a reduction of 58 percent. Although the percent removals were lower, the half-lives of total PNAs and phenanthrene, 7.07 days for the former and 7.99 days for the latter, were about four times faster than the half-lives achieved by the solid-phase system.

Total Organic Carbon. Total organic carbon removal was 19 percent in the solid-phase system and 22 percent in the slurry-phase system (Table 2). TOC half-lives were an order of magnitude different; 310.86 days for the solid-phase demonstration and 21.66 days for the slurry-phase demonstration. The low percent removal and long half-lives indicate that some component of the total organic carbon is not readily biodegradable. During the slurry-phase demonstration, solvent-extractable organic carbon (total extractable organic carbon or TEOC) was monitored. The initial TEOC concentration of 9,380,000 μg/kg was reduced to 2,464,000 μg/kg over the 10-day treatment period, a 74 percent reduction with a half-life of 5.18 days (Table 3). This indicates that some but not all of the total organic carbon is biodegradable. The nondegradable TOC may be the high molecular weight fraction of the organic residues, naturally occurring organics such as humic or tannic acids or some combination of the two.

Demonstration of Biodegradation

Volatile organic compounds were removed by the physical process of air stripping in both bioremediation demonstrations. A comparison of microbial activity and phenanthrene concentration indicates that semi-volatile organics were removed by biodegradation and not physical processes. This conclusion is supported by the occurrence of the highest microbial activity with the largest decrease in phenanthrene concentration in both biodegradation demonstrations (Figure 1). Additional support for this conclusion comes from a comparison of daily TEOC and ATP measurements made during the slurry-phase bioremediation demonstration (Table 4). The occurrence of high microbial activity with greatest decrease in TEOC concentration indicates removal by biological, rather than physical, processes (Figure 2).

TABLE 3. Comparison of total extractable organic carbon to total organic carbon degradation in the slurry-phase bioremediation demonstration, Brio Refining Superfund Site, Houston TX.

Solid-Phase (µg/kg)		Day 0	Day 4	Day 10	Removal (percent)	Half-Life (days)
Total Extractable Organic Carbon	Mean	9,380,000	5,681,000	2,464,000	73.73	5.18
	Std. Dev.	11,061,000	7,357,000	2,625,000		
Total Organic Carbon	Mean	15,804,000	14,721,000	11,477,000	27.38	21.66
	Std. Dev.	10,338,000	10,357,000	8,145,000		

Notes:
(1) Reported concentrations are the average of all samples collected.
(2) Half-life = 0.693[(lnCo/C)/t], where: Co = initial concentration, C = final concentration, and t = treatment period.
(3) Slurry-phase results represent high-carbon slurry and low-carbon slurry data combined.

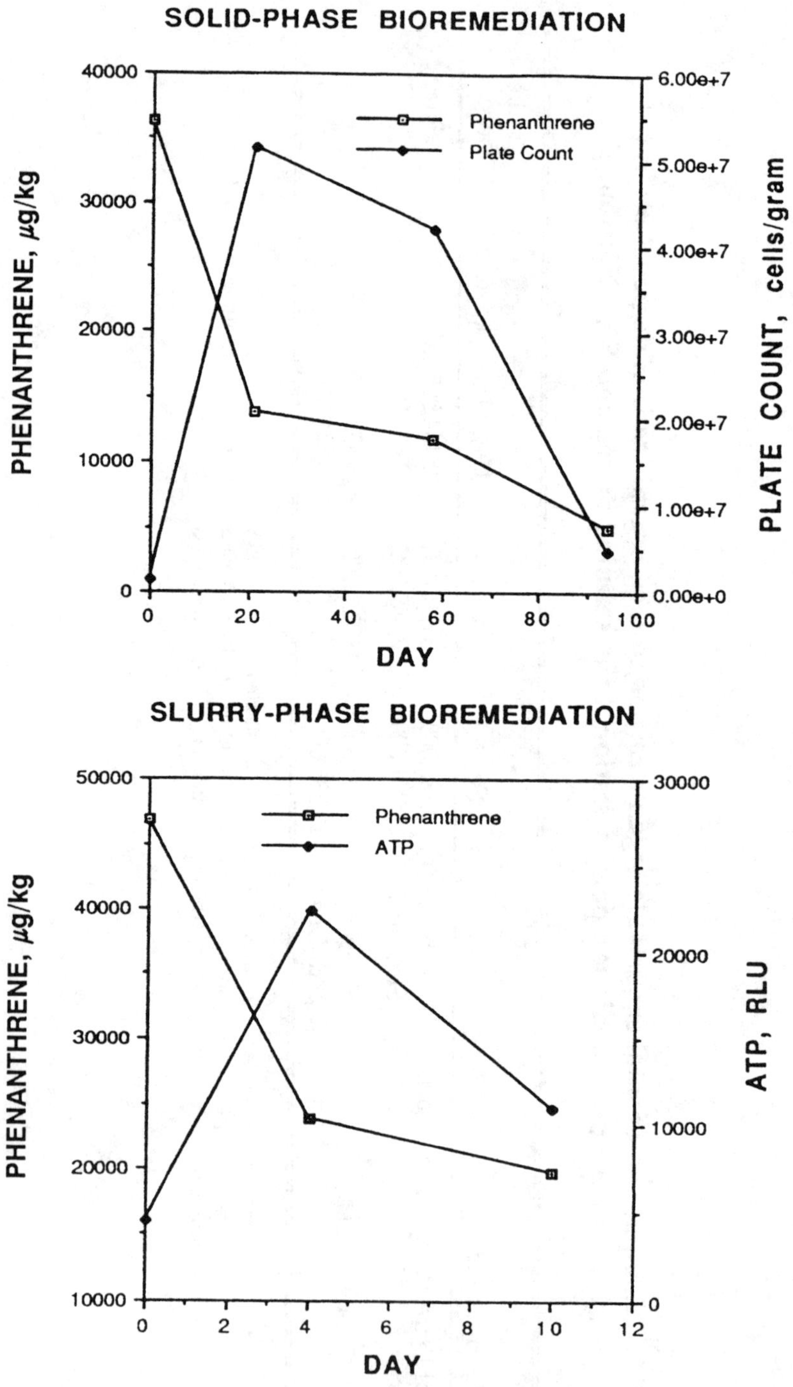

FIGURE 1. Comparison of phenanthrene removal and microbial activity, Brio Refining Superfund Site, Houston TX.

TABLE 4. Demonstration of biological removal of total extractable organic carbon, Brio Refining Superfund Site, Houston TX.

Day	TEOC (µg/kg)	ATP (RLU)
0	9,380,000	4,576
1	4,540,000	55,339
2	4,104,000	45,327
3	3,874,000	24,278
4	5,681,000	22,402
5	3,985,000	18,966
6	5,010,000	18,631
7	2,847,000	18,528
8	4,785,000	15,965
9	4,233,000	15,843
10	2,464,000	10,911

CONCLUSIONS

Volatile organics were effectively removed by air stripping in both systems with a half-life of 7.57 days in the solid-phase demonstration and 1.21 days in the slurry-phase demonstration. Semivolatile organics were degraded biologically in both demonstrations; however, a substantially faster degradation rate was achieved in the slurry-based system. Phenanthrene had a half-life of 7.99 days in the slurry-phase system, four times faster than the 32.41-day half-life achieved by the solid-phase system. Total organic carbon half-life, although nearly 15 times faster in the slurry-phase system than in the solid-phase system (21.66 days versus 310.86 days), was still higher than the rate achieved for semivolatile organics. However, the solvent-extractable portion of the total organic carbon had a half-life of 5.18 days in the slurry-phase system, indicating the organic carbon available to microorganisms could be degraded at a relatively rapid rate.

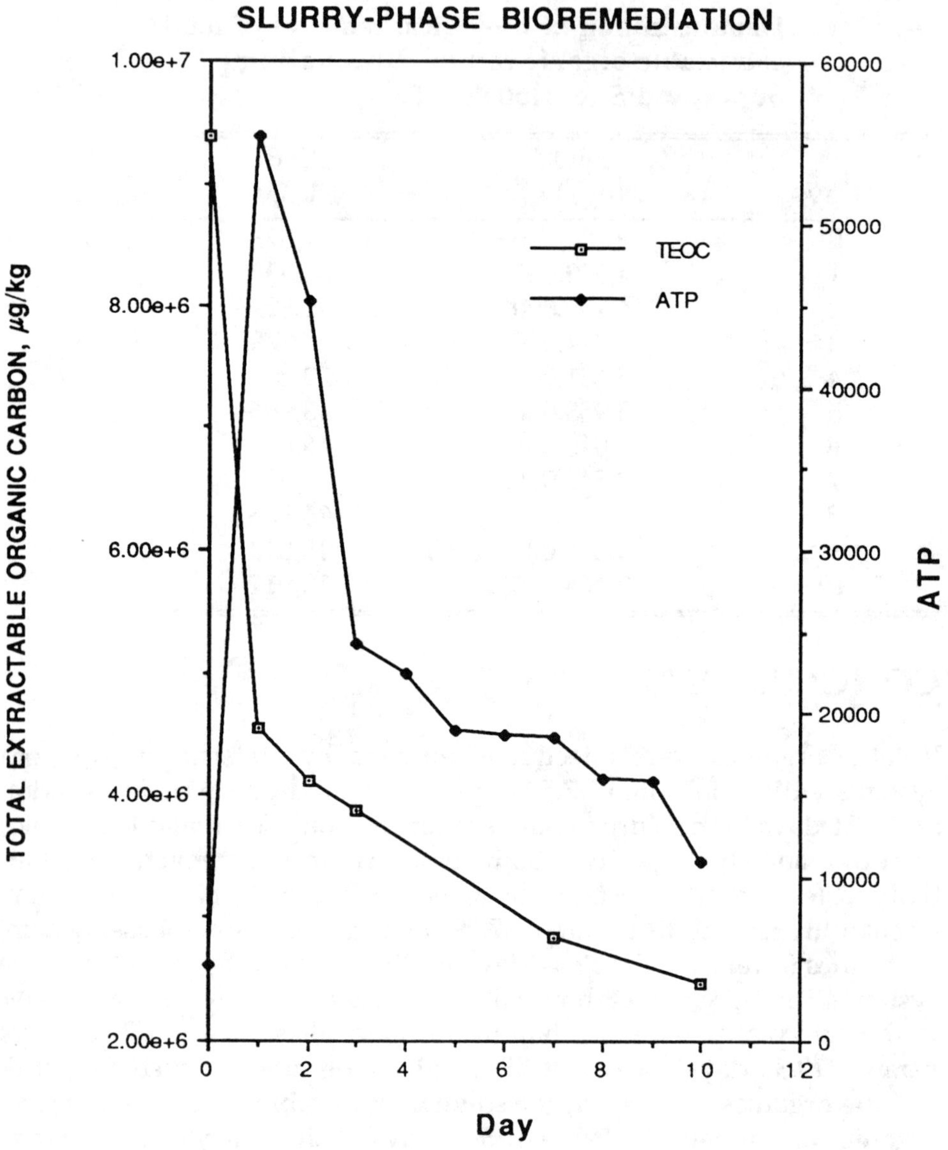

FIGURE 2. Comparison of total extractable organic carbon removal and microbial activity, Brio Refining Superfund Site, Houston TX.

Acknowledgments

The efforts of Derek Ross and David Ashcom on the solid-phase bioremediation demonstration and Ed Valines, Bill Adams, Martina Bianchini-Akberg and Will Renaudette on the slurry-phase demonstration were vital to the successful completion of these projects.

Biological Treatment of Oil-Contaminated Soils in Bioreactors

R. Stegmann, S. Lotter, J. Heerenklage*
Technical University of Hamburg-Harburg

INTRODUCTION

In Germany biological soil treatment is practiced in most cases by using the windrow technique, where more and more the process takes place inhouse. Dependent upon the specific situation and the experience of the operating company, the process works more or less successfully, but quite a few open questions remain to be answered.

Biological soil treatment in closed reactors is more frequently considered, and one company in Germany uses this technique already. The reactors are more costly to run and should be used in those cases where the windrow technique is not satisfactory. Closed reactor treatment is appropriate when

- Higher standards are needed for the emission control of the volatile components
- Some organics are difficult to degrade

* Technical University of Hamburg-Harburg, Arbeitsbereich Umweltschutztechnik, Eißendorfer Str. 40, D-2100 Hamburg 90, Federal Republic of Germany

- Soil with a higher clay content must be treated
- Naturally occurring biological processes must be enhanced

To operate biological soil treatment under optimum conditions and to discover the treatment limits, the specific factors of influence have to be determined in advance by using a series of laboratory-scale experiments. Since each case is different, this step is obligatory. Using these tests the following factors of influence can be optimized: oxygen supply, temperature, water content, amount of compost (soil:compost ratio), compost quality (i.e., stage of maturation), addition of microbes, nutrients, and other additives.

Tests have been performed in respirometers (Sapromat by Messrs. Voith, Germany), test jars, and forced air aerated test reactors (volume about 3 L). These and other investigations are subjects of an interdisciplinary research project funded by the Deutsche Forschungsgemeinschaft (DFG) called "Treatment of Contaminated Soils" ("Sonderforschungsbereich:Reinigung kontaminierter Böden"). During the first phase of this research program, mainly reproducible oil-contaminated soils are used for investigation. One research project of this program deals with the "Treatment of Contaminated Soils in Bioreactors" and is investigated by the authors.

Soil, Oil, Compost, Test Facilities, and Analytics

Most of the projects in the research center used the same kind of soil (A_h = horizon) with a relatively high proportion of sand and humic substances (Miehlich & Wagner 1990; Goetz *et al.* 1990). For respirometer investigations, only the fraction ≤ 2 mm is used. The artificial contamination was performed by mixing the soil with 1 percent (by weight) of diesel fuel that had been analyzed in detail (Steinhart *et al.* 1989). The used diesel fuel has the following composition: *n*-alkane, 31.9; rest aliphatics, 39.8; monoaromatics, 16.4; diaromatics, 8.0; and polyaromatics, 3.9 percent.

The elementary analysis of the oil resulted in a carbon content of 86.1 percent and a hydrogen content of 12.8 percent (Francke 1990). Based on these findings, a stoichimetric amount of oxygen for complete oxidation of the oil to CO_2 and H_2O can be calculated to 3,370 mg O_2/g hydrocarbon (HC). This calculation respects only the oxidation of the carbon and the hydrogen, for which the following relationship had been used (Sontheimer *et al.* 1980):

$$C_xH_yO_z + s \bullet O_2 \rightarrow x \bullet CO_2 + y/2 \bullet H_2O$$

The oxygen number, s, is calculated by the following equation:

$$s = x + y/4 - z/2$$

where z was set to zero for diesel fuel.

In the experiments, compost had been added that originates from biowaste that was composted at a small windrow plant in Hamburg-Harburg. In the respirometer, compost samples ≤ 4 mm were used. A scheme explaining the principle of the respirometer operation is shown in Figure 1. The reaction vessels in which the contaminated soil samples had been placed were kept at a temperature of 22 C. The oxygen consumption values were transferred to a personal computer so that the biological activity could be continuously monitored by the

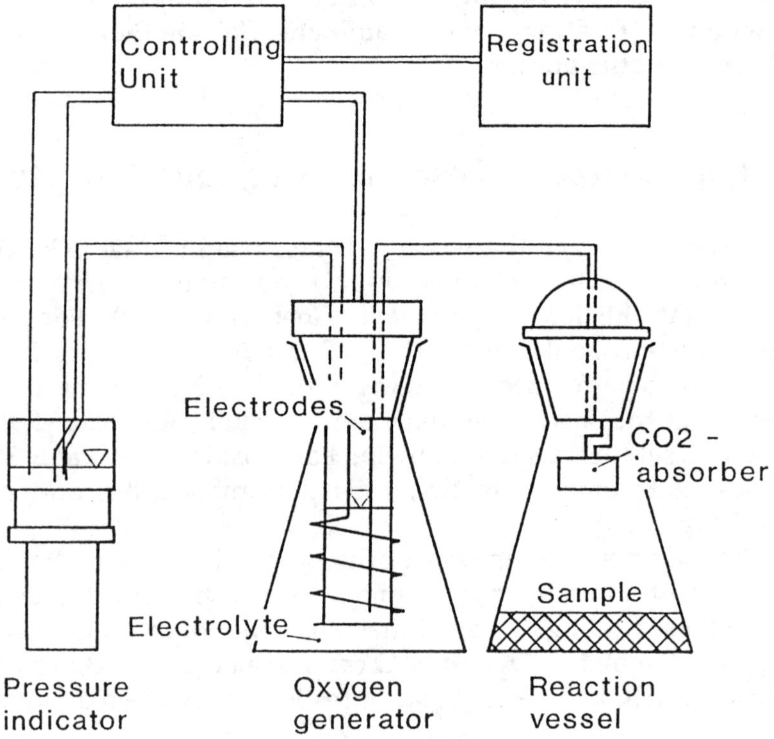

FIGURE 1. Operation principle of the respirometer.

oxygen uptake curves. A high reproducibility of the results is obtained by using this method so that usually tests were run in parallel. In case of the tests investigating the influence of different amounts of compost addition, only single samples were used. For the determination of the final oil concentration in the soil, samples from the respirometer vessels had been analyzed at the end of the experiment.

The test reactors (see Figure 2) are made of glass and have a volume of 3 L. They are installed in a 30 C heated room. In the lower part of the reactor, a sieve is installed on which the soil-compost mixtures are placed. The reactor is equipped with sampling devices at three different levels. Forced air aeration is practiced from the bottom; the air is water-saturated to avoid drying the soil samples. Samples are taken from the gas that leaves the reactor and are immediately injected into a gas chromatograph. To measure the CO_2 concentrations, a molecule sieve combined with a WLD detector is used. The soil samples are removed from the reactor at the sampling points.

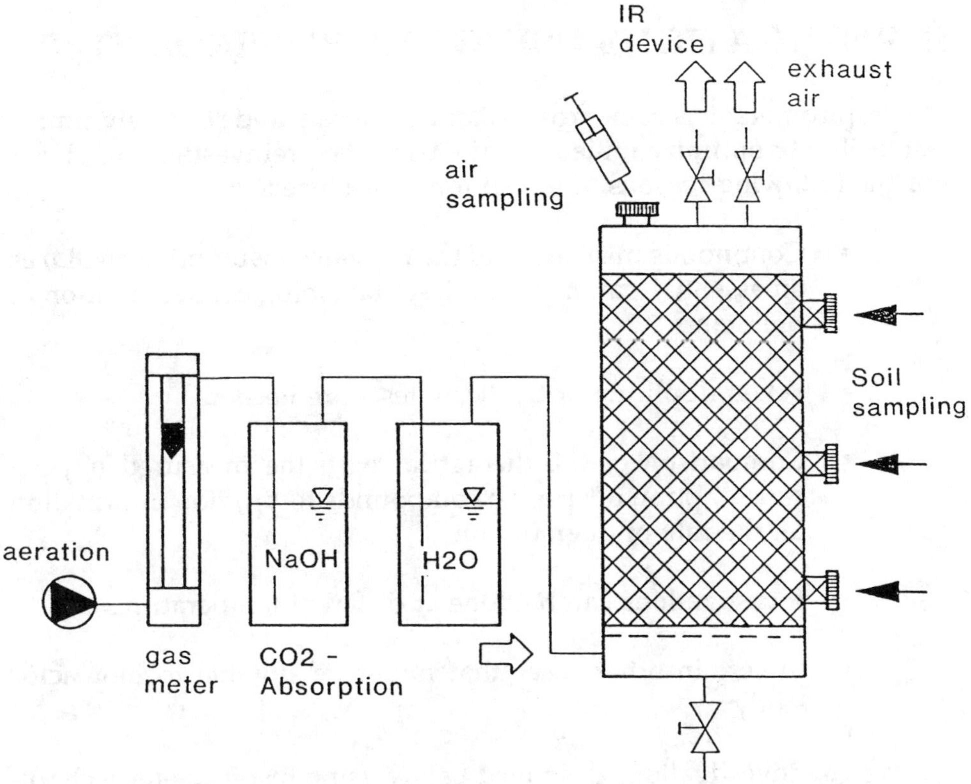

FIGURE 2. Test reactors for the biological treatment of contaminated soils (volume 3 L).

The total amount of hydrocarbons was measured using infrared spectrography (procedure DIN H18, Anonymous 1981), where two different extraction methods were used. In the beginning, an 8-h Soxhlet extraction with trichlor-trifluro-ethane was practiced (25 mg soil with 250 mL trichloro-trifluoro-ethane). The extract was contacted with aluminum oxide to remove polar compounds.

To reduce the amount of trichlor-trifluro-ethane for analytical purposes, an extraction method using ultrasound was developed (Bundt *et al.* 1990), in which 3 g soil were extracted with 10 mL trichloro-trifluoro-ethane. The extraction rate was higher using ultrasound, but because of the low sample volume, extraction was done in duplicate.

Biomass was indirectly determined using the SIR (substrate-induced respiration) method. Glucose is added to the soil samples, and the resulting CO_2 production is related to the biomass (Anderson & Domsch 1978). A modified method by Beck (1984) was used: instead of CO_2 production, oxygen uptake in the respirometer was measured.

INVESTIGATIONS USING THE RESPIROMETER

The respirometer has been proven as an important and relatively simple test facility to optimize milieu conditions in the preinvestigation phase. For the following reasons, the respirometer is used:

- Continuous measuring of the oxygen consumption (BOD) as an essential parameter to describe biological degradation of pollutants
- When duplicate or triplicate tests are needed
- Comparison of the theoretical with the measured oxygen consumption of specific compounds to provide information on the rate of degradation
- Investigations can be done at different temperatures
- In certain other cases that might be suitable to bioreactor study

The investigations described below using respirometer technology have been evaluated by presenting the cumulative oxygen uptake (mg O_2/g hydrocarbons); the oxygen consumption caused by the degradation of the oil is calculated by subtracting the oxygen consumption of the noncontaminated sample from that of the contaminated one.

Optimization of the Water Content

In this project, oil-contaminated soils at high total solid content (no suspension) are treated. The optimum water content of the soil-compost mixtures was selected on the basis of the results of a pretest series during which the activity of oil-contaminated soils at different water contents was measured in a respirometer over a period of 4 days. As an example, the results from the soil-compost mixture of 8:1 (based on dry weight, dwt) are presented. Both contaminated and noncontaminated soil samples were investigated, and the maximum water capacity of the soil-compost mixture was determined (Kretzschmar 1986). For the noncontaminated sample, a value of 48.1 g H_2O/100 g dwt was measured (contaminated sample: 43.9 g H_2O/100 g dwt). Figure 3 presents oxygen uptake in relation to different percentages of maximum water capacity.

In both cases the maximum oxygen consumption (or biological oxygen demand, BOD) is reached at about 60 percent of the total water capacity; this result corresponds well with the recommendations for the determination of the biomass in soils (Jenkinson & Powlson 1976). The significantly higher BOD of the contaminated soil-compost mixture is caused by the strong biological degradation of the oil components, whereas the oxygen consumption in the test vessel with the noncontaminated soil-compost mixture is constant during the 4 days. This reflects basal respiration.

Selection of the Age of the Compost

The addition of additives like bulking material (wood chips, etc.) and/or compost to improve the biological degradation of oil-contaminated soils is already practiced. When compost is added, the amount and the quality of the added material often are not mentioned or not known. The compost characteristics depend on the kind of original organic material (*e.g.*, straw, garden, and yard waste). The compost used in this project was produced from biowaste.

Since compost quality changes with time of biological degradation, the selection of material becomes important. Using the

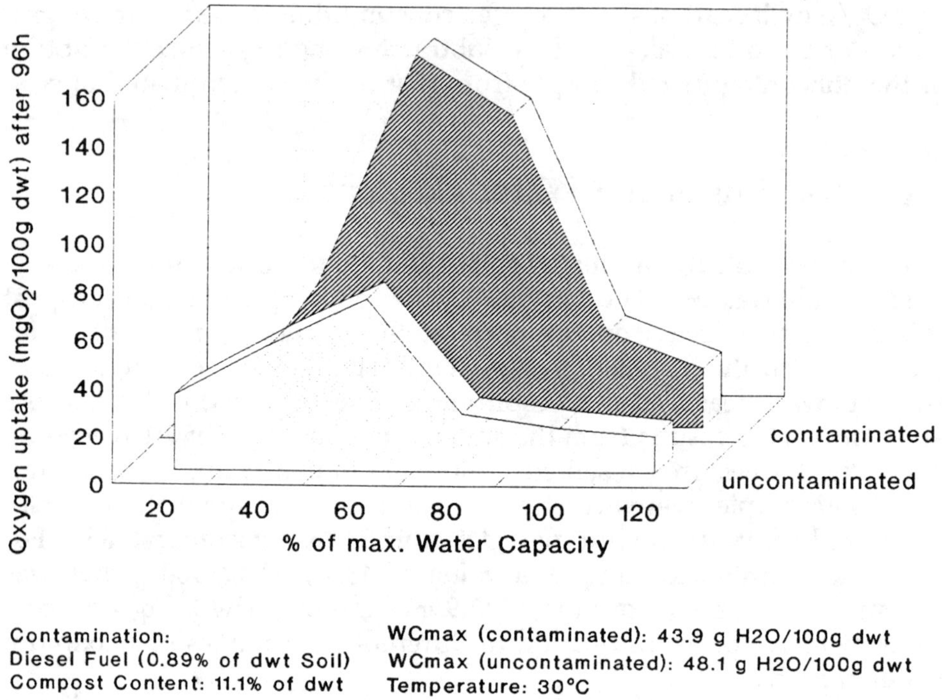

FIGURE 3. Oxygen uptake in 4 days of a soil-compost mixture of 8:1 dependent upon the soil water content (as percentage of the maximum water capacity).

respirometer (Sapromat), compost materials of different "age" (degree of maturation) have been investigated (½, 2, 4, and 6 months old). Initial biowaste composition differed also with time of year.

Diesel fuel polluted soil was mixed at 1 percent (by weight) with the compost material at a ratio of 2:1 (dry weight). Table 1 shows some characteristics of the different kinds of compost used in the respirometer tests. The only 2-week-old compost material could not be used since the self-heating was so significant that it influenced the pressure inside the vessel.

The biological parameters of biomass and respiration activity were quite similar for 4- and 6-month-old compost, but 1.7 higher for 2-month-old material. The relatively low organic content (measured as lost on ignition) indicates a fairly high degree of mineralization.

Figure 4 shows the influence of compost age on degradation of diesel fuel in the soil-compost mixture. The hydrocarbon concentration after 300 h reflects a significantly higher rate of biological degradation

TABLE 1. Description of the characteristics of the used composts of different age.

Age of Compost	(Months)	2	4	6
Biomass	(mgC/100g dwt)	218	120	126
O_2 – uptake	(mgO_2/100g dwt)	289	181	172
Organic dry matter	(%)	41	42	30

when the 6-month-old compost had been added; this can also be seen from the oxygen consumption values. After a lag phase of 2 days, the respiration values of the 6-month-old compost test increased to a value of 1,330 mg O_2/g hydrocarbon (4 resp.; 2-month-old compost = 730 resp.; 350 mg O_2/g hydrocarbon). For the hydrocarbon analysis of the soil-compost mixture, the extraction was done in the 2- and 4-month-old compost by the Soxhlet method and in the 6-month-old material by means of ultrasound.

Although the biomass was higher in the young compost, the oldest material enhanced degradation the most. It seems that the kind of microorganisms in the old compost resulted from an advanced humification process is more appropriate for the degradation of oil components.

Investigations Regarding the Volume of Compost Added

It could be shown that the addition of one part of compost to two parts of soil has positive effects on the degradation of the oil in the soil, but this is a large amount of compost to be added. Common practice in Germany is a 9:1 soil:compost ratio. For this reason, the influence of different soil:compost ratios on biological degradation had been investigated at four ratios (dwt) 2:1, 4:1, 8:1, and 16:1. Twenty-five grams (dwt) of soil were artificially contaminated with diesel fuel (1% by weight), and the different amounts of compost (6 months old) were added and mixed with the soil.

In additional tests, 25 mg N (as KNO_3) were added to the different soil-compost mixtures to obtain a molar C:N:P ratio of 73.2:9.2:1. The following biomass composition was used for the calculation: C_{250} H_{611} O_{77} P_6 S. The nutrient composition of the used compost is listed in Table 2.

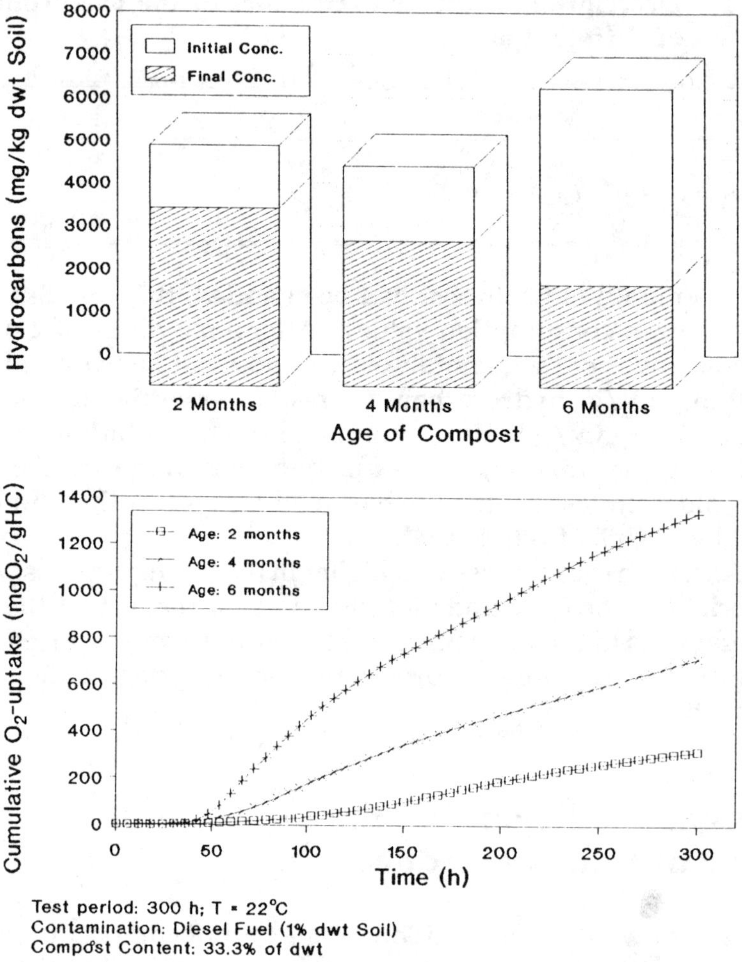

FIGURE 4. Influence of different compost age additions on hydrocarbon degradation and cumulative oxygen consumption (substrate: oil contaminated soil).

TABLE 2. Nutrient concentrations in the compost.

TKN	NH_4-N (gN/kg dwt)	NO_3-N	P_{tot} (gP/kg dwt)	PO_4-P
5.6	0.4	0.2	3.2	2.35

In addition, the pH should have the same value independent of the different soil-compost mixtures. For this reason, different amounts of $Ca(OH)_2$ were added; because of the analytical principle of the respirometer, CO_2 production by means of $Ca(OH)_2$ addition does not affect the results. Table 3 shows the test program.

TABLE 3. Soil-compost mixtures, nutrient addition, and pH regulation.

Soil/ Compost	Soil (g dwt)	Compost (g dwt)	N – Addition (mg $KNO_3 - N$/ 25g dwt Soil)	$Ca(OH)_2$ (mg/ kg dwt Soil)	pH	
					Start	End
2:1	25	12.5	25.0	0.014	6.96	7.02
4:1	25	6.25	25.0	0.066	6.82	6.87
8:1	25	3.12	25.0	0.088	6.82	6.76
16:1	25	1.56	25.0	0.190	6.92	6.83

In Figure 5 the hydrocarbon concentrations in the soil-compost mixtures after 300 h are presented, where the results are related to 1 kg (dwt) of soil. From the initially installed concentration of 10,000 mg HC/kg (dwt) soil, 70 to 100 percent could be detected analytically at the start of testing. The final hydrocarbon concentrations decreased with increasing compost content, where nutrient addition had no positive effect.

The biological degradation of the oil can be well documented by the oxygen consumption measured in the respirometer. The upper part of Figure 6 presents cumulative oxygen consumption, and the lower part, oxygen consumption with time. The lag phase is again 2 days, where those tests without nutrient addition start earlier with the biological degradation of the oil. Table 3 presents a summary of the final and maximum oxygen consumption rates.

Figure 6 demonstrates that with decreasing compost content the cumulative and the maximum oxygen uptake also decrease; in addition, it takes longer to reach the maximum. Nitrogen addition results in comparatively lower maximum oxygen consumption rates. One advantage of nitrogen addition cannot be seen: a slightly negative effect that might be the case for the lower compost addition rates. Perhaps the increased salt content is of influence.

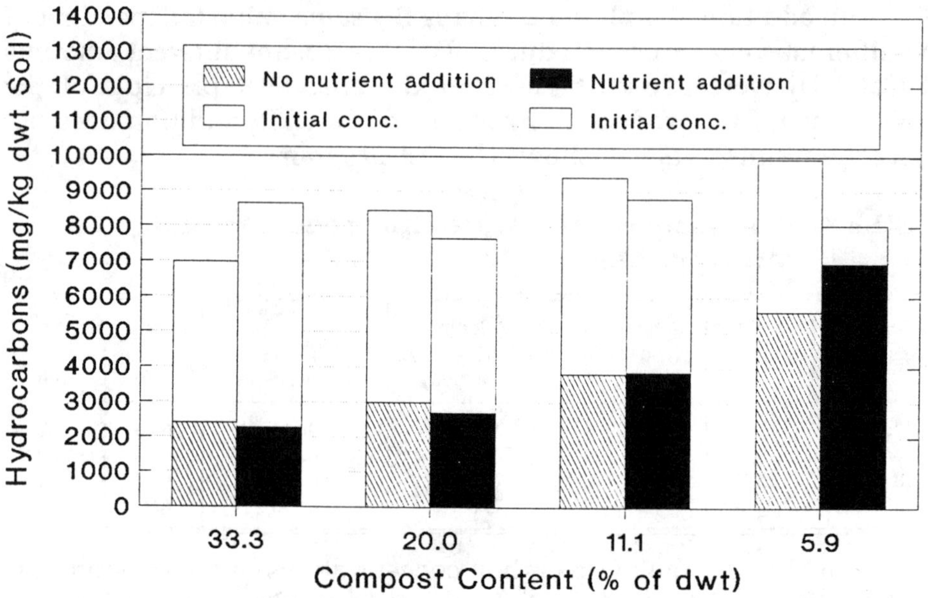

Test period: 300 h; T = 22°C
Contamination: Diesel Fuel (1% dwt Soil)
Nutrients: 25 mg KNO_3-N

FIGURE 5. The effect of different compost volume addition rates and nutrient supply on the degradation of oil (respirometer tests).

Relationship between Theoretical Oxygen Demand and Actual Oxygen Consumption

To what extent pollutants are biologically degraded and whether complete mineralization was achieved must always be confirmed. The stoichiometric oxygen demand of the oil (3.370 g O_2/g HC) is compared with the measured oxygen consumption values after 300 h (see Table 4); the oxygen consumption may proceed, but with decreasing tendency if the test would have been prolonged. This point will be discussed later.

Cumulative oxygen consumption after 300 h (ratio 2:1) was related to hydrocarbon reduction to calculate the ratio BOD/ThOD (Table 5). As a result of this calculation, only 47.9 percent of the theoretical oxygen demand has been measured when a 2:1 soil-compost mixture was tested. Calculated from the used oxygen, only 36.4 percent of the initial hydrocarbon concentration was biologically degraded. The last column in Table 4 shows the percentage of hydrocarbons that are biologically degraded (i.e., 36.4%) plus the final concentration of

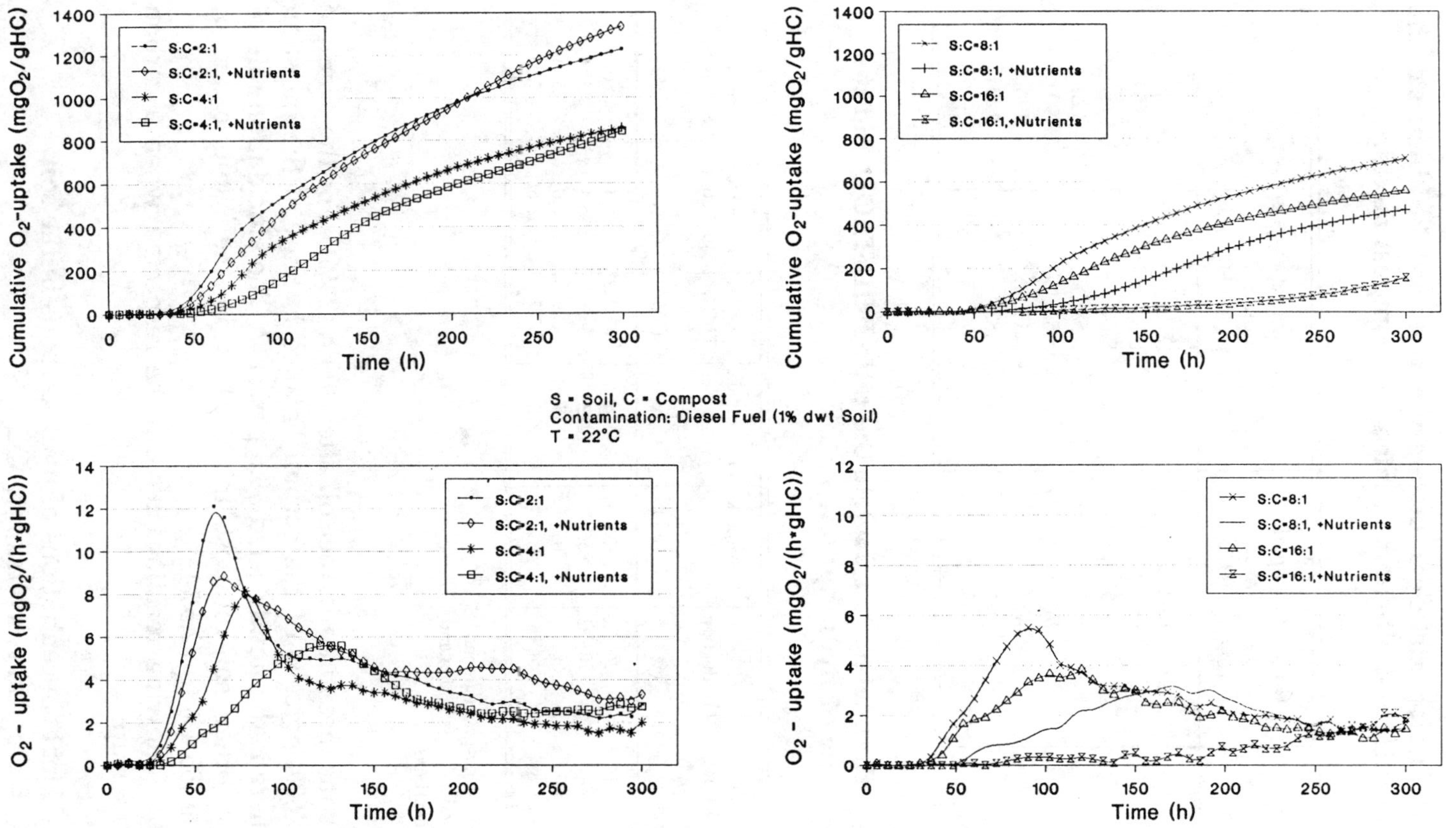

FIGURE 6. Cumulative and time-dependent oxygen uptake kinetics of the different oil-contaminated soil-compost mixtures (respirometer tests).

TABLE 4. Characteristic parameters of oxygen uptake kinetics of the different oil-contaminated soil-compost mixtures.

Mixture Soil/ Compost	Cum. O_2 – uptake after 300 h (mgO_2/gHC)	Max. O_2 – uptake Amount ($mgO_2/(h \cdot gHC)$)	Time (h)
2:1	1226	12.10	60
2:1 + Nut.	1334	8.84	66
4:1	856	8.17	78
4:1 + Nut.	841	5.58	126
8:1	707	5.50	90
8:1 + Nut.	473	3.17	174
16:1	561	3.83	120
16:1 + Nut.	158	–	–

TABLE 5. Parameters to calculate the ratio BOD/ThOD and the fate of hydrocarbons.

Mixture Soil/ Compost	HC – Residue (mg/kg Soil)	HC – Reduction (%)	O_2 – uptake (mgO_2/ gHC add.)	O_2 – uptake (mgO_2/ gHC red.)	BOD/ ThOD (%)	HC – De – gradation (%)	Sum of Residue + Degradation (%)
2:1	2403	76	1226	1614	47.9	36.4	60.4
4:1	3000	70	856	1223	36.3	25.4	55.4
8:1	3817	62	707	1143	33.9	21.0	59.1
16:1	5583	44	561	1270	37.7	16.6	72.5

BOD: Biological oxygen demand
ThOD: Theoretical oxygen demand
HC: Hydrocarbons

hydrocarbons measured in the soil after 300 h (24%). For the example 2:1 ratio, 60.4 percent of the carbon balance is known; the question arises, where is the nondetected 39.6 percent? The elucidation rate of tests with other soil-compost mixtures is in the same order of magnitude (55.4% to 72.5%).

The following reactions have to be respected when a carbon balance is set up:

- Final concentrations of hydrocarbons in the soil
- Biological conversion to CO_2

- Production of new biomass
- Strip effects
- Reactions with humic and soil substances like clay
- Basic oxygen consumption of the bacteria

The authors found some evidence that the reaction and/or incorporation of the hydrocarbons into the humic matrix may be of temporary and/or general influence. If this is true, hydrocarbons can be extracted only to a certain degree, and therefore some will not be analyzed. Other factors, such as stripping, also influence the hydrocarbon balance. With increasing compost addition, this effect decreases (i.e., 2:1 ratio about 3%, no compost addition about 20%; see also Lotter *et al.*, 1990). The amount of carbon incorporated into the biomass is difficult to predict since measuring biomass concentrations in a soil-compost mixture is problematic, but it will not explain the gap. From other experiments it is known that the gap in the carbon balance varies during the experiment and is at its highest after 10 to 20 days (Lotter *et al.* 1990). The inexplicable carbon in the balance can be in the range of 40 percent and decreases with increasing test period. One must remember that these test results are obtained after a period of 300 h (12.5 days) and that the biological degradation processes are not terminated; this is apparent from the still-rising cumulative oxygen uptake curves in Figure 6. If these curves are mathematically extrapolated, the total amount of oxygen consumption can be calculated. Under the assumption that the entire oxygen consumption is due to hydrocarbon degradation, the possible total hydrocarbon degradation can be calculated. This procedure was not very successful using the data from Figure 6, where the test period was only 300 h and the temperature was 22 C. But evaluating another test that was run in the respirometer at 30 C over a period of 500 h, where again a 2:1 soil:compost ratio and an initial 1 percent oil contamination of the soil (dwt) were used, the extrapolation was most successful (see Figure 7). The regression curve was calculated on the basis of the data measured between 150 and 500 h, during which the following function was optimized:

$$y = a \bullet e^{-bx} + c$$

The regression coefficient reached an R^2 value of 1.

The maximum oxygen consumption was evaluated from the regression curve to 3.415 g O_2/g HC, which corresponds well with the theoretical value of 3.370 g O_2/g HC. This value was reached after

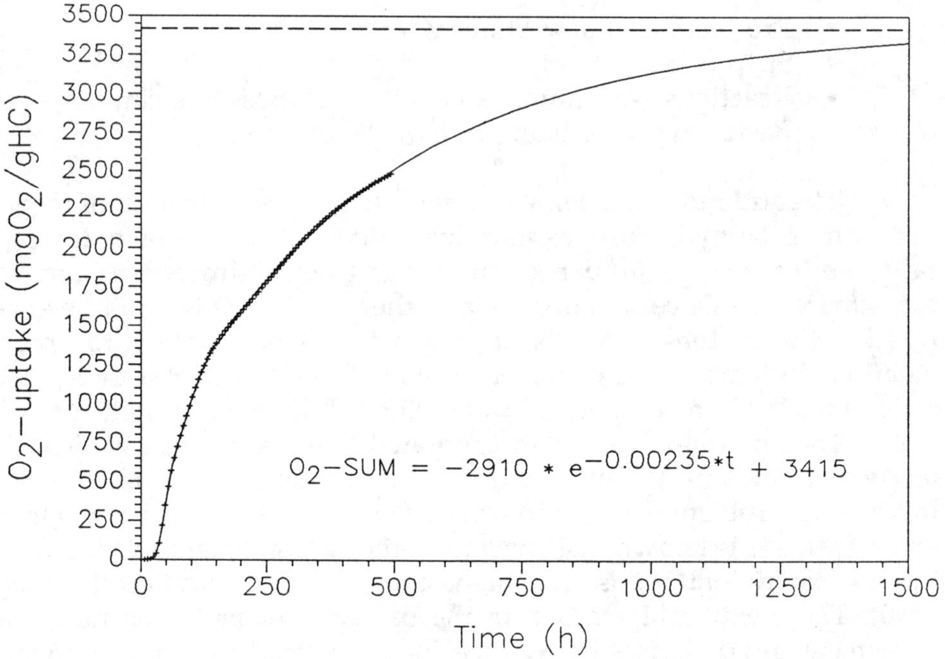

FIGURE 7. Extrapolation of a cumulative oxygen consumption curve. Experimental data: soil-compost = 2:1 (dwt), compost age 7 months, regression curve calculated on the basis of the data between 150 and 500 h, initial oil contamination 1 percent (dwt soil).

1,500 h (62.5 days). This test is a first approach and has to be validated by other tests of at least 65 days' duration.

COMPARISON OF TEST SYSTEMS (RESPIROMETER, JAR, AND BIOREACTOR TESTS)

The three different test systems are used for the following purposes:

- *Respirometer Tests*: The continuous measurement of the oxygen consumption is the main advantage of this system; in addition, setup is relatively simple and a variety of tests can be run in parallel.

- *Jar Tests*: Since this test unit is relatively simple and inexpensive, it can be used for long-term testing. In the commercially available gas-tight jars, a glass filled with NaOH is placed to absorb the produced CO_2; this CO_2 can be determined at certain times and represents the consumed oxygen. Soil samples can also be taken and analyzed. It is important that enough oxygen is always available.

- *Bioreactor Tests*: Since this is a test procedure where, in contrast to jar tests, gas samples can be analyzed on line, more information can be obtained. The stripped hydrocarbons can be monitored continuously, as can oxygen and CO_2 concentrations. The oxygen supply may also be continuous by forced air aeration. Soil samples can be taken at any time from different sampling points at the side of the reactor without disturbing the process.

One test series investigated how far the results from these three test systems can be compared. The following test parameters were installed: water content—60 percent of maximum water capacity, compost age—6 months, soil:compost ratio—2:1, and soil contamination by oil—1 percent (dwt). The following analytical parameters were compared: oxygen consumption in the respirometer, CO_2 production in the jar and reactor, hydrocarbon concentrations in all test facilities, and biomass in the respirometer and reactor.

Figures 8 and 9 show the results of hydrocarbon concentration and biomass with time. Because of optimized milieu conditions and 30 C reaction temperature, enhanced degradation could be achieved. After 21 days, the HC concentrations in all three test systems were between 400 and 500 mg/kg (dwt) soil; if the results are based on the total solid soil-compost mixture, values between 280 and 330 mg/kg have been reached, which are comparable to background levels. The HC concentrations measured on the specific days in the different test systems correspond satisfactorily (considering the problems associated with soil analysis); only at day 2 were hydrocarbon concentrations in the respirometer significantly higher.

Good correspondance of the values could also be reached with the parameter biomass. Because of the easily degradable fractions in the oil, the biomass increased rapidly during the first 2 days; after this time biomass concentrations decreased in all three test systems to about the same values as at the beginning of the test. It should be emphasized (as already mentioned) that the method used to determine

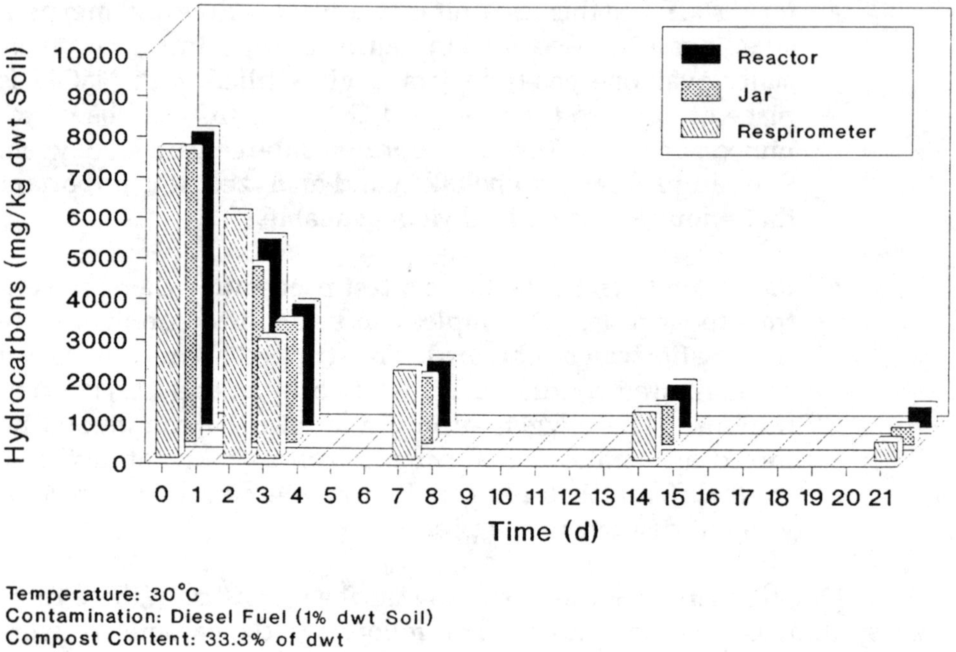

FIGURE 8. Hydrocarbon concentrations at different times in three different test systems: respirometer, jar and reactor.

biomass is not very accurate, in which glucose is added as a substrate. It is doubtful that all bacteria are activated when glucose is added; however, the biomass values described quantitate quite well the development of the biomass.

If CO_2 production in the jar is compared with that in the reactor, good correspondence is found. Figure 10 shows the molar relation of the CO_2 concentrations in the used air of the reactor and the oxygen consumption in the respirometer. Based on a stoichiometric calculation respecting the oil composition, a relation of CO_2:O_2 of 0.7 results; tests obtained from the contaminated samples are in the range of 0.5 to 0.7. This is not true for the uncontaminated samples. To understand the order of magnitude, the CO_2:O_2 ratio calculated for sewage sludge, in the range of 1.25 (Bidlingmaier 1983), might be used.

As an overall conclusion, the results from the three different test systems can be compared to a relatively high degree.

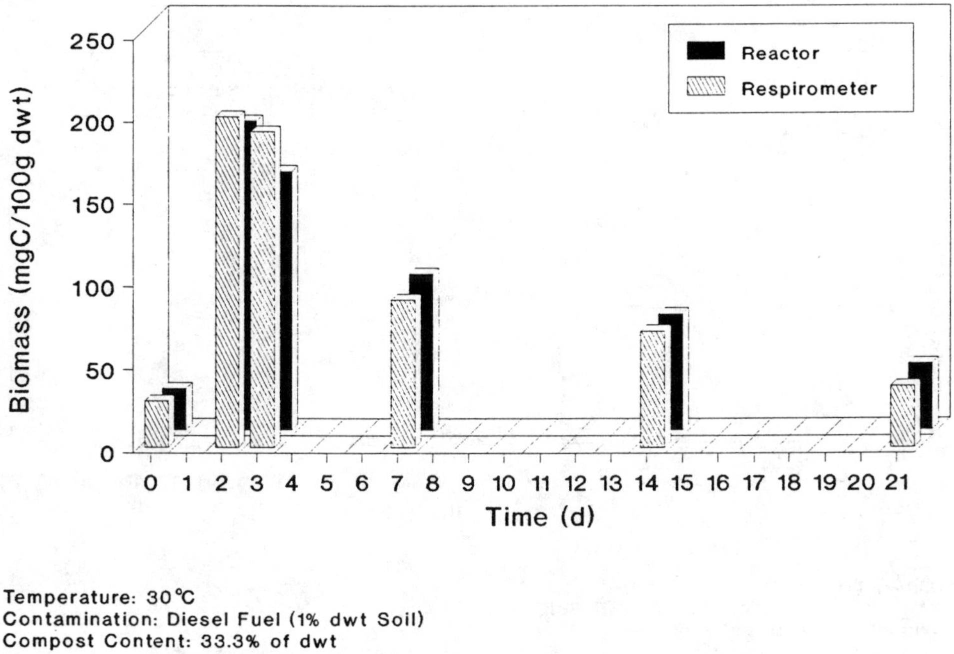

FIGURE 9. Biomass concentrations at different times in respirometer and reactor test systems (method: glucose addition, SIR [Anderson, Domsch, 1978]).

SUMMARY AND CONCLUSIONS

The optimum water content for the biological treatment of oil-contaminated soils could be selected by means of test results determined in the respirometer where the maximum oxygen consumption took place at 60 percent of the maximum water capacity. Since high oxygen consumption could also be found at 80 percent maximum water capacity and low oxygen consumption at 40 and 100 percent, an acceptable range might be between 50 and 80 percent of the maximum water capacity. Of course, this range has to be determined for each individual soil.

Compost addition has proven advantageous for the biological degradation of oil-contaminated soils where preferably a highly matured material (>6 months old) should be used. The biomass determination using glucose addition (SIR method) resulted in 1.7 more biomass for the 2-month-old compost than for 4- and 6-month-old material. It might be the case that not only the biomass concentration but also the

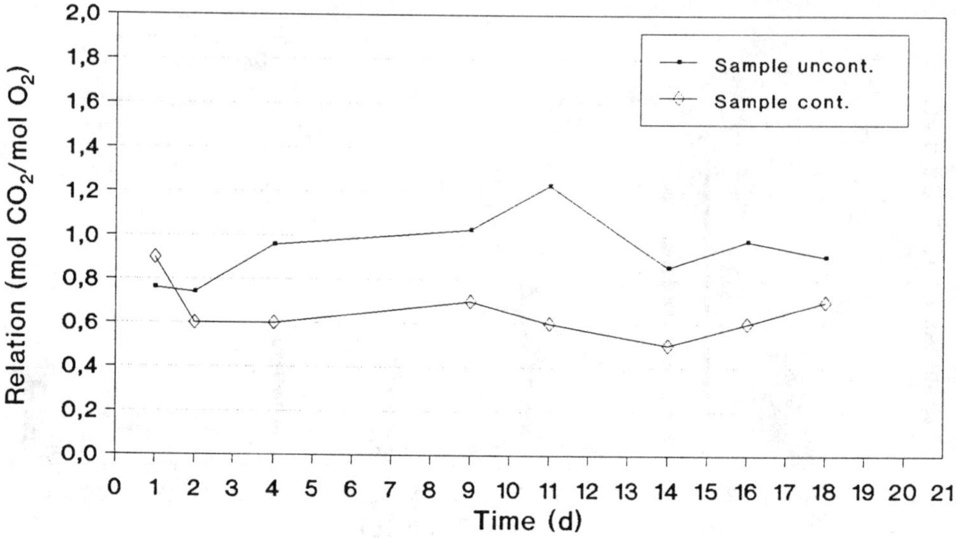

Temperature: 30°C
Contamination: Diesel Fuel (1% dwt Soil)
Compost Content: 33.3% of dwt

FIGURE 10. Molar relation of mol CO_2/mol O_2 of bioreactor and respirometer (CO_2 measured in the exhaust air of the bioreactor, O_2 consumption measured in the respirometer).

kind of organism influence the enhancement of degradation. With proceeding humification, specific organisms may develop that are more appropriate for degrading oil components.

Different soil:compost ratios—2:1, 4:1, 8:1, and 16:1 (dwt)—were investigated to enhance oil degradation. The 2:1 ratio gave the best results for maximum and total oxygen consumption; consequently, also the lowest final hydrocarbon concentrations could be measured at this ratio.

The ratio of actually consumed oxygen (BOD) to theoretical oxygen consumption potential (ThOD) of the added oil resulted for the 2:1 soil:compost ratio in 48 percent, compared with 35 percent for other ratios of less compost material.

As a first approach to setting up a carbon balance, residual hydrocarbon concentrations and biologically degraded HC were summarized; using this procedure only about 60 to 72 percent of the total added HC could be identified (test period, 300 h), where the stripped proportion and the hydrocarbons used for biomass production have not been included in the calculation. From other experiments it can be concluded that these factors do not explain the 40 resp. 28 percent of undetected hydrocarbons. Reactions between oil components and the humic matrix seem to be of significant influence, and it is important to know whether the incorporation in the humic substances is reversible or irreversible.

Future investigations should measure all factors of influence on the carbon balance, especially where a reliable biomass determination and an extraction procedure for the hydrocarbons incorporated into the humic matrix have to be developed.

Comparison of the different test systems showed good correspondence. These test procedures are available for different purposes: respirometer tests can be used to optimize the milieu condition; jar tests are simple to handle, inexpensive, and can favourably be used for long-term degradation tests; and reactors are more sophisticated and should be chosen when processes must be described more in detail.

Acknowledgments

This project was funded by the Deutsche Forschungsgemeinschaft. It is a subproject of Research Center 188, Hamburg, "Treatment of Contaminated Soils." Results from the intensive discussions in the different working groups of the research center are to a certain extent incorporated in this paper; their input is most appreciated.

REFERENCES

Anderson, J.P.E.; Domsch, K. H. (1978): "A Physiological Method for the Quantitative Measurement of Microbial Biomass in Soils." *Soil Biol. Biochem., 10,* 215–221.

Anonymous. (1981): "Deutsche Einheitsverfahren zur Wasser-, Abwasser- und Schlammuntersuchung" *Bestimmung von Kohlenwasserstoffen,* H18.

Beck, Th. (1984): "Mikrobiologische und biochemische Charakterisierung landwirtschaftlich genutzter. Böden, I. *Mitteilung: Die Ermittlung einer Bodenmikrobiologischen Kennzahl." Zeitschrift für Pflanzenernährung und Bodenkunde, Band 147,* H1, 456–467.

Bidlingmaier, W. (1983): "Das Wesen der Kompostierung von Siedlungsabfällen." In *Müll- und Abfallbeseitigung,* Hrsg. Straub, H., Hösel, G., Schenkel, W., *Band 4,* Kennziffer 5306; Berlin.

Bundt, J.; Herbel, W.; Steinhart, H. (1990): "Mineralölanalytik von Bodenkontaminationen unter beson derer Berücksichtigung von aromatischen Kohlenwasserstoffen." In *Reinigung kontaminierter Böden*, Hrsg. Stegmann, Franzius; Economica Verlag; Bonn.

Francke, W. (1990): "Teilprojekt 'Mikroanalytische Untersuchungen an kontaminierten Böden: Erfassung und Quantifizierung ausgewählter Schadstoffe in unbehandeltem und behandeltem Material'" in *SFB 188: Mitteilungen in der SFB–Arbeitsgruppe "Analytik," unveröffentlicht.*

Goetz, D.; Wiechmann, H.; Berghausen, M. (1990): "Teilprojekt 'Einfluß von Ölkontaminationen auf bodenphysikalische und -mechanische Eigenschaften von kontaminierten Standorten'" in *SFB 188: Mitteilungen in der SFB-Arbeitsgruppe "Boden,"* unveröffentlicht.

Jenkinson, D.S.; Powlson, D.S. (1976): "The Effects of Biocidal Treatments on Metabolism in Soil — V. A Method for Measuring Soil Biomass." *Soil Biol. Biochem.*, *8*, 209–213.

Kretzschmar, R. (1986): *Kulturtechnisch-Bodenkundliches Praktikum*; Institut für Wasserwirtschaft und Landschaftsökologie, Christian-Albrechts-Universität, Kiel.

Lotter, S.; Stegmann, R.; Heerenklage, J. (1990): "Basic Investigations on the Optimization of Biological Treatment of Oil-Contaminated Soils." In *Altlastensanierung '90*, Hrsg. Arendt, Hinsenveld, van den Brink, Third International KfK/TNO, Conference on Contaminated Soil; Kluwer Academic Publishers: Drodrecht, Boston, London.

Miehlich, G.; Wagner, A. (1990): "Teilprojekt 'Veränderung bodenchemischer Eigenschaften durch Ölverunreinigung und -dekontamination'" in *SFB 188: Mitteilungen in der SFB – Arbeitsgruppe "Boden,"* unveröffentlicht.

Sontheimer; Spindler; Rohmann (1980): *Wasserchemie für Ingenieure*, DVGW–Forschungsstelle am Engler-Bunte-Institut, Universität Karlsruhe (TH): Eigenverlag, Karlsruhe.

Steinhart, H.; Herbel, W.; Bundt, J.; Paschke, A. (1989): "Teilprojekt 'Mineralöllasten – ihre Identifikation und Analytik zur Optimierung ihres biochemischen und chemischen Abbaus unter besonderer Berücksichtigung mehrcyclischer Aromaten'" in *SFB 188: Mitteilungen in der SFB – Arbeitsgruppe "Analytik,"* unveröffentlicht.

Design of a Packed Column Bioreactor for On-Site Treatment of Air Stripper Off Gas

Robert H. Douglass, John M. Armstrong, William M. Korreck*
The Traverse Group, Inc.

INTRODUCTION

The use of biological filters for control of odorous organic and some inorganic vapor emissions from industrial, agricultural, solid waste, and wastewater treatment facilities has been employed since the 1960s. Its success and practicality is well documented (Carlson & Leiser 1966). Air quality concerns regarding risk and human exposure to toxic pollutants have resulted in the increasing and broadening role of biological treatment methods for control and elimination of not only nuisance odors, but also airborne toxics from a variety of sources.

Waste gases from contaminated soil and groundwater treatment processes are sources of potentially toxic vapors. Conventional groundwater remediation involves purging aqueous contaminants from the aquifer and treating the purged water by various physical, chemical, or biological methods (i.e., activated carbon, bioreactor, air stripping, or incineration). The principal disadvantage of nonbiological methods is the creation or persistence of undesirable end products. Either contaminants are transferred to another medium, or the process results in potentially hazardous air emissions. Of the available groundwater treatment technologies, air stripping is one of the most cost-effective techniques, but it results in a waste gas stream that requires treatment

* The Traverse Group, Inc., 2525 Aero Park Drive, Traverse City MI 49684

and control. Traditionally, air stripper off gas is released without treatment or controlled by use of incineration or activated carbon, perpetuating the waste stream cycle. Biological treatment of air stripper off gas is an economical and practical method of controlling vapor emissions, because such processes offer the prospects of converting the contaminant to benign products.

Design of a vapor phase bioreactor is dependent on the characteristics of the waste stream. Chemical composition, contaminant solubility, contaminant concentration, and vapor stream flow rate control the applicability of biological treatment. Based on these criteria, an appropriate reactor packing material must be selected to provide the necessary conditions for biological oxidation of the contaminants.

Traditional bioreactors for control of odors from wastewater treatment systems have consisted of soil or peat/compost packing media (Pomeroy 1982). The importance of media selection is twofold. First, the material must have good pneumatic conductivity to minimize pressure drop and air movement power requirements. Second, the media should have high surface area, good wetting characteristics, and good sorptive capacity to enhance the capture and retention properties of the reactor. Because the microbes reside in or on the wet stationary phase (biolayer), the retention and distribution of a thin water and biolayer film over the media surface(s) is important to a large and well-distributed microbial population.

This study of the applicability of biofiltration to the treatment of contaminant streams typically found in air stripper off gas was conducted in two phases. The initial phase of this study evaluated the properties and effectiveness of eight readily available and inexpensive packing materials for application to a simple, portable bioreactor. The second phase focused on bioreactor performance when using soil, peat, and fiberglass wool packings, which were identified as the better media during the first phase.

METHODS AND MATERIALS

Experimental Setup

Six columns were constructed from 2.1-m lengths of 20.3-cm diameter polyvinyl chloride (PVC) pipe patterned after previous studies (Carlson & Leiser 1966). Each column was mounted vertically and filled first with a layer of washed pea gravel, then 1.5 m of test packing. The remaining column volume was left as headspace (Figure 1). The

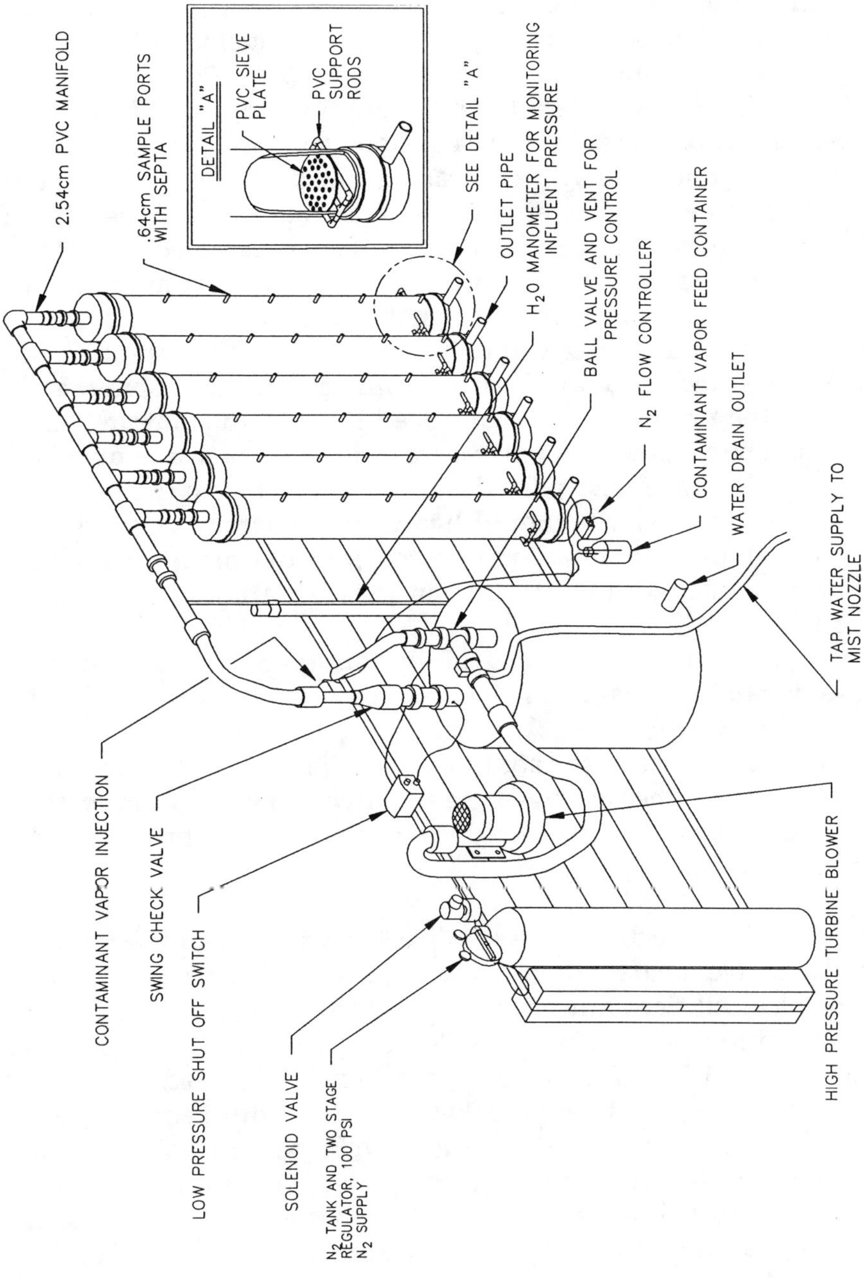

FIGURE 1. Bioreactor test system.

reactors were tapped with septum-capped sample ports located at 0.3-m intervals along the bioreactor tube for vapor sampling, water sampling, and drainage. A 2.54-cm vapor inlet line was installed at the top and bottom of each column.

The air supply system for inducing flow through the columns was designed to provide air with a constant contaminant concentration and 100 percent humidity. Air was supplied to the columns by a regenerative type of blower with a maximum discharge rate of 1.1 m^3/min. Because test run flow rates never exceeded 0.1 m^3/min per column, excess air was bled off through a vent. From the blower, air flowed through a humidification chamber to raise the moisture content of the air (Ottengraf & van den Oever 1983). The humidification chamber consisted of a 213-L drum with a fine mist spray nozzle installed in the top and a float-controlled drain in the bottom below a water seal. The air flowed countercurrent through the water mist in the air space of the drum and passed out the top to the columns. Before entering the columns, a stream of hydrocarbon vapors was fed to the airstream. Hydrocarbon vapor was fed into the main airstream by bubbling a regulated flow of nitrogen through a liquid hydrocarbon reservoir (Figure 1).

To evaluate the removal of hydrocarbon vapors within the columns, vapor samples were collected on tenex tubes and analyzed by GC/FID and GC/PID.

The first phase of the study was conducted during 4 months of system operation. The second phase required 3 months of operation.

The packings were tested to evaluate flow properties and substrate removal.

Phase I Packing. Each column was packed with one of the following materials: native sand with chemical nutrient amendments, sawdust with chemical nutrient amendments, peat buffered with calcium carbonate to control pH, a mixture of 20 percent manure and 80 percent sawdust, a mixture of 20 percent aerobically digested wastewater sludge and 80 percent vermiculite, or a mixture of 20 percent aerobically digested wastewater sludge and 80 percent sawdust. These packings were selected to provide a comparison of traditional materials (i.e., soil and peat) with other materials exhibiting favorable properties.

The packings were assured an active microbial community by seeding each first with fresh activated sludge liquor from the local municipal wastewater treatment plant (Ottengraf *et al.* 1986) and later with groundwater from a site known to have hydrocarbon-degrading microbes.

Phase II Packing. After completion of the initial testing, the six columns were repacked for further testing. The columns were packed in duplicate with washed 20/40 mesh Ottawa sand, buffered peat, and fiberglass wool (insulation). The columns were also reconfigured for downflow operation. Inoculation of the columns this time was accomplished by saturating to field capacity with contaminated groundwater known to contain hydrocarbon-degrading microbes.

Calculation of Compound Retention Times

As part of the second phase of the study, the degree of hydrocarbon vapor retention due to sorption and solubility was investigated. Contaminants are only available to the bacteria when dissolved in the water film coating the packing surface or adsorbed onto the packing surface. Contaminants in the air-filled pore space are not available for degradation, so a long retention alone does not assure good treatment. Performance of the bioreactor is influenced by the mass transfer from vapor to the biolayer.

A series of tests was run using methane, toluene, and *n*-heptane as the contaminant gases. Because of its low solubility and sorption, methane was used as a measure of minimum retention time in the column. Toluene, which is both soluble and easily adsorbed, was used to identify the composite effect on retention; *n*-heptane, which can be adsorbed but has very low solubility, was used to quantify the contribution of sorption to contaminant retention.

Using simple partitioning theory and a chromatographic analogy, theoretical retention times were calculated using equation (1) and compared to measured values.

$$R = R_{tm}\,(1 + K_d/\beta) \qquad (1)$$

where:
- R = retention time
- R_{tm} = minimum retention time
- K_d = vapor/liquid partition coefficient
- β = phase ratio

The minimum retention time, R_{tm}, in the column is that of the air flowing through the reactor with no solubility or sorption. However, if a compound is soluble and spends half its time in the water film, its retention in the column will be twice R_{tm}. The amount of a compound in the water and air at any given time can be predicted by its partition coefficient, K_d. For air and water at any given temperature and pressure, K_d is the ratio of the equilibrium concentration in the water divided by the concentration in the air. Because the bioreactors do not have equal amounts of water and air, a phase ratio β (the volume of water divided by the volume of air), is calculated to correct K_d for the reactor in question. The corrected partition coefficient, K, is calculated by dividing K_d by β.

The values for K_d were estimated by dividing the water solubility in mg/L at 25 C by the calculated saturated air concentration in mg/L (vapor pressure at 25 C divided by one atmosphere of pressure, multiplied by mg/M, and divided by 22.4 L/M). Using the K_d values for water to predict K will model the air/water partitioning, but will not account for any adsorption onto the solid packing materials. The packed columns are actually a three-phase system consisting of air/water/solid support. The value of K determined in the test runs can exceed the values calculated by water partitioning due to adsorption onto the solid support, which is then also part of the stationary phase.

For the three packings, β was calculated from the ratio of the air-filled porosity divided by the water-filled porosity. A value for K was determined for all three packings for both toluene and *n*-heptane. The results are summarized in Table 1.

TABLE 1. Calculated mass transfer parameters.

Packing Material	β	Toluene K_d	*n*-Heptane K_d	Toluene K	*n*-Heptane K
20/40 Ottawa Sand	1.7	3.33	0.010	1.96	0.0059
Peat/$CaCO_3$	2.0	3.33	0.010	1.67	0.0050
Fiberglass	7.2	3.33	0.010	0.46	0.0014

To determine the actual K value and retention time, the column effluent gas was monitored with a Bacharach TLV sniffer after 20 cc of contaminated air was injected into the headspace above the column packing. Retention times were determined by an HP 3390A integrator. It should be noted that peak widths at times were over a minute, which could result in errors in retention time values. From the retention time data, the actual K value can be compared to the value calculated by the air/water partitioning.

TEST RESULTS

Packing Properties

The following packing properties were determined prior to placing in the columns: total porosity, water-filled porosity, water content, wet bulk density, and specific surface area (Table 2).

TABLE 2. Packing material physical characteristics.

Packing Material	Wet Density (kg/m^3)	Water Content (kgH_2O/m^3)	Porosity (%)	Specific Surface Area (m^2/m^3)
Sawdust	718.8	425.9	80	2,160
Vermiculite/Sludge	746.1	634.0	90	2,900–3,300
Sawdust/Sludge	714.1	494.7	75	2,160
Peat/$CaCO_3$	760.5	581.0	90	5,000–10,000
Sawdust/Cow Manure	576.4	385.8	90	2,160
Native Sand	1,761.0	224.1	40	11,600
Ottawa Sand Col (1 and 2)	1,769.0	104.0	27	–
Peat/$CaCo_3$ Col (3 and 4)	612.0	296.0	90	5,000–10,000
Fiberglass Col (5 and 6)	183.0	120.0	98	20,000

Packing densities, water contents, and porosities for the second phase of the study were determined after the columns were packed by weighing the columns before and after loading to represent more accurately the expected conditions. Many of the estimates for density and porosity may be subject to error caused by settling and compressibility. The wet and dry densities for the buffered peat columns were 760.5 kg/m^3 and 179.3 kg/m^3, respectively, in the first phase and 589.2 kg/m^3 and 294.6 kg/m^3, respectively, in the second phase.

Estimates of specific surface area for each packing were made from different sources. The estimate for peat was based on approximate fiber size and density, so the surface area is reported as a range.

Fiberglass surface area data were obtained from manufacturer specifications, and the granular packings, sand, and sawdust were based on grain size analysis and the assumption that the grains were spherical.

Flow Properties

For the first phase of the study, the blower and manifold did not allow attainment of velocities (loadings) of greater than 0.31 m/min ($m^3/m^2/min$). A comparison of the influent pressure necessary to maintain a velocity of 0.3 m/min for the packings of both tests is presented in Table 3. As will be demonstrated later, a velocity of at least 1.55 m/min will be necessary for a practical bioreactor. Higher velocities and their corresponding pressure drops were achieved in the second phase of the study (Table 4). The fiberglass packed columns had the lowest pressure drop.

A higher pressure drop in the peat column during the second phase of testing probably resulted from higher density in the column. Comparison of the native sand column in the first phase and the frac sand column in the second phase show the importance of grain size distribution and angularity in flow performance. Fine particles in the native sand may have effectively plugged flow paths through the column.

Another factor evaluated during testing was flow configuration. The first phase was conducted with the columns operated in a downflow configuration. Pressure measurements at points across the column indicated water blocking near the bottom of several columns. Reversing the flow direction in the second phase improved the pressure profiles.

During the tests, bed temperature remained relatively constant, varying from 22 to 26 C as measured by effluent air temperature. Influent air temperatures were comparable to effluent, indicating little temperature increase within the columns.

Contaminant Retention

The first phase of the study investigated only the overall removal of contaminant vapors from the flow stream. Initially, a contaminant stream of unleaded gasoline vapor was fed to the columns, with influent and effluent concentrations measured. Later test runs were conducted with single-compound vapor streams of toluene and

TABLE 3. Pressure drop across 5 feet of packing.

Packing Material	Column Influent Pressure[(1)] (mm Hg)
Sawdust	0.022
Vermiculite/Sludge	0.168
Sawdust/Sludge	0.030
Peat/$CaCO^3$	0.075
Sawdust/Cow Manure	0.071
Native Sand	0.336
20/40 Ottawa Sand	0.168
Peat/$CaCO_3$	0.243
Fiberglass	0.084

(1) For a flow loading rate of 0.3 m/min.

TABLE 4. Column flow loss, Phase II packings.

Packing Material	Flow Loading (m/min)	Pressure Loss across Bed (cm Hg)
Peat/$CaCo_3$	0.2	1.7
	0.4	3.3
	1.1	7.2
Ottawa Sand	0.3	1.7
	0.6	3.3
	1.6	7.2
Fiberglass	0.6	1.7
	1.3	3.3
	2.6	7.2

benzene. Results of this work indicated that, at the velocities attainable (0.06 to 1.55 m/min), concentrations up to 60 mg/m^3, and after an acclimation period of approximately 2 weeks, the removal of aromatic compounds was very good, whereas many of the gasoline constituents were recalcitrant. The peat and sand columns provided the greatest removal efficiency (Table 5).

TABLE 5. Substrate removal.

Packing Material	Substrate	Flow Loading (m/min)	Feed Concentrations (mg/m^3)	Removal (%)
Sawdust	Gasoline	0.31	59.4	21.6
Vermiculite/Sludge	Gasoline	0.31	52.8	8.5
Sawdust/Sludge	Gasoline	0.31	63.2	5.2
Peat	Gasoline	0.31	55.5	30.6
Sawdust/Cow Manure	Gasoline	0.31	57.9	0
Sand	Gasoline	0.03	58.3	63.1
Sawdust	Toluene	0.3	271	11.8
Vermiculite/Sludge	Toluene	0.3	321	45.1
Sawdust/Sludge	Toluene	0.3	191	0
Peat	Toluene	0.3	304	92.7
Sawdust/Cow Manure	Toluene	0.3	237	13.9
Sand	Toluene	0.06	308	94.3
Peat	Benzene	2.1	76,400	70.1
Sand	Benzene	0.9	76,400	81.5

Once the packing was seeded with appropriate organisms, acclimation took from 1 to 4 weeks. This was verified by collecting samples of influent and effluent vapors until effluent concentrations had stabilized.

The retention calculations based on air/water partitioning do not fit the observed data very well for the peat column (Table 6). Only one data point for *n*-heptane was available. On all other *n*-heptane or toluene injections, the mass injected was too low to saturate all the adsorption sites. Clearly, the adsorption and retention resulting from the organic carbon of the peat causes a much longer compound retention than the simple air/water partitioning model would predict.

For the Ottawa sand, the retention time predicted by air/water partitioning was greater than the observed value (Table 7). The

TABLE 6. Packing retention times, peat/$CaCO_3$.

Flow (L/min)	Substrate	m/min	Methane R_t (min)	Observed R_t (min)	Calculated R_t (min)
53.8	Toluene	1.2	1.05	Adsorbed	2.80
20.9	Toluene	0.4	2.58	Adsorbed	6.69
10.8	Toluene	0.2	4.46	Adsorbed	11.9
36.8	*n*-Heptane	1.2	1.05	1.19	1.06
13.9	*n*-Heptane	0.4	2.58	Adsorbed	2.59
6.5	*n*-Heptane	0.2	4.46	Adsorbed	4.48

TABLE 7. Packing retention times, 20/40 Ottawa sand.

Flow (L/min)	Substrate	m/min	Methane R_t (min)	Observed Rt (min)	Calculated R_t (min)
53.8	Toluene	1.7	0.53	0.85	1.57
20.9	Toluene	0.65	1.15	1.80	3.40
10.8	Toluene	0.34	2.14	2.86	6.33
53.8	*n*-Heptane	1.7	0.53	0.51	0.53
20.9	*n*-Heptane	0.65	1.15	1.17	1.16
10.8	*n*-Heptane	0.34	2.14	2.15	2.15

residence time for a compound in the stationary phase can be calculated by $K(R_{tm})$ or determined experimentally by subtracting the methane retention time (R_{tm} estimate) from the observed retention time. For example, at a flow loading of 1.66 m/min in Ottawa sand, the expected residence time for toluene in the stationary phase was 1.04 min, but only 0.32 min was observed. It is possible that estimates for porosity or moisture content were in error or that mass transfer limitations prevented the predicted air/water partitioning from occurring. No retention of *n*-heptane was calculated, which matches the measured value of only 1.2 sec at 1.66 m/min very well. The Ottawa sand is a very inert silica sand, so most of the retention is expected to be simple air/water partitioning.

For the fiberglass insulation, both toluene and *n*-heptane exhibited retention times in the column that were much higher than air/water partitioning would predict. Although the glass fibers are very similar to the minerals in the Ottawa sand column, fiberglass insulation is sprayed with a plastic resin binding agent. For *n*-heptane, at a loading of 2.7 m/min, the column retention is 0.21 min longer than the methane retention time, a value 214 times higher than that predicted for the water film alone.

For soluble or slightly soluble organic solvents, an inert packing coated by a thin water film will result in an adequate retention time. For relatively insoluble compounds such as *n*-heptane, adsorption onto the solid support packing may be the only way to retain compounds in the column. The key question then becomes whether or not the adsorbed compounds are available to bacteria as a substrate. Water solubility is a key factor in compound retention, as illustrated by the test results from the fiberglass column at an air loading of 2.7 m/min (Table 8). The retention of *n*-heptane is comparable to methane, suggesting little sorption. Toluene, by virtue of its comparatively high solubility, had a retention time nearly twice that of *n*-heptane.

TABLE 8. Packing retention times, fiberglass insulation.

Flow (L/min)	Substrate	m/min	Methane R_t (min)	Observed R_t (min)	Calculated R_t (min)
84.9	Toluene	2.7	0.70	1.27	1.02
42.5	Toluene	1.3	1.24	2.24	1.81
19.8	Toluene	0.6	2.84	5.29	4.15
84.9	*n*-Heptane	2.7	0.70	0.91	0.70
42.5	*n*-Heptane	1.3	1.24	1.62	1.24
19.8	*n*-Heptane	0.6	2.84	3.54	2.84

The assumption that methane is not dissolved or adsorbed can affect the results. Methane is actually more soluble in water than *n*-heptane, which could explain why *n*-heptane had a shorter retention time than methane in some of the runs.

Contaminant Transport and Removal

The column tests indicated that removal of the target aromatic compounds was not diffusion limited at the loadings and concentrations tested. However, in the test using gasoline vapors, diffusion limitations appear to control removal. Although some sorption occurred, the packing material capacity was reached during the acclimation period.

The degradation rates of gasoline vapor, toluene, and benzene were studied in the first phase using the six original packings. Initial tests were conducted with gasoline and toluene. Except for sand and peat columns, removal of these substrates was poor (Table 5). Because benzene is the aromatic compound of greatest regulatory concern, a third test run using benzene was conducted to evaluate its removal in the sand and peat columns. The degradation of benzene, as expected, was similar to that of toluene, even though much higher concentrations of benzene were used (Table 5). A relatively high influent of 76.4 mg/L benzene was fed to both columns; the peat achieving about 70 percent removal and the sand about 80 percent. Accounting for flow rates, the elimination rate per volume of packing and time was 74.5 $mg/min\text{-}m^3$ for the peat and 35.6 $mg/min\text{-}m^3$ for the sand. These rates are essentially the same as those expected for toluene degradation. Tests using *ortho*-xylene as the substrate were also conducted. Removal in the sand and peat columns was again best, at 96 and 87 percent, respectively.

In the second phase of the study, the duplicate columns of 20/40 Ottawa sand, peat/$CaCO_3$, and fiberglass insulation were prepared, treated with nutrient solution and inoculum, then fed unleaded gasoline vapors for 1 month. Samples were collected and analyzed for benzene, toluene, and total gasoline. Results are presented in Table 9.

TABLE 9. Comparative compound removal, Phase II testing.

Packing Material	Column #	Toluene % Removal	Toluene (μg/min-m³)	Benzene % Removal	Benzene (μg/min-m³)	Gasoline % Removal	Gasoline (μg/min-m³)
Ottawa Sand	1	98	1660	95	1235	40	11,225
Ottawa Sand	2	>99	3180	95	1305	54	15,850
Peat/CaCO³	3	85	885	95	777	21	3,600
Peat/CaCO³	4	96	705	97	845	61	11,545
Fiberglass	5	90	3920	70	2365	20	14,365
Fiberglass	6	83	3810	69	2470	-	-

Notes: Total gasoline feed concentration = 61.4 μg/L; toluene concentration = 3.6 μg/L; benzene concentration = 2.8 μg/L.

Although the constituent concentrations of benzene and toluene fed to the columns during the second phase of the study were low, treatment was excellent for all columns, ranging from 69–99 percent removal. The results for total gasoline removal ranged from 20 to 61 percent.

Also in the second phase of the study, one column of each of the three packings was chosen to test the effect of airflow loading on removal of toluene. The experimental plan was to keep the mass loading rate steady while varying the airflow rate. Unfortunately, the mass flow loading did not remain constant, making interpretation of the data more difficult. At the low-contaminant vapor concentrations tested, ranging from 54 to 124 µg/L, a three- to sevenfold increase in flow rate does not appear to affect treatment efficiency greatly. Table 10 presents results from this test.

TABLE 10. Effect of flow loading on removal rate.

Packing Material	Flow (L/min)	Mass Loading Toluene (mg/min)	Toluene Conc. (µg/L)	% Removal	Removal Rate (µg/min-m^3)
Ottawa Sand	59.4	3.22	54.1	99.2	64.6
Ottawa Sand	21.2	1.94	91.4	99.9	39.2
Ottawa Sand	7.6	0.956	124	99.9	19.4
Fiberglass	99.1	5.36	554.1	>1.5	77.3
Fiberglass	53.8	4.92	91.4	>7.9	77.3
Fiberglass	28.3	3.54	83.1	83.1	59.3
Peat	39.6	2.14	54.1	99.9	43.1
Peat	15.0	1.3>	91.4	99.9	27.5
Peat	6.5	0.814	125	99.9	16.4

Since the rate of substrate removal was related primarily to mass loadings and not airflow, the design choice for airflow will probably be made from pressure drop design criteria rather than biological rate parameters.

Although the second phase of the study yielded less degradation rate data, the patterns for treatment efficiency versus contaminant loading were similar. There appeared to be a maximum biological reaction rate that was approached as mass loading of contaminants increased. When mass loading exceeds this level, the removal declined.

Because the biological rate of degradation appears to be the limiting factor for treatment efficiency, the chemical composition of the contaminants can be very important since rates of biodegradation vary by substrate.

For the three packings studied in the second phase, maximum degradation rates for benzene and toluene were about 53 to 78 mg/min-m^3 versus about 18 to 28 mg/min-m^3 for *o*-xylene.

DISCUSSION

The challenge in treating the air stream from an air stripping tower is handling the high airflows with generally low-contaminant vapor concentrations. Typically, air strippers are operated at air-to-water ratios of approximately 50:1, depending on the contaminant stream. Therefore, for every cubic meter of water treated, approximately 50 m^3 of contaminated air is created. Packing selection and reactor sizing are critical for providing the contact time necessary to attain sufficient mass transfer of contaminants from the vapor phase to the biolayer, where they become available to the microorganisms for oxidation. This availability is a combination of contact and retention time, solubility, diffusion, and packing sorptive capacity. The results validate the use of biotreatment for air stripper off treatment by demonstrating the removal of those contaminants most commonly found in groundwater at fuel spill sites. However, as shown by the poor removal of total gasoline vapor, treatment of a vapor stream from other sources (i.e., soil venting) may not be feasible.

Ideally a packing should have high porosity, high specific surface area, low bulk density, and good moisture retention (Fischer & Bardtke 1984). These properties are the keys to a packing's performance. High porosity assures low pressure drop (assuming good permeability), high surface area assures a well-distributed water film and microbial community, low bulk density minimizes bioreactor construction and transportation constraints, and high moisture retention assures optimal conditions for microbial growth.

One of the major considerations in choosing packing material is pressure drop across the treatment bed. Fibrous materials with 30 to 50 percent air-filled porosity give relatively low pressure drops across a treatment bed, but are somewhat prone to settling or uneven packing distribution. A granular material such as Ottawa sand has an air-filled porosity of less than 20 percent, but because it packs evenly and has relatively large pore channels it has a very low pressure drop across the treatment bed. Perhaps the high-quality effluent of peat/$CaCO_3$ could be combined with the stability and even packing of the Ottawa sand by using a granular packing in the 10 to 40 mesh range that has an adsorptive capacity for the organics being treated. Such a material has yet to be identified.

Pneumatically, downflow operation of the reactor improves the pressure profile across the bed by eliminating water hold-up and countercurrent flow.

If a compound has very low solubility, its retention time and exposure to the bacteria then becomes dependent on its sorption to the packing. If sorption is high, it may be retained long enough to be degraded. If sorption is low, it may not reside in the reactor long enough to be exposed to the bacteria.

Since the calculated values for K were so low for *n*-heptane in water, it seems that any significant retention in the stationary phase is due to adsorption on the solid support material rather than the water film of the packing. On several of the runs with the peat column, no value could be determined because of irreversible adsorption onto the packing (Table 6).

Removal of the substrate in a bioreactor is both diffusion and reaction limited depending on substrate concentration and location in the reactor (Ottengraf & van den Oever 1983). A high substrate removal rate did not necessarily ensure a high percentage of removal or a low concentration in the effluent. The fiberglass packing had a very high capacity for removal, but did not exceed 80 percent removal efficiency because of the higher mass loading rate compared to the other columns. This could result from channelling or problems with uneven packing distribution. The Ottawa sand performed better perhaps because the nearly spherical particles packed evenly. The peat/$CaCO^3$ packing is fibrous and could also have problems with uneven distribution or channelling; however, its effluent quality was very good, with removal rates exceeding 99 percent. This may result from the strong adsorption of organic contaminants onto the packing.

Flow loadings of 1.66 to 3.32 m/min did not seem to lower removal rates significantly. Removal of toluene in the fiberglass column did improve slightly from 72 to 83 percent when the flow rate was lowered 3.5-fold, at roughly the same mass loading.

Packing cost is an important consideration. Peat/$CaCO_3$ costs \$35 to \$70/m^3, depending on quantity. At the packing density used in the second phase of the study, fiberglass insulation will cost from \$105 to \$140/m^3. The Ottawa sand is relatively expensive at \$245 to \$280/m^3.

Design of a bioreactor based on the described criteria and the experimentally obtained parameters has been presented in the literature (Eitner & Gethke 1987; Ottengraf 1987; Prokop & Bohn 1985). For design and application to a mobile treatment system, peat buffered with $CaCo_3$ is the lightest, least expensive, and among the best biological support media. Ideally, the reactor should be transportable by trailer, which limits its size to approximately 3 m wide, 10 m long, and 2 m deep. Using a conservative removal capacity of 40 mg/min-m^3, the maximum mass loading on the bed will be 2,400 mg/min. Achieving a desired bed velocity of 1.55 to 3.1 m/min will yield a maximum flow rate of 93 m^3/min. The inlet vapor concentration will be 26 mg/m^3. A vapor stream with this concentration coming from an air stripper operating at a 50:1 air/water ratio (and assuming complete transfer) would result from a water flow rate of 1.9 m^3 with a concentration of 1,300 mg/m^3. A reactor of this size could easily treat air stripper off gas from most groundwater remediation systems.

REFERENCES

Carlson, D. A.; Leiser, C. P. *Journal Water Pollution Control Federation* **1966,** May, 829–840.

Eitner, D.; Gethke, H. G. Presented at the 80th Annual Meeting of APCA, New York, NY, June 21–26, 1987.

Fisher, K.; Bardtke, D. Presented at the International Symposium on Characterization and Control of Odoriferous Pollutants in Process Industries, Society Belge De Filtration, Louvain-La-Neuve, Belgium, April 25–27, 1984.

Ottengraf, S.P.P. *TIBTECH* **1987,** May 5, 132–136.

Ottengraf, S.P.P.; van den Oever, A.H.C. *Biotechnology and Bioengineering;* Wiley: New York, 1983; pp 3089–3102.

Ottengraf, S.P.P.; Meesters, J.J.P.; van den Oever, A.H.C.; Rozema, H. R. *Bioprocess Engineering I* **1986,** 61–69.

Pomeroy, R. D. *Journal Water Pollution Control Federation* **1982,** *54,* 1541–1545.

Prokop, W. H.; Bohn, H. L. *Journal of the Air Pollution Control Association* **1985,** *35,* 1332–1338.

Anaerobic Treatment of *o*-Xylene

*Kun Mo Lee**
Ajou University
Ihor R. Melnyk
Burns and McDonalds
D. F. Bishop
U.S. Environmental Protection Agency

INTRODUCTION

Recently, the presence of volatile organic compounds (VOCs) in receiving waters has been of great concern because of their potential toxicity and carcinogenicity. Some VOCs can accumulate in the food chain through the biomagnification process.

Most toxic VOCs enter receiving water after being discharged into sewers where they are mixed with domestic wastewater and treated by publicly owned treatment works (POTWs). Although some toxic VOCs can be removed effectively by the activated sludge process (Kirsch *et al.* 1984; Kirsch & Wukasch 1985), studies have indicated that the principal removal mechanism for most VOCs is air stripping from the aeration basin (Petrasek *et al.* 1983).

Hannah *et al.* (1986) and Weber and Jones (1985) suggested that volatilization from the aeration basin can be a significant removal mechanism of VOCs in activated sludge processes. Since volatilization is simply a transfer from the aqueous to the vapor phase, this should not be regarded as removal. Thus, there is a strong need to treat VOCs in a closed environment such as in anaerobic digesters.

* Ajou University, Department of Environmental Engineering, Suwon, Korea 441-749

o-Xylene is a common VOC found in municipal wastewater. The concentration of *o*-xylene in wastewater cannot exceed its solubility, which is 175 mg/L at 20 C. When waste streams containing *o*-xylene are not discharged into sewers, but collected for later disposal, the *o*-xylene concentration in the waste stream can be significantly higher than its solubility in water.

The use of existing anaerobic digesters in POTWs as reactors for treatment of waste toxic organic compounds, especially VOCs, can be a viable alternative for toxics treatment. The toxic waste could be collected and transported to POTWs and fed into anaerobic digesters for treatment. The closed environment of an anaerobic digester would prevent the VOCs from escaping into the atmosphere. Furthermore, anaerobic metabolism appears to achieve better removal of toxic organics than aerobic metabolism does.

Therefore, this study was designed to investigate the feasibility of removing *o*-xylene in mesophillic anaerobic digesters under both the batch and semicontinuous modes of operation. The specific objectives include the determination of the *o*-xylene concentration that does not affect the digester operation adversely and removal efficiencies of *o*-xylene.

Experimental Approach

Three different experiments were performed in this study. Firstly, a preliminary serum bottle test using unacclimated seed was conducted with *o*-xylene concentrations from 0 to 5,000 mg/L to estimate the toxic concentrations to the anaerobic digestion process. The total gas production was monitored and used to estimate the inhibition coefficient (K_i) for *o*-xylene. Secondly, bench-scale mesophillic anaerobic digesters were operated with o-xylene concentrations from 0 to 750 mg/L to verify the findings from the serum bottle test. Finally, a second serum bottle test was performed to compare performance of acclimated versus unacclimated sludges when spiked with *o*-xylene.

EXPERIMENTAL PROCEDURE

Sludge Source. A mixture of waste-activated sludge and primary sludge with a 2:1 volume ratio from the gravity thickeners at the U.S. EPA Test and Evaluation (T&E) Facility was used. The T&E Facility is located inside the Mill Creek wastewater treatment plant in

Cincinnati, Ohio. The average total solids (TS) and volatile solids (VS) contents of the sludge were 3.2 and 2.1 percent, respectively. The average pH was 6.4 and the average *o*-xylene concentration was 4.1 mg/L.

Bench-Scale Anaerobic Digester and Digester Operation. Four conventional single-stage anaerobic digesters were fabricated; each was a Plexiglas® cylinder, 20 cm in diameter and 61 cm in height. The reactor volume was 20 L with 15 L of active culture. A ½-HP motor with a rotational speed of 80 rpm provided complete mixing to each digester. All four digesters were operated at a solids residence time (SRT) of 15 days at 35 C.

A draw-and-fill mode of operation eliminated contamination of withdrawn sludges with feed sludge. Digester operation was divided into five phases (Table 1), and sludge withdrawal and feeding was done once a day. A control without *o*-xylene spiking was in operation for all phases. *o*-Xylene was spiked into the top of the digester through a septum via a 1-mL syringe.

Serum Bottle Tests. Serum bottles with capacities of 125 mL were set up in triplicate. Placed in each bottle was 30 mL of a 4:1 ratio of fresh feed sludge to acclimated digested sludge. Nitrogen gas was injected just above the mixture to purge the air from the headspace of the bottle. The bottle was sealed using Teflon®-coated, silicon septum and plastic screw seal.

The bottles were kept in a water bath at 35 C for 45 min, and excess gas production from heating was vented using a syringe. The bottles were then spiked with *o*-xylene using a microliter syringe, and time zero was established. During incubation, gas production volume was measured daily using a syringe. Methane content of the gas also was measured periodically.

Sampling and Analytical Methods. Digested sludges withdrawn daily from four anaerobic digesters were analyzed for pH, alkalinity, TS, VS, and chemical oxygen demand (COD), all according to *Standard Methods* (1985). Measurement of volatile fatty acids (VFA) followed Jenkins *et al.* (1983). Gas production volume was measured using a wet test meter, and composition of the digester gas was determined using a Porapak column with a thermal conductivity detector (TCD) in a Varian 3700 gas chromatograph (GC).

o-Xylene determination was made using a Chromosorb-W, SE 30 percent column with a flame ionization detector (FID) in the same

TABLE 1. Digester operating conditions.

Reactor ID	Phase I	Phase II	Phase III	Phase IV	Phase V
A	Control	1,000 mg/L	Recovery	Acclimation[(a)] 250 mg/L	Spiking[(b)] 250 mg/L
AR	Control w/ gas recycle	1,000 mg/L w/gas recycle	Recovery	Acclimation[(a)] 500 mg/L	Spiking[(b)] 500 mg/L
B	Control	Control	Control	Acclimation[(a)] 750 mg/L	Spiking[(b)] 750 mg/L
BR	Control w/ gas recycle	Control w/ gas recycle	Control	Control	Control
Operating Period (days)	1–32 (32)	33–60 (28)	61–109 (49)	110–124 (15)	125–161 (37)

(a) Spike *o*-xylene gradually over 1 SRT (15 days) period to attain 250 mg/L at the end of 1 SRT.
(b) A given *o*-xylene concentration is maintained all the time.

GC. For gas-phase *o*-xylene, 0.5 mL of the digester gas collected in a Tedlar® gas sampling bag was injected into a GC.

The sludge-phase *o*-xylene concentration was determined by shaking 20-mL sludge samples thoroughly in a 250-mL amber glass bottle fitted with a Mininert® valve. The bottle was then heated in a water bath at 90 C for 10 min. Headspace gas of 2 mL was injected into a GC. Standards were prepared by injecting the known amount of *o*-xylene into the 250-mL bottle and followed the same procedure. The aqueous-phase *o*-xylene concentration was determined using the same method as the sludge-phase analysis, except the sample was centrifuged at 5,000 rpm for 10 min.

RESULTS AND DISCUSSION

Preliminary Serum Bottle Test

Figure 1 shows the cumulative gas production at 100 to 5,000 mg/L *o*-xylene concentrations and control. Gas production decreased as *o*-xylene concentration increased. The bottles spiked at 100 mg/L produced a similar amount of gas as the control. However, 40 percent reduction in total gas production occurred at 1,000 mg/L *o*-xylene concentration compared with the control.

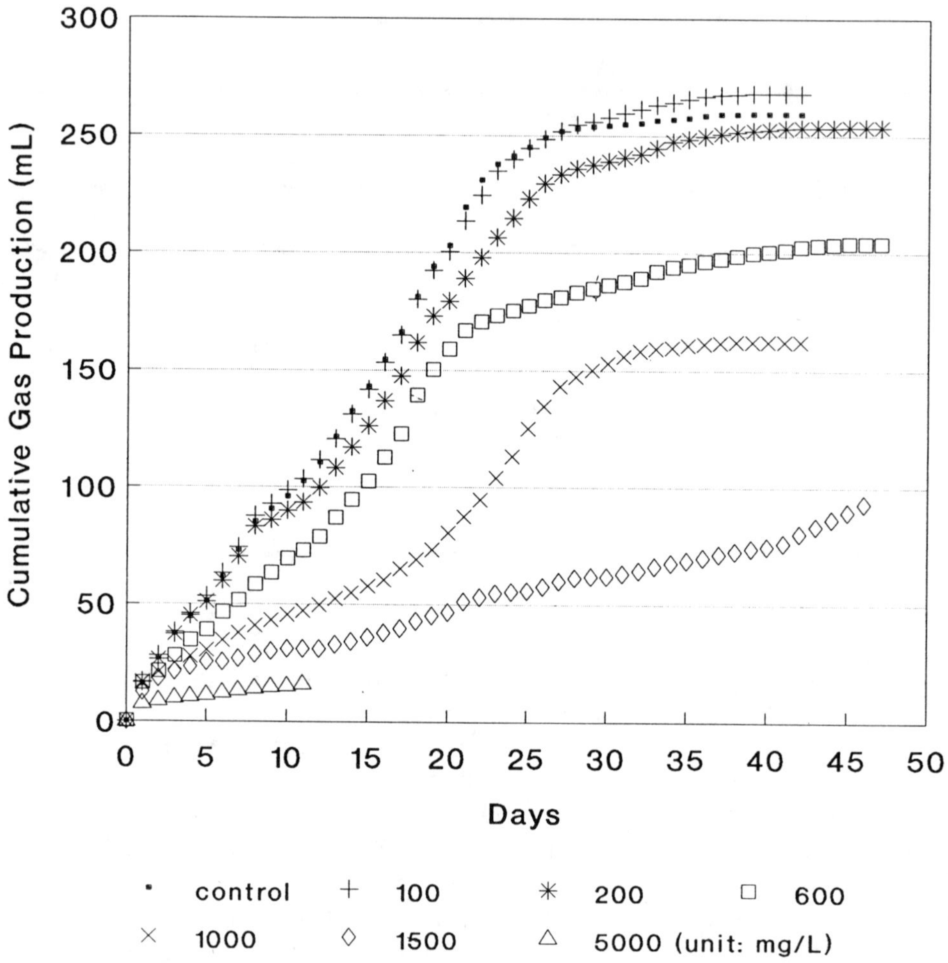

FIGURE 1. Cumulative gas production: preliminary serum bottle test.

Gas production data were used to calculate the half-kill dose or inhibition coefficient (K_i) as described by Vargas and Ahlert (1987). The calculated K_i was 1,014 mg/L at correlation coefficient (R^2) of 0.90. Using the limits set by Blum *et al.* (1986), a 50 percent reduction in total gas production as toxic, the *o*-xylene dose at 1,000 mg/L would exert a toxic effect on anaerobic digestion. This corresponds with the inhibition coefficient calculated above.

Maximum rate ratio (MRR) values were calculated for the time period from day 1 to day 21 according to Owen *et al.* (1979). An *o*-xylene dose of greater than 200 mg/L indicates probable inhibition. Overall, preliminary serum bottle test results indicate that the degree of inhibition increases with increasing doses of *o*-xylene above 100 mg/L with a toxic concentration over 1,000 mg/L.

Second Serum Bottle Test

Because the preliminary serum bottle test results were from the unacclimated sludges, the second serum bottle test was conducted to compare performance of acclimated versus unacclimated sludges when spiked with *o*-xylene. Sludges acclimated with 250 mg/L of *o*-xylene spiking produced a slightly greater amount of total gas than the controls (Figure 2). In addition, they did so at a much quicker rate. The unacclimated sludge spiked at 250 mg/L and the acclimated sludge at 500 mg/L performed almost identically. The total gas production by these bottles was within 0.1 mL of that of the control. However, the unacclimated sludges spiked at 500 mg/L produced gas rather slowly, and the total gas produced was 81 percent of that produced by the controls.

These results indicate that acclimation can overcome the toxic effect of *o*-xylene up to 500 mg/L. Furthermore, the fact that acclimated sludges spiked at 250 mg/L produced more gas than the controls indicates that the spiked portions of *o*-xylene may be utilized as substrate for the anaerobes or biodegraded in the anaerobic digester, provided the anaerobes are acclimated to *o*-xylene.

Digester Operational Results

Operation of four bench-scale digesters was divided into five phases (Table 1). Digester off-gas recycling was attempted to achieve better mixing as well as to increase contact time between *o*-xylene and anaerobic organisms. The gas recycling rate was held constant at 10 L/min; however, gas recycling was abandoned after Phase II because there was no apparent benefit of recycling on digester operation or on *o*-xylene degradation.

Phase I. The purpose of Phase I operation was to collect baseline data for the anaerobic digesters. The pHs of all four digesters were stable near 6.9. Effluent TS and VS were also held fairly constant at 1.5 to 1.6 percent and 0.7 to 0.8 percent, respectively. Percent VS destruction

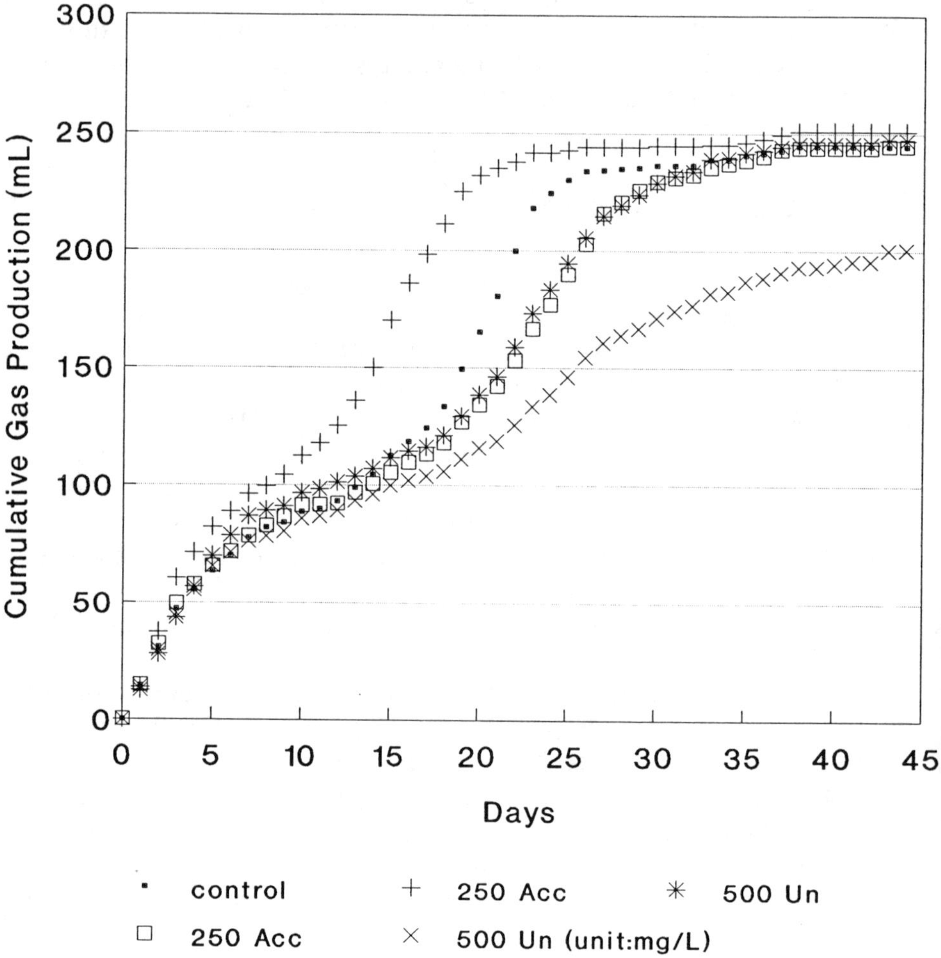

FIGURE 2. Cumulative gas production: secondary serum bottle test.

was stable ranging from 45 to 52 percent, and no differences among digesters were noted at the 95 percent confidence level. Methane content also was similar, the average ranging from 54 to 62 percent.

The above results showed that all four digesters were operating similarly. Furthermore, gas recycling did not improve digester performance.

Phase II. Digesters A and AR were given an initial dose of 15,000 mg/L of *o*-xylene to attain 1,000 mg/L concentration and a daily dose of 1,000 mg/L thereafter. The other two digesters served as controls.

After 10 days, the spiked digesters had pH values below 6.4, which is not acceptable for proper operation. Effluent TS and VS concentrations of the spiked digesters were higher than those of the controls, and daily gas production dropped almost immediately (within 1 day) after spiking. The percent VS destruction dropped from the control value of 40 to 50 percent to 10 to 15 percent by the end of Phase II. Methane content of the digester gas also dropped from the control value of 45 to 70 percent to 30 to 40 percent.

All of these results clearly showed that the spiked digesters failed completely, suggesting that a 1,000 mg/L concentration is toxic to anaerobic digester. Furthermore, unacclimated sludges had no chance of breaking *o*-xylene.

One important observation was that the digesters performed no better with gas recycling than without. In some cases gas recycling yielded detrimental effects on the digesters. These may be deduced from the gas production data. Cumulative gas production in this period for the digesters A, AR, B, and BR were 53.9, 38.9, 394.7, and 369.3 L, respectively. Other parameters, although not so obvious, showed similar trends. Therefore, gas recycling was abandoned in Phase III operation.

Phases III and IV. By the end of the recovery period, Phase III, two digesters, A and AR, appeared to return to normal as verified by the pH, gas production, methane content, and other parameters comparable with those of the controls. Phase III lasted over three SRTs; thus, simple hydraulics would have flushed the digesters of any *o*-xylene.

After Phase III, three digesters were acclimated with *o*-xylene to give final concentrations of 250, 500, and 750 mg/L at the end of one SRT. All digester performance did not differ significantly from that of the control. This is in contrast with the shock spiking of Phase II where spiked digesters failed.

Phase V. The three digesters received a constant amount of *o*-xylene daily, and initially the digesters appeared to operate identically. The percent VS destruction, however, became significantly different by the third week for the digester with 750 mg/L *o*-xylene spiking (Figure 3). The pH of this digester dropped below critical levels (as low as 6.2)

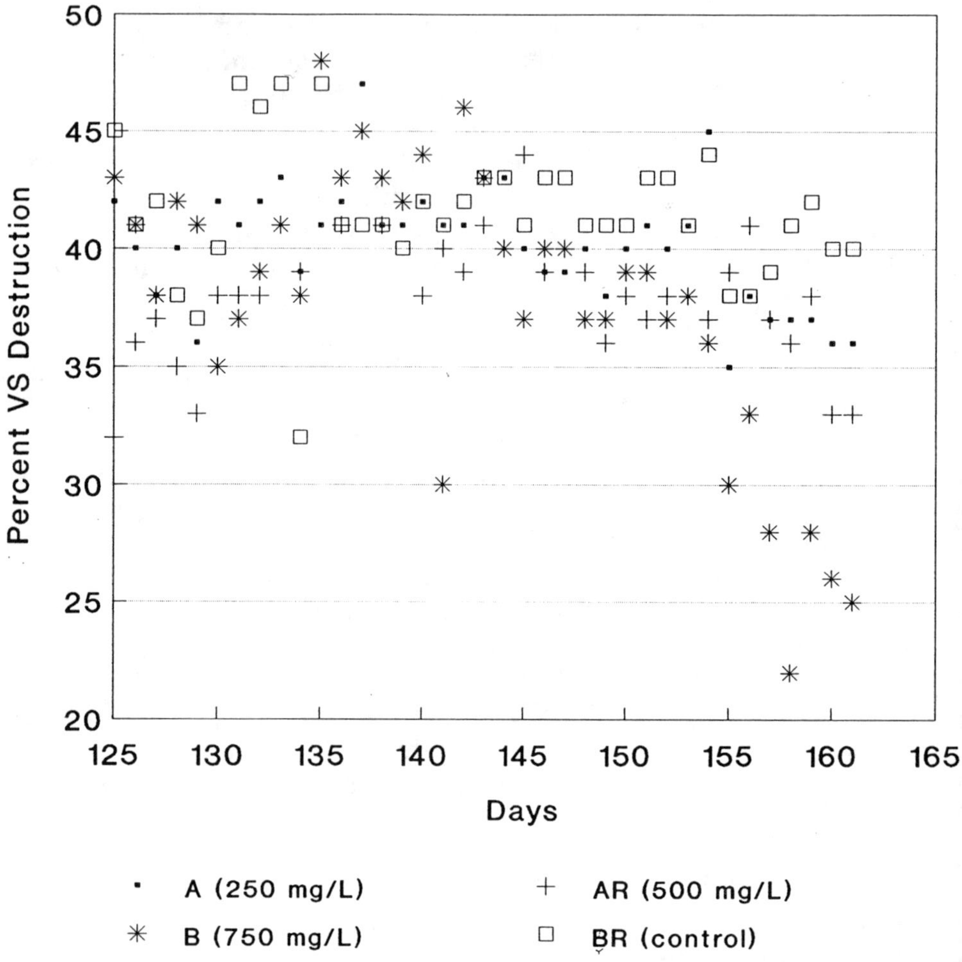

FIGURE 3. VS destruction efficiency: Phase V.

during the fourth and fifth weeks; methane concentration also dropped to approximately 45 percent, as compared with 60 percent for the 250 and 500 mg/L spiked digesters and 70 percent for the control.

Cumulative gas production for the four digesters (Figure 4) indicates possible failure of the 750 mg/L spiked digester. The greater gas production of the 250 mg/L spiked digester compared with the control suggests possible utilization of *o*-xylene as a substrate.

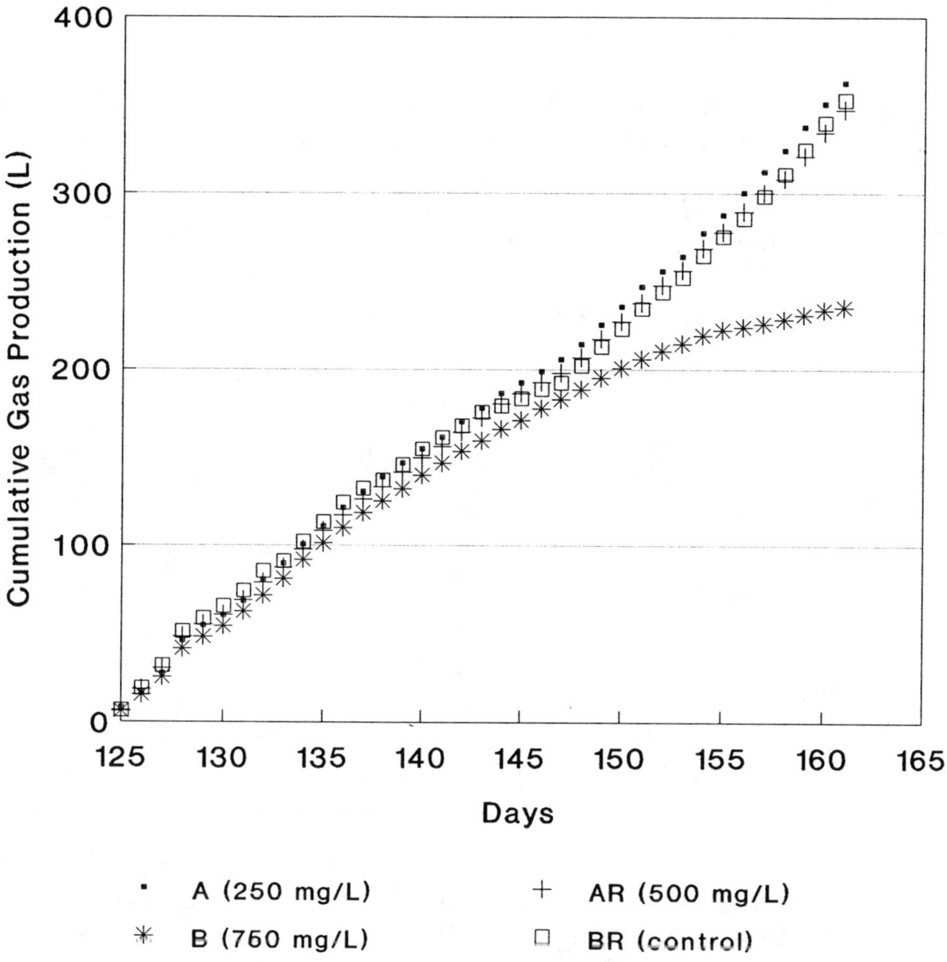

FIGURE 4. Cumulative gas production: Phase V.

No significant difference in performance among the digesters A, AR, and the control was observed. Thus, digesters acclimated to *o*-xylene may be able to tolerate the toxic effect of *o*-xylene up to 500 mg/L. Digesters, though acclimated, spiked with more than 750 mg/L *o*-xylene would fail, and digesters spiked with less than or equal to 250 mg/L would biodegrade portions of *o*-xylene.

o-Xylene Removal

o-Xylene removals were calculated based on the difference between total mass spiked plus mass fed into the feed sludge and the total mass in the withdrawn sludge as well as in the digester gas (Table 2). Table 3 shows the weekly average of total *o*-xylene mass in and out from each digester during Phases II and V, respectively.

TABLE 2. *o*-Xylene concentration statistics.

		Feed Sludge mg/L	A		AR		B		BR	
			Gas mg/m^3	Sludge mg/L	Gas mg/m^3	Sludge mg/L	Gas mg/m^3	Sludge mg/L	Gas mg/m^3	Sludge mg/L
Phase II:										
	AVE	4.1	3,455	763	2,835	804				
	MIN	0.0	563	444	691	149				
	MAX	15.8	7,336	1,061	6,054	1,086				
	SD	5.4	2,124	213	1,877	262				
	C.I	3.0	1,350	129	1,193	158				
Phase V:										
	AVE		4,074	149	4,689	201	6,344	303	11	3
	MIN		2,192	87	2,486	97	2,831	93	0	0
	MAX		7,624	284	6,668	623	10,504	823	90	13
	SD		1,652	63	1,480	155	2,387	229	26	4
	C.I		1,269	53	1,138	120	1,996	176	16	2

Mass balance data in Table 3 indicate that *o*-xylene removal in the control reactor during Phase II operation was close to 100 percent because the feed sludge contained very low levels of *o*-xylene. For Phase V operation where BR served as a control, feed sludge contained a significant amount of *o*-xylene.

Digesters A and AR, which were shock spiked at 1,000 mg/L *o*-xylene concentration, showed significantly reduced *o*-xylene removal efficiencies. The removals for these digesters dropped quickly from 74 and 68 percent during the first week to 11 and 14 percent for the fourth week, respectively. As was discussed earlier, all of the operational parameters showed a failure of digester operation. This means shock spiking at 1,000 mg/L exerted significant toxic effects to these digesters that caused complete failure.

Up to 62 percent removals of *o*-xylene were attained in the acclimated digesters, and increasing the *o*-xylene spiking increased the removal efficiencies. Although this could result simply from adsorption of *o*-xylene with the solids, it is unlikely since the removal of *o*-xylene

TABLE 3. *o*-Xylene mass balance during Phases II and V operation.

		A		AR		B		BR	
		Total in	Total out	Total in	Total out	Total in	Total out	Total in	Total out
	Day	(mg)		(mg)		(mg)		(mg)	
Phase II:									
	33–39	21,000	5,402	21,000	6,701				
	40–46	7,000	4,797	7,000	4,450				
	47–53	7,002	6,351	7,002	6,138				
	54–60	7,004	6,208	7,004	6,019				
Phase V:									
	127–133	4,637	2,265	9,137	2,278	13,637	3,466	137	102
	134–140	3,692	2,886	7,192	2,409	10,692	5,128	192	152
	141–147	3,682	2,321	7,182	2,768	10,682	3,798	182	160
	148–154	3,673	1,909	7,123	2,691	10,673	2,393	173	107
	155–161	3,672	1,919	7,172	3,482	10,672	2,257	172	34

during Phase II has decreased as the digesters failed. Table 3, therefore, suggests that acclimated sludges with *o*-xylene could biodegrade *o*-xylene in the anaerobic digester.

The only inexplicable result is the continued high removals by digester B (spiked at 750 mg/L), which had appeared to fail by several operating parameters, notably a significant reduction in gas production. This may be such a case as that mentioned by Yang *et al.* (1980) in which an absence of gas production does not necessarily indicate a lack of biological degradation in anaerobic digester. However, digester operating parameters are major criteria in judging the digester performance; thus digester B should be considered a failure and spiking at 750 mg/L should be considered toxic to anaerobic digestion.

SUMMARY AND CONCLUSIONS

Preliminary Serum Bottle Test. Total gas production decreased with increasing *o*-xylene concentrations at 200 mg/L and above, and MRR values indicated probable inhibition at and above 200 mg/L. The inhibition coefficient, K_i, for *o*-xylene was 1,000 mg/L.

Second Serum Bottle Test. The acclimated sludge spiked at 250 mg/L produced more gas than the control and the other spiked sludges. This indicates the biodegradation of *o*-xylene in the anaerobic digesters if anaerobes are acclimated to *o*-xylene.

Bench-Scale Digester Test. Acclimated digesters were operated without difficulty at 250 and 500 mg/L of *o*-xylene, whereas the digester spiked at 750 mg/L failed after 2 SRTs. *o*-Xylene removals up to 41 and 62 percent were achieved for the acclimated digesters spiked at 250 and 500 mg/L, respectively.

REFERENCES

Blum, D.J.W.; Hergenroeder, R.; Parkin, G. F.; Speece, E. *Jour. Water Poll. Control* **1986,** *58,* 122.

Hannah, S. A.; Austern, B. M.; Eralp, A. E.; Wise, H. W. *Jour. Water Poll. Control* **1986,** *58,* 27.

Jenkins, S. R.; Morgan, J. M.; Sawyer, C. L. *Jour. Water Poll. Control* **1983,** *55,* 448.

Kirsch, E. J.; *et al.* WERL, U.S. EPA, 1984, EPA CR 805858.

Kirsch, E. J.; Wukasch, R. F. WERL, U.S. EPA, 1985, EPA CR 8076380-10.

Owen, W. F.; Stuckey, D. C.; Healy, J. B.; Young, L. Y.; McCarty, P. L. *Water Research* **1979,** *13,* 485.

Petrasek, A. C.; *et al.* *Jour. Water Poll. Control* **1983,** *55,* 1286.

Standard Methods for the Examination of Water and Wastewater, 16th Ed. AWWA, WPCF, 1985.

Vargas, C.; Ahlert, R. C. *Jour. Water Poll. Control* **1987,** *59,* 964.

Weber, W. J.; Jones, B. E. WERL, U.S. EPA, 1985, EPA CR 806030.

Yang, J.; *et al.* *Prog. Water Technol.* **1980,** *12,* 977.

Bioremediation of Seleniferous Soils

W. T. Frankenberger, U. Karlson*
University of California, Riverside

INTRODUCTION

On the west side of the San Joaquin Valley, California, drainage waters from an estimated 25,000 ha of farmland have elevated Se levels, and in many cases the element has been accumulating in regional and on-farm evaporation ponds (Deverel & Millar 1986; Letey *et al.* 1986). Although problems with agronomic crops or livestock have not been encountered, aquatic wildlife has been affected seriously. At Kesterson Reservoir, California, bioaccumulation has led to toxic levels of Se at the high end of the food chain, causing widespread death and deformities of indigenous and migratory bird species (Letey *et al.* 1986; Ohlendorf *et al.* 1986). After draining the reservoir, sediment Se concentrations in the top 0.15 m ranged from 1 to 209 mg/kg (Frankenberger & Karlson 1988). Conventional methods of removal or containment of the accumulated Se were found to be environmentally unacceptable or extremely costly, with estimates being as high as U.S. $100 million for the Kesterson site alone. Microbial volatilization of Se is being considered as an alternative method of detoxifying the contaminated sediments (California WRCB 1988). The purpose of this paper is to review laboratory experiments on the control of Se methylation by microbial processes and to present the results of a field study on the effects of various treatments on Se emission from soil.

Biosynthesis of gaseous alkylselenides has a pronounced effect on Se cycling (Craig 1986). Microbial volatilization of Se from soils and

* Department of Soil and Environmental Sciences, College of Natural and Agricultural Sciences, University of California, Riverside, Riverside CA 92521-0424

sediments and leaf-mediated volatilization from plants are considered to be major sources of atmospheric Se. The global biogenic flux has been estimated at 30 × 10^8 g Se/year, which comprises approximately one-fifth of the total atmospheric emission of Se (MacKenzie *et al.* 1979). Gaseous Se has a very high volatility and readily escapes from the soil into the atmosphere (Ross 1985). The volatile Se reported to evolve from soils is mainly dimethylselenide (DMSe), along with small quantities of dimethyldiselenide (DMDSe) and possibly dimethylselenone (Jiang *et al.* 1983). Methylation of Se is thought to be a detoxification mechanism for microorganisms, animals, and humans, since DMSe is at least 500 to 700 times less toxic than the inorganic Se species (Franke & Moxon 1936; Ganther *et al.* 1966; McConnell & Portman 1952).

Kinetics of Atmospheric Reactions of Dimethylselenide

Volatile Se compounds are released into the environment from both anthropogenic and natural sources (MacKenzie *et al.* 1979). One of the most important vapor-phase Se species found in the atmosphere appears to be DMSe. Biomethylation is a major source for this metalloid-carbon bond formation in the natural environment. Although the biological methylation of Se has been poorly quantified in the past and was once considered to be less important than the generation of volatile Se compounds from combustion of fossil fuels and refuse, biological transformations are now believed to constitute an important link in the global cycling of this element. Higher levels of Se are detected in atmospheric and rainfall samples collected in Denmark, Scotland, and the United States during the warmer spring and summer months, when biological activity is at its peak, than in the winter months when combustion of fossil fuels is at its highest.

The photolysis and kinetics of the gas-phase reactions of DMSe with OH and NO_3 radicals and ozone (O_3) were investigated at 296 ± 2 K and atmospheric pressure of air (Atkinson *et al.* 1990). No evidence for photolysis of DMSe was observed. Rate constants for reactions with the OH radical and O_3 of (6.78 ± 1.70) × 10^{-11} and (6.80 ± 0.72) × 10^{-17} cm^3/molecule/s, respectively, were obtained (Table 1). The rate constant for reaction with the NO_3 radical was observed to increase with increasing NO_2 concentration, with an extrapolated value of 1.4 × 10^{-11} cm^3/molecule/s at zero NO_2 concentration. Combining these rate constants with estimated ambient tropospheric concentrations of OH

TABLE 1. Rate constants at room temperature and atmospheric pressure for the gas-phase reactions of dimethylselenide and dimethylsulfide with OH and NO_3 radicals and O_3, and calculated lifetimes of dimethylselenide in the lower troposphere following reaction with these species.

Reactant Species	Rate Constant, cm^3/molecule/s $(CH)_3)_2Se$	 $(CH_3)_2S$	$(CH_3)_2Se$ Lifetime[(a)]
OH	$(6.78 \pm 1.70) \times 10^{-11}$	6.3×10^{-12}	2.7 h
NO_3	1.4×10^{-11}	1.0×10^{-12}	5 min
O_3	$(6.80 \pm 0.72) \times 10^{-17}$	$<1 \times 10^{-18}$	5.8 h

(a) Based on assumed atmospheric concentrations as follows: OH, $1.5 \times 10^6/cm^3$; NO_3, $2.4 \times 10^8/cm^3$; O_3, $7 \times 10^{11}/cm^3$. Lifetime is defined as the length of time a substance exists in its original chemical structure.

Source: Reprinted with permission from *Environmental Science and Technology*, 1990, Vol. 24, pp. 1326–1332.

radicals (1.5×10^6 molecule/cm^3), NO_3 radicals (2.4×10^8 molecule/cm^3), and O_3 (7×10^{11} molecule/cm^3) leads to the following calculated lifetimes of DMSe following reaction with these species: OH radicals 2.7 h, NO_3 radicals 5 min, and O_3 5.8 h (Table 1). The products of these reactions of DMSe are presently unknown, but are likely to be involved in gas-to-particle conversions and aerosol scavenging during the global cycling of Se.

Although DMSe reacts relatively rapidly with OH and NO_3 radicals and with O_3 and has a calculated lifetime in the lower troposphere of a few hours or less, the resulting Se-containing products may be sorbed onto submicrometer particulates that can have a relatively long residence time in the atmosphere and can travel considerable distances. For example, MacKenzie *et al.* (1979) estimated from mass balance calculations that the residence times for particulates are generally on the order of 2 to 10 days.

Accelerated biomethylation of Se on the west side of the San Joaquin Valley occurs at optimal levels in the daytime during the warm spring and summer months (Frankenberger & Karlson 1988). This is a period of very little precipitation, and the prevailing winds blow mainly in a southeasterly direction, toward the Se-deficient areas on the east side of the valley. Average wind speeds are usually >6 km/h (weather data collected July 1988 through March 1989 at the Tranquility 7W Weather Station, Westlands Water District, 3130 N. Fresno Street, P.O. Box 6056, Fresno CA 93703). Although some dry deposition of Se-containing product compounds may be expected, these data indicate that during the lifetime of gaseous DMSe, the gas will be dispersed and diluted by air currents directly away from the contaminated areas, with deposition possibly occurring in the Se-deficient areas.

Further work is needed to characterize the products of gas-phase reactions of DMSe as well as other organoselenium compounds that make up the atmospheric vapor phase. Although we know that sub-micrometer aerosol particulate Se is the dominant phase in the atmosphere, the mechanisms of Se association with particulates are not well known. Further experiments are needed to investigate gas-to-particle conversions and aerosol scavenging, as well as the mechanisms that cause Se compounds to leave the atmosphere in rainfall.

Determination of Gaseous Selenium Evolved from Soil

Currently, most researchers are employing gas chromatography (GC) for determination of volatile Se, with the detection modes being flame ionization, electron capture, or atomic absorption spectrometry. In these techniques, a special trap column is required for sampling. Stationary phases used for traps include Porapak QS and Carbosieve B, OV-1, and alusil. The sample is introduced into the gas chromatograph by applying reverse flow of a carrier gas through the trap column into the injection port. A severe drawback with these techniques is the time required per analysis (>30 min), which severely limits the number of samples to be processed for one experiment, even if full-time labor can be dedicated to the analyses. Another possible disadvantage with the GC methods is that nonspecific detectors are employed and total Se emissions may not be accounted for. Other Se gases besides DMSe and DMDSe may evolve from soil and could remain undetected.

In a few reports, ^{75}Se labelling has been employed. Acidic mercuric chloride has been used for trapping Se volatilized from soil (Ganje & Whitehead 1958). Another method was developed to determine Se volatilized by soil microorganisms using nitric acid as a trapping agent (Ganje & Whitehead 1958). Lewis *et al.* (1966) employed activated C to trap ^{75}Se evolved from plants requiring a cold trap as part of their experimental apparatus.

In this laboratory, a novel method was developed for large-scale measurements of Se volatilization from soils (Karlson & Frankenberger 1988a) (Figure 1). Nonspecificity, limited sensitivity, and time constraints excluded GC as a feasible detection method, hence labeling with ^{75}Se was adopted. Using nitric acid as a trapping agent was abandoned after several attempts because of the corrosiveness of the reagent. Activated C was found to be an efficient adsorbant for Se evolved from soils. We developed an experimental apparatus with C traps that required minimum sampling time and no sample processing (Figure 1). Sampling only involved replacing a specially designed C cartridge that could be counted directly without the need for sample transfer. Besides being rapid, the procedure was also safe, precise, and inexpensive. The limit of detection was 8.2 ng Se using a specific activity of 5 Ci/kg Se. The method may be applicable to other microbially produced volatiles, such as mono- and dimethylmercury, di- and trimethylarsine, dimethylsulfide, methylgermanium, tetramethyltin, dimethyltelluride, methylantimony, and tetraalkyl-lead compounds.

ALKYLSELENIDE PRODUCTION FROM SOIL

In this work, the efficiency of different C materials, the interaction of N fertilization and C amendments, and the influence of trace elements on Se alkylation were evaluated. Soil amendments with ground wheat (*Triticum aestivum* L.) grain had been found to stimulate Se evolution (Doran & Alexander 1976). Other organic C sources reported to enhance this process include glycerol (Doran & Alexander 1977a) and glucose (Francis *et al.* 1974). Field-moist Los Banos clay loam, obtained from an Se-affected area, was treated with 100 mg/kg Se(IV) labeled with $^{75}Se(IV)$. Gaseous Se was determined by direct gamma counting of activated charcoal traps (Karlson & Frankenberger 1988a). Without C amendments, the volatilization potential was 2.9 percent in 13 days; the addition of 2 g C (pectin)/kg soil increased that potential to 8.8 percent. With other C sources volatilization in 13 days was 6.6 percent [corn (*Zea mays* L.) leaves], 5.2 percent (starch), 3.7 percent (dairy manure), 3.3 percent (cellulose), and 2.7 percent (sewage sludge) (Karlson & Frankenberger 1988b). Volatilization rates decreased as C was depleted, but increased again upon C reload. The addition of molybdenum (Mo), mercury (Hg), chromium (Cr), and lead (Pb) at

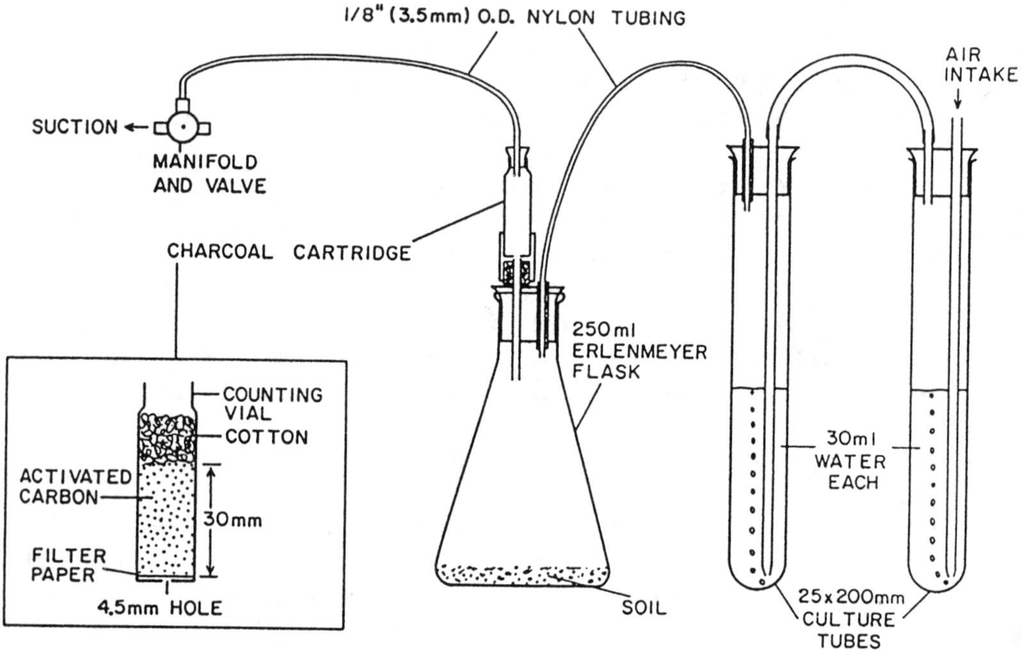

FIGURE 1. Schematic representation of the experimental apparatus for trapping ^{75}Se volatilized from soil.

25 mM/kg soil greatly inhibited Se volatilization, but arsenic (As), boron (B), and manganese (Mn) had little effect. The presence of cobalt (Co), zinc (Zn), and nickel (Ni) (5 to 25 mM/kg soil) increased Se volatilization rates up to 150 percent after 37 days of incubation. Cobalt in the form of methylcobalamin (methyl-B_{12}) has been known as a cofactor in numerous alkylation reactions, in particular alkyl-transfer to Tl(III), As(III), and Hg(II) (Halpern 1982). Wood *et al.* (1968) reported that methylcobalamin influenced the synthesis of both monomethyl and dimethylmercury in enzymatic as well as nonenzymatic systems. Our findings indicate that methyl-B_{12} may have a similarly important function in the alkylation of Se in soils.

Low levels of N in the presence of added C slightly enhanced alkylselenide production, but without C amendments, N had a slightly inhibitory effect.

The addition of specific C amendments selectively promoted the methylating organisms in soil. Fungal growth is favored by the addition of complex C sources, such as pectin and plant residues. Fungi have been suggested to be the primary producers of methyl-Se (Ab-Erreish *et al.* 1968). We have isolated and identified the following fungi from seleniferous sediments with the methyl-Se capacity: *Acremonium, Alternaria, Fusarium, Penicillium,* and *Ulocladium* spp. (Karlson & Frankenberger 1989).

ENVIRONMENTAL FACTORS AFFECTING MICROBIAL PRODUCTION

The objective of this study was to determine the environmental factors that stimulate biomethylation of Se from an actual problem site, i.e., a soil with high salinity (>20 dS/m) and with the native Se species being comprised of both inorganic and organic fractions. Attempts were made to maximize Se volatilization. The parameters that were evaluated included pH, moisture content, temperature, organo-Se substrates, and amendments such as L-methionine, carbohydrates, and proteins.

A saline (>20 dS/m) seleniferous soil collected from Pond 4, Kesterson Reservoir, was assayed for DMSe production upon incubation for up to 120 h at 22 C under various treatments. The DMSe was monitored by gas chromatography and identified by gas chromatography/mass spectrometry.

Effects of pH, Moisture, and Temperature

Chemically, the formation of DMSe, i.e., the reduction of Se(VI) and Se(IV) to the selenide state, should be favored at a low pH (Geering *et al.* 1968). In this study, however, the pH optimum observed for DMSe production (pH 8) coincides with the native pH of the seleniferous sediment. The indigenous microflora apparently had the highest activity when exposed to the pH to which it had presumably adapted.

A decrease in Se evolution under waterlogged conditions was expected since obligate aerobic fungi are thought to be the predominant Se-methylating organisms among the soil microbiota (Karlson & Frankenberger 1989). However, it should be noted that considerable DMSe is evolved from submerged soils. Chau *et al.* (1976) also found substantial amounts of DMSe being produced by sediment samples saturated with water (3:1) in the laboratory. Cooke and Bruland (1987) reported the presence of methylated Se compounds (DMSe, DMDSe, and $DMSe^{+}$-R) in diverse natural aquatic systems and suggested that biomethylation of Se is a widespread process in aquatic systems.

Volatilization of Se is highly temperature-dependent. Increasing the temperature to 35 C promoted volatilization of Se. The average Q_{10} calculated for DMSe production was 2.60. Zieve and Peterson (1981) reported similar findings with ^{75}Se added to soil samples as Se(IV). The total amount of Se being evolved after 4 days at 20 C was more than three times that at 10 C and more than 15 times that at 4 C.

Effects of Organic Substrates

Stimulation of DMSe production upon the addition of selenomethionine is most probably caused specifically by the methionine constituent. It has been speculated that methionine or its metabolites (*e.g.*, S-adenosylmethionine, SAM) could serve as a methyl donor in transferring a methyl group directly to the Se atom during the microbial methylation reaction (Doran 1982). Doran and Alexander (1977b) reported that SAM enhanced production of DMSe from Se(IV) and elemental Se when catalyzed by cell-free extracts of *Corynebacterium* sp. To obtain a better understanding of the mechanism involved in the formation of DMSe, further work is needed to determine if SAM added to soil also stimulates DMSe production.

Increased Se volatilization in response to the addition of various carbohydrates was expected, since previous studies have shown that plant residues, comprised of many of these constituents, also stimulate the evolution of DMSe from soils through enhanced microbial activity (Abu-Erreish *et al.* 1968; Doran & Alexander 1977a; Karlson & Frankenberger 1988b). Substrate-specific effects apparently play a role, since some carbohydrates stimulated DMSe production strongly, whereas others had no effect. Stimulation of methylating versus nonmethylating microflora by specific carbonaceous soil amendments has been reported upon the addition of pectin and proteins (Frankenberger & Karlson 1989; Karlson & Frankenberger 1988b). The application of albumin causes the largest enhancement in Se volatilization. Albumin applied to soil enhanced gaseous Se production 19.3-fold over the unamended control. Other proteins, including casein and gluten, also had a strong stimulatory effect on DMSe production. Further studies are underway to identify the active ingredients of these proteins so that lower organic loads may be applied.

SELENIUM VOLATILIZATION AS A BIOREMEDIATION TREATMENT

Microbial methylation of Se can be stimulated in Se-contaminated sediments by adding specific organic amendments and following basic principles that favor aerobic microbial processes (Figure 2). Field applications of these findings may be possible, provided that similar conditions can be maintained on a large scale over a sufficient period of time. A remediation program would consist of adding a specific C source with frequent tillage and irrigation and intensive management practices in the warm summer months.

Dissipation of Soil Selenium by Microbial Volatilization at Kesterson Reservoir

The scope of this project was to assess Se biovolatilization as a bioremediation technique to deselenify sediments at Kesterson Reservoir. This study was initiated in 1987 with the goals of identifying the most effective practices for accelerating volatilization and obtaining information necessary to determine the time and cost factors associated with this technology. The field study was conducted on cattail-enriched sediments of Pond 4.

This site was selected to represent one of the more highly contaminated areas at Kesterson Reservoir. Selenium concentrations in Pond 4 ranged from 10 to 209 mg/kg (median = 39 mg/kg). The field experiment was initiated on July 28, 1987. Sixteen subplots were amended with various treatments, including application of water alone or combined with cattail straw, cattle manure, citrus (orange) peel, and proteins (casein and gluten). Some subplots were also treated with fertilizers such as ammonium nitrate and zinc sulfate. The application rates of inorganic and organic amendments are specified in Table 2. All subplots were rototilled 15 cm deep on a weekly basis. Gaseous Se was sampled approximately biweekly for 1-h periods, typically during the early afternoon to detect the peak of activity. Background Se evolution was measured on nontilled and nonirrigated sediments of the experimental site. Sediment samples were collected monthly in a five-point pattern at the 0–15 cm depth to account for depletion of soil Se with time.

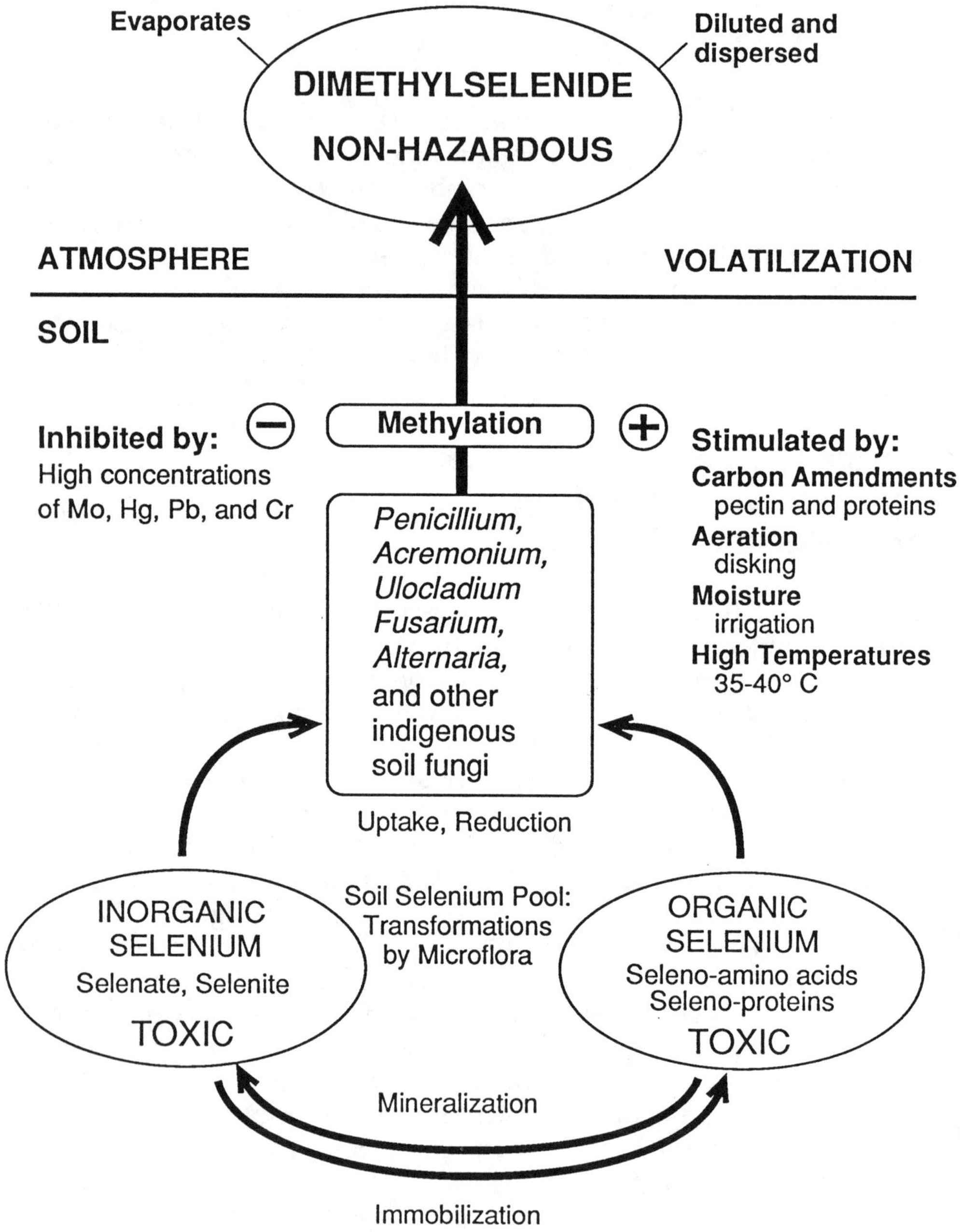

FIGURE 2. Factors affecting volatilization of selenium from soil.

TABLE 2. Organic and inorganic materials used as amendments in the field study.

Material	Source	Application Rate
Citrus (orange) peel	Sunkist Growers, Tipton CA	67 Mg/ha
Nitrogen + orange peel	ammonium nitrate	5 g N/kg orange peel
Zinc + orange peel	zinc chloride	1 g Zn/kg orange peel
Cattle manure	Harris Feed Col., Coalings CA	200 Mg/ha
Barley straw (chopped)	local	22 Mg/ha
Typha latisolia (cattail straw) + N	on site	11 Mg/ha plus $(NH_4)_2SO_4$ at 1 g N/kg cattail straw

Volatilization was monitored in the field with a flux chamber system as designed by Lawrence Berkeley Laboratory. The equipment required for this operation consisted of gas sampling chambers, a compressor vacuum pump, gas washing bottles (40 to 60 µm porosity), a gas manifold, and vinyl and surgical tubing. The air above the sediment was pulled into a trap of alkaline hydrogen peroxide (85 percent recovery) at 2 L/min. The trapped volatile Se was oxidized to Se(VI) and analyzed by atomic absorption spectrometry/hydride generation. Results are reported as volatile emission rates expressed in µg Se/m^2/h (Table 3).

Seasonal variation of gaseous Se emissions was evident, with the highest emissions recorded in the late spring and summer months (Figure 3). The greatest emissions of gaseous Se with all treatments occurred in July through September 1988, which correlates with the high soil temperatures and a relatively high Se inventory. Less volatile Se was released in the fall and winter months. The cumulative hourly gas emission flux with different seasons of the year showed plateaus during the winter season and increased rates in the early spring and summer months of each year. As time proceeds, the Se inventory available for methylation appears to be decreasing with each season of warmer temperatures.

TABLE 3. Evolution of DMSe from Pond 4 sediments, Kesterson Reservoir.

Treatments	Subplots	Average Emission	Highest Emission
		-----μg Se/m²/h-----	
Moist	56, 57, 62	16.1	140
Cattail straw +N	66	10.7	39
Cattle manure	51, 53, 60	11.2	116
Citrus peel	54, 58, 61	32.5	297
Citrus peel + N + Zn	63, 64	110.1	808
Casein	65	49.8	749
Gluten	52, 55, 59	44.7	391
Background		2.6	8

Irrigation with tillage alone resulted in an average volatile Se emission of 16.1 μg $Se/m^2/h$ during this 2-year field study. This rate is considerably higher (6.2-fold) than the average background level (2.6 μg $Se/m^2/h$). The effectiveness of applied C sources on volatilization of Se is shown in Table 3. The application of citrus peel plus N plus Zn promoted Se volatilization more than any other treatment. The highest emission rate recorded with this treatment was 808 μg $Se/m^2/h$. The average emission reading with citrus peel plus N plus Zn was 110.1 μg $Se/m^2/h$, which is 42 times the background level.

Soil samples collected at different depths in Pond 4 indicated that 90 percent of the total Se was within the upper 15 cm of the soil profile. Composite soil samples from each subplot were analyzed monthly for total Se during the course of this field study. The measured values for residual sediment Se best fit a logarithmic regression equation that approximates a first-order reaction. The least effective treatment in promoting Se volatilization was cattle manure, with 30 percent Se removal in 23 months (Figure 4). Proper soil water management (moist alone combined with tillage) led to a considerable decline in the Se inventory with 49 percent removed after 23 months of investigation. Two promising amendments that promoted biomethylation were casein and citrus peel plus N plus Zn. Approximately 69 percent and 62 percent of the initial Se inventory was removed in the Ap (plow) layer after 2 years of investigation with these two amendments,

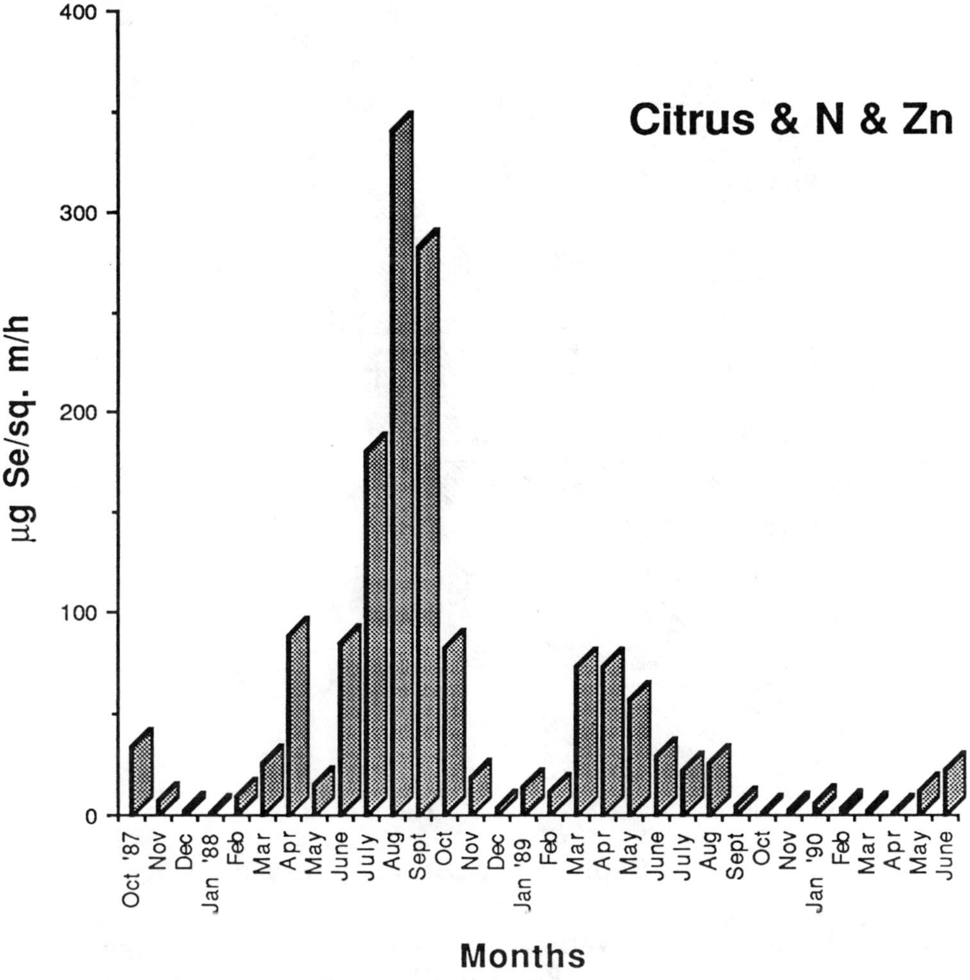

FIGURE 3. Seasonal flux of the average monthly volatile Se emission from Pond 4 sediments upon treatment with citrus peel + N + Zn.

respectively. Figure 5 shows the dispersion of the collected data and the striking decline in the Se inventory in the Ap horizon upon rototilling and adding water.

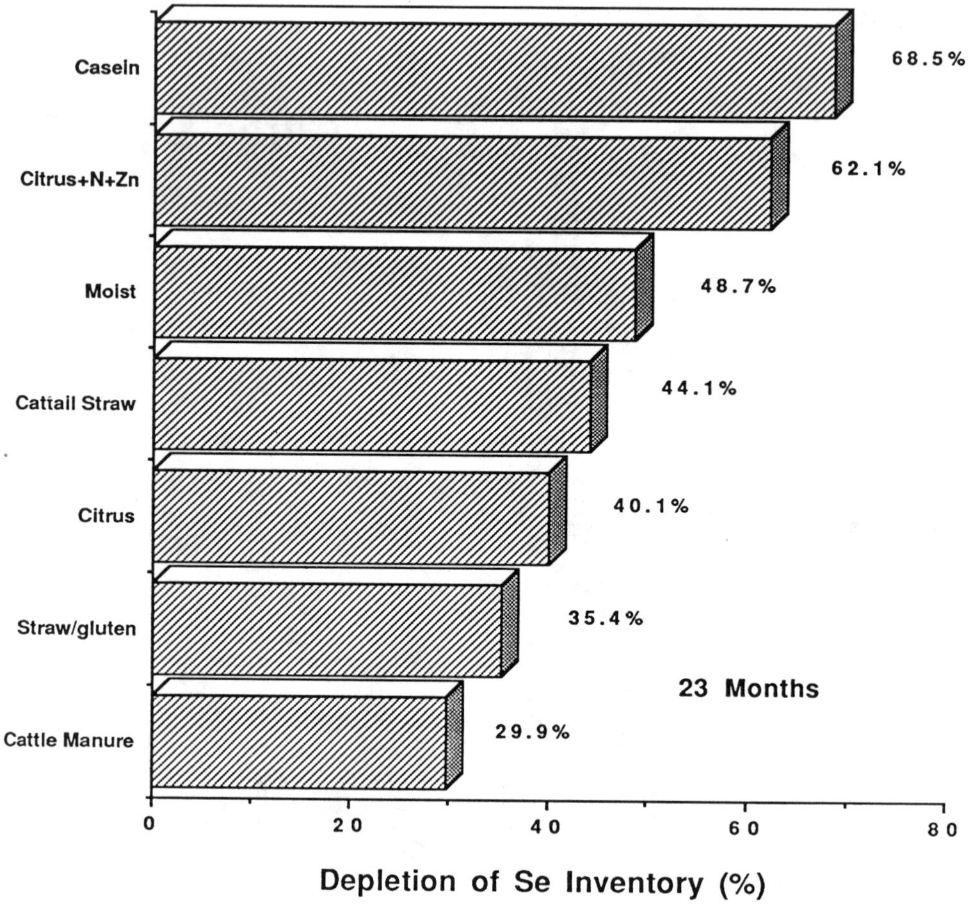

FIGURE 4. Influence of various treatments on the percentage of Se dissipated from Pond 4 sediments, Kesterson Reservoir, based on sediment analysis. Sampled from 3 independent subplots over a 23-month period.

CONCLUSION

Selenium contamination from agricultural wastewater has been blamed for wildlife deaths and grotesque deformities of hatchlings at Kesterson Reservoir (Merced County CA). Some approaches in reducing toxic Se concentrations have been proposed, but most of these tend to be costly and/or ineffectual (*e.g.*, physical removal and disposal). One approach, based on biomethylation of Se, is being considered as a remedial alternative. This study involves the development of an *in situ*

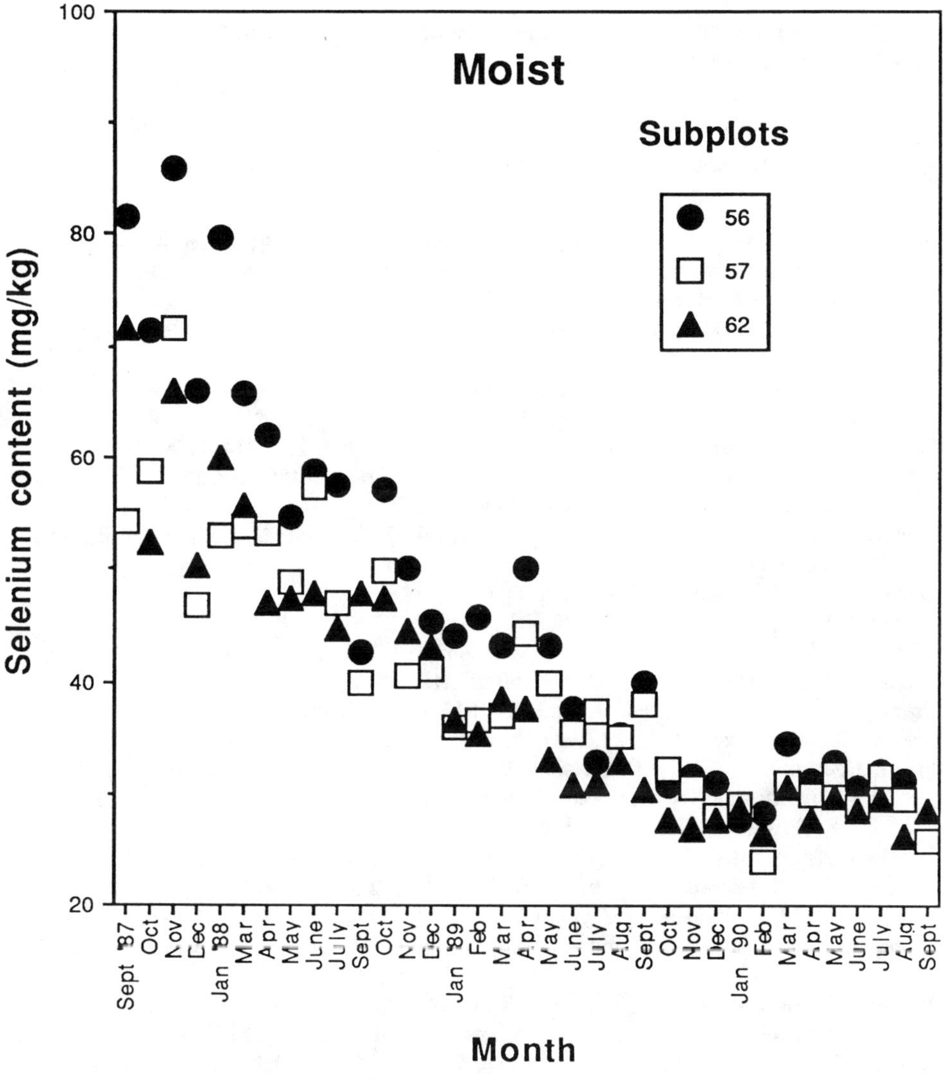

FIGURE 5. Selenium content of Pond 4 sediments, Kesterson Reservoir, versus time of treatment with water and

biomethylation process based on landfarming principles to detoxify seleniferous sediments. Optimum management practices were identified to accelerate Se volatilization, a naturally occurring component of Se cycling in the biosphere. This bioremediation technique is highly dependent on specific carbon amendments, temperature, moisture, aeration, and activators (cofactors). The biotechnology prototype developed could be applicable for the cleanup of polluted sediments on irrigated farms throughout the western United States where Se is a growing problem.

REFERENCES

Abu-Erreish, G. M.; Whitehead, E. J.; Olson, O. E. *Soil Sci.* **1968**, *106*, 415–420.

Atkinson, R.; Aschmann, S. M.; Hasegawa, D.; Thompson-Eagle, E. T.; Frankenberger, W. T. *Environ. Sci. Technol.* **1990**, *24*, 1326–1332.

California WRCB (Water Resources Control Board). Order No. WQ 88-7; Sacramento, CA, July 5, 1988.

Chau, Y. K.; Wong, P.T.S.; Silverberg, B. A.; Luxon, P. L.; Bengert, G. A. *Science* **1976**, *192*, 1130–1131.

Cooke, T. D.; Bruland, K. W. *Environ. Sci. Technol.* **1987**, *21*, 1214–1219.

Craig, P. J. In *Organometallic Compounds in the Environment*; Craig, P. J., Ed.; Longman Group: Harlow, Essex, U.K., 1986.

Deverel, S. J.; Millar, S. P. Open-File Report 86-538, 1986; U.S. Geological Survey; Sacramento, CA.

Doran, J. W. *Adv. Microb. Ecol.* **1982**, *6*, 1–32.

Doran, J. W.; Alexander, M. *Soil Sci. Soc. Am. J.* **1976**, *40*, 687–690.

Doran, J. W.; Alexander, M. *Soil Sci. Soc. Am. J.* **1977a**, *41*, 70–73.

Doran, J. W.; Alexander, M. *Appl. Environ. Microbiol.* **1977b**, *33*, 31–37.

Francis, A. J.; Duxbury, J. M.; Alexander, M. *Appl. Microbiol.* **1974**, *28*, 248–250.

Franke, K. W.; Moxon, A. L. *J. Pharmacol. Exp. Ther.* **1936**, *58*, 454–459.

Frankenberger, W. T., Jr.; Karlson, U. *Dissipation of Soil Selenium by Microbial Volatilization at Kesterson Reservoir*; U.S. Dept. of the Interior, Bureau of Reclamation; Sacramento, CA, 1988.

Frankenberger, W. T., Jr.; Karlson, U. *Soil Sci. Soc. Am. J.* **1989**, *53*, 1435–1442.

Ganje, T. J.; Whitehead, E. I. *Proc. S.D. Acad. Sci.* **1958**, *37*, 81–84.

Ganther, H. E.; Levander, O. A.; Saumann, C. A. *J. Nutr.* **1966**, *88*, 55-60.

Geering, H. R.; Cary, E. E.; Jones, L.H.P.; Allaway, W. H. *Soil Sci. Soc. Am. Proc.* **1968**, *32*, 35–40.

Halpern, J. In B_{12}; Dolphin, D., Ed.; Wiley Interscience: New York, 1982.

Jiang, S.; Robberecht, H.; Adams, F. *Atmos. Environ.* **1983**, *17*, 111–114.

Karlson, U.; Frankenberger, W. T., Jr. *Soil Sci. Soc. Am. J.* **1988a**, *52*, 678–681.

Karlson, U.; Frankenberger, W. T., Jr. *Soil Sci. Soc. Am. J.* **1988b**, *52*, 1640–1644.

Karlson, U.; Frankenberger, W. T., Jr. *Soil Sci. Soc. Am. J.* **1989**, *53*, 749–753.

Letey, J.; Roberts, C.; Penberth, M.; Vasek, C. *An Agricultural Dilemma: Drainage Water and Toxics Disposal in the San Joaquin Valley*; Publication No. 3319; Div. of Agric. and Natural Res., University of California, 1986.

Lewis, B. G.; Johnson, C. M.; Delwiche, C. C. *J. Agric. Food Chem.* **1966**, *14*, 638–640.

MacKenzie, F. T.; Lontey, R. J.; Paterson, V. *Math. Geol.* **1979**, *11*, 99–141.

McConnell, K. P.; Portman, O. W. *Proc. Soc. Exp. Biol. Med.* **1952**, *79*, 230–231.

Ohlendorf, H. M.; Hoffman, D. J.; Saiki, M. K.; Aldrich, T. W. *Sci. Total Environ.* **1986**, *52*, 49–63.

Ross, H. B. *Tellus, Ser. B.* **1985**, *37*, 78–90.

Wood, J. M.; Kennedy, F. S.; Rosen, C. G. *Nature* **1968**, *220*, 173–174.

Zieve, R.; Peterson, P. J. *Sci. Total Environ.* **1981**, *19*, 277–284.

Possibilities and Limitations of Landfarming for Cleaning Contaminated Soils

*Joop Harmsen**
The Winand Staring Centre

INTRODUCTION

The biological treatment of soil by landfarming is a relatively simple and cheap method of cleaning soil contaminated by different organic compounds. The polluted soil is spread out on a specially constructed site where the microbiological degradation is stimulated by cultivation of the soil and addition of nutrients. On-site treatment is possible if the pollution is at the surface or if the location is large enough to allow the polluted soil to be spread out. Landfarming has often been used for treating soils containing different types of mineral oil and for soils contaminated with polycyclic aromatic hydrocarbons or pentachlorophenol. Results of experiments in the Netherlands have been summarized by Soczo and Staps (1988) and by Soczo *et al.* (1990).

The success of landfarming is dependent on several factors. In general an organic compound will degrade under favourable conditions, i.e., the presence of microorganisms, oxygen, nutrients, and the bioavailability of the organic compound.

In this paper results obtained from different types of experiments are described. The first part deals with results of laboratory experiments on the optimization of the landfarming process, starting

* The Winand Staring Centre for Integrated Land, Soil and Water Research, P.O. Box 125, 6700 AC Wageningen, The Netherlands

with the original polluted soil. Next the laboratory results were used in (semi)field experiments on landfarming of mineral oil and pentachlorophenol. In these experiments it was found that a residual concentration was left in the soil. To investigate why this residual concentration was hardly degraded, research into bioavailability was carried out. Again laboratory experiments, now focused on the desorption mechanism of model compounds in soil, were done. Furthermore, the use of detergents to increase the solubility was investigated.

In the discussion all these separate investigations are evaluated and related to results from available literature. This leads to a better understanding of the possibilities and limitations of the landfarming method.

LABORATORY EXPERIMENTS ON OPTIMIZATION OF DEGRADATION

Effects of Nutrients

Materials and Methods. Although the Sapromat manufactured by Voith (Germany) was made to measure oxygen consumption in water, we used it to measure the oxygen consumption in soil. A layer about 1 cm thick of a crude oil contaminated soil sample mixed with nutrients (NH_4NO_3 and KH_2PO_4) was put into a reaction vessel. This was connected to a pressure sensor and an electrolytic cell. Carbon dioxide produced in the soil was adsorbed with soda lime pellets. The drop in pressure caused by oxygen consumption was compensated by producing oxygen with the electrolytic cell. The electric current required to produce the oxygen was monitored, and a graph of the cumulative oxygen consumption versus time was derived. Because the Sapromat contained six reaction vessels, more nutrient combinations could be measured simultaneously.

Results. Biodegradation could be optimized at the commencement of a landfarming process by adding varying quantities of nutrients to a soil polluted with crude oil. Results of the Sapromat experiment are presented in Figure 1 (Ten Holder, personal communication, 1980). The oxygen consumption increased in the presence of higher amounts of nutrients.

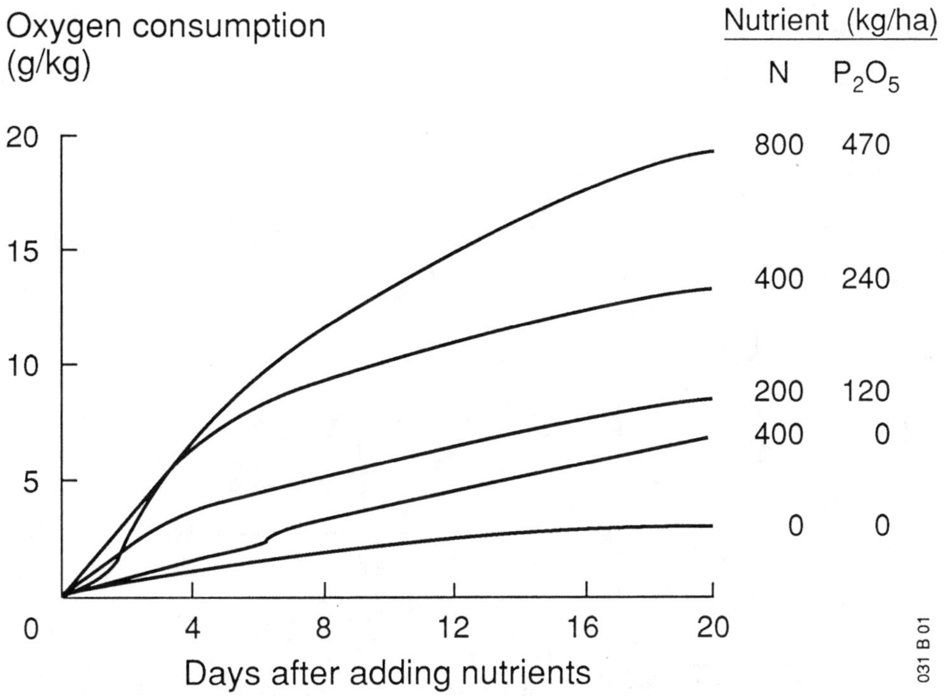

FIGURE 1. Effect of nutrients added in varying quantities on the biodegradation of crude oil in soil. The cumulative oxygen consumption of soil polluted with crude oil as a function of time is given. The amount of nutrients was based on a layer 25 cm thick.

Effect of Cultivation

Materials and Methods. The effect of cultivation was studied by leading air through a reaction vessel containing 2.5 kg of soil contaminated with crude oil. The flow velocity of the oxygen was 10 mL/min. The oxygen concentration of the ingoing and outcoming air was measured using an oxygen meter. The effect of cultivation was measured after thoroughly mixing the soil by hand every 4 days.

Results. Mixing the soil resulted in a higher oxygen consumption (Figure 2), not only just after mixing, but also several days later (Hoeks *et al.* 1988). Cultivation is important for providing essential aeration of the soil. The introduction of new oxygen caused little effect, because this amount can normally be ignored compared to the total amount of oxygen required to clean the soil. It is assumed that cultivation of the

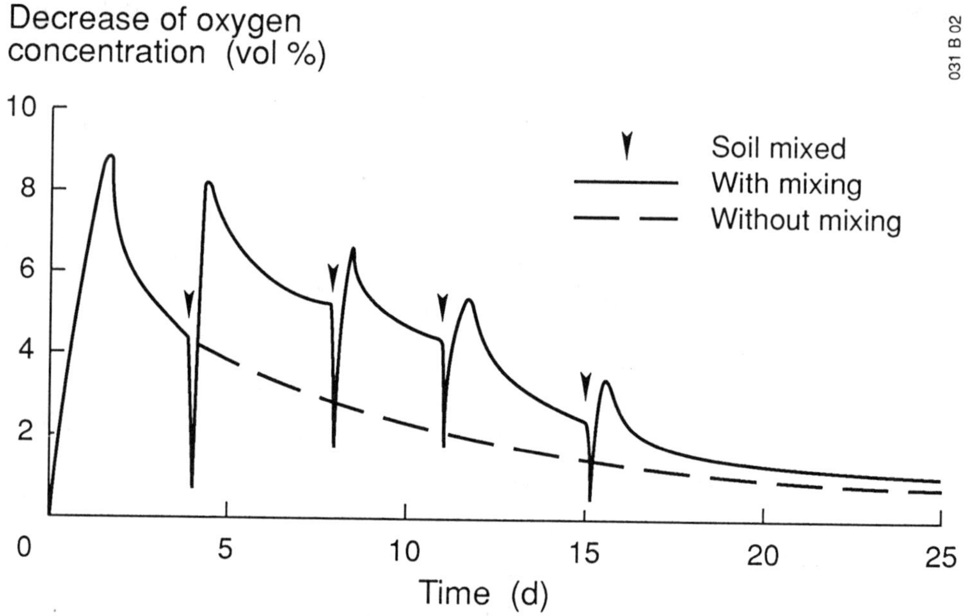

FIGURE 2. Effect of soil mixing on the oxygen consumption measured as decrease in oxygen concentration at a depth of 25 cm in a vessel with crude oil polluted soil at a temperature of 20 C.

soil redistributes the oil, nutrients, and microorganisms, creating new points of attack for the microorganisms.

FIELD EXPERIMENTS

Oil Contaminated Soil

Materials and Methods. To test the results of the laboratory experiments, an experimental plot (32 × 43 m) for landfarming was constructed on the site of the Waste Disposal Company in Wijster (Hoeks *et al.* 1988). The landfarming field consisted of a nonpermeable layer of polyethene (0.27-mm Valeron) covered with a layer of 70-cm permeable sandy soil containing a drainage system. The polluted soil was spread over this field in a layer 35 to 40 cm thick. Two types of polluted soils were investigated, one containing gas oil and the other crude oil. Half of the experimental site was treated as bare soil so that cultivation of the soil was possible; the other half was covered with grassy vegetation.

This resulted in four experimental fields: gas oil noncultivated (1A) and cultivated (1B), and crude oil noncultivated (2A) and cultivated (2B). During the experiment the mineral oil contents of soil were measured using an extraction with hexane followed by gas chromatography. By this method it was possible to analyse the hydrocarbons, producing retention times corresponding to the *n*-alkanes with 10 to 40 carbon atoms. In the Netherlands in 1986, a new extraction procedure using acetone and pentane was recommended (DHV 1986). This method gave comparable results for gas oil, but in the case of crude oil the extraction efficiency of the acetone/pentane method was higher.

Organic carbon (TOC, measured with the Dohrman DC180 carbon analyser) and nutrients (measured by ion chromatography) were analysed in the water of the drainage system. Oxygen in the soil pores was measured *in situ* using an oxygen meter.

Results. The mineral oil content of both fields 1A and 1B (Figure 3A) decreased from about 1,500 mg/kg dry soil in 5 months to a level of 400 mg/kg dry soil. Any further decrease could not be established. The only difference between the cultivated soil and soil with a grassy cover was that the homogeneity in the cultivated soil was better, visible as a slight deviation in Figure 3A. One summer period was enough for the degradation of about 75 percent of the pollution. (The Netherlands lies in the moderate maritime climate zone.) Because it improved soil structures, cultivation had a good effect on the average oxygen content in the soil (Table 1). Oxygen consumption was high in the summer period of 1985 (low O_2 concentration) and low in 1986 (high O_2 concentration), indicating that degrading of the residual concentration in the soil was difficult.

TABLE 1. Average oxygen concentration in the soil as a function of depth at a site with soil polluted by gas oil for cultivated and noncultivated (grassy cover) fields (the open air contains 21 percent oxygen).

	Oxygen Concentration (vol %)					
	Cultivated			Grassy Cover		
Period	15 cm	30 cm	50 cm	15 cm	30 cm	50 cm
Summer 1985	18	13	8	10	5	2
Summer 1986	21	21	20	21	20	18

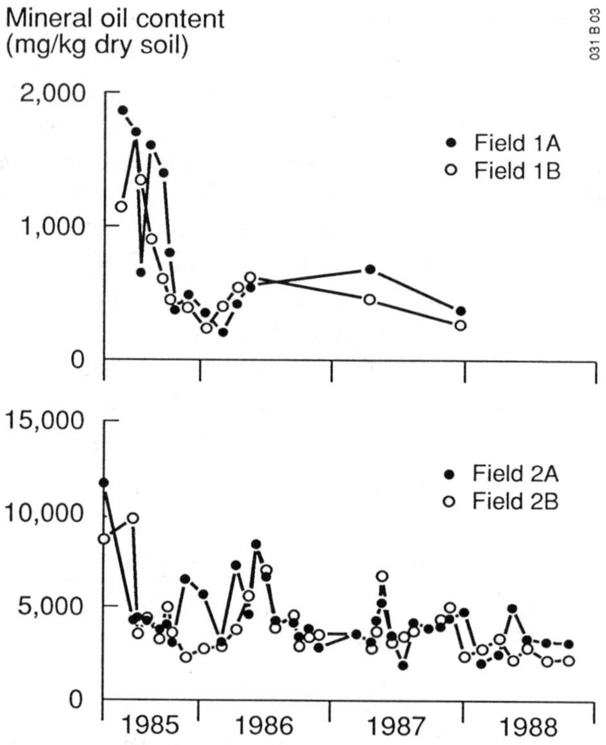

FIGURE 3. Oil concentration (mg/kg dry soil) in soils polluted with oil as a function of time: (A) Soil contaminated with gas and oil. The oil concentration was measured with the hexane method. Field 1A had a (grassy) vegetation cover, and field 1B was cultivated. (B) Soil contaminated with crude oil. The oil concentration was measured using the acetone-pentane method. Concentrations measured in 1985 were originally measured with the hexane method and adjusted using the ratio of both methods measured in 1986. Field 2A had a (grassy) vegetation cover, and field 2B was cultivated.

No oil was detected (<0.2 mg/L) in the drainage water. The TOC level in the drainage water was 110 mg/L C, which could be products of degradation. These components had the same polarity as natural occurring organic carbon compounds. Nitrate was present in concentrations of between 40 and 200 mg/L and phosphate in concentrations of 1 mg/L or less (<0.1 mg/L).

The presence of crude oil complicated the picture (Figure 3B). After a decrease measured in the summer period, the amount of oil in the soil increased during the winter period. This phenomenon can be related to the ambiguous definition of mineral oil. From an analytical point of view it is possible to analyse part of the crude oil—i.e., the part that can pass chromatographic column—producing peaks with retention times between the *n*-alkanes with 10 and 40 carbon atoms. This part decreased during the summer period from biological degradation. It is assumed that during the next winter and spring, compounds with higher boiling points would be partly degraded, forming compounds whose boiling points fall within the range of the analytical procedure defined. These compounds were measured, but disappeared yet again during the following summer period (Figure 3B). The same effect occurred the following year, but to a lesser extent.

Although it was difficult to measure the "actual" concentration of mineral oil, it appeared that degradation occurred from the decreased oxygen concentration in the pores. In contrast to the gas oil soil, even after a period of 3 years, oxygen concentrations in the crude oil soil remained low (15 to 21% at 15 cm, 10 to 19% at 30 cm, and 2 to 17% at 50 cm).

The composition of the crude oil changed drastically during the landfarming process (compare Figures 4A and 4B). The easy degradable *n*-alkanes disappeared (separate peaks), and the relatively volatile part (up to C_{20}) decreased more than the relatively nonvolatile part (above C_{20}).

The TOC level and the nitrate and phosphate concentrations in the drainage water were comparable with those found in the gas oil experiment.

Pentachlorophenol-Contaminated Soil

Materials and Methods. The degradation of pentachlorophenol was investigated using lysimeters (140 × 120 × 120 cm) on an experimental farm (Verschoor & Scheffer, personal communication, 1988). The lysimeters were filled with a layer of sand 70 cm deep (0.1% organic carbon) and covered with a polluted layer of 40 cm. The soil was taken from a small wood-conserving factory. The base of each lysimeter was

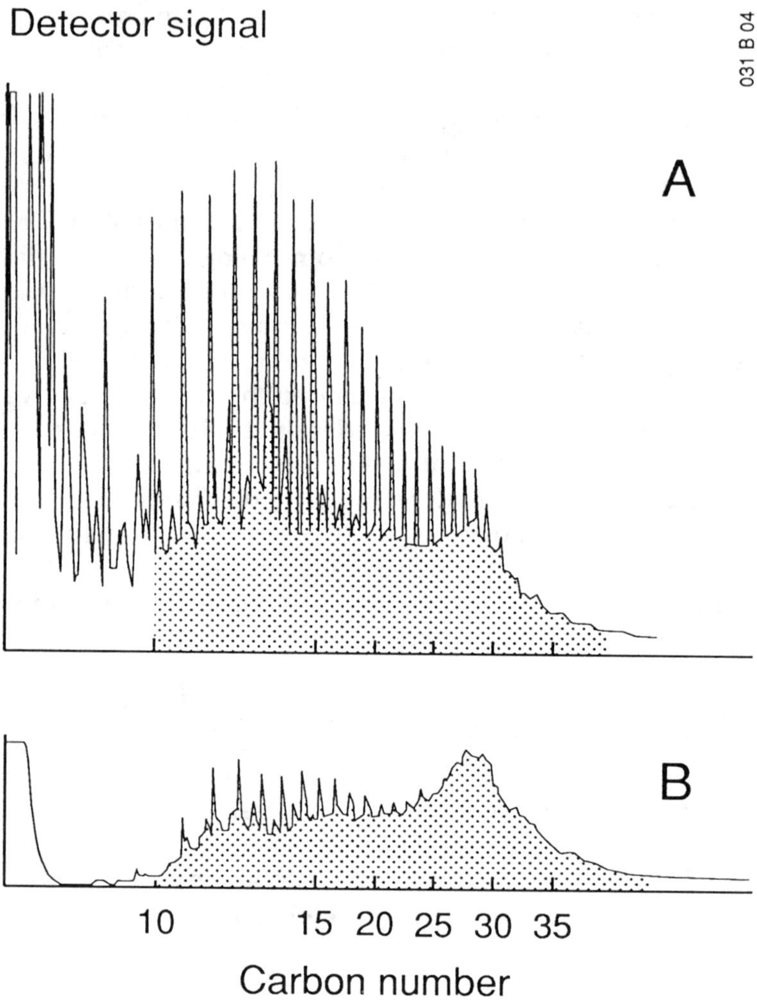

FIGURE 4. Gas chromatogram of crude oil. The shaded part is measured as oil: (A) Original crude oil, content *ca.* 20 g/kg soil. (B) Crude oil after 423 days of landfarming, content *ca.* 5 g/kg soil.

fitted with a drainage system, the lower 5 cm being kept saturated with water. Four different treatment procedures were used:

- No treatment.

- Addition of nutrients: nitrogen and phosphate were added 8.5 and 11 weeks after start, 6.7 g and 2.5 g per lysimeter, respectively.

- Cultivation: lysimeters were cultivated by hand every 3 weeks, starting at week 5.

- Anaerobic treatment: after 5 weeks one lysimeter was saturated with water to investigate whether the degradation would continue under anaerobic conditions.

Soil samples and water samples from the drainage system were analysed for pentachlorophenol using a gas chromatographic method after derivatization with acetic anhydride (Abrahamson & Xie 1983). To isolate pentachlorophenol, the soil was first mixed with 1 M HCl, then extracted twice using toluene.

After sterilization of the soil (γ rays), the equilibrium concentration in water was measured after shaking the soil (40 g) with water (100 mL) for 24 h.

Results. Pentachlorophenol was degraded rather rapidly. The concentration in the soil decreased from 100 to 20 mg/kg dry soil within a period of 10 to 20 weeks (Figure 5). It was remarkable that the only treatment procedure that influenced the degradation of this polluted soil was saturation with water. The degradation stopped in the anaerobic circumstances created. Neither cultivation of the soil nor addition of nutrients favoured the degradation velocity.

The concentration of pentachlorophenol in the drainage water was low during the entire experiment; between 0.0002 and 0.0005 mg/L. Higher concentrations were measured in the drainage water of the lysimeter where the degradation had been stopped by saturation with water (*ca.* 0.05 mg/L). Water in equilibrium with sterilized soil contained much more pentachlorophenol, in concentrations varying from 0.5 mg/L for a soil containing 20 mg/kg, up to 4.8 mg/L for a soil containing 70 mg/kg.

In the drainage water of the unfertilized lysimeters, nitrate concentrations were between 20 and 80 mg/L, indicating that the soil contained enough nutrients. Adding fertilizers did indeed increase the nitrate concentrations to values between 70 and 400 mg/L. TOC values in the drainage water of all the lysimeters were between 20 and 50 mg/L C.

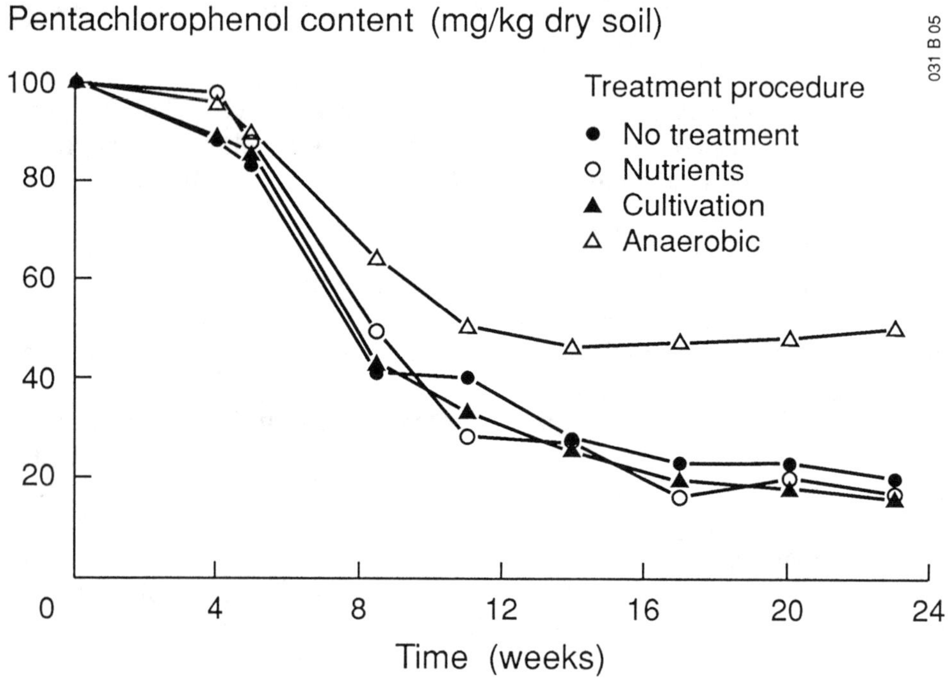

FIGURE 5. Effect of different treatment procedures on the biodegradation of pentachlorophenol. The pentachlorophenol concentration is measured as a function of time. The experiment was started in May 1988.

LABORATORY EXPERIMENTS ON BIOAVAILABILITY

Effect of Sorption Kinetics

Materials and Methods. Bioavailability is not only related to the partition of a compound between the water and solid phase in the soil, but the kinetics of the desorption process are also of importance. To study the kinetics of adsorption and desorption, a small column was used (25 cm × 4.6 mm i.d.). The column was filled with soil. Water with or without model components was pumped through the column using HPLC pumps (Kratos SF400) at a flow velocity of 0.1 mL/min. In the research presented, aromatic compounds with one alkyl substitute (Velthorst, personal communication, 1990) were used as model components. The columns were percolated with water containing model components to measure the adsorption onto the soil. Next, the desorption was studied by pumping water not containing model components through the column. The effluent of the column passed a small HPLC preconcentration column on line, filled with C18 reversed-phase material (1 cm × 2 mm i.d.). The analytical procedure was started after 30 min of concentration, and all components were separated on a Chromsper C18 column (20 cm × 3 mm i.d., Chrompack). A gradient of water and acetonitril was used (Varian 5000 pump), and the components were detected on a UV detector at 210 nm (Kratos SF773). The results were registered on a personal computer using Nelson 2600 software. The whole system was automated in such a way that an effluent sample was analyzed every 75 min and an influent sample after six runs. A single adsorption/desorption experiment took 1 month to complete, and in this period the water-filled pore volume of the column was replaced 2,200 times.

Results. Figure 6 gives the breakthrough curves of three of the six model components. Considering the adsorption curves, the aromatic hydrocarbons were retarded by the soil. The highly soluble compounds, such as benzene and toluene, had the smallest breakthrough volumes (measured as volume at c/c0 = 0.5); they were adsorbed the least. The more strongly adsorbed hexylbenzene had a larger breakthrough volume, and the c/c0 value did not reach one. The phenomenon of the equilibrium not being reached was more clearly visible in the desorption curves. With desorption the compounds with lower solubility were present much longer in the system. Adsorption volumes and desorption volumes (measured at half-height of the curve) were the same. Again, for highly soluble compounds like benzene and toluene equilibrium was reached rapidly, but the relative concentrations of butylbenzene and hexylbenzene did not reach the value of zero.

Although all curves showed a tailing effect, the tailing from benzene to hexylbenzene was greater. To measure the tailing, the value of c/c0 at three times the desorption volume (measured at c/c0 = 0.5) was taken and related to the solubility (Figure 7). This figure not only provides the results derived from the experiment presented in Figure 6, but also those obtained from a second soil.

Effect of Detergents

Materials and Methods. The effect of detergents on the residual concentration in soil after landfarming was studied on soil from a landfarming experiment polluted with gas oil (Nieveen & Verheijen, personal communication, 1988). Even though the soil had already been

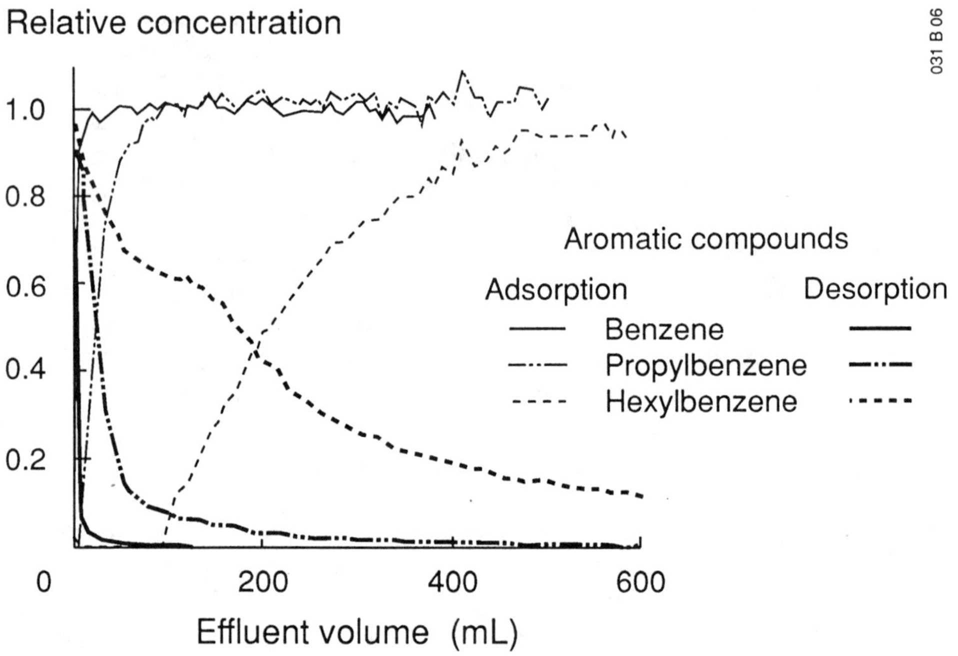

FIGURE 6. Breakthrough curves of alkylbenzenes in a column containing a sandy soil with an organic carbon content of 0.74 percent. The relative concentration, c/c0 (i.e., the concentration of the effluent divided by the concentration of the influent), as a function of the effluent volume, is given. The pore volume in the column was 1.8 mL. Adsorption process. Concentration of each alkylbenzene in the influent was 0.6 mg/L. Desorption process. No alkylbenzenes in the influent.

treated with the landfarming method for 2 years, it still contained 400 mg oil/kg dry soil. Measurement of the effect of detergents on the solubility was done by shaking the soil with water containing different amounts of detergents. Detergents used were Serqua 710 and the Dobanols 91-5, 91-6, and 91-8. Dobanol contains C_9, C_{10}, and C_{11} alcohols with, respectively, 5, 6, and 8 M of ethoxylate. Serqua 710 is an alkylphenolpolyglycolether. The mineral oil content was measured using gas chromatography, and the results were adjusted for the signal caused by the detergents.

The biodegradability of the residual concentration of the remaining oil, the detergents, and mixtures of the two was tested in water cultures inoculated with a suspension of the polluted soil at a

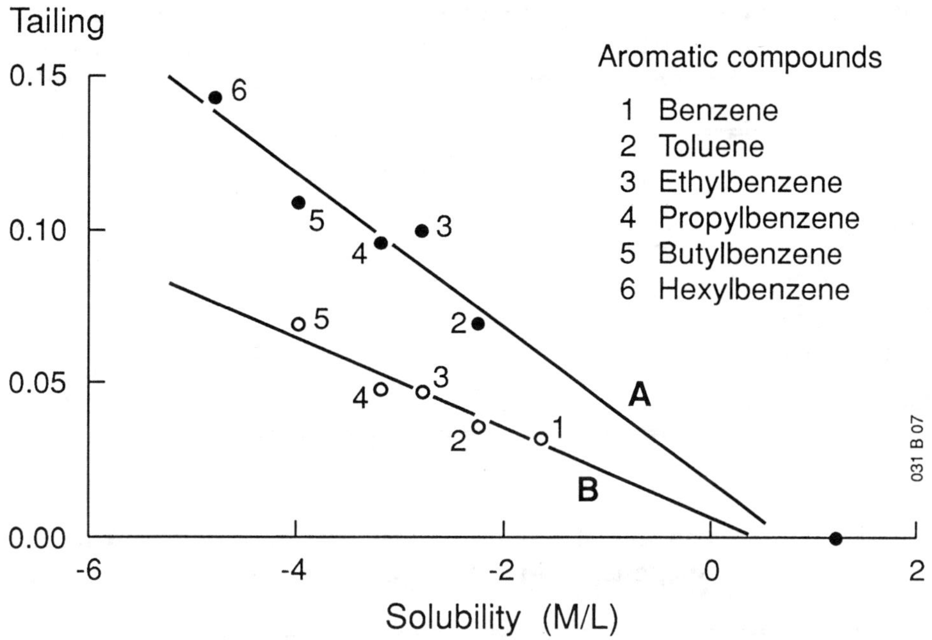

FIGURE 7. Relation between tailing (c/c0 value at three times the desorption volume measured at c/c0 = 0.5) and the solubility in water. (A) Values from the desorption experiment presented in Figure 6. (B) Values from a desorption experiment with a sandy soil containing 1.51 percent organic carbon.

temperature of 30 C. The oil was obtained by extracting the polluted soil with hexane. An hexane extract was added to a water culture, producing a concentration of about 300 mg/L, which was too much to be dissolved. The hexane was removed by purging with air. The oil content was measured after the detergents had been added at the start and after several weeks.

Results. The solubility in water of the residual concentration of the remaining gas oil in soil had been increased by the detergents (Table 2). The highest solubility was obtained with Dobanol 91-6.

The residual concentration of oil in a water culture disappeared under "ideal" circumstances (presence of oxygen, nutrients, and a temperature of 30 C). Starting with 320 mg/L the concentration reduced to 300 mg/L after 23 days and to 68 mg/L after 63 days (Table 3). In the presence of detergents the degradation rate was higher.

TABLE 2. Effect of detergents (0, 0.2, and 2.0 g/L) on the solubility of a residual concentration of gas oil in soil. One liter of detergent solution has been shaken with 250 g of soil.

	Mineral Oil in Water (mg/L)		
Detergent	0	0.2	2
Serqua 710	0.4	0.5	4.3
Dobanol 91-5	0.4	1.4	11.1
Dobanol 91-6	0.4	1.6	39.4
Dobanol 91-8	0.4	1.2	15.1

TABLE 3. Oil concentrations in a water culture inoculated with a soil suspension in the presence of detergents (250 mg/L).

	Oil Concentration (mg/L)		
Detergent	t = 0	t = 9 days	Oil Disappeared (%)
None	255	188	26
Serqua 710	200	20	90
Dobanol 91-5	228	40	82
Dobanol 91-6	275	83	70
Dobanol 91-8	315	43	86

DISCUSSION

Cleaning soil using the landfarming methods is possible for several compounds. The treatment of soil polluted by mineral oil is often practised, and the results achieved vary. Light mineral oils are relatively easily degradable by microorganisms; a residual concentration of about 500 mg/kg dry soil can be obtained (Figure 3A; Marel *et al.* 1988) in one summer. More difficult to degrade is mineral oil with a higher boiling part; it takes several seasons to lower significantly the concentration (Figure 3B; IWACO 1989), and even then a high residual concentration remains in the soil. Another degradable compound, pentaclorophenol, can be removed to the extent of 80 to 90 percent in one summer period (Figure 5; Valo & Salkinoja-Salonen 1986). Polycyclic aromatic hydrocarbons (PCA) are much more difficult to degrade, and again several summer periods may be necessary to reduce PCA contents to a level of 100 mg/kg dry soil (Berends & Kloeg 1987; Marel *et al.* 1988). Results of various projects are, however, difficult to compare because different analytical procedures are used. Mineral oil specially could be analysed in substantially different ways because it is poorly defined and a standard mineral oil does not exist. Furthermore, natural occurring components can appear to testing instruments as mineral oil. Often different extraction procedures have been used, producing different results. Therefore, it would be unwise to compare absolute amounts from different investigations, but comparing changes over time can be useful.

Much effort is put into optimizing the landfarming process. In particular the right circumstances at the commencement produce high degradation rates. The addition of nutrients and the cultivation of the soil increase the rate of degradation; this is shown in Figures 1 and 2. Similar results to those shown in Figure 1 are presented by Verheul *et al.* (1988). Further optimization of the process does not improve the results because, compared to a bioreactor, more time is needed to accomplish the landfarming process. In many cases, using more than one summer for the process is not a problem. If it is possible to reduce a certain compound in one summer by 80 percent, there is no reason to try to reach the same level of degradation in 1 month, specially if no further degradation can be realized during the remainder of the period. Figure 5 clearly shows that the presence of oxygen is sufficient to degrade pentachlorophenol. The other circumstances for the microorganisms in this soil are already more or less optimal; the effect of other activities is therefore marginal.

Another example is adding active microorganisms, which produces good results in a laboratory situation (Oldenhuis *et al.* 1989). The benefit however is small in a landfarm situation. Follow-up research has shown that adding microorganisms acted as a booster at the start of the degradation, but after 2 months the amount of degraded oil was the same as without added microorganisms (IWACO 1989). It may be concluded that biodegradation occurring under aerobic conditions is optimized rather quickly: optimum landfarming conditions are obtained if the soil is well aerated and enough nutrients are present.

In all cases a residual concentration is left in the soil. This concentration is relatively low with respect to the original concentration for the easily degradable pentachlorophenol (Figure 5) and light mineral oil products (Figure 3A; Soczo *et al.* 1990). Higher residual concentrations are found with crude oil (Figure 3B) and other compounds of higher boiling points like PCAs (Berends & Kloeg 1987; Marel *et al.* 1988).

There are two explanations for these differences in residual concentrations:

- In a mixture of compounds like mineral oil, first the easily degradable compounds disappear and nondegradable compounds are left.
- At the end of the landfarming process, the concentration is low and compounds are no longer available for microorganisms.

For a single compound like pentachlorophenol, only the second explanation can be true because the molecular structure of pentachlorophenol does not change. With respect to mineral oil the composition changes during the landfarming process (Figure 4) and less degradable compounds are left. The cluster in the chromatogram is composed of several branched hydrocarbons. The experiments with detergents show that these hydrocarbons are degradable under "ideal" circumstances, but in practice the degradation in landfarming will be very slow. To improve the landfarming method into an operational method for soil restoration, it is necessary to find methods to decrease residual concentrations.

It is assumed from the experiments with detergents and with pentachlorophenol that bioavailability is the most important parameter. Bioavailability can be increased using detergents. However, high concentrations of detergents are necessary, and it is not possible to dissolve the entire residual concentration in one step because of the limited amount that can be put into a solution. Hurtig *et al.* (1988) also found that, even if a detergent was used, a residual concentration remained in the soil. Furthermore, the detergents producing the highest solubility of oil (the Dobanols, Table 2) are the most easily degradable, resulting in a high oxygen consumption and leaving less oxygen available for oil degradation (Table 3). It is questionable whether these detergents work long enough in landfarming.

A more fundamental experiment found that part of the adsorbed components desorbed relatively fast (Figure 6). The partition between the solid phase and the water phase in the soil of this same part can be predicted from the solubility and the amount of organic carbon in the soil (Harmsen *et al.* 1988; Schwarzenbach & Westall 1981). The second part desorbed very slowly (the tailing in Figure 6) and enlarged with decreasing solubility (Figure 7). It is assumed that this is the part adsorbed inside soil particles. To degrade this part, it is necessary to transport it from the inside of a soil particle or small soil pore to bigger pores where the microorganisms are present. During transportation the same retardation occurs as in a soil column: highly soluble compounds are transported rapidly, and low soluble compounds very slowly. For this phenomenon Wu and Gschwend (1986) introduced an effective diffusion coefficient that includes the solubility. Figure 6 shows that this diffusion coefficient is important for the desorption of the second part. Pentachlorophenol also follows the general rule, as presented in Figure 7, and 20 percent of the pentachlorophenol is likely to be desorbed slowly (Harmsen & Pennings 1990).

The soil should contain sufficient water for the desorption process of the second part, i.e., for diffusion into the mobile water phase. In the Netherlands, soil water content is high during winter and spring. In the spring the temperature may be high enough to effect biodegradation. In summer the degradation slows in spite of higher temperatures, and water content decreases because evaporation exceeds the amount of rain.

It is assumed that during spring the readily available part of the pollution is degraded. Although in summer the less available part ought to be degraded, this part will not be released because of the insufficient amount of the transport medium, water. During the following winter the sufficient amount of water present enables redistribution of the pollutants. Consequently, the newly available components can be degraded during next spring. Not only is water necessary for the redistribution, but time as well. From Figure 7 and the theory of Wu and Gschwend (1986), it may be concluded that the lower the solubility, the slower the redistribution.

CONCLUSIONS

Landfarming is a biological method used to clean soils contaminated by organic compounds. It is relatively simple to stimulate the degradation both under laboratory and field conditions. Microorganisms are not able to remove all the contaminants from the soil; depending on the properties of the polluted substance, some residual concentration is left in the soil. It is assumed that this residual concentration becomes only slowly available for degradation. Desorption experiments have shown that the velocity of desorption from the soil matrix can be related to the water solubility of the compound of interest. Use of detergents can help to increase the water solubility.

The use of landfarming is limited at the moment by the residual concentrations found in the cleaned soil. Methods to improve landfarming should no longer be focused on the stimulation of the process at the beginning. More effort is necessary to increase the bioavailability of residual concentrations.

Acknowledgments

This research is part of The Netherlands Integrated Soil Research Programme.

REFERENCES

Abrahamson, K.; Xie, T. M. *J. Chromatography* **1983**, *279*, 199–208.

Berends, J.; Kloeg, D. *Land + Water/Milieutechniek* **1987**, *12*, 72–75.

DHV Consultancy. Soil protection series of the Ministry of Housing, Physical Planning and Environment nr 55 B; State Publishers: The Hague, The Netherlands, 1986.

Harmsen, J.; Pennings, M. *Proceedings*, Symposium "Biotechnologische Bodemsanering"; 12 June, NIAIA: The Hague, The Netherlands, 1990.

Harmsen, J.; Bouter, P.; Bransen, F.; Versteeg, G. J. In *Contaminated Soil '88*; Wolf, K.; Van den Brink, W. J.; Colon, F. J., Eds.; Kluwer: Dordrecht, 1988; pp 109–111.

Hoeks, J.; Harmsen, J.; Pennings, M. ICW Report 33; The Winand Staring Centre: Wageningen, The Netherlands, 1988.

Hurtig, H. W; Knacker, T.; Schalnass, H.; Arendt, G. In *Contaminated Soil '88*; Wolf, K.; Van den Brink, W. J.; Colon, F. J., Eds.; Kluwer: Dordrecht, 1988; pp 921–928.

IWACO Consultancy. Report 20.358; IWACO: Groningen, The Netherlands, 1989.

Marel, M. van der; Zewald, C. A.; Veerkamp, W. *Land + Water/Milieutechniek* **1988**, *10*, 115–119.

Oldenhuis, R.; Kuijk, L.; Lammers, A.; Jansen, D. B; Witholt, B. *Appl. Microbiol. Biotechnol.* **1989**, *30*, 211–217.

Schwarzenbach, R. P.; Westall, J. *J. Environ. Sci. Tech.* **1981**, 1360–1367.

Soczo, E. R.; Staps, J. J. M. In *Contaminated Soil '88*; Wolf, K.; Van den Brink, W. J.; Colon, F. J., Eds.; Kluwer: Dordrecht, 1988; pp 663–670.

Soczo, E. R.; Staps, J. J. M; Visser, W. J. F. RIVM Report No. 736102002; National Institute of Public Health and Environmental Hygiene: Bilthoven, The Netherlands, 1990.

Valo, R.; Salkinoja-Salonen, M. *Appl. Microbiol. Biotechnol.* 1986, *25*, 68–75.

Verheul, J.H.A.M.; Van den Berg, R.; Eikelboom, D. H. In *Contaminated Soil '88*; Wolf, K.; Van den Brink, W. J.; Colon, F. J., Eds.; Kluwer: Dordrecht, **1988**; pp 707–715.

Wu, S.; Gschwend, P. M. *J. Environ. Sci. Tech.* **1986**, 717–724.

Treatment of Toxic or Refractory Aromatic Compounds by Chemical Oxidants

Soon Haing Cho[*]
Ajou University
Alan R. Bowers
Vanderbilt University

INTRODUCTION

A variety of organic compounds are refractory, toxic, or inhibitory in nature. Wastewaters containing these organics must be pretreated before being introduced to conventional biological treatment systems. The purpose of pretreatment is to remove or destroy these compounds, or alternately to convert them to less toxic and more readily biodegradable intermediate products which can be more easily treated by conventional biological treatment methods. One of the possible pretreatment methods is partial oxidation of these chemicals by chemical oxidants.

For the efficient and economical treatment of refractory organic compounds, treatability studies must be conducted on a variety of individual compounds, since each compound reacts differently with different oxidants. Therefore, in this study the effect of chemical oxidants on the biodegradability and the toxicity of various nonbiocompatible compounds was tested. The oxidants were compared based on reactivity with the original compounds, biodegradability of the by-products, and toxicity of the by-products.

---

[*] Department of Environmental Engineering, Ajou University, Suwon, Korea

Hydrogen peroxide has long been known to be a powerful oxidant to treat sulfur compounds (Eisenhauer 1965; Kibbel *et al.* 1972; Satterfield *et al.* 1954). Typically, when combined with a ferrous iron salt known as Fenton's reagent, hydrogen peroxide is used as a catalyst and is very effective in treating phenolic wastes over a wide range of temperatures and concentrations (Eisenhauer 1964; Sosnovsky & Rawlinson 1971).

Permanganate is another vigorous oxidant that has long been used in the laboratory. It also has been commonly used as a water treatment chemical since 1913. Permanganate is very effective in removing tastes and odors caused by the decomposition of organic matter. Ozone is also known to be a powerful oxidant readily available for water and wastewater treatment. It has been used for disinfection, chemical oxidation, and preoxidation in preparation for biological treatment. However, ozone is much more selective in its oxidation reactions than hydrogen peroxide and permanganate (Hoigne & Bader 1983; Lalezary *et al.* 1986; Peyton *et al.* 1982).

In this study, hydrogen peroxide, potassium permanganate, and ozone were considered as the chemical oxidants to treat the refractory and toxic organic compounds. Chlorine was not included in this study because of the possible formation of chlorinated hydrocarbon reaction products which can be carcinogenic to humans.

MATERIALS AND METHODS

Test Compounds

Test compounds were obtained from Fisher Scientific Company, Inc., or Aldrich Chemical, Inc. All chemicals used in this study were of reagent grade or better. Stock solutions, 0.1 M, of the test compound were prepared with distilled water. For compounds that had a low solubility, such as dinitrobenzene, naphthalene, and diphenylamine, stock solutions were prepared at lower concentrations depending on their solubilities. Table 1 shows the compounds and their known solubilities.

Oxidants and Reaction Conditions

Hydrogen peroxide was obtained from Interox America, Inc., Deer Park, Texas, at a concentration of 50 percent by weight. Fenton's reagent ($FeSO_4 \cdot 7H_2O$) was prepared before the start of each new set of

TABLE 1. Solubility and molecular weight of test compounds.

Compound	Solubility (mg/L)	Molecular Weight
Benzaldehyde	3,000	106.12
Sulfanilic Acid	8,000	173.19
Vanillin	10,000	152.15
Indole	soluble (hot)	117.15
Resorcinol	soluble	110.11
Catechol	–	110.11
Pyrogallol	–	126.11
Hydroquinone	–	110.11
Phthalic Acid	7,000	166.13
Salicylic Acid	2,000	138.12
Coumarin	3,000	146.15
Pyrrolidine	infinity	71.12
Diphenylamine	300	169.23
Naphthalene	30	128.17
Dinitrobenzene	100	168.11
Skatole	500	131.18

experiments at a concentration of 50 mg/L (as Fe). Fermcolase-1000, manufactured by Finnsugar Biochemicals Inc., Schaumburg, Illinois, was used to destroy the hydrogen peroxide remaining after 24 h of oxidation. H_2O_2 test strips, manufactured by E. M. QUANT, EM Science, Inc., Cherry Hill, New Jersey, were used to verify the complete destruction of residual H_2O_2 in the test solutions.

Potassium permanganate ($KMnO_4$) and sodium sulfite (Na_2SO_3) were also obtained from Fisher Scientific, Inc. Permanganate was added to the test solution in solid form. Sodium sulfite was used to remove the residual permanganate in the test solutions.

Ozone (O_3) was produced in the laboratory using an Ozone Generator (LINDE Model SG-4060). Ozone gas was generated each time a reaction was performed. Figure 1 shows the flow schematic for the ozone oxidation experiments.

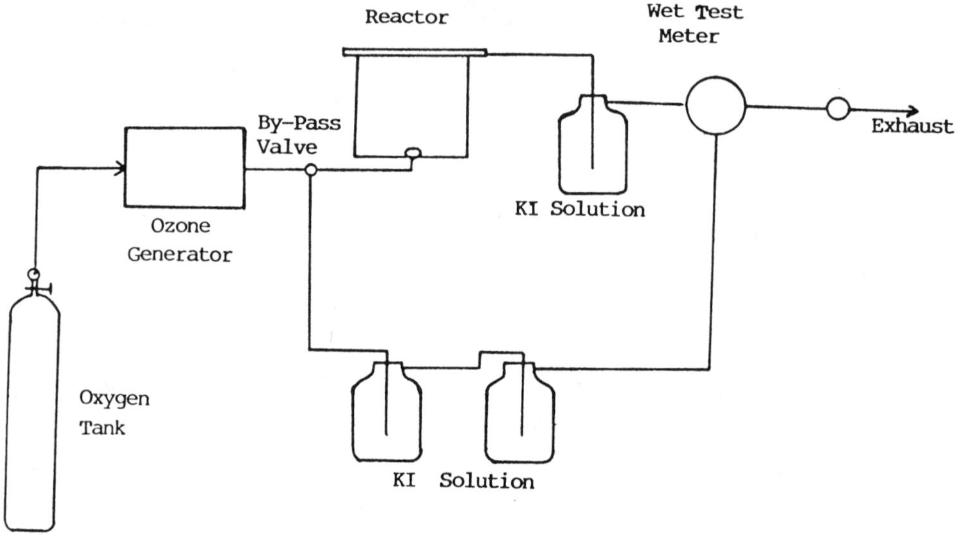

FIGURE 1. Flow schematic for the ozone oxidation experiments.

Oxidation of compounds by catalyzed hydrogen peroxide and potassium permanganate was carried out in batch shake flasks. Stoichiometric amounts of oxidants were used in hydrogen peroxide and permanganate oxidations. The dosages of oxidants were calculated based on the assumption of ultimate degradation, which is the complete destruction of a compound to CO_2, H_2O_2, and/or other organics. For oxidation of compounds by ozone, ozone gas was introduced to the synthetic wastewater containing organics.

An initial pH of 3.5 was found to be the most suitable for catalyzed hydrogen peroxide from previous research (Bowers *et al.* 1989). The optimum pH for permanganate oxidation varies widely depending on the nature of organic compounds. The reactivity of permanganate depends on the types of organic compounds present and to a great degree on the reaction pH conditions (Wieberg 1965). Extreme pH conditions are also more favorable for permanganate oxidation than neutral pH. The general factors governing the treatment efficiency are the reactivity of ozone with the target compound, the rate of reactivity, extent of incidental stripping associated with ozone dispersion, temperature, and pH (Masschelein 1982). The residence time of ozone gas in water was reported to be the longest at a pH of about 3.5 to 4. At high pH values, ozone self-decomposition accelerates, resulting in a higher net ozone consumption. Because of these reasons and in order to normalize the standard experimental conditions, the oxidation of organic compound was carried out at a solution pH of 3.5.

Experimental Procedures

The oxidation reaction was carried out in batch shake flask tests. The oxidation time was 24 h for both catalyzed hydrogen peroxide and potassium permanganate. The time required for the stoichiometric amount of ozone infusion was less than 1 min in most cases. This reaction time was very short compared with the oxidation time of catalyzed hydrogen peroxide or potassium permanganate. In addition, the solubility of ozone is reported to be very low. The solubility of ozone is 8.92 mg/L at 20 C (Rice & Browning 1981). For these reasons, the ozonation time for the synthetic wastewater was extended to 10 times the time required for stoichiometric amount of ozone infusion (i.e., 10 min in most cases).

After 24 h of oxidation by peroxide or permanganate, the amount of residual oxidants in solution was measured. Generally there was no significant oxidant remaining in the solution after 24 h. Any residual peroxide was decomposed by adding Fermcolase-1000. Residual permanganate was removed by adding sodium sulfite to reduce Mn (VII) to Mn (IV). Any residual precipitates, such as MnO_2, were separated by filtration using 0.45-um filters (Gelman No. GN-6, 47mm) and a vacuum apparatus. No residual ozone was detected after 4 h of oxidation (based on preliminary tests). Therefore, no further destruction or removal procedure was applied to eliminate ozone from the sample solution after oxidation.

COD (chemical oxygen demand), TOC (total organic carbon), and phenolic residual concentration were measured to monitor the decomposition of the organic compound by chemical and biological oxidation. For the COD test, the closed reflux, titrimetric method was used. For TOC analysis of samples, a Dohrman (Model DC 90) total organic carbon analyzer was used. The residual phenolic compound was measured using the 4-amino antipyrine direct photometric method. All of the analytical procedures were conducted in accordance with the methods prescribed in *Standard Methods for the Examination of Water and Wastewater* (APHA 1985). Toxicity of each organic was measured before and after chemical oxidation to compare the degree of toxicity using different oxidants. Toxicity was expressed in terms of the EC_{50} using a Beckman Microtox (Model 2055) toxicity analyzer. The EC_{50} is the effective concentration causing a 50-percent diminution in light output of luminescence bacteria under stated conditions of exposure time and test temperature. In this toxicity test, 15 min and 15 C were selected. Test procedures are described in the *Beckman Microtox System Operating Manual* (Kasler & Greene 1982).

Biodegradability was measured using municipal activated sludge obtained from the Metro Nashville Central Wastewater Treatment Facility. A 14-day batch test was performed with the unoxidized compound and oxidized compound adjusted to the same initial value (if the COD of the oxidized compound is less than 100 mg/L) or 100 mg/L (as COD). The reason for this dilution is to shield the organism from toxic and inhibitory effects of the original compound.

The pH of the test solution was maintained at 7.0 using a phosphate buffer. This phosphate buffer also served as the phosphorus nutrient source for bacterial growth. In addition, nitrogen was added. The calculation of the amount of N and P required was based on COD:N:P = 100:5:1 (Eckenfelder 1980). Other trace minerals were not added, but tap water was used as the dilution water and it was assumed that sufficient amounts of trace elements were available.

COD and TOC were measured at 0, 2, 7, and 14 days of biooxidation. These data were used to characterize the extent of biodegradation of organic compounds. Toxicity was also measured after 14 days of biodegradation. Figure 2 shows the flow chart of the overall experimental procedure.

RESULTS AND DISCUSSION

Hydrogen Peroxide Oxidation

Fe-catalyzed hydrogen peroxide reacted well with most of the 16 compounds tested. The percent reduction of COD and TOC and the change of toxicity after 24 h of peroxide oxidation compounds are summarized in Table 2 for each of the compounds tested. Based on TOC data, 17 to 79 percent of the compounds were converted to CO_2. Among the compounds tested, benzaldehyde showed the maximum degradation, and skatole did not show any ultimate degradation. The COD reduction of compound with hydrogen peroxide was generally high. It ranged 39 to 95 percent after chemical oxidation. However, dinitrobenzene did not exhibit any reduction after 24 h of oxidation in terms of COD.

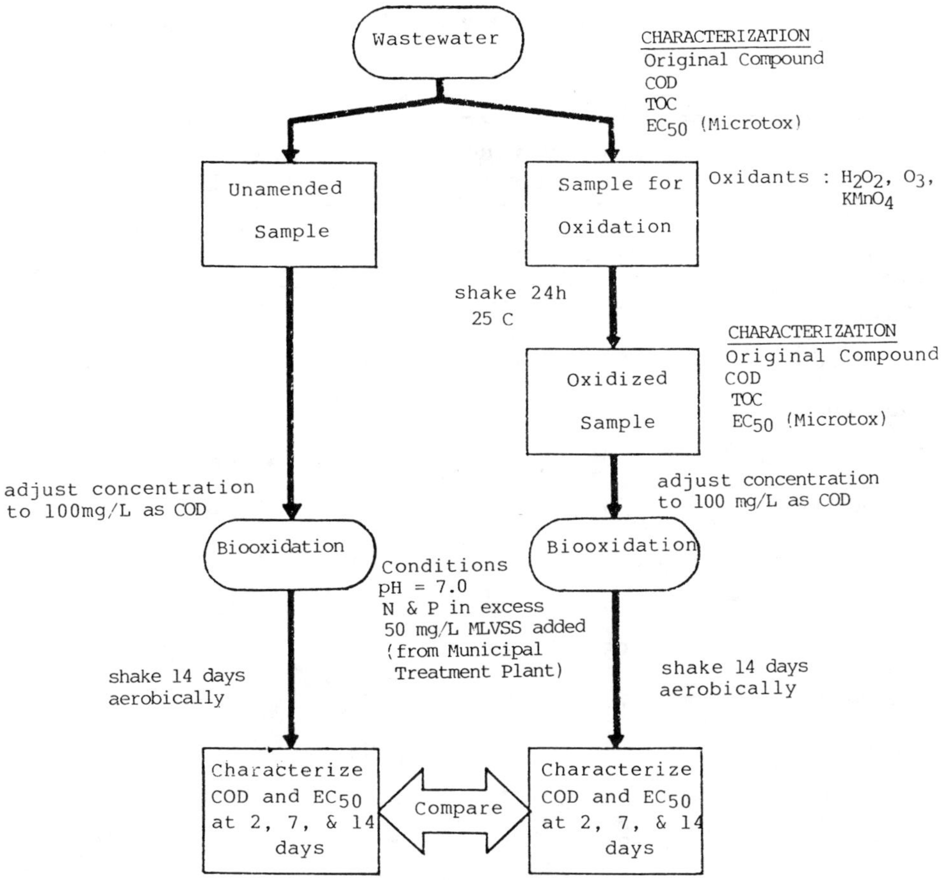

FIGURE 2. Flow chart for experimental procedure.

Based on the Microtox® test, toxicities of the compounds were reduced to a certain extent after chemical oxidation. However, resorcinol, vanillin, and salicylic acid did show an increase in toxicity after oxidation when considering the 24-h toxicity data on an equivalent TOC basis. COD and TOC reduction of these three organics ranged from 75 to 88 percent and 29 to 70 percent, respectively. Based on these data, the toxicity increase after oxidation might be attributed to the toxicity of intermediate oxidation by-products.

The oxidation products of coumarine and naphthalene did not exhibit any detectable toxicity. The degree of toxicity reduction of other compounds varied. The toxicity reduction of catechol, hydroquinone, benzaldehyde, diphenylamine, sulfanilic acid, and dinitrobenzene after 24-h hydrogen peroxide oxidation was improved by a factor of about

TABLE 2. Hydrogen peroxide oxidation of aromatic organic compounds.

	% Reduction after Oxidation		Change of Toxicity (EC_{50})		
Compound	COD	TOC	Before Oxidation	After Oxidation	Adjusted
Catechol	80	57	8.1	30	13
Resorcinol	80	57	55	64	28
Hydroquinone	78	31	0.3	1.6	1.1
Pyrogallol	75	45	2.8	61	34
Benzaldehyde	93	79	4.8	54	11
Coumarine	65	26	2.0	NT[(a)]	NT
Vanillin	88	70	22	53	16
Indole	95	62	1.3	56	21
Pyrrolidine	72	35	NT	NT	NT
Skatole	39	0	0.9	4.2	(b)
Salicylic Acid	75	29	27	25	18
Phthalic Acid	71	37	NT	NT	NT
Diphenylamine	88	70	14	83	25
Sulfanilic Acid	75	46	14	62	34
Dinitrobenzene	0	17	4	5	4.2
Naphthalene	80	44	41	NT	NT

(a) NT = Nondetectable toxicity.
(b) Toxicity cannot be adjustable since TOC is 0.

1 to 3. For pyrogallol and indole, the toxicity was improved by the order of magnitude.

Biodegradability tests indicated that hydrogen peroxide by-products of sulfalinic acid, hydroquinone, diphenylamine, dinitrobenzene, phthalic acid, and pyrrolidine (based on COD or TOC data) were generally more amenable to biological treatment than the original compounds. Other organics showed opposite results—chemical oxidation by-products were not more amenable to biological treatment than the original compound at low concentration (100 mg/L as COD). This would imply that some of the oxidation by-products may be more difficult to biodegrade than their original compounds at low concentrations. Or it can be speculated that the chemical oxidation by-product might be the final refractory end product which will not be biodegraded any further.

Another speculation can be explained as follows. Since both of the initial samples for the biodegrabality test were prepared based on the COD, the unoxidized sample was diluted to the COD value of 100 mg/L of original compound. The oxidized sample was used without any dilution (if the COD was less than 100 mg/L) or diluted to 100 mg/L (if the COD was over 100 mg/L), while the original unoxidized sample was diluted to 100 mg/L as COD.

Therefore the dilution factor was far less than that of the original unoxidized sample. Thus the amount of unoxidizable portion might be more in the oxidized sample than in the unoxidized sample.

Permanganate Oxidation

Among the 16 compounds tested, 8 compounds were not reactive with permanganate at the experimental condition of this study. These compounds are coumarine, pyrrolidine, skatole, diphenylamine, sulfanilic acid, phthalic acid, dinitrobenzene, and naphthalene. Compounds other than these showed 22 to 68 percent conversion of the original compound to CO_2 based on TOC data. COD reduction of these organics ranged from 50 to 91 percent. As in the case of hydrogen peroxide oxidation, benzaldehyde showed maximum degradation in terms of TOC. A summary of data is given in Table 3.

The degree of toxicity reduction of organic compounds by permanganate oxidation was better than that of hydrogen peroxide oxidation when comparing the 24-h toxicity data on an equivalent TOC basis. Even though the toxicity was substantially reduced by permanganate oxidation, every oxidized product was observed to retain some degree

TABLE 3. Permanganate oxidation of aromatic organic compounds.

	% Reduction after Oxidation		Change of Toxicity (EC_{50})		
Compound	COD	TOC	Before Oxidation	After Oxidation	Adjusted
Catechol	66	52	8.1	50	24
Resorcinol	73	28	55	42	30
Hydroquinone	71	27	0.3	63	46
Pyrogallol	78	22	2.8	50	39
Benzaldehyde	79	68	4.8	11	14
Coumarine	–	–	2.0	–	–
Vanillin	55	53	22	62	29
Indole	91	60	1.3	25	10
Pyrrolidine	–	–	(a)	NT	NT
Skatole	–	–	0.9	–	–
Salicylic Acid	50	32	27	26	18
Phthalic Acid	–	–	NT	NT	NT
Diphenylamine	–	–	14	–	–
Sulfanilic Acid	–	–	14	67	–
Dinitrobenzene	–	–	4	–	–
Naphthalene	–	–	41	NT	NT

(a) NT = Nondetectable toxicity.

of toxicity. Resorcinol and salicylic acid showed an increase in toxicity after oxidation.

Same trends were observed in the biodegradability test in terms of permanganate oxidation products and hydrogen peroxide oxidation products. According to the COD and TOC data, the degree of biodegradation rapidly occurred for 2 days and slowed during the rest of the experimental period. Vanillin, pyrogallol, and hydroquinone (based on TOC or COD data) were more amenable to biological treatment than the original compounds. Benzaldehyde, indole, resorcinol, catechol, and salicylic acid showed opposite results.

Ozone Oxidation

For the 16 compounds tested, all compounds were reactive with ozone except coumarine. Ozone oxidation data are summarized in Table 4. Even though the reaction time may not be sufficient, 0 to 74 percent of the organics were converted to CO_2 (based on TOC reduction). The COD reduction of organic compounds after oxidation ranged from 31 to 100 percent. Among the 16 compounds, benzaldehyde showed maximum degradation. This was the same as in the case of both hydrogen peroxide and permanganate oxidation. Coumarine, skatole, dinitrobenzene, and naphthalene did not show degradation in terms of TOC. For skatole, dinitrobenzene, and naphthalene, however, there was a substantial decrease in the COD value. Coumarine did not show degradation both in TOC and COD data. Benzaldehyde and diphenylamine showed the highest percentage of reduction in terms of TOC and COD, respectively.

The degree of toxicity reduction of compounds by ozone oxidation was generally better than that of the other two oxidants when individually comparing toxicity data on an equivalent TOC basis. Most of the residual products from ozone oxidation showed some detectable toxicity except naphthalene. Oxidized sulfalinic acid, hydroquinone, phthalic acid, diphenylamine, and naphthalene were more amenable to biological treatment than the unoxidized original compound.

Biooxidation and Overall Removal Efficiencies

Based on the results of biological oxidation tests, organics can be categorized by the following three groups: (1) Organics that showed the same degradation patterns whether pretreated with chemical oxidants or untreated. Figure 3 shows the biodegradation of resorcinol treated with chemical oxidants and of untreated resorcinol. Organics such as vanillin, catechol, pyrogallol, salicylic acid, pyrrolidine, coumarine, phthalic acid, skatole, diphenylamine, hydroquinone, and sulfanilic acid showed similar trends. (2) Organics treated with chemical oxidants did not show the effect of biodegradation, while the untreated organics showed decreasing trends. Figure 4 shows the biodegradation of indole treated with chemical oxidants and of untreated indole. Naphthalene

TABLE 4. Ozone oxidation of aromatic organic compounds.

	% Reduction after Oxidation		Change of Toxicity (EC_{50})		
Compound	COD	TOC	Before Oxidation	After Oxidation	Adjusted
Catechol	31	22	8.1	39	30
Resorcinol	50	29	55	48	34
Hydroquinone	45	17	0.3	48	40
Pyrogallol	48	29	2.8	42	30
Benzaldehyde	74	74	4.8	39	10
Coumarine	–	–	2.0	–	–
Vanillin	64	64	22	65	23
Indole	77	61	1.3	45	18
Pyrrolidine	59	32	(a)	NT	NT
Skatole	38	0	0.9	8	8
Salicylic Acid	42	31	27	28	19
Phthalic Acid	52	31	NT	NT	NT
Diphenylamine	90	31	14	88	61
Sulfanilic Acid	57	57	14	70	30
Dinitrobenzene	35	0	4	11	11
Naphthalene	100	0	41	NT	NT

(a) NT = Nondetectable toxicity.

and benzaldehyde showed similar trends. (3) Organics that did not show the effect of biodegradation whether chemically pretreated or untreated. Figure 5 shows the biodegradation of chemically treated dinitrobenzene and of chemically untreated dinitrobenzene. As can be seen, biological oxidation may not be a good approach to treat these types of organics.

The average oxidation state of organic carbon in the waste mixture may be expressed using the following equation (Stumm & Morgan 1981):

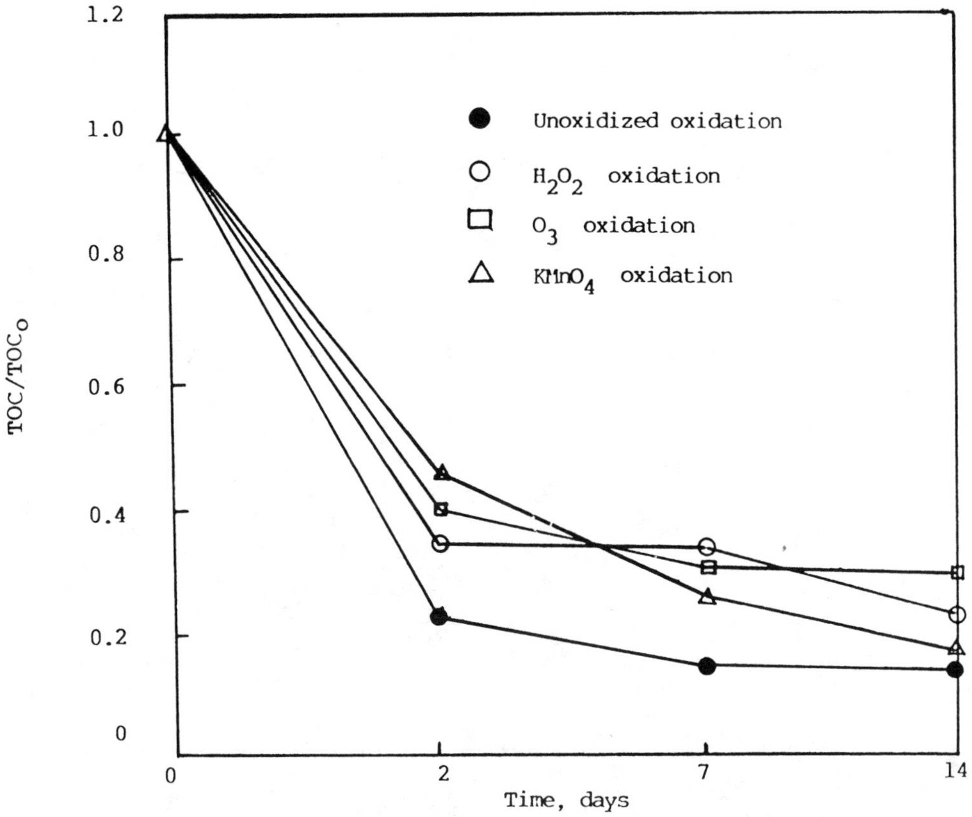

FIGURE 3. Normalized TOC/TOC_0 for resorcinol.

$$OX\ (\text{Oxidation State}) = 4\ (TOC - COD)\ /\ TOC \tag{1}$$

The change in oxidation state for each of the organic compounds by H_2O_2 oxidation, $KMnO_4$ oxidation, and ozone oxidation is compared in Figures 6, 7, and 8, respectively. The data show that the original compounds are drastically altered even though all of the oxidants were not used for ultimate conversion of organic carbon. Indole exhibited the greatest change ($\Delta OX = 4.28$ for H_2O_2 oxidation, $\Delta OX = 3.75$ for $KMnO_4$ oxidation) in both peroxide and permanganate oxidation. Naphthalene exhibited the greatest change ($\Delta OX = 6.04$) by ozone oxidation. The smallest change was for dinitrobenzene ($\Delta OX = 0.71$) for H_2O_2 oxidation, vanillin ($\Delta OX = 0.16$) for $KMnO_4$ oxidation, and sulfanilic acid and vanillin ($\Delta OX = 0.01$) for ozone oxidation. Based on the change of oxidation state value (ΔOX), organic compounds that were oxidized by hydrogen peroxide showed comparable values. This does not necessarily mean that hydrogen peroxide is the best oxidant among the three

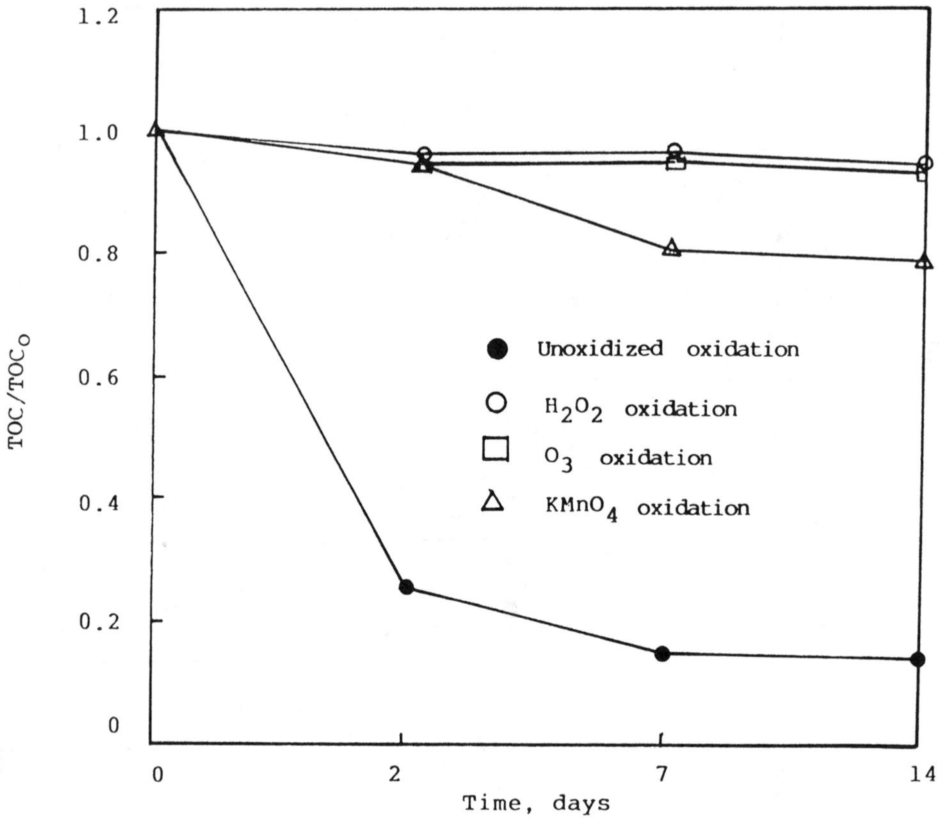

FIGURE 4. Normalized TOC/TOC$_0$ for indole.

oxidants tested. However, it can be said that most of the organic compounds tested in this study reacted well with hydrogen peroxide.

The overall COD removal efficiency of both chemical oxidation and biological oxidation can be expressed as the following equation:

$$\text{Overall Removal Efficiency, \%} = (1 - \alpha) \times 100 \qquad (2)$$

where,

$$\alpha = (COD_{RB} / COD_{OB}) \quad (COD_{RC} / COD_O)$$

COD_{RB} = COD remaining after 14 days of biooxidation, mg/L

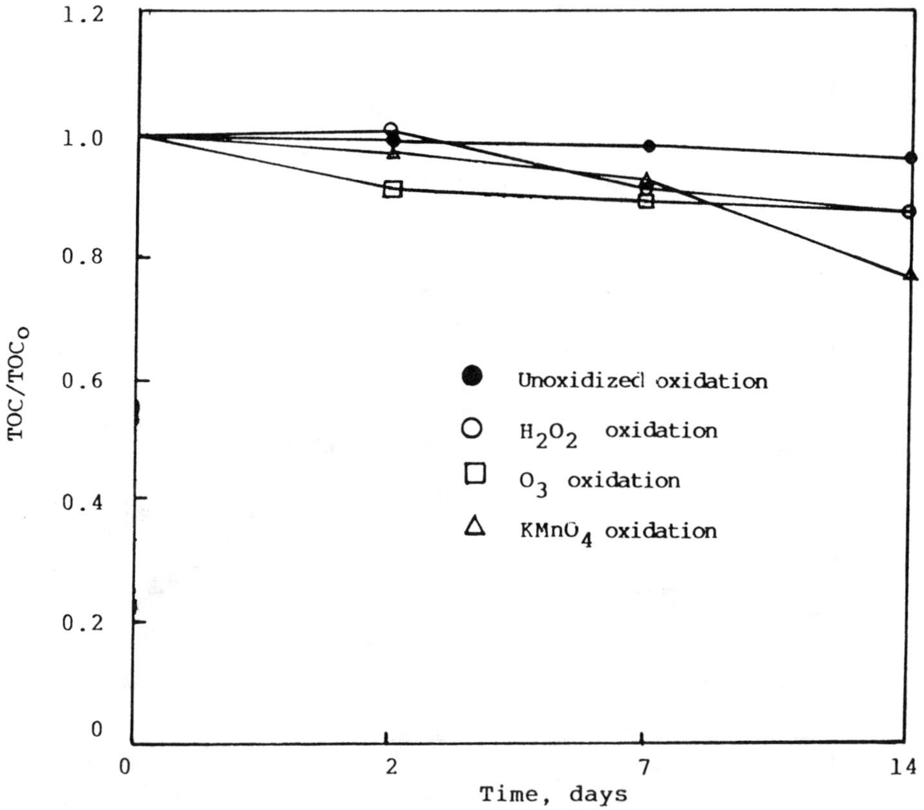

FIGURE 5. Normalized TOC/TOC$_o$ for dinitrobenzene.

COD_{OB} = initial COD in the biooxidation reactor, mg/L

= 100 mg/L in most cases of this work

COD_O = initial COD before any treatment, mg/L

COD_{RC} = COD remaining after chemical oxidation, mg/L.

Table 5 shows the overall removal efficiencies of organic compounds by each oxidant. The overall removal efficiencies showed promising results for using combined treatment of chemical oxidation and biological oxidation for most of the organics tested. Based on these data, catalyzed hydrogen peroxide showed the best results among the three oxidants compared. However, if the oxidation reaction was conducted at optimal pH for permanganate oxidation, and a sufficient amount of time was allowed for ozonation, these two oxidants would

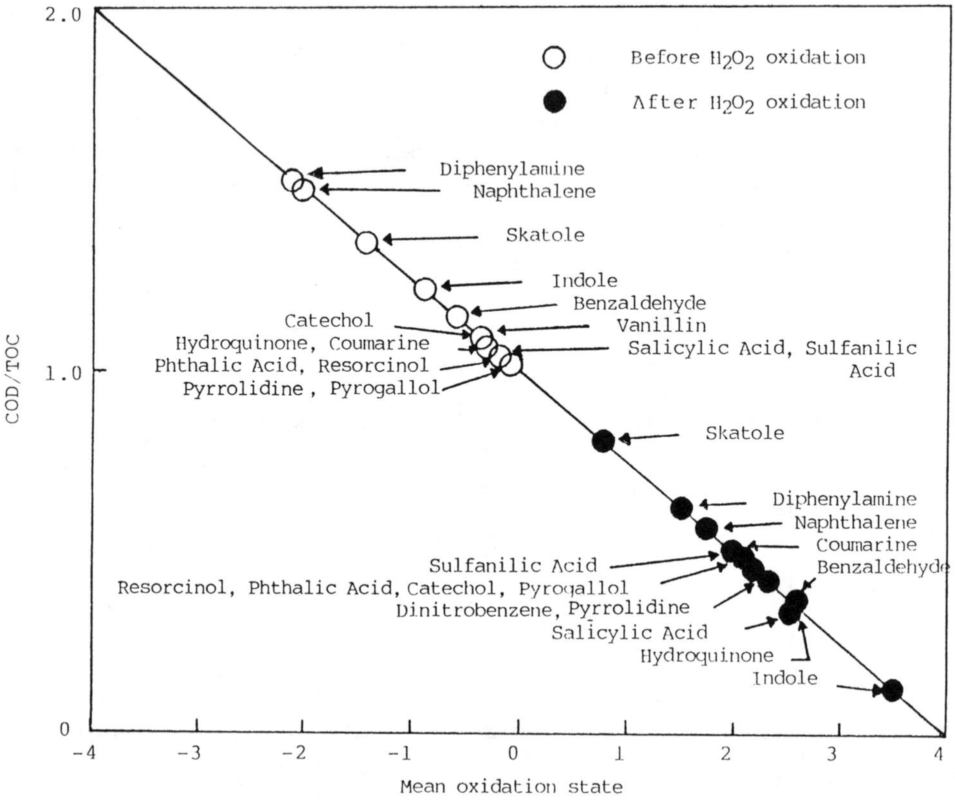

FIGURE 6. The COD/TOC versus mean oxidation state of carbon in the system (H_2O_2 oxidation).

show results similar to those of hydrogen peroxide. Further research is recommended for these two oxidants (permanganate and ozone) to determine optimal oxidation conditions and feasibility.

SUMMARY AND CONCLUSIONS

Among the three oxidants tested with 16 compounds, catalyzed hydrogen peroxide and ozone proved to be effective oxidants. Even though permanganate did not oxidize half of the 16 chemicals tested, it cannot be concluded that permanganate is not a good oxidant. For example, hydroquinone and pyrogallol oxidized by permanganate showed the best results in the biodegradability test among the three oxidants. The reason for this selective oxidation of permanganate might be due to the wide range of optimum pH conditions for permanganate oxidation.

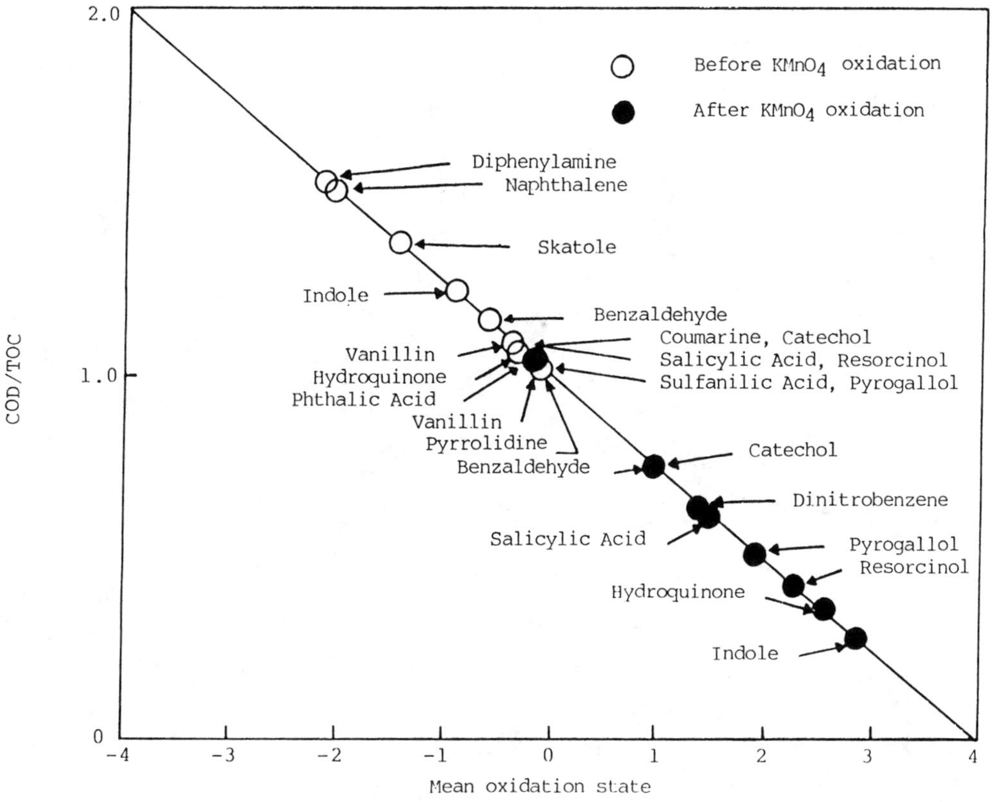

FIGURE 7. The COD/TOC versus mean oxidation state of carbon in the system ($KMnO_4$ oxidation).

However, a fixed pH (3.5) was used as the standard oxidation condition in this experiment. Further study is recommended for the unoxidized compound by permanganate under different pH conditions.

Comparing toxicity data on an equivalent TOC basis, toxicity reduction of compounds by ozone oxidation generally shows better results than by the other two oxidants. The percentage reduction with catalyzed hydrogen peroxide showed the highest value in most of the compounds tested.

Most of the oxidation products were biodegraded rapidly in 2 days regardless of which oxidant was used. In general, applying a chemical pretreatment to the difficult aromatic compounds has proven to be an effective technique for reducing the toxicity and improving the overall reduction of COD and TOC. Since the sudden loading of wastewater containing toxic nonbiodegradable compounds can cause the biological treatment process to fail, this chemical pretreatment technique

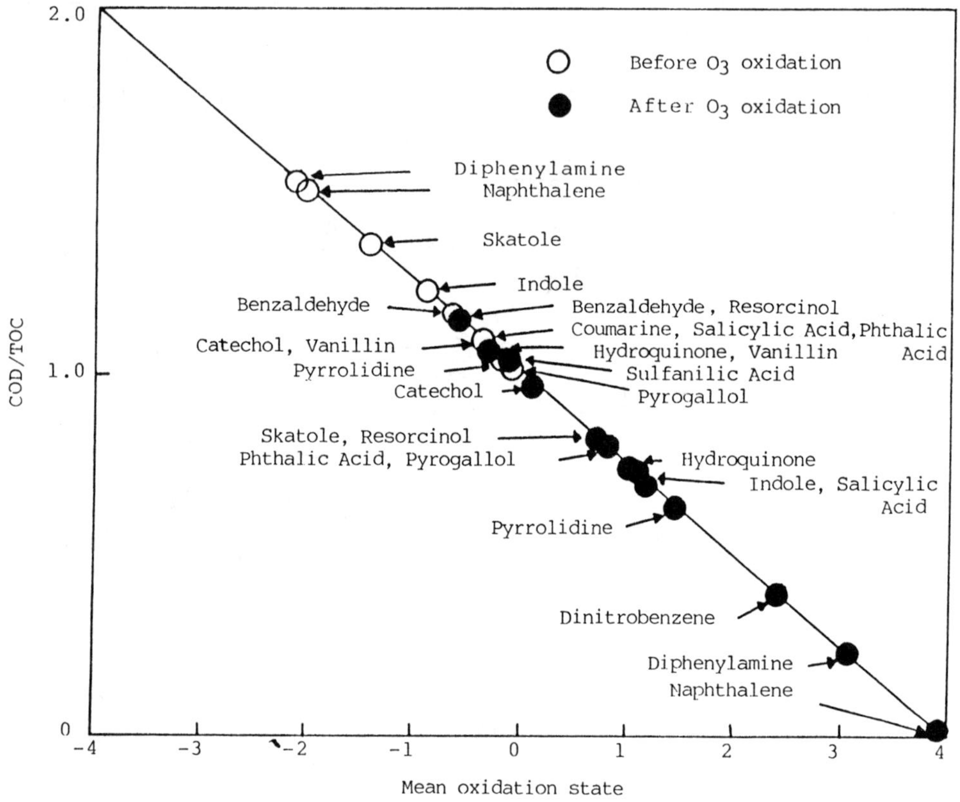

FIGURE 8. The COD/TOC versus mean oxidation state of carbon in the system (O_3 oxidation).

will be especially valuable for treating such wastewater. A feasibility study would be recommended depending on the organic compounds to be treated.

TABLE 5. Overall removal efficiency of chemical oxidation and biooxidation by each oxidant.

	Removal Efficiency, %		
Compound	Hydrogen Peroxide	Permanganate	Ozone
Catechol	94	77	90
Resorcinol	89	86	88
Hydroquinone	70	62	69
Pyrogallol	62	66	64
Benzaldehyde	79	78	77
Coumarine	81	–	–
Vanillin	97	96	93
Indole	100	87	71
Pyrrolidine	90	–	84
Skatole	90	–	73
Salicylic Acid	91	75	81
Phthalic Acid	88	–	81
Diphenylamine	98	–	98
Sulfanilic Acid	96	–	92
Dinitrobenzene	(a)	–	51
Naphthalene	100	–	100

(a) COD increased after chemical oxidation.

REFERENCES

APHA (American Public Health Association). *Standard Methods for the Examination of Water and Wastewaters,* 16th Ed.; American Public Health Association: Washington, DC, 1985.

Bowers, A. R.; Gaddipati, P.; Eckenfelder, W. W., Jr.; Monsen, R. M. "Treatment of Toxic or Refractory Wastewaters with Hydrogen Peroxide." *Wat. Sci. Tech.* **1989,** *21,* 477–486.

Eckenfelder, W. W., Jr. Principles of Water Quality Management;, CBI Publishing Company: Boston, 1980.

Eisenhauer, H. R. "Oxidation of Phenolic Wastes, Part I, Oxidation with Hydrogen Peroxide and a Ferrous Salt Reagent." *J. Wat. Pol. Con. Fed.* September **1964,** 1116–1128.

Eisenhauer, H. R. "Sweetening of Petroleum Distillates with Hydrogen Peroxide." *Industrial Engineering Chemistry,* Product Research and Development **1965,** *4*(3), 185–189.

Hoigne, J.; Bader, H. "Rate Constants of Reactions of Ozone with Organic and Inorganic Compound in Water, I Non-Dissociating Organic Compounds." *Water Research* **1983,** *17,* 173.

Kasler, L.; Greene, M. *Beckman Microtox System Operating Manual,* Beckman Instrument: Microbics Operations: Carlsbad, CA, 1982.

Kibbel, W. H., Jr.; Raleigh, C. W.; Shaperd, J. S. "Hydrogen Peroxide for Industrial Pollution Control." *Eng. Bull.,* Purdue Univ., Eng. Ext. Ser., **1972,** *141,* 824.

Lalezary, S.; Pirbazari, M.; McGuire, M. J. "Oxidation of Five Earthy-Musty Taste and Odor Compounds." *J. AWWA.* **1986,** *78*(3), 62.

Masschelein, W. J. *Ozonization Manual for Waste and Wastewater Treatment;* John Wiley and Sons: New York, 1982.

Peyton, G. R.; Huang, F. Y.; Burleson, J. L.; Glaze, W. H. "Destruction of Pollutants in Water with Ozone in Combination with Ultraviolet Radiation. 1. General Principles and Oxidation of Tetrachloroethylene." *Environ. Sci. & Technol.* **1982,** *16,* 448.

Rice, R. G.; Browning, M. E. *Ozone Treatment of Industrial Wastewater;* Noyes Data Corporation: Park Ridge, NJ, 1981.

Satterfield, C. N.; Reid, R. C.; Briggs, D. R. "Rate of Oxidation of Hydrogen Sulfide by Hydrogen Peroxide." *J. American Chemical Society* **1954,** *76,* 3922.

Sosnovsky, G.; Rawlinson, D. J. *Organic Peroxides,* Vol. II; Wiley-Interscience: New York, 1971; pp 269–336.

Stumm., W.; Morgan, J. J. *Aquatic Chemistry;* Wiley and Sons: New York, 1981.

Wieberg, K. B. *Oxidation in Organic Chemistry;* Academic Press, Inc.: New York, 1965.

Use of Liposomes for Combating Oil Spills and Their Potential Application to Bioreclamation

Shimon Gatt, Herve Bercovier, Yechezkel Barenholz*
Hebrew University

INTRODUCTION

Oil is a poor substrate for most microorganisms. Being a hydrocarbon it contains only carbon and hydrogen; an external source of nitrogen, phosphorus, and other essential nutrients is needed to support growth of microorganisms (Atlas 1984; Gould & Corey 1980).

Being very hydrophobic, petroleum and hydrophobic wastes do not interact with water and form a separate phase. The forces that prevent their interaction with water also reduce, very considerably, interaction of the oil or waste with other surfaces including those of most microorganisms (Atlas 1984; Beachey 1980; Gould & Corey 1980; Jann & Jann 1990; Racke & Coats 1990; Savage & Fletcher 1985).

Being composed of phospholipids (Szuhaj 1989), liposomes (Lichtenberg & Barenholz 1988) are a source of carbon, hydrogen, phosphorus, and nitrogen, and because of their vesicular structure, they can be used as a reservoir of encapsulated nutrients, minerals, and growth factors. We propose that liposomes modify the physical properties of oil or hydrophobic wastes, increasing their availability to microbial degradation, while concomitantly supplying nutrients essential for microbial growth.

* Department of Biochemistry, Hadassah School of Medicine, Hebrew University, Jerusalem 91010, Israel

Phospholipids

Phospholipids, which are the raw materials of liposomes (Lichtenberg & Barenholz 1988; Szuhaj 1989), are naturally occurring membrane-lipids in which glycerol is linked to two long chain fatty acids, a phosphoric acid residue and a nitrogen-containing base (*e.g.* choline). For large-scale environmental application, availability and low cost of the phospholipids are two major prerequisites. Plant phospholipids fulfill both requirements; their potential yield is close to 500,000 tons/year, and as by-products of the edible oil industry, they are included in animal feed, human food products, cosmetics, pharmaceuticals, and such technical applications as tanning and components of paints (Szuhaj 1989).

The phospholipids are composed of three major classes: (1) neutral, zwitterionic phosphatidylcholine (PC), referred to as lecithin; (2) the lesser hydrophilic, amphotheric phosphatidylethanolamine (PE), and (3) negatively charged phospholipids such as phosphatidylglycerol (PG) or phosphatidylinositol (PI).

Liposomes

Liposomes that can be multilamellar, oligolamellar, or unilamellar are sealed vesicles (microsacs) having one or more bilayered membrane, composed of phospholipids and containing water in their interior (Lichtenberg & Barenholz 1988). In the latter, one can encapsulate water-soluble compounds such as minerals, sugars, amino acids, vitamins, hormones, drugs, and growth factors (Becher 1983; Gregoriadis 1988). The membrane of the vesicle, which is composed of two phospholipid monolayers, is hydrophobic in its interior and hydrophilic in its two respective exteriors, thereby permitting entrapment of a variety of lipophilic or amphiphilic compounds, whether neutral or charged. Numerous methodologies have been developed for preparing liposomes (Lichtenberg & Barenholz 1988). As a general rule, the preparation requires two major steps: hydration of the lipid and bringing the liposomes to the desired size. We have studied the effect of the composition of various natural or modified soybean phospholipids on the size, physical properties, and stability of the liposomes, as well as their ability to withstand the action of high ionic strength and divalent ion concentration present in seawater and oil wells. Our results suggested that high proportions of PI and PE in the raw material reduce the stability of the liposomes in the above salt-containing media.

We have therefore developed procedures for increasing the proportion of the neutral phospholipids (i.e., PC) without considerably increasing the cost.

FACILITATION OF BIOREMEDIATION BY LIPOSOMES

Liposomes could facilitate cleanup of contaminated soils or of oil spilled on water by a combined effect, i.e., a concerted operation that enhances bacterial growth by supplying essential nutrients while concomitantly modifying the physical state of oil or wastes, thereby making them accessible to interaction with, and degradation by, microorganisms.

Modification of the Physical State of Petroleum or Hydrophobic Wastes by Liposomes

We have shown that liposomes modify the physical state of oil spilled on water; they also dislodge and further modify the physical state of oil entrapped in underground reservoirs or of oil or wastes spilled on soil (discussed below). In the latter cases, the oil or hydrophobic waste is entrapped in small capillaries, and because of the high interfacial tension between the respective hydrophobic molecules and water, the latter can neither detach oil or waste molecules, nor facilitate their moving out of the capillaries. Addition of phospholipids in the form of liposomes causes a 10,000- to 50,000-fold reduction of the interfacial tension, enabling oil or waste molecules to break out of the capillaries, coalesce, and be pushed by the water to the surface.

The interaction of phospholipid liposomes with oil spread on water is of a different nature. Dissimilar to oil attached to soil, there are no capillaries and no need for detachment of any sort from an external surface. The oil spread on the water has two interfaces, oil-water and oil-air. Whereas in the latter the interfacial tension is relatively low (less than 20 dynes/cm^2), that of the oil-water interface is high (above 50 dynes/cm^2). Following the law of gravity, molecules of oil move toward and spread on large areas of the water surface. But, because of their hydrophobic nature, there is no interaction even between very thin layers of oil molecules and those of the water underneath them.

Addition of liposomes causes an immediate spreading of phospholipid molecules on the water surface, forming a monomolecular layer at the oil-water interface. The phospholipid molecules mediate between the oil and water by associating their hydrophobic fatty acid residues with the oil phase while their polar headgroups are in contact with the water. The next step results in breaking up the continuous oil phase into droplets, each of which is surrounded by a monomolecular or bimolecular layer of phospholipid. A likely arrangement of the phospholipid molecules at the respective oil-water and oil-air interfaces is shown in Figure 1. These oil droplets are thus surrounded by a layer of phospholipid molecules that prevents the formation of a continuous oil layer on the water surface. It should be emphasized that when liposomes are used for interacting with oil on water, those phospholipid molecules that coat the oil droplets constitute a minor portion of the total liposomal content. Thus, the vast majority of the liposomes remain intact and serve as a reservoir for coating more oil once it reaches them.

We have observed that pouring oil on water pretreated with liposomes indeed resulted in its instant conversion to droplets (as shown in Figure 1), thereby preventing its spreading, as a continuous phase, on the water surface. The oil stayed as condensed, nonspreadable drops of oil on the surface of phospholipid-water interface of the liposome-containing water. Oil spread on water adheres poorly to solid surfaces, but when converted to droplets surrounded by phospholipid molecules, it adheres to numerous surfaces (*e.g.*, Figure 2B), thereby facilitating its removal from the water by skimming. The precise molecular organization of the phospholipid layer, surrounding the drop of oil or hydrophobic waste at the solid surface, probably depends on whether the latter is hydrophobic (Figure 3A) or hydrophilic (Figure 3B). The observations on the effect of liposomes on oil on water suggest that the use of liposomes might broaden the spectrum of microorganisms that can adhere to oil or wastes and thereby facilitate their biodegradation.

Supply of Nutrients

The phospholipids that make up the liposomes are composed of glycerol, fatty acid, and phosphorylated bases such as phosphocholine or phosphoethanolamine, the latter contain phosphorus and nitrogen. The phospholpids thus contain about 2 percent nitrogen and 4 percent phosphorus and can, by themselves, supply nutrients for growth of

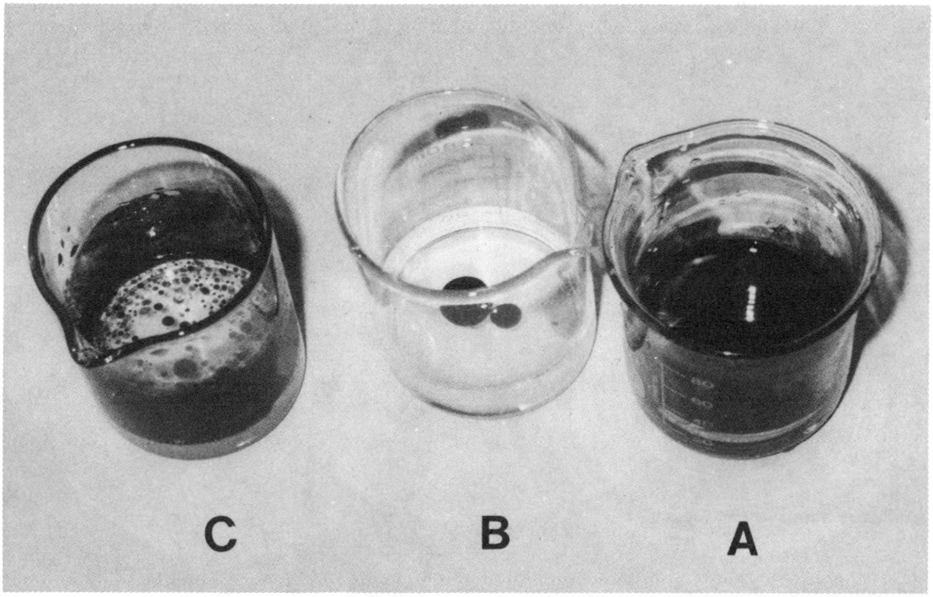

FIGURE 1. Effect of liposomes on petroleum on the surface of water. Alaskan crude oil was pipetted onto seawater without (A) or with (B) S45-M small, unilamellar liposomes (SUV). In beaker C, 20 µL of 0.5 percent S45-M were added to oil spread on water as in beaker A.

oil-degrading microorganisms. Furthermore, being sealed capsules, a broad spectrum of nutrients, minerals, and growth factors can be entrapped in the liposomes and thereby supplied to the microorganisms. These nutrients and growth factors could be divided into two main categories. The first category is hydrophilic compounds (including minerals) encapsulated in the internal aqueous compartment of the sealed liposomes. These will remain entrapped and their spontaneous release, which could be controlled by the lipid composition, will occur continuously into the microbial environment. In addition, the release of these compounds will occur when the microorganisms attack or digest the lipids that compose the liposomes. In the second category, hydrophobic or amphiphilic growth factors could be incorporated into the bilayered, lipidic membrane of the liposomes and thereby delivered to the microorganisms. This could permit supplying essential fatty acids, fat-soluble vitamins, and growth factors, as well as a naturally occurring sulfolipid (cerebroside sulfate, sulfatide) that would supply sulfur to microorganisms requiring it. These properties should

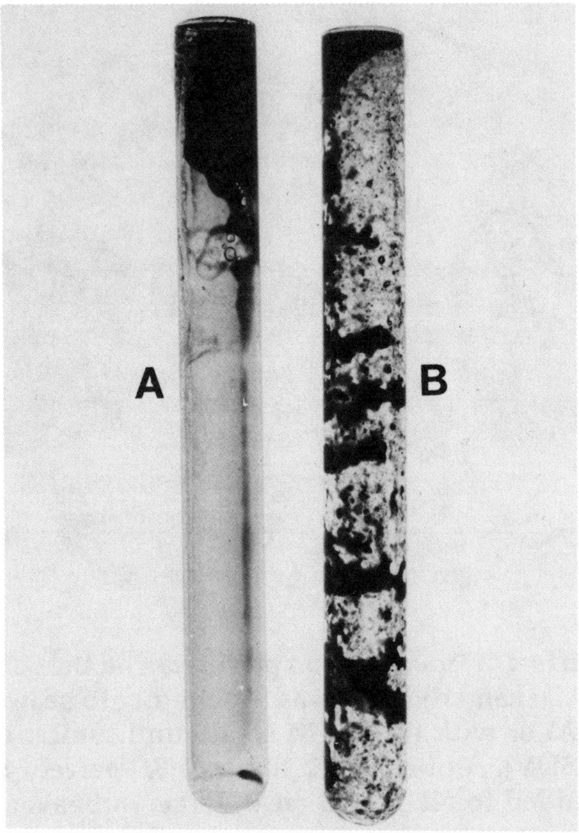

FIGURE 2. Adherence of liposome-treated petroleum to a solid surface. Alaskan crude oil was layered over seawater containing 0.1 percent S45-M SUV. The tubes were tumbled gently and the liquids were poured out. A = no liposomes. B = with liposomes.

permit liposomes to be custom-tailored to deliver suitable nutrients required for the proliferation of microorganisms capable of degrading specific oils (Atlas 1984) or wastes (Racke & Coats 1990) in a broad spectrum of environmental conditions.

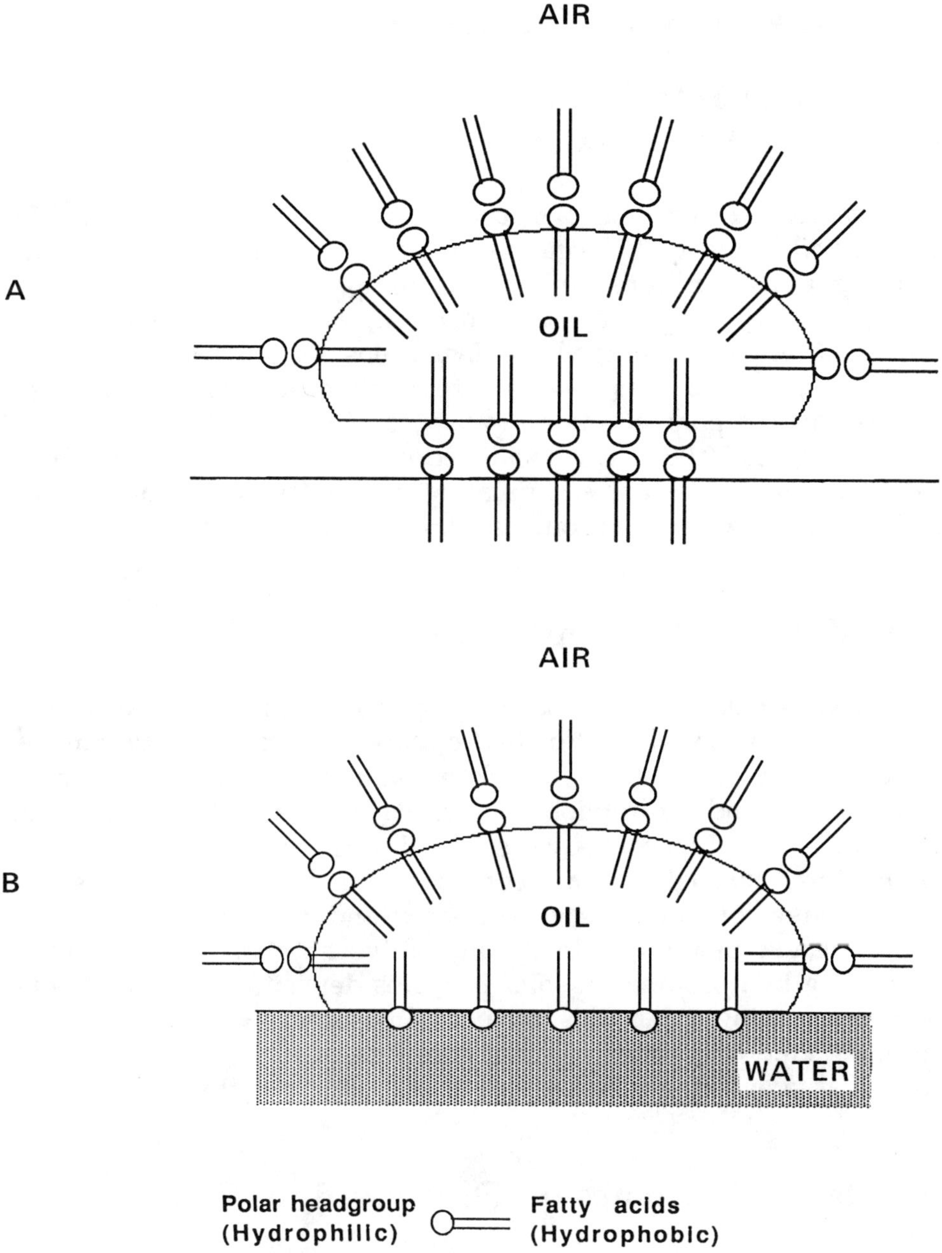

FIGURE 3. **Physical state of oil droplets, surrounded by molecules of phospholipid originating from liposomes, facilitating its adherence to hydrophobic (A) or hydrophilic (B) surfaces.**

LIPOSOMAL APPLICATIONS

Decrease of Interfacial Tension between Oil and Water

We have studied the effect of liposomes on the interfacial tension (IFT) between oil and water. The interfacial tension was measured at varying concentrations of phospholipids, salts, temperatures, and types of oil. Using a variety of liposomal compositions, we obtained four preparations that, at 60 C, provided interfacial tensions between hexadecane and 15 percent NaCl as low as 1 to 3 millidynes/cm^2, indicating a decrease of the parallel value in absence of liposomes by a factor of about 20,000 to 50,000. This facilitated efficient recovery of petroleum from Barea sandstone cores, simulating tertiary (enhanced) oil recovery (EOR) from underground sources (U.S. Patent 4 811 791).

Dislodging of Oil Spilled on Soil

Tests were carried out on a laboratory scale to simulate treatment of soils contaminated with oil or hydrophobic wastes. For this purpose various oils (*e.g.*, light oils, Alaskan crude oil, heavy petroleum or machine oil) and hydrophobic materials were adsorbed onto silica, sand, or earth. Liposomes were added without or with shaking, or were filtered through columns containing the above. In all cases, part of the above contaminants was released and recovered in the water phase (patent pending). In an alternative approach, we simulated dislodging by liposomes of solid paraffins deposited in oil well bores or tubing string. For this purpose liposomes were employed to emulsify solid polywaxes, canning paraffins, or tubing string, scraping as well as preventing their redeposition (patent pending).

Effects of Liposomes on Oil Spread on Water

Two basic approaches are currently used for cleanup of oil spilled on a water surface, dispersal by detergents, or containment and recovery. For the latter, progress of the oil is prevented by a barrier of floating booms, and collection is achieved by skimmers, sorbents, or manual methods (Fingas *et al.* 1979).

We have conducted numerous laboratory experiments to define the effects of applying phospholipids to oil spread on water. For this purpose we have compared liposomes of a wide variety of phospholipids, from the crudest to the purest; from polyunsaturated to fully hydrogenated, and both unilamellar and multilamellar. We used a series of pure paraffins, ranging from hexane to solid tetracosane ($C_{24}H_{50}$), as well as several natural petroleums. Liposomes were prepared in distilled, tap, seawater, and salt solutions, as well as various solutions of divalent ions. We have also compared the effects of the liposomes with those produced by numerous detergents, including nonionic surfactants, anionic sulfonates, and commercial dispersants (*e.g.*, Corexit 9527), as well as sodium oleate.

The results can be summarized as follows:

1. Addition of oil onto water results in its even spreading on the water surface (Figure 1A). Addition of traces of the liposomal dispersion onto the water surface immediately pushes away and disrupts the oil layers, forming viscous, condensed oil droplets, which adhere to the walls (*e.g.* glass, polyethylene, or ceramic) of the container (Figure 1C).

2. Addition of oil onto water that already contains liposomes results in coalescence of the oil into viscous, condensed droplets (Figure 1B). The lecithin-treated droplets adhere to and can be collected by many objects, such as wetted paper, cloth, ceramics, and glass.

3. Gentle shaking of thick layers of oil on liposome-containing water results in formation of small droplets that rise to the surface, but do not coalesce into a uniform oil layer. Vigorous shaking results in smaller droplets that rise to the surface and do not coalesce. Extensive ultrasonic irradiation results in formation of much smaller droplets that stay dispersed on the aqueous phase for extended periods. This conforms with the hypothesis illustrated in Figure 4.

4. Treating oil on water with liposomes, followed by gentle shaking or stirring, results in tight adherence of the condensed, viscous droplets to the surface of the container, whether plastic, ceramic, or glass. Nearly 100 percent of the oil can thus be adsorbed onto the surface of the

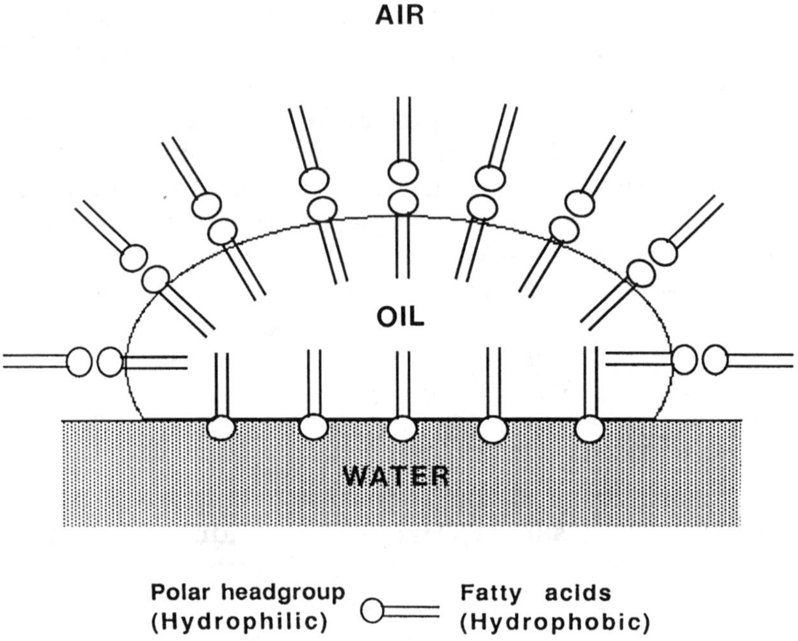

FIGURE 4. Physical state of oil droplet on water, surrounded by phospholipid originating from liposomes.

container (Figure 2B). Droplets that do not adhere to the surface of the container can be collected by paper, cloth, glass, plastic, or ceramic materials. The oil that adheres to the surface of the container can be dislodged easily by applying a light hydrocarbon, such as hexane.

5. The above effects can also be reproduced when oil spread on water is sprayed with a solution of phospholipids in a solvent such as a light hydrocarbon, an alcohol, and numerous other solvents, miscible or immiscible with water.

The above laboratory-scale experiments indicate that application of liposomes to an actual oil spill would result in the following beneficial effects: stopping the progressing oil layer, pushing it back, and breaking it up into individual drops, thereby preventing its reaching the beach, while concomitantly facilitating its collection by booms, sorbents, or skimmers. Because of the increased adherence of the liposome-treated oil drops, they will be more susceptible to interaction with, and degradation by, microorganisms. Finally, by disrupting the continuous oil layer, liposomes clear large areas of the water surface, thereby enabling penetration of oxygen, essential for the survival of marine organisms.

ENHANCEMENT BY LIPOSOMES OF MICROBIAL GROWTH AND BIODEGRADATION OF OIL

Facilitation by liposomes of growth of oil-degrading bacteria was tested in the following two sets of experiments. In the first, soil contaminated with petroleum was collected from a local gas station and incubated in minimal medium, to which Alaskan crude oil was added. Bacteria that grew in the medium were collected and used to inoculate minimal medium to which Alaskan oil and unilamellar liposomes were added (final concentration of phospholipid 0.15 percent). Several liposomal preparations were employed, composed of partially or fully purified PC of soybean or egg yolk, as well as fully hydrogenated phosphatidylcholine. In all cases, the presence of the liposomes increased the bacterial count by 2 to 3 logs after incubating for 3 days at 30 C. To determine the optimal and minimal inhibitory concentrations (MIC) of liposomes for bacterial growth, an aliquot (3×10^3) of the above bacteria was incubated in the presence of the petroleum-contaminated soil in 1 mL minimal essential medium to which had been added unilamellar liposomes composed of soybean lecithin (S45-M) modified to contain mostly PC. The data of Figure 5 indicate an optimal value of 0.1 percent and a minimal inhibitory concentration of about 0.4 percent liposomal phospholipids.

In the second extended series of experiments, soils contaminated with petroleum were obtained from the oil-loading terminal at Israel's southern port of Eilat and used as a source for bacteria. Two soils were obtained, a dry soil (ECS-A) and a wet, oily, rocky sludge (ECS-B). These respective soils were incubated at 30 C in distilled, tap, or seawater in the absence or presence of liposomes. In some experiments ^{14}C-hexadecane was also added and the $^{14}CO_2$ collected on hyamine-hydroxide and counted. The results are summarized as follows:

1. In all cases, addition of liposomes stimulated the bacterial growth by several logs. Figure 6 indicates a 5 to 6 log increase in the microbial count in the presence of small, unilamellar vesicles of S45-M. When incubation of the wet,

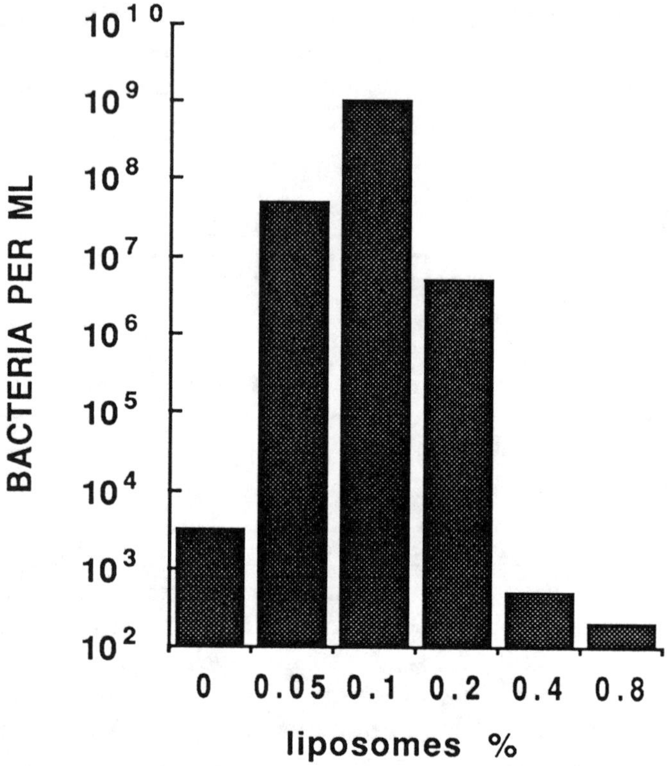

FIGURE 5. Growth of bacteria isolated from gasoline-contaminated soil on varying concentrations of liposomes. Bacteria were isolated from a soil contaminated with petroleum collected from a gasoline station. Inoculums (5×10^3) of bacteria isolated from the soil were then incubated in 0.5 mL minimal medium to which 0.15 mL Alaskan crude oil was added. Increasing concentrations of S45-M liposomes were then added and the flasks incubated for 5 days at 30 C.

oil-contaminated soil (ECS-B) was continued for 2 weeks or more, a whitish opaque emulsion appeared in the liposome-containing flasks, but not in their counterparts from which liposomes were omitted. In the flasks in which soil ECS-A was incubated with liposomes, a complete cleanup of the soil was obtained.

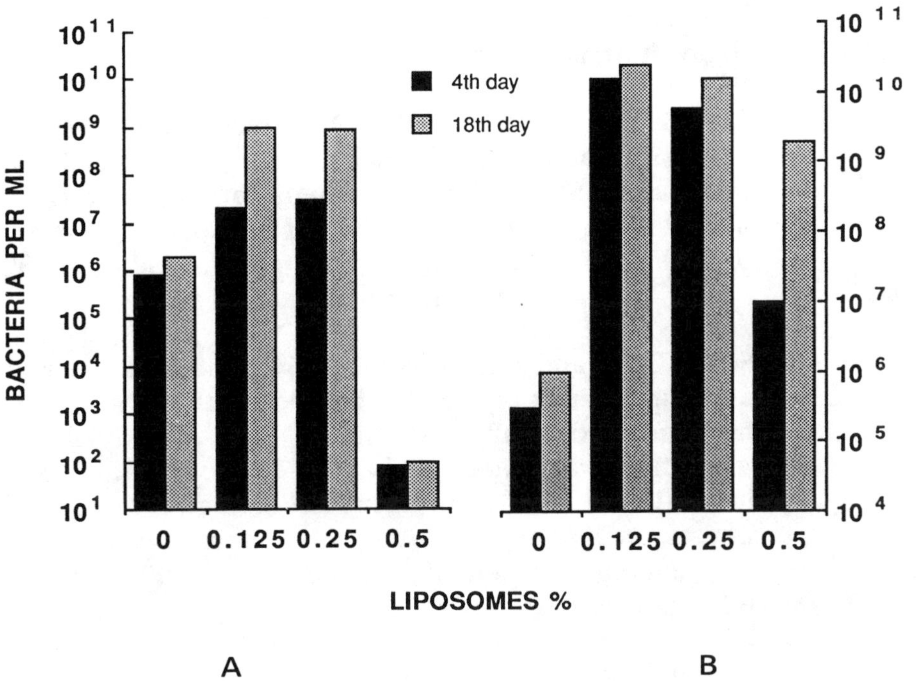

FIGURE 6. Enhancement by liposomes of bacterial growth on petroleum-contaminated soils. One gram each of the petroleum-contaminated soils ECS-A (A) and ECS-B (B) was incubated in 20 mL tap water at 30 C for 4 days (black bars) or 18 days (dotted bars). Various concentrations of S45-M SUV were included as specified in the abcissa, which indicates the final percentage of liposomal phospholipid.

2. Experiments were done to ascertain that bacteria grown on the contaminated soils indeed degraded hydrocarbons. In a typical experiment, 2 g of the above soils were incubated in 2 mL of distilled water in the absence or presence of S45-M liposomes (final phospholipid concentration 0.25 percent); 40 µL of ^{14}C-hexadecane were also included. After 5 days of incubation at 30 C, the bacterial count per mL, in the presence of liposomes, was 9×10^8 and 4×10^9 for ECS-A and ECS-B, respectively. The $^{14}CO_2$ collected was equivalent to 0.36 µL and 0.5 µL of oxidized ^{14}C-hexadecane for the above two soils, respectively. In the absence of liposomes only half of the disintegrations per minute of

$^{14}CO_2$ was obtained. This indicated that the soil bacteria chain-shortened or fully oxidized the ^{14}C-hexadecane.

3. Bacteria obtained from soils ECS-A and ECS-B were mixed in equal numbers and further incubated in 5 mL distilled water and 0.5 mL Alaskan crude oil for 4 days at 30 C. In the presence of unilamellar S45-M liposomes (0.3 percent phospholipid), the bacterial count per mL increased from 2×10^3 to 3×10^8, that is, more than 5 logs.

4. Soils ECS-A and ECS-B were also incubated in water with other liposomal preparations composed of egg, soybean, or hydrogenated fully saturated phosphatidylcholines. Although some variations have been observed, especially in the initial rates of proliferation of the bacteria, after a 2-week incubation, all the respective liposomes resulted in a similar enhancement of bacterial growth (10^{10} bacteria per mL, Figure 7).

5. An experiment for testing enhancement of bacterial growth by liposomes formed *in situ* was conducted as follows: soils ECS-A and ECS-B were sprayed evenly with 0.4 percent by weight of soybean phosphatidyl-choline solubilized in hexane; this solvent was then evaporated in an oven. The phospholipid-coated soil was wetted with water (10 or 30 percent by weight), incubated for 3 weeks at 30 C, and the bacterial count in 1-g samples of soil was determined. Under these conditions hydration and swelling of the dry phospholipids to a liposome-like lamellar phase is obtained (Small 1986). Figure 8 exemplifies the results obtained using the wet ECS-B soil. Soil was sprayed with phospholipid but not wetted served as control and had 2×10^3 bacteria per gram. The corresponding value for the hydrated soil, which had not been sprayed with phospholipid, was about 10^5 bacteria per gram; in either case the bacterial count remained at the same level during the experiment. In contrast, in the soil that had been coated with phospholipids and hydrated with water, the bacterial count increased continuously, reaching 2×10^{11} bacteria per gram after 24 days of incubation (Figure 8).

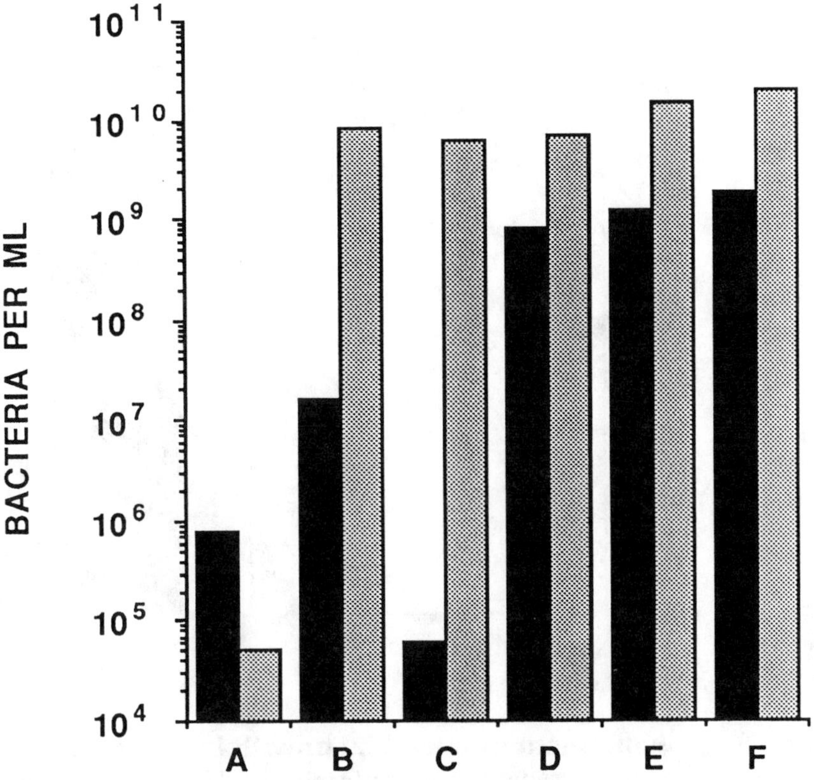

FIGURE 7. Effect of the type of liposome on enhancement of bacterial growth on petroleum-contaminated soil. One gram of wet oil-contaminated soil (ECS-B) was incubated in 3 mL distilled water with each of several unilamellar liposomal preparations (final concentration of phospholipids in the incubation medium 0.1 to 0.2%, w/v). The flasks were incubated at 30 C for 8 days (black bars) or 13 days (dotted bars). A = no liposomes. B = fully saturated phosphatidylcholine (PC). C = partially purified soybean phospholipids. D = S45-M. E = soybean PC. F = egg PC.

6. An experiment for testing the effect of encapsulation of nutrients in liposomes on the enhancement of growth of bacteria on oil-contaminated soil was done as follows: SUVs of soybean lecithin were prepared in the presence of yeast extract and separated from excess yeast extract by filtration through Sepharose 6B under sterile conditions.

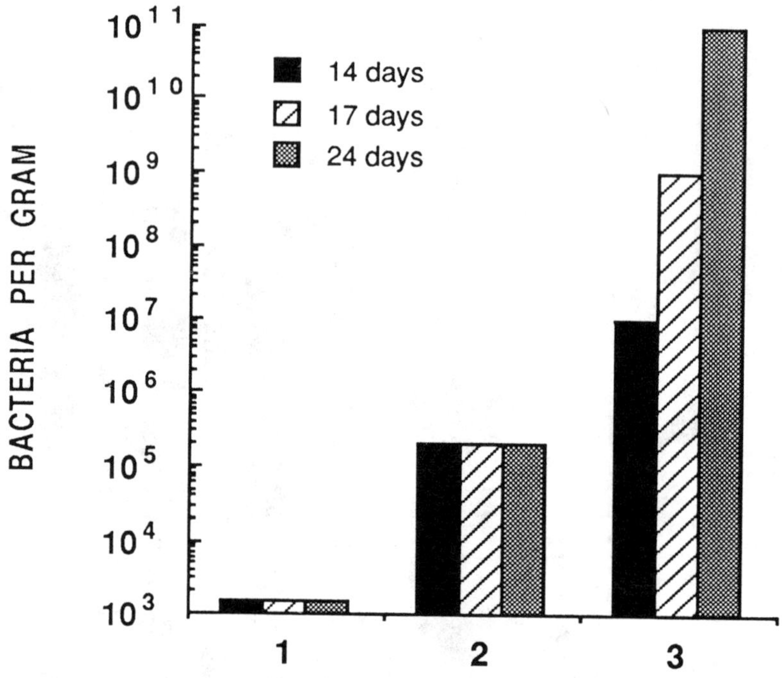

FIGURE 8. Enhancement of microbial growth by liposomes formed *in situ* on petroleum contaminated soil. Fifty grams of petroleum-contaminated soil (ECS-B) were sprayed with 0.2 g of soybean phosphatidylcholine in 10 mL hexane (final concentration 0.4% PC). The hexane was evaporated in an oven, and each 10-g portion of soil was hydrated with 1 mL water and incubated at 30 C. After 13, 16, and 24 days, 1-g samples were collected, suspended in minimal medium, and the bacterial count determined. 1 = soil sprayed with phospholipid but not hydrated with water. 2 = soil *not* sprayed with phospholipid but hydrated with water. 3 = soil sprayed with phospholipid and hydrated with water.

When added to soils ECS-A or ECS-B in distilled water, they induced a very rapid increase in bacterial proliferation (Figure 9). Thus, after 2 days at 30 C, the bacterial count per mL had already increased from 10^3, in absence of liposomes, to values as high as 4×10^{10} in their presence, an increase of more than 7 logs.

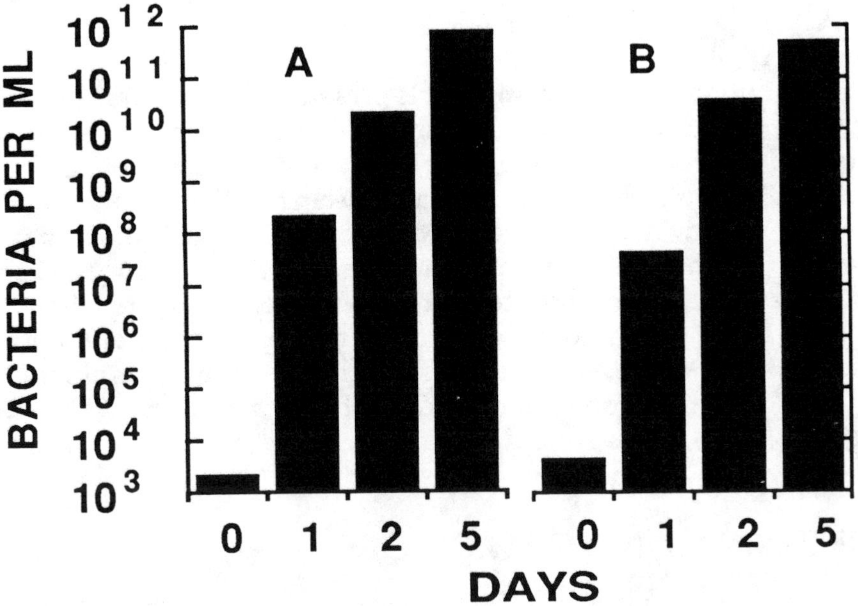

FIGURE 9. Enhancement of bacterial growth on petroleum-contaminated soil by liposomes in which nutrients have been encapsulated. Yeast extract was encapsulated in SUV as described in the text. One gram each of soils ECS-A and ECS-B was incubated at 30 C in 10 mL distilled water without and with 0.1 to 0.4 percent liposomal phospholipid. At days 0, 2, and 5, samples were removed for bacterial count.

In summary, the above preliminary experiments indicate that liposomes induced a very considerable (up to 7 logs) enhancement of growth of bacteria present in a natural environmental hazard, i.e., soil contaminated with petroleum. When these bacteria were collected and further incubated with crude petroleum, liposomes again increased the bacterial count by at least 5 logs. The bacteria, whose growth was stimulated by the liposomes, oxidized ^{14}C-hexadecane. The increased bacterial prolfieration induced visible changes in the appearance of the soil, converting the nearly black color of the oil-contaminated soil to light brown. Darkening of the medium and in some cases very heavy emulsions were also observed. When oil-contaminated soil was coated with phospholipids (by spraying a solution thereof) and wetted with water, the liposomes thus formed *in situ* increased bacterial growth on this soil up to 8 logs.

TOXICITY

Phospholipids are natural compounds present in all membranes, including those of humans, animals, plants, or microorganisms. They are employed in routine clinical use for intravenous infusion of emulsified lipids and for drug delivery in humans (Becher 1983; Gregoriadis 1988). They are very widely used for feeding animals and fish. Thus, it is expected that they will have no toxic effects on natural ecosystems and could therefore be applied to soil or water contaminated by oil or wastes without the need for prolonged toxicological studies. Excess phospholipids will be biodegraded by microorganisms, thus eliminating any possibility of unforeseen hazards to the flora or fauna of the sea, to its shores, or to underground water.

SUMMARY AND PROSPECTS FOR THE FUTURE

In our studies related to enhanced oil recovery and cleanup of oil spills, we have used liposomes composed of commercial preparations of soybean phospholipids that were sorted, selected, partially purified, and modified chemically. This permitted us to obtain various combinations of phospholipid composition and varying degrees of purity at a reasonable cost. Additional studies should define the precise compositions of phospholipids suitable for cleanup and/or biodegradation of specific oils spilled on water or of oils or wastes contaminating soils.

We have used various phospholipids to prepare liposomes and have demonstrated their effects on decreasing the interfacial surface tension and increasing oil-wettability in relation to enhanced oil recovery and cleanup of oil spills on the ground and on water. Also, current experiments clearly indicated considerable enhancement by liposomes of growth of oil-degrading microorganisms. Further experimentation should permit adapting the use of liposomes for enhancing biodegradation of oil and hydrophobic wastes on an industrial scale.

Although natural phospholipids are nontoxic and biodegradable, direct toxicity tests still have to be done.

Liposomes can be prepared on a small scale by sonic irradiation or extrusion; for large-scale preparations, dairy homogenizers could be employed, permitting preparation of many thousands of liters of small, unilamellar liposomes in a day. Nutrients, minerals, and other growth factors can be added during the preparation of the liposomes, resulting in their encapsulation in the aqueous interior or, alternatively, incorporation into the membrane of the liposomes.

One could also prepare a concentrated dispersion of liposomes (*e.g.*, 20% phospholipid in water) to facilitate delivery and on-the-spot dilution with water. Alternatively, in the presence of cryoprotectant agents, the liposomes could be lyophilized or spray-dried and delivered or stored as a dry powder. Based on our stability studies, concentrated stocks of liposomes can be stored for several years and used when needed.

Being aqueous dispersions, the liposomes can be applied to the ground or water surface by any means used for delivering water, such as spraying, flooding, etc. For treating oil- or waste-contaminated soil, they could be dissolved in a light kerosene (*e.g.*, hexane or another suitable solvent) and sprayed on the oily surface, thereby combining two processes, that is, delivery of phospholipids and softening of the weathered oil or waste. Following evaporation of the kerosene, application of water will cause hydration and swelling of the phospholipids *in situ*, thereby forming liposomes at the site of the contaminated area.

Once optimal conditions are established for cleanup by physical means and/or enhanced biodegradation, a large-scale supply of suitable liposomes can be secured for use in emergencies (*e.g.*, combatting and cleanup of oil spills on water or beaches), or for an in-depth bioreclamation of soils contaminated with petroleum or numerous hydrophobic wastes.

Acknowledgments

We thank Dr. L. Bar for assistance in preparing liposomes and Ms. S. Cherbu and R. David Pour for expert technical assistance.

REFERENCES

Atlas, R. M., Ed. *Petroleum Microbiology*; MacMillan, 1984.
Beachey, E. H., Ed. *Bacterial Adherence*; Chapman and Hall, 1980.
Becher, P., Ed. *Encyclopedia of Emulsion Technology*; Marcel-Dekker: N.Y., 1983.
Fingas, M. F.; Duval, W. S.; Stevenson, G. B. *The Basis of Oil Spill Cleanup*; Environmental Protection Service: Canada: 1979.
Gould, G. W.; Corey, J. L., Eds. *Microbial Growth and Survival in Extremes of Environment*; Academic Press, 1980.
Gregoriadis, G., Ed. *Liposomes as Drug Carriers*; Wiley: N.Y., 1988.

Jann, K.; Jann, B. *Bacterial Adhesions*; Springer-Verlag, 1990.
Lichtenberg, D.; Barenholz, Y. In *Methods of Biochemical Analysis*; Glick, D., Ed.; Wiley. **1988**; Vol. 33, pp 337–462.
Racke, K. D.; Coats, J. R., Eds. *Enhanced Biodegradation of Pesticides in the Environment*; American Chemical Society, 1990.
Savage, D. C.; Fletcher, M., Eds. *Bacterial Adhesion: Mechanisms and Physiological Significance*; Plenum, 1985.
Small, D. M. *The Physical Chemistry of Lipids: Handbook of Lipid Research*; Plenum, 1986; Vol. 4.
Szuhaj, B. F., Ed. *Lecithin: Sources, Manufacture, Uses*; American Oil Chemical Society, 1989.

Use of Indigenous Microorganisms for the Disposal of Cattle Dip Waste

Daniel R. Shelton[*], *Cathleen J. Hapeman-Somich*
U.S. Department of Agriculture

INTRODUCTION

Coumaphos [*O,O*-diethyl *O*-(3-chloro-4-methyl-2-oxo-2*H*-1-benzopyran-7-yl)-phosphorothioate] is used as an acaricide for control of the southern cattle tick (*Boophilus microplus*) and the cattle tick (*Boophilus annulatus*) by the Animal and Plant Health Inspection Service (APHIS), U.S. Department of Agriculture (USDA), in its Tick Eradication Program. Several hundred thousand head of cattle are dipped annually in one of approximately 60 vats along the U.S.-Mexican border. The vats contain approximately 15,000 L of coumaphos solution, 0.15-0.30 percent active ingredient (a.i.), in the form of Co-Ral Wettable Powder or Co Ral Flowable Liquid formulations (Rogers 1990). Vats are emptied, cleaned, and recharged every 6 to 12 months, depending on extent of usage or length of service, generating approximately 10^6 L of coumaphos waste per year. Coumaphos wastes are stored on-site; there is currently no demonstrated disposal technology available.

The levels of coumaphos in vats are generally stable over time such that any decreases in coumaphos concentrations are due to removal by cattle. However, dramatic losses in coumaphos have been observed over the past few years in several high-use vats (greater than 1,000 cattle dipped per year). These vats accumulate significant

[*] Pesticide Degradation Laboratory, U.S. Department of Agriculture, Agricultural Research Service, Building 050, BARC-West, Beltsville, MD 20705

amounts of soil and cattle waste resulting in enhanced levels of microbial activity. These observations suggested that coumaphos losses were due to microbial degradation and that disposal of waste coumaphos could be achieved by exploiting this phenomenon.

Methods and Materials

Samples of cattle dip were obtained by APHIS personnel at selected sites by immersing 250-mL or 1-L containers into vats immediately after vat contents were mixed, then shipped via surface mail (Shelton & Karns 1988). Coumaphos, potasan [*0,0*-diethyl *0*-(4-methyl-2-oxo-2*H*-1-benzopyran-7-yl)-phosphorothioate], chlorferon (3-chloro-4-methyl-2-oxo-2*H*-1-benzopyran-7-ol), and [*benzo ring*-U-^{14}C]coumaphos were gifts from the Animal Health Division, Mobay Corporation (Shawnee, Kansas). Co-Ral Flowable Liquid and Co-Ral Wettable Powder were purchased from Mobay. 4-Methylumbelliferone was purchased from Aldrich. The synthesis of diethylthiophosphoric acid (DETP) has been previously described (Shelton & Somich 1988). The structures of α-chloro-β-methyl-2,3,4-trihydroxy-*trans*-cinnamic acid (CMTC) and β-methyl-*trans*-cinnamic acid (MC) were elucidated based on mass spectral and nuclear magnetic resonance (NMR) data (Shelton & Somich 1988).

Coumaphos, potasan, chlorferon, 4-methylumbelliferone, DETP, CMTC, MC, tyrosine, and unidentified metabolites were analyzed by high performance liquid chromatography (HPLC) (Shelton & Karns 1988; Shelton & Somich 1988). Evolution of $^{14}CO_2$ and production of radio labelled soluble metabolites were quantified by liquid scintillation (Shelton & Karns 1988; Shelton & Somich 1988). Experiments involving inhibition of growth by chlorferon were monitored via HPLC (substrate depletion) or turbidimetrically (Shelton & Somich 1988).

Isolates of pure cultures and estimates of population densities were obtained by performing appropriate serial dilutions of enrichment cultures and spread plating them onto diluted nutrient broth (1 g/L) with 1.6 percent agar. Pure cultures were routinely maintained on mineral salts (Shelton & Somich 1988) amended with 0.1 percent arginine or 0.1 percent acetate, with or without a vitamin solution (Balch *et al.* 1979). Consortia were maintained on mineral salts amended with 0.1 percent coumaphos, 0.1 percent 4-methylumbelliferone, or 0.1 percent DETP (with 10 mg/mL yeast extract).

Fermentation experiments were conducted using a 1-L BioFlow fermentor (New Brunswick Scientific, Edison, New Jersey) with pH control. One-liter quantities of cattle dip were supplemented with 4 mM phosphate buffer and 100 mg/mL yeast extract and inoculated with 200 mL of enriched culture; samples from low-use vats were also supplemented with ammonium and trace metals (Shelton & Somich 1988). Incubations were conducted at 25 to 35 C with the pH buffered at 6.8 to 7.2.

RESULTS AND DISCUSSION

Preliminary Observations

Experiments were initiated to determine if the potential for microbial degradation of coumaphos existed in both low-use and high-use vats. Under aerobic conditions, coumaphos was degraded in all vat samples tested, although rates were variable; degradation of coumaphos in Laredo (LAR) was inhibited due to a decrease in pH (Figure 1) (Shelton & Karns 1988). In addition, experiments with [*benzo ring*-U-^{14}C]coumaphos demonstrated that a significant portion of the aromatic moiety was mineralized to $^{14}CO_2$ (27 to 53 percent depending on the vat) (Figure 2). Coumaphos was reductively dechlorinated under anaerobic conditions to potasan in the high-use vats as a result of the accumulation of organic matter (cattle waste). Potasan was subsequently degraded under aerobic conditions.

Enrichment cultures were initiated using the Flowable Liquid (42 percent coumaphos, 58 percent inert ingredients) as the carbon and energy source. Despite the potential for coumaphos degradation in all of the vats tested, only enrichments with LAR, one of the high-use vats, were consistently successful.

Isolation of Coumaphos-Degrading Bacteria

After several transfers (10 percent inoculum), enrichment cultures were serially diluted, plated, and isolates screened for organophosphate hydrolase activity. Three morphologically distinct bacteria (designated B-1, B-2, and B-3) possessed with the ability to hydrolyze coumaphos to chlorferon (3-chloro-4-methyl-2-oxo-2*H*-1-benzopyran-7-ol) and diethylthiophosphoric acid (DETP) (Shelton & Somich 1988). Of the three isolates, however, only B-1 significantly metabolized chlorferon (Figure 3). B-1 was a gram-negative rod and required no organic nutritional supplements. Experiments with [*benzo ring*-U-^{14}C]coumaphos or

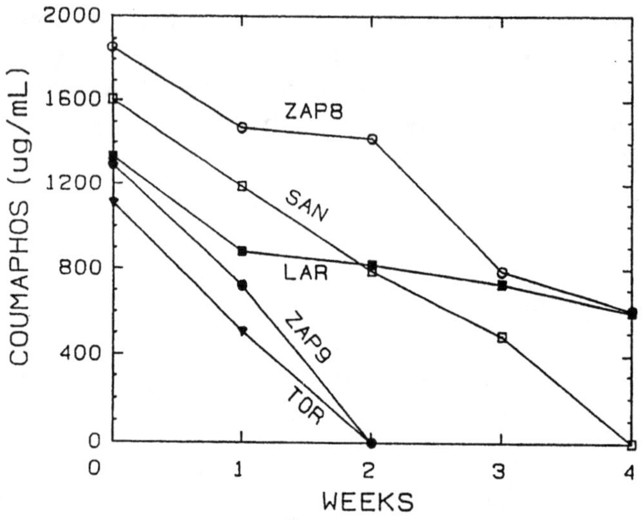

FIGURE 1. Degradation of coumaphos under aerobic conditions in unbuffered vat dip solutions. LAR and TOR were high-use vats (>1,000 cattle dipped/year); SAN, ZAP8, and ZAP9 were low-use vats. (Reproduced with permission from Shelton & Karns 1988, American Chemical Society.)

chlorferon indicated that B-1 was capable of both mineralizing and incorporating into biomass the aromatic portion of the molecule, although the majority of the label was recovered as soluble polar products, as determined by HPLC, including α-chloro-β-methyl-2,3,4-trihydroxy-*trans*-cinnamic acid (CMTC).

Repeated attempts to grow B-1 with coumaphos or chlorferon as sole growth substrates, however, were unsuccessful. Inhibition experiments showed that the addition of 50-100 μg/mL of chlorferon to an actively growing culture of B-1 resulted in partial inhibition, while complete growth inhibition was observed with quantities of chlorferon greater than 200 μg/mL (Figure 4). In comparison, 4-methylumbelliferone, the hydrolysis product of potasan, was utilized by B-1 as a growth substrate. A nonchlorinated analogue of CMTC, β-methyl-*trans*-cinnamic acid (MC), accumulated in the culture solution. These data indicated that B-1 produced intermediates from chlorferon which were inhibitory to its growth and that other bacteria were presumably present in the enrichment culture that could metabolize these intermediates.

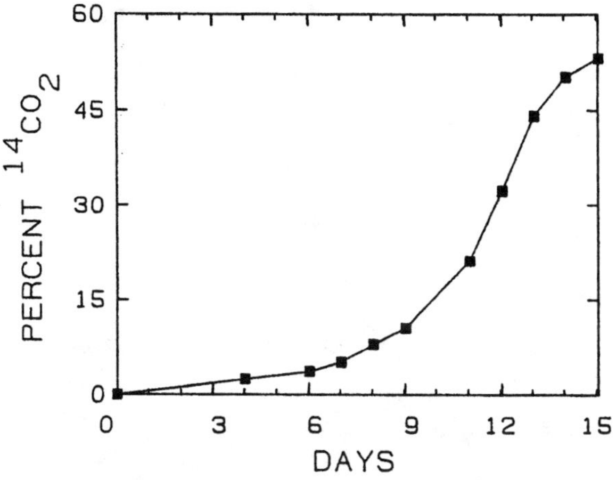

FIGURE 2. Production of $^{14}CO_2$ from [*benzo ring*-U-^{14}C] coumaphos in buffered (pH = 7.0) LAR vat dip solution. (Reproduced with permission from Shelton & Karns 1988, American Chemical Society.)

Characterization of Coumaphos-Mineralizing Consortia

Experiments were initiated in which bacteria, previously isolated from enrichment cultures, were combined with B-1. An active consortium growing on analytical-grade coumaphos was observed during the course of these incubations. This undefined consortium degraded both chlorferon and DETP and did not require any organic nutritional supplements. The consortium consisted primarily of B-1 (60 percent of all colonies) and an additional bacterium, designated B-4 (35 percent of all colonies); the remaining 5 percent of assorted colonies were not characterized. B-4 was a gram-negative rod that required a vitamin supplement for growth and was inhibited by 50 μg/mL of chlorferon. The recombination of B-1 and B-4 from pure cultures resulted in a stable two-membered consortium able to utilize coumaphos as a growth substrate with the addition of a vitamin supplement (Figure 5); DETP accumulated stoichiometrically, while CMTC accumulation was 15 to 20 percent on a molar basis. Apparently, the 5 percent of assorted isolates in the undefined consortium were responsible for DETP metabolism and providing vitamins for B-4.

Enrichment cultures were also initiated from Laredo vat dip with DETP as the sole carbon and energy source. A consortium of microorganisms was subsequently obtained which mineralized DETP

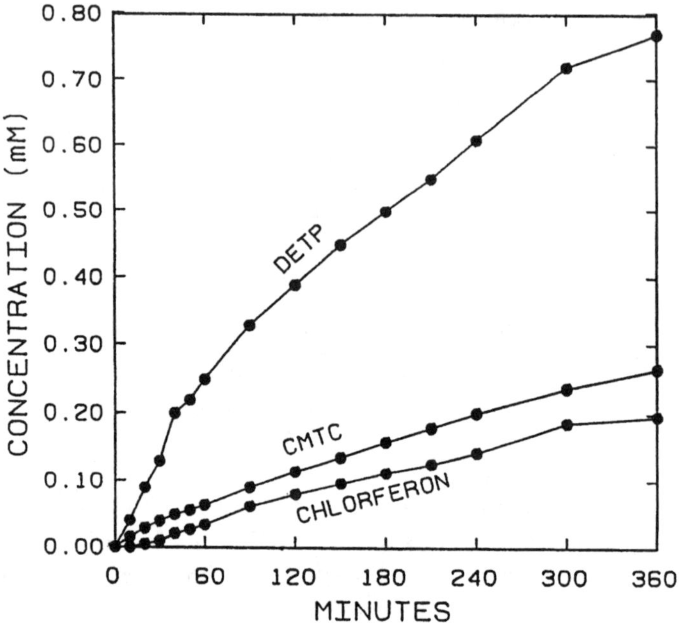

FIGURE 3. Time course of metabolism of coumaphos (as Co-Ral Flowable Liquid) to DETP, chlorferon, and an aromatic metabolite (CMTC) by B-1. (Reproduced with permission from Shelton 1988, American Society for Microbiology.)

to phosphate, sulfate, and CO_2, with the addition of yeast extract as a nutritional supplement (10 mg/mL) (Figure 6) (Shelton 1988). The consortium was composed of at least six morphologically distinct bacteria, two of which could metabolize ethanol (the hydrolysis product of DETP); however, attempts to reconstitute the consortium from pure cultures were unsuccessful.

These data collectively indicated that the initial step in coumaphos metabolism in the dip vats was hydrolysis to chlorferon and DETP by one or more bacteria, followed by the metabolism of chlorferon by B-1 and B-4 (or biochemically comparable microorganisms); DETP was metabolized by a separate consortium (Figure 7). Potasan was hydrolyzed to DETP and 4-methylumbelliferone; 4-methylumbelliferone was metabolized by B-1 alone. CMTC and MC were end products of B-1 metabolism but were degraded by one or more unidentified microorganisms in the original enrichment culture.

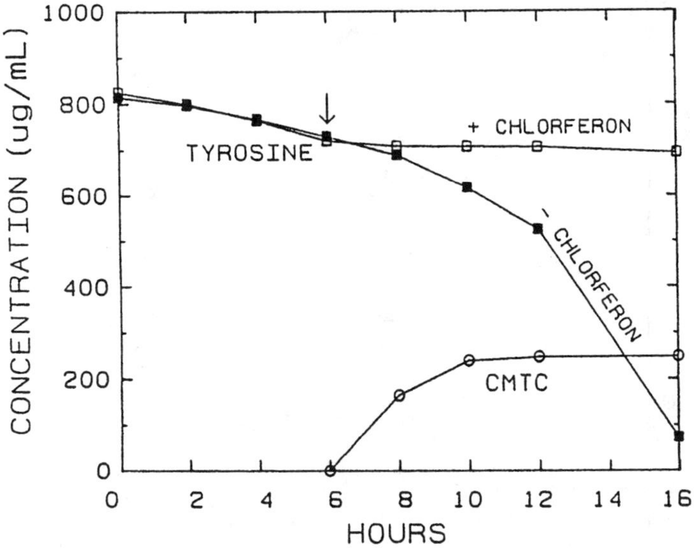

FIGURE 4. Inhibition of tyrosine consumption by B-1 as a result of the addition of 400 µg/mL chlorferon. Chlorferon was added at the time indicated by the arrow and the production of CMTC monitored. (Reproduced with permission from Shelton & Somich 1988, American Society for Microbiology.)

Use of Consortia for Disposal of Cattle Dip Waste

Experiments were conducted to determine if the various microbial consortia could be used to dispose of coumaphos in cattle dip waste. Reconstituted pure cultures of B-1 and B-4 were unable to degrade coumaphos when formulated as either the Wettable Powder (25 percent coumaphos, 75 percent inert ingredients) or the Flowable Liquid (42 percent coumaphos, 58 percent inert ingredients). The undefined consortium, from which B-1 and B-4 had been isolated, was able to utilize coumaphos formulated as Flowable Liquid as a growth substrate but did not metabolize the coumaphos when formulated as the Wettable Powder. This suggested that some of the formulating reagents (inert ingredients) were inhibitory to the growth of B-1 and B-4 but that these compounds may be degraded or eliminated by one or more unidentified microorganisms.

Incubations were initiated with samples of fresh cattle dip solution from Laredo City Vat (containing Wettable Powder) or Laredo Import Vat (containing Flowable Liquid). Degradation of coumaphos

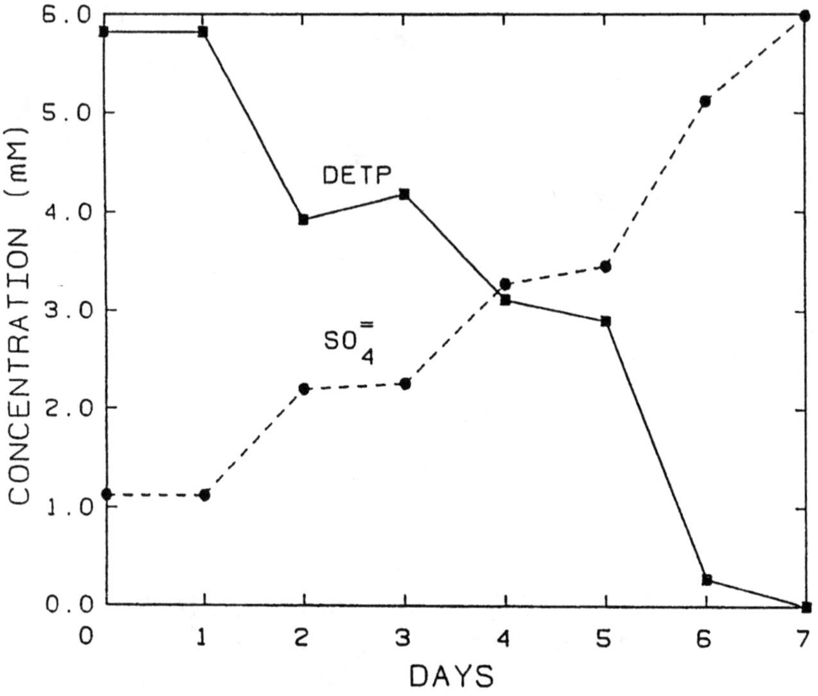

FIGURE 5. Time course of DETP disappearance and sulfate appearance by an enriched microbial consortium. (Reproduced with permission from Shelton & Somich 1988, American Society for Microbiology.)

was observed within 1 to 2 weeks, indicating that the inhibitory compound(s) were dissipated in the vats, due either to microbial or chemical degradation, or physical removal. Experiments to determine if cattle dip waste from other vats (containing different formulations) could be degraded were initiated using a combination of the Laredo City Vat and Laredo Import Vat enrichment cultures as inocula. Extensive coumaphos degradation occurred in all vat samples with degradation consistently exceeding 99 percent (Table 1); degradation was lower in two instances because of equipment failure which resulted in suboptimal growth.

CONCLUSIONS AND FUTURE STUDIES

Based on these studies, it appears that disposal of coumaphos from cattle dip waste can be effectively accomplished using microbial

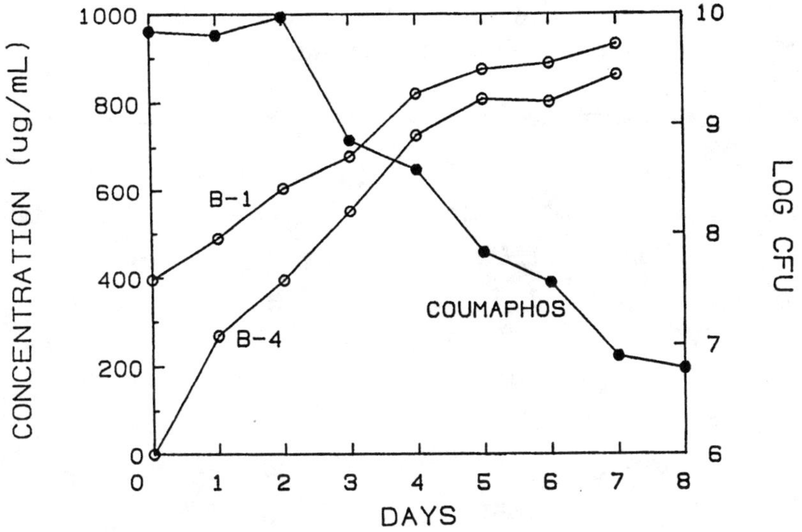

FIGURE 6. Utilization of coumaphos as a growth substrate by a two-membered consortium consisting of B-1 and B-4 (CFU = colony forming units).

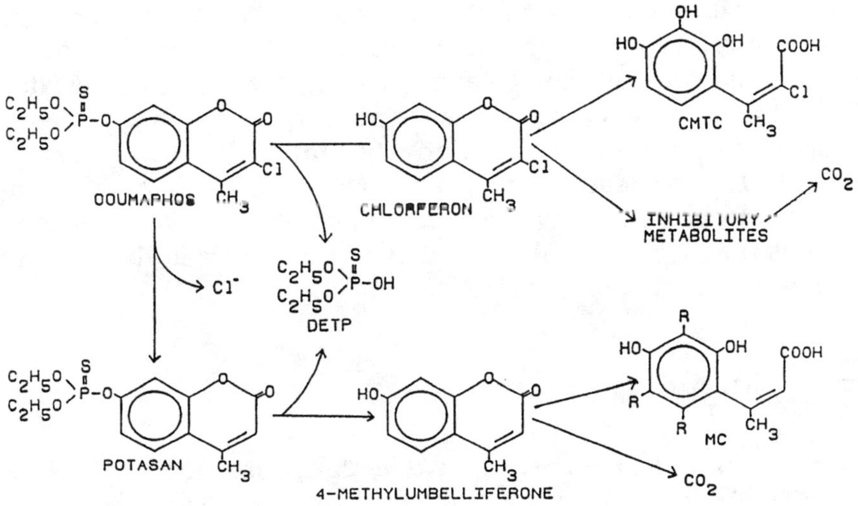

FIGURE 7. Partial pathway of potasan (by B-1) and coumaphos (B-1 and B-4) metabolism. Potasan is produced as a result of the reductive dechlorination of coumaphos. DETP is mineralized by a separate bacterial consortium. CMTC and MC are metabolized by one or more unidentified microorganisms.

TABLE 1. Degradation of coumaphos in vat dip by an enriched microbial consortium.

Vat (Formulation)[a]	Cattle Dipped	Final Concentration (mg/mL)	Percent Degradation
Laredo Import (FL)	11,923	3	99.8
Juarez (FL)	32,252	3	99.7
Laredo City (WP)	951	4	99.6
Zapata (WP)	375	4	99.7
Calaboz (WP)	—	4	99.8
San Andres[b] (WP)	387	12	99.9
Pinto[c] (WP)	451	10	97.4

(a) FL = Flowable Liquid, initial coumaphos concentration approximately 0.15%. WP = Wettable powder, initial coumaphos concentration approximately 0.1%.
(b) pH probe failed resulting in low pH.
(c) Aeration and agitation failed.

consortia. The advantage of this process is that complete disposal (mineralization) of coumaphos can be achieved at relatively low cost. Although the microbial ecology and biochemistry of coumaphos degradation is extremely complex, the actual fermentation system for coumaphos disposal should be relatively simple. Pilot-scale fermentations (400 L) are currently in progress to determine if disposal of larger quantities of coumaphos waste is feasible. The ultimate goal of this research is the disposal of all spent coumaphos waste in the field using a microbial fermentation process.

REFERENCES

Balch, W. E.; Fox, G. E.; Magrum, L. J.; Woese, C. R.; Wolfe, R. S. "Methanogens: reevaluation of a unique biological group." *Microbiol. Reviews.* 1979, *43*:260–296.

Rogers, B., Animal and Plant Health Inspection Service, U.S. Department of Agriculture, personal communication, 1990.

Shelton, D. R. "Mineralization of diethylthiophosphoric acid by an enriched consortium from cattle dip." *Appl. Environ. Microbiol.* 1988, *54*:2572–2573.

Shelton, D. R.; Karns, J. S. "Coumaphos degradation in cattle-dipping vats." *J. Agric. Food Chem.* 1988, *36*:831–834.

Shelton, D. R.; Somich, C. J. "Isolation and characterization of coumaphos-metabolizing bacteria from cattle dip." *Appl. Environ. Microbiol.* 1988, *54*:2566–2571.

Development and Application of a Multilevel Respirometric Protocol to Determine Biodegradability and Biodegradation Kinetics of Toxic Organic Pollutant Compounds

*Henry H. Tabak**
U.S. Environmental Protection Agency
Sanjay Desai, Rakesh Govind
University of Cincinnati

INTRODUCTION

The main objective of this study was to develop a protocol based on oxygen consumption to generate a database on biokinetic constants of representative priority pollutants, RCRA and CERCLA toxic organics. Experiments were conducted using electrolytic respirometry to generate oxygen consumption data from which biokinetic parameters were developed. The representative compounds included 20 benzenes, 7 phthalates, 5 polycyclic aromatic hydrocarbons, 4 nitrosoamines, 8 phenols, 4 ketones, *p*-nitroaniline, and benzyl alcohol, with aniline as the reference compound.

The environmental factors investigated were temperature, concentration of biomass and substrate, and source of biomass. These effects on biodegradation kinetic rates will subsequently be related to the structural properties of the compound. This in turn will facilitate

* U.S. Environmental Protection Agency, Office of Research and Development, Risk Reduction Engineering Laboratory, 26 W. Martin Luther King Dr., Cincinnati OH 45268

prediction of the extent and rate of biodegradation of organic pollutant chemicals from the knowledge of their structural properties. The specific organic compounds used for this part of the study were benzene, toluene, ethylbenzene, phenol, resorcinol, 2,4-dimethyl phenol, dimethyl phthalate, diethyl phthalate, butyl phthalate, and butyl benzyl phthalate.

The results presented in this paper are applicable for aqueous aerobic degradation of organic compounds. A similar approach can be developed for studying aerobic degradation in soil systems. Furthermore, in this study, each compound was the sole carbon source for the microorganisms. Respirometric studies involving mixed substrate systems are currently under way.

Background

An extensive review of the use of oxygen consumption measurements in routine examination of sewage, in control of sewage treatment processes, and in application to specific wastes or compounds has been presented by several researchers including King and Dutka (1986). Techniques used to evaluate biodegradation kinetics were reviewed in detail by Grady (1985). These techniques utilize continuous, fed-batch, and batch-type reactors for providing data from which kinetic parameters can be evaluated.

The success of batch techniques in obtaining intrinsic kinetic parameters (dependent only on the nature of the compound and the degrading microbial community, and not on the reactor system used) was indicated in the studies of Simkins and Alexander (1985), Braha and Hafner (1987), Tabak *et al.* (1981), and Paris and Rogers (1986).

Measurement of oxygen consumption through electrolytic respirometry is a very promising batch-type technique for automating data collection associated with biodegradation and intrinsic kinetic parameters. Studies performed by several researchers including Tabak *et al.* (1984), Oshima *et al.* (1985), and Grady *et al.* (1989a,b,c) have utilized the electrolytic respirometry approach to generate biodegradation kinetic data.

Tabak *et al.* (1984) and Oshima *et al.* (1985) sought to capitalize on Dojlido's method of identifying inflection points to quantify more fundamental rate coefficients. Substrate removal was divided into exponential and declining phases, separated by an inflection point, and the endogenous phase. They coupled substrate removal and cell growth to oxygen consumption by imposing an electron balance and consequently were able to evaluate μ_m from oxygen consumption data up to the inflection point. They proposed using the lag time to indicate the difficulty of achieving acclimation to a test compound, and their respirometric studies were carried out with an unacclimated biomass.

The justification for using respirometry to obtain intrinsic kinetic coefficients lies in the concept of oxygen consumption as an energy balance (Gaudy & Gaudy 1971). This concept states that all of the electrons available in a substrate undergoing biodegradation must either be transferred to the terminal acceptor or be incorporated into new biomass or soluble microbial products. Furthermore, since biomass growth and product formation are proportional to substrate removal, this suggests that an oxygen uptake curve can provide the same information as either a substrate removal curve or a biomass growth curve. Specific growth rates obtained from growth studies as slopes of plots of Ln (biomass concentration) or Ln(X) versus time at different substrate concentrations compared favorably with those obtained from the exponential phase of respirometric oxygen uptake curves as slopes of plots of Ln (d oxygen uptake/d time) or Ln (dO_u/dt) versus time.

Studies of Grady *et al.* (1989a,c) have demonstrated that it is possible to determine intrinsic kinetics of single organic compounds by using only measurements of oxygen consumption in respirometric batch reactors. With the use of computer simulation techniques and nonlinear curve fitting methods, intrinsic kinetic parameters were obtained from oxygen consumption data and were shown to be in agreement with those obtained from more traditional measurements of substrate removal (DOC, SCOD, ^{14}C) or cell growth.

EXPERIMENTAL APPROACH

The electrolytic respirometry studies were conducted using an automated continuous oxygen uptake and BOD measuring Voith Sapromat B-12 (12-unit system) electrolytic respirometer-analyzer.

The nutrient solution used in these studies was an OECD synthetic medium consisting of measured amounts per liter of deionized distilled water of mineral salts solution, trace salts solution, and a solution (150 mg/L) of yeast extract as a substitute for a vitamin solution.

A comprehensive description of the procedural steps of the respirometric tests and of the experimental design employing test and control systems is presented elsewhere (Tabak *et al.* 1984).

In this study, biodegradation was measured by three approaches: the first, as the ratio of the measured BOD values in mg/L (oxygen uptake values of test compound minus inoculum control, endogenous oxygen uptake values) to the theoretical oxygen demand (ThOD) of substrate as a percent; the second, as a percentage reduction of the original amount of test compound as measured by dissolved organic carbon (DOC) changes; and the third, as a percentage reduction of the original amount of test compound as measured by specific substrate analysis.

The biodegradation of the substrate and cell growth rate was modeled using the Monod equations, as summarized in Table 1. Based on the concept of oxygen uptake as energy balance (Gaudy & Gaudy 1971), if the concentrations of the substrate, products, and biomass are all expressed in BOD units, then the oxygen uptake (O_u) at any time in the batch reactor may be calculated (Grady *et al.* 1989a) from

$$O_u = (S_{so} - S_s) - (X - X_o) - (S_p - S_{po}) \quad (1)$$

The above equation was applied in conjunction with the Monod equations in Table 1 to obtain the values of substrate, biomass, and product concentrations over time.

The methodology for quantitation of the Monod kinetic parameters requires the use of specific methods for estimating them initially (Tabak *et al.* 1989a,b,c), subsequently followed by computer simulation methods coupled with nonlinear curve fitting techniques, and is based on the use of measured values of initial growth and growth yield. The method requires the use of the kinetic equation relating growth rate of biomass in the presence of substrate, substrate utilization rate, product formation rate, and rate of oxygen consumption from O_2 uptake (BOD) curves to calculate and use the theoretical oxygen consumption data to quantitate biokinetic parameters.

A more comprehensive description of the procedural steps in determining rates of exponential and declining growth, as well as in estimating and quantitating Monod kinetic parameters for toxic organic compounds from oxygen consumption data, has been given elsewhere by Tabak *et al.* (1989a,b,c).

TABLE 1. Summary of equations for the batch respirometer. Specific equations have been given for volatile compounds and sparingly soluble compounds.

$$dS_s/dt = -(\mu_m/Y)S_sX/K_s + S_s) \quad (2)$$

$$dS_p/dt = Y_p(-dS/dt) = (Y_p\mu_m/Y)S_sX/(K_s + S_s) \quad (3)$$

$$dX/dt = \mu_m S_sX/(K_s + S_s) - K_sbX/(K_s + S_s) \quad (4)$$

where S_s is the soluble substrate concentration, S_p is the soluble product concentration, X is the biomass, Y_p is the product yield, and b is the decay coefficient.

Volatile Compounds

$$dS/dt = -(\mu_m SX/Y)/(K_s' + S) \quad (5)$$

where $K_s' = K_s(1 + HV_g/V_L)$. Similarly, equations (3) and (4) change to

$$dS_p/dt = (Y_p\mu_m SX/Y)/(K_s' + S) \quad (6)$$

$$dX/dt = [(\mu_m SX)/(K_s' + S)] - [(K_s'bX)/(K_s' + S)] \quad (7)$$

Sparingly Soluble Compounds

$$X = X_o \exp[(\mu_m S_s - bK_s)t/(K_s + S_s)] \quad (8)$$

$$S = S_o - [(\mu_m S_s/Y)/(\mu_m S_s - bK_s)](X - X_o) \quad (9)$$

$$S_p = [(Y_p\mu_m S_s/Y)/(\mu_m S_s - bK_s)](X - X_o) \quad (10)$$

Thus, equations (8), (9), and (10) should be used when the total concentration of the substrate, S, is greater than $S_s(1 + HV_g/V_L)$; and equations (5), (6), and (7) should be used when the total substrate concentration, S, is less than or equal to $S_s(1 + HV_g/V_L)$.

Modifications for Volatile and Sparingly Soluble Compounds

Volatile Compounds. Equations (2) through (4) (Table 1) are valid for a nonvolatile compound present at a concentration below its solubility limit, so they have to be modified for volatile and sparingly soluble compounds. Because respirometer reactors are sealed, when compound is added it partitions itself between gas and liquid phases. As biodegradation proceeds and the substrate concentration decreases in the liquid phase, it is transferred from the gas phase to the liquid phase to maintain equilibrium in accordance with Henry's law. The Monod equations were modified to include the effect of vapor-liquid equilibrium; these equations have been presented in Table 1.

Sparingly Soluble Compounds. A sparingly soluble compound will dissolve into the solution phase up to its solubility limit, and the undissolved part will exist as a different phase. As the compound undergoes biodegradation in solution phase, more of the undissolved part dissolves into the solution, keeping its concentration in solution phase constant at the solubility limit. Therefore, degradation proceeds at a constant substrate concentration as long as total substrate concentration in the system is greater than its solubility limit. If the compound is also volatile, then depending on Henry's law constant, it will partition itself into the gas and liquid phases. When the microorganisms degrade this compound from the solution phase, compound from the undissolved part dissolves into solution phase. This keeps the compound concentration in the solution and gas phases constant. If S is the total substrate concentration (per unit volume of liquid phase) in the system and S_s is its solubility limit, then the concentration in solution and gas phases will be constant as long as $S > S_s/(1 + HV_g/V_L)$. Therefore, if substrate concentration is held constant at S_s in equations (2) through (4), then they can be solved analytically, resulting in equations (8) through (10), as presented in Table 1.

DEVELOPMENT OF MULTILEVEL RESPIROMETRIC BIODEGRADATION TESTING PROTOCOL

The oxygen consumption data generated with the use of electrolytic respirometry have been adequately utilized for assessing the biodegradative activity of sludge microbiota, the degradability and toxicity of toxic organic compounds, as well as for determining the intrinisic kinetic parameters of biodegradation. Methodologies have been developed for quantitating the biodegradability and biodegradation kinetics of representative classes of RCRA toxic organics, with the resultant development of a comprehensive multilevel respirometric biodegradation testing protocol based on oxygen consumption data.

The methods of each successive testing level of the respirometric protocol are characterized by increased complexity and a consequently higher cost for performing the tests pertaining to each testing level. Accordingly, testing level can be selected to match research needs.

Studies to assess the biodegradability and biodegradation rates of toxic organics by sludge microbiota with the use of respirometric oxygen uptake data can involve the use of one or more levels of the protocol, depending on the amount of information needed for assessing the biodegradability of the organic toxic pollutant or the toxicant-bearing waste to determinate its fate and rate of biotreatment in municipal or industrial waste treatment systems. A logic flow diagram of the respirometric biodegradation testing protocol, providing a brief description of each of the testing levels, is shown in Figure 1.

RESULTS AND DISCUSSION

Respirometric biodegradability, biokinetic, and Monod kinetic data for selected RCRA alkyl benzenes, phenols, phthalates, and ketones were generated in these studies. The quantitative kinetic parameters for these compounds have been summarized in Table 2 as the averages of two replicates (except for aniline, which had eight replicates). A representative oxygen uptake curve is given in Figure 2, which shows the agreement between the experimental (asterisk) and the calculated (solid line) data.

The effects of temperature, substrate concentration, biomass concentration, and source of biomass were studied for 10 compounds selected from the 32 compounds that were found to biodegrade. The criterion for selection was that they had to be present in at least three of the four U.S. EPA lists: RCRA list, Superfund list, Priority pollutant list, and Best Demonstrated Available Technology (BDAT) list. The ten compounds selected were benzene, toluene, ethyl benzene, phenol, resorcinol, 2,4-dimethylphenol, dimethyl phthalate, diethyl phthalate, dibutyl phthalate, and butyl benzyl phthalate. One way analysis of variance (ANOVA) was used to find whether a kinetic constant was

LEVEL I BIODEGRADATION SCREENING METHOD

DETERMINATION OF PRIMARY DEGRADATION FROM OXYGEN UPTAKE DATA

* ACCLIMATION TIME
* PERCENT BIODEGRADATION
* MICROBIAL INHIBITION/TOXICITY

LEVEL II STUDIES ON BIOKINETICS OF BIODEGRADATION BASED ON RESPIROMETRIC OXYGEN UPTAKE DATA

DETERMINATION OF SPECIFIC GROWTH RATES AND INITIAL ESTIMATES OF MONOD KINETICS

* RATES OF EXPONENTIAL AND DECLINING GROWTH PHASE OF BOD CURVE
* MONOD KINETIC PARAMETERS FROM THE OXYGEN UPTAKE CURVE

LEVEL III STUDIES TO QUANTITATE MONOD KINETIC PARAMETERS FROM RESPIROMETRIC OXYGEN UPTAKE DATA

USE OF INITIAL ESTIMATES OF MONOD KINETIC PARAMETERS TO DETERMINE BEST FIT VALUES THROUGH COMPUTER SIMULATION AND NONLINEAR CURVE FITTING TECHNIQUES

LEVEL IV STUDIES TO DETERMINE THE EXTENT OF BIODEGRADATION

TESTS FOR ASSESSMENT OF ULTIMATE BIODEGRADABILITY DURING RESPIROMETRIC EXPERIMENTS

* RADIOISOTOPE RESPIROMETRY
* RESPIROMETRY COUPLED WITH SPECIFIC SUBSTRATE ANALYSIS

FIGURE 1. Logic flow diagram for the respirometric biodegradation testing program.

similar or different (Walpole & Myers 1972) and Cochran's test was used to check the equality of variances for different treatments. The value of f, which is the value of the random variable F having the F distribution, was calculated and compared with the critical value at the 95 percent confidence level, denoted by the variable F. The Cochran's

TABLE 2. Quantitative Monod kinetic parameter values.

Compound	μm (h^{-1})	K_s mg/L	Y mg/mg	Avg. % error
Aniline	0.253± 0.015	28.93± 2.61	0.43± 0.05	6.2 ± 2.4
<u>Benzenes</u>				
Benzene	0.282	6.57	0.27	5.32
Ethyl benzene	0.216	10.07	0.34	8.77
Toluene	0.523	7.75	0.36	5.62
o-Xylene	–	–	–	–
m-Xylene	0.123	0.75	0.26	12.94
p-Xylene	0.140	2.47	0.36	13.43
Cumene	0.194	10.03	0.23	19.21
Butyl benzene	0.109	10.00	0.23	18.29
sec-Butyl benzene	0.110	10.00	0.62	0.32
tert-Butyl benzene	0.021	10.00	0.22	18.91
1-Phenyl hexane	0.135	7.47	0.53	8.52
<u>Phthalates</u>				
Dimethyl phthalate	0.317	7.00	0.49	8.87
Diethyl phthalate	0.281	7.49	0.36	7.89
Dipropyl phthalate	0.220	6.50	0.23	6.65
Dibutyl phthalate	0.180	7.89	0.38	6.95
<u>Phenols</u>				
Phenol	0.283	3.00	0.30	4.51
o-Cresol	0.294	1.75	0.53	9.52
m-Cresol	0.542	6.50	0.44	6.39
p-Cresol	0.321	6.00	0.39	8.04
2,4-Dimethyl phenol	0.303	6.99	0.32	5.21
Catechol	0.495	5.40	0.28	8.19
Resorcinol	0.392	5.87	0.28	5.32
<u>Ketones</u>				
Acetone	0.217	1.00	0.55	5.62
2-Butanone	0.239	9.99	0.45	3.06
4-Methyl-2-pentanone	0.240	7.47	0.43	4.55
Isophorone	0.102	10.00	0.54	12.26

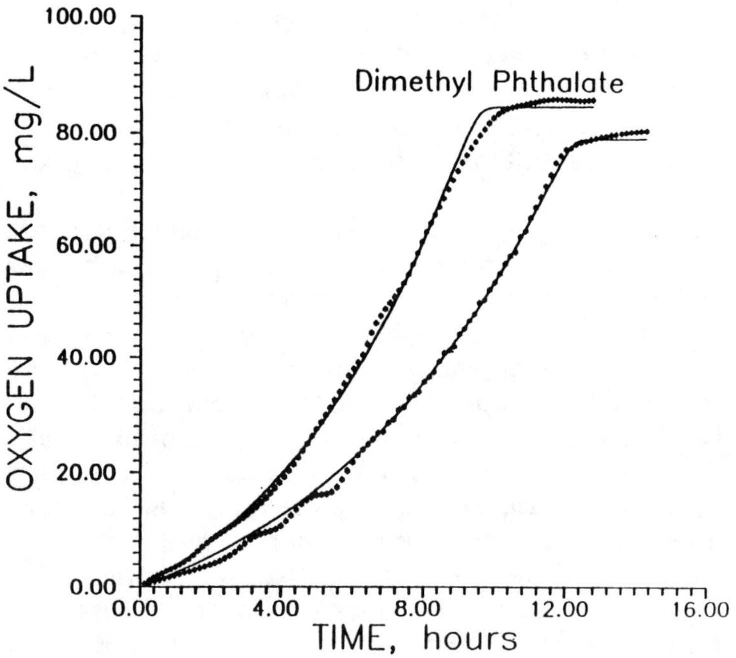

FIGURE 2. Oxygen uptake curves for dimethyl phthalate at 100 mg/L.

equality of variance test requires calculating the parameter g and comparing its value with the critical value G at the 95 percent confidence level (Walpole & Myers 1972).

Effect of Substrate Concentration. The experiments were conducted at 25 C and at substrate concentrations of 20, 40, 60, 80, and 100 mg/L, using sludge from the Little Miami wastewater treatment plant (Cincinnati OH) as inoculum. The duration of each of the experiment was about 20 days. Duplicate runs were made for each of the five concentrations for all the compounds.

The F test showed that μ_m values are similar at all concentration levels for all compounds. This suggests that μ_m is independent of initial substrate concentration. Similar trends have been confirmed for ethylene dibromide, aniline, and *m*-nitro phenol (Aelion *et al.* 1989), at low concentrations of phenol (Scott *et al.* 1983), and for phthalates and polyaromatics (Grady *et al.* 1989a). Fannin *et al.* (1981) reported that the degradation rate decreased with increase in concentration for phenol. They studied phenol degradation in the concentration range 100 to 400 mg/L and found phenol to inhibit microbial degradation beyond

100 mg/L (Grady *et al.* 1989a,b,c), so degradation rate decreases with increasing concentration beyond 100 mg/L. Variation in lag time is not more than 1 day for a substrate concentration range of 20 to 100 mg/L, which indicates that at least in this concentration range lag time is independent of substrate concentration.

Effect of Biomass Concentration. The experiments were carried out at 25 C with a substrate concentration of 100 mg/L and biomass concentrations of 20, 30, and 40 mg/L. Activated sludge from the Little Miami wastewater treatment plant (Cincinnati OH) was used as inoculum. The duration of experiments was 10 to 20 days. Experiments used duplicates for all the compounds at each concentration level of biomass.

The F tests show that μ_m is independent of initial biomass concentration. Aichinger (1989) has found a similar trend for phthalates and polyaromatic hydrocarbons. He used a method similar to the one used in this study to estimate kinetic parameters from oxygen uptake data. Fannin *et al.* (1981) reported that increasing the amount of bacterial inoculum caused a significant linear increase in the degradation rate. They had calculated the rate by dividing percentage degraded by time period (%/h).

Effect of Temperature. To study the effect of temperature, experiments were carried out at substrate concentration of 100 mg/L and at 15, 25, and 35 C, using activated sludge from the Little Miami wastewater treatment plant (Cincinnati OH) as inoculum. The duration of experiments was 10 to 20 days. Experiments used duplicate samples of all the compounds at each temperature. Microorganisms were not acclimated to the temperatures of 15 or 35 C before the start of the experiment. The results of these experiments are presented in Table 3. At 15 C benzene, toluene, and ethyl benzene were not mineralized completely at the cessation of oxygen consumption. The residual materials were probably the transformation products, although no studies were done to identify them. These products may have inhibited further degradation of three benzenes.

The maximum specific rate constants for all the compounds increased with an increase in temperature from 15 to 35 C. The effect of temperature on the rate constant was described by the Arrhenius equation

$$\mu_m = A \exp(-E/RT) \qquad (11)$$

TABLE 3. Results of effect of temperature.

Compound	Parameter	Temperature (C) 15	25	35	A (h^{-1})	E kcal/M
Benzene	μ_m	0.263	0.282	0.292	0.14	0.94
	Y	0.22	0.27	0.29		
	K_s	7.44	6.57	7.44		
	lag	8.5	3.5	2.5		
Toluene	μ_m	0.514	0.523	0.530	0.83	0.28
	Y	0.32	0.36	0.35		
	K_s	7.07	7.75	6.27		
	lag	6.5	2.2	2.0		
Ethyl benzene	μ_m	0.210	0.216	0.221	0.46	0.44
	Y	0.32	0.34	0.36		
	K_s	10.51	10.07	9.15		
	lag	6.5	3.0	2.3		
Phenol	μ_m	0.220	0.283	0.332	0.52	0.37
	Y	0.28	0.30	0.34		
	K_s	3.50	3.00	4.87		
	lag	2.5	1.0	0.8		
2,4-Dimethyl phenol	μ_m	0.267	0.303	0.330	0.64	0.64
	Y	0.28	0.32	0.27		
	K_s	7.01	7.00	5.50		
	lag	8.5	2.2	1.0		
Resorcinol	μ_m	0.360	0.395	0.420	3.27	0.77
	Y	0.22	0.28	0.29		
	K_s	5.00	5.87	5.50		
	lag	2.5	1.3	1.0		
Dimethyl phthalate	μ_m	0.298	0.317	0.331	1.43	0.89
	Y	0.43	0.49	0.41		
	K_s	6.00	7.00	6.89		
	lag	5.50	2.50	2.00		
Diethyl phthalate	μ_m	0.229	0.282	0.334	1.72	0.93
	Y	0.30	0.36	0.35		
	K_s	8.98	7.49	7.37		
	lag	5.50	2.00	1.50		
Dibutyl phthalate	μ_m	0.172	0.180	0.188	1.93	0.97
	Y	0.39	0.37	0.38		
	K_s	9.04	7.89	9.27		
	lag	6.0	2.0	1.5		

where A is the frequency factor (h^{-1}), E is the activation energy (cal/M), R is the gas constant (1.986 cal/K × g M), and T is the absolute temperature (K). An Arrhenius plot of the temperature data showed good agreement for different compounds. Table 3 lists the frequency factor and the activation energy of some of the test compounds. These activation energies of test compounds, except for phenol and diethyl phthalate, are in the range reported for ordinary enzyme reactions, indicating that the degradation of these compounds in wastewater treatment is not thermodynamically hampered (Larson & Perry 1981). Several researchers have reported that degradation rates are directly proportional to temperatures (Hollis 1976; Johnson & Heitkamp 1984; Stephenson *et al.* 1983). Furthermore, the results in Table 3 indicate that lag time decreases with increase in temperature.

Effect of Sludge Source. The experiments were conducted with sludges from four different sources: Little Miami wastewater treatment plant (Cincinnati OH), Test and Evaluation Facility of the U.S. EPA (Cincinnati), Monsanto plant (Addyston OH), and BP America plant (Lima OH). The substrate and biomass concentrations used were 100 and 30 mg/L, respectively. Duplicate samples were run for all the compounds with sludge from each source. The results of the experiments are given in Table 4.

Statistical analysis showed that the kinetic constants obtained with sludges from domestic, domestic-industrial, and industrial sludge sources were, with few exceptions, in fairly good agreement and were shown to fall within a similar range of microbial activity. In this study correlations were made between microbial populations representing domestic and industrial waste treatment sludge, and exceptions could suggest that kinetic constant may be unique to the biomass of a particular industrial waste treatment system. In such exceptions, the history and composition of the microbial community may play an important role in the rate of biodegradation, but the extent of biodegradation does not depend on the culture used. Data support that in most cases, however, there was a fair correlation between specific rate constant for any compound biodegraded by microbiota from different sources of sludge, and that the percent biodegraded for all the test compounds was similar. An even better correlation between kinetic constants was observed in studies (not reported here) relating sludges from different domestic waste treatment systems only. Furthermore, these data corroborate the work of Levandowski (1988) on POTW treatment of industrial organic wastes.

TABLE 4. Results of effect of sludge source.

Compound	Parameter	Little Miami	T&E	Monsanto	BP
Benzene	μ_m	0.282	0.116	0.122	0.672
	Y	0.27	0.23	0.22	0.22
	K_s	6.57	8.77	7.64	9.24
	extent	54.0	53.0	56.5	52.0
Toluene	μ_m	0.523	0.376	0.141	0.776
	Y	0.36	0.32	0.26	0.29
	K_s	7.75	6.52	4.51	8.13
	extent	67.5	62.0	60.5	64.0
Ethyl benzene	μ_m	0.216	0.130	0.240	0.499
	Y	0.34	0.30	0.26	0.25
	K_s	10.07	2.90	7.97	10.12
	extent	93.0	95.5	99.0	99.0
Phenol	μ_m	0.283	0.156	0.183	0.316
	Y	0.30	0.39	0.39	0.22
	K_s	3.00	9.50	5.00	8.99
	extent	95.0	93.0	94.0	97.0
2,4-Dimethyl phenol	μ_m	0.303	0.179	0.158	0.113
	Y	0.32	0.34	0.30	0.21
	K_s	7.00	3.00	8.50	7.00
	extent	90.0	95.0	87.0	88.0
Resorcinol	μ_m	0.398	0.159	0.351	0.394
	Y	0.28	0.39	0.31	0.30
	K_s	5.87	4.88	8.77	6.50
	extent	87.5	95.5	90.0	92.0
Dimethyl phthalate	μ_m	0.317	0.152	0.118	0.460
	Y	0.49	0.41	0.40	0.47
	K_s	7.00	2.00	5.25	5.00
	extent	99.8	99.0	99.9	99.5
Diethyl phthalate	μ_m	0.282	0.140	0.114	0.413
	Y	0.36	0.47	0.32	0.32
	K_s	7.50	2.00	10.00	7.48
	extent	89.5	90.0	86.0	84.5
Dibutyl phthalate	μ_m	0.180	0.219	0.265	0.487
	Y	0.37	0.51	0.45	0.30
	K_s	7.89	7.35	6.12	5.23
	extent	80.0	84.5	86.0	85.5

Microbial adaption to the compounds tested was quite rapid, resulting in a two- to fivefold increase in the average degradation rates after only the second exposure in a batch reactor. When multiple substrates were tested, the individual compound degradation rates in the mixture were very similar to the degradation rates as sole carbon source.

CONCLUSIONS

The electrolytic respirometric approach is a very successful, accurate, cost-effective, and less labor intensive method for generating biodegradability, inhibition, and biokinetic data. The values of the kinetic parameters for single organic compounds in synthetic media determined from oxygen consumption data are similar to those obtained from traditional measurement of substrate removal (DOC, SCOD, ^{14}C) or cell growth.

From this study the following conclusions can be drawn: (1) Monod parameters were calculated for 32 compounds that degraded in the respirometer. (2) The Monod kinetic equations were modified to handle volatile and slightly soluble compounds. (3) The rate constant for degradation was found to increase with temperature and was described by the Arrhenius equation. (4) The rate constant were found to be independent of the initial substrate and biomass concentrations. (5) The kinetic constants were found to be independent of the source of biomass (with exceptions). (6) In the concentration range of 20 to 100 mg/L, lag time proved independent of substrate concentration and decreased with an increase in initial biomass concentration, as suggested by Wiggins *et al.* (1987) and Ventullo and Larson (1986).

Acknowledgments

The authors wish to thank Suzanne Chase, Secretary in the U.S. Environmental Protection Agency's Risk Reduction Engineering Laboratory (Cincinnati OH) for her excellent and timely wordprocessing skills in preparing this manuscript.

REFERENCES

Aelion, C. M.; Dobbins, R. C.; Pfaender, F. K. "Adaptation of Aquifier Microbial Communities to the Biodegradation of Xenobiotic Compounds: Influence of Substrate Concentration and Preexposure." *Environ. Toxicol. Chem.* **1989**, *8*, 75.

Aichinger, G. "Determination of Biodegradation Kinetics of Selected Phthalates and Polyaromatic Hydrocarbons Using Electrolytic Respirometry." M.S. Thesis, Environmental Systems Engineering, Clemson Univ., Clemson, SC, 1989.

Braha, A.; Hafner, F. "Use of Lab Batch Reactors to Model Biokinetics." *Water Res.* **1987**, *21*, 73.

Fannin, T. E.; Marcus, M. D.; Anderson, D. A.; Bergman, H. L. "Use of a Fractional Design to Evaluate Interactions of Environmental Factors Affecting Biodegradation Rates." *Appl. Environ. Microbiol.* **1981**, *42*, 936.

Gaudy, A. F., Jr.; Gaudy, E. T. *Biological Concepts for Design and Operation of the Activated Sludge Process*; U.S. Environmental Protection Agency: Cincinnati, OH; EPA-17090 FQJ.

Grady, C.P.L., Jr. "Biodegradation: Its Measurement and Microbiological Basis." *Biotech. Bioengineer* **1985**, *27*, 660.

Grady, C.P.L., Jr.; Lim, H. C. *Biological Wastewater Treatment: Theory and Application*; Mercel Dekker: New York, 1980.

Grady, C.P.L., Jr.; Dang, J. S.; Harvey, D. M.; Jobbagy, A.; Wang, X.-L. "Determination of Biodegradation Kinetics through Use of Electrolytic Respirometry." *Wat. Sci. Tech.* **1989a**, *21*, 957.

Grady, C.P.L., Jr.; Dang, J. S.; Harvey, D. M.; Jabbagy, A. "Evaluation of Biodegradation Kinetics with Respirometric Data." Submitted to *Water Pollut. Control Fed.* **1989b.**

Grady, C.P.L., Jr.; Aichinger, G.; Cooper, S. F.; Naziruddin, M. "Biodegradation Kinetics for Selected Toxic/Hazardous Organic Compounds." In *Hazardous Waste Treatment: Biosystems for Pollution Control*; AWMA, 1989c.

Hollis, C. G. "Thermal Effects on Biodegradation of Pollutants in Water"; Research Report No. 53; Dept. of Biology, Memphis State Univ.: Memphis, TN, 1976.

Johnson, B. T.; Heitkamp, M. A. "Environmental and Chemical Factors Influencing Biodegradation of Phthalic Acid Esters in Fresh Water Sediments." *Environ. Pollut.* (Series B) **1984**, *8*, 101.

King, E. F.; Dutka, B. I. "Respirometric Techniques." In *Toxicity Testing Using Microorganisms*; Britton, G.; Dutka, B. J., (Eds.); CRC Press: Boca Raton, FL, Vol.I, p 76.

Larson, R. J.; Perry, R. L. "Use of Electrolytic Respirometer to Measure Biodegradation in Natural Waters." *Wat. Res.* **1981**, *15*, 697.

Levandowski, G. "POTW Treatment of Industrial Organic Wastes"; final report, Project BICM-6; New Jersey Institute of Technology: Newark, NJ; p 104.

Oshima, A.; Tabak, H. H.; Lewis, R. F. "The Evaluation of Biological Treatability and Removability of Toxic Organic Chemicals by Respirometry"; in-house report; U.S. Environmental Protection Agency, Water Engineering Research Laboratory: Cincinnati, OH, 1985.

Paris, D. F.; Rogers, J. E. "Kinetic Concepts for Measuring Microbial Rate Constants: Effects of Nutrients on Rate Constants." *Appl. Environ. Microbiol.* **1986**, *51*, 221.

Scott, H. D.; Wolf, D. C.; Lavy, T. L. "Adsorption and Degradation of Phenol at Low Concentrations in Soil." *J. Environ. Quality* **1983**, *12*, 91.

Simkins, S.; Alexander, M. "Nonlinear Estimation of the Parameters of Monod Kinetics That Best Describe Mineralization of Several Substrate Concentrations by Dissimilar Bacterial Densities." *Appl. Environ. Microbiol.* **1985**, *50*, 816.

Stephenson, T.; Lester, T. N.; Perry, R. "The Influence of Transient Temperature Changes on the Biodegradation of Nitrilotriacetic Acid in the Activated Sludge Process: A Pilot Plant Study." *Environ. Pollut.* (Series A) **1983**, *32*, 1.

Tabak, H. H.; Quave, S. A.; Mashi, C. I.; Barth, E. F. "Biodegradability Studies with Organic Priority Pollutant Compounds." *J. Water Poll. Control Fed.* **1981**, *53*, 1503.

Tabak, H. H.; Lewis, R. F.; Oshima, A. "Electrolytic Respirometry Biodegradation Studies, OECD Ring Test of Respiration Method for Determination of Biodegradability"; in-house report, U.S. Environmental Protection Agency, Municipal Environmental Research Laboratory: Cincinnati, OH, 1984.

Tabak, H. H.; Desai, S.; Govind, R. "The Determination of Biodegradability and Biodegradation Kinetics of Organic Pollutant Compounds with the Use of Electrolytic Respirometry." *Proceedings*, EPA 15th Annual Research Symposium: Remedial Action, Treatment and Disposal of Hazardous Waste, 1989a.

Tabak, H. H.; Desai, S.; Govind, R. "Determination of Biodegradation Kinetics of RCRA Compounds Using Respirometry for Structure-Activity Relationships." *Proceedings*, 44th Annual Purdue Industrial Waste Conference, West Lafayette, IN, May 9–11, 1989b.

Tabak, H. H.; Desai, S.; Govind, R. "Biodegradability and Biodegradation Kinetics Assessment for Toxic Organics Using Oxygen Uptake Measurements." *Proceedings*, 12th U.S./Japan Conference on Sewage Treatment Technology, Cincinnati, OH, October 12–13, 1989c.

Urano, K.; Kato, Z. "Evaluation of Biodegradation Ranks of Priority Organic Compounds." *J. Hazardous Materials* **1986**, *13*, 147.

Ventullo, R. M.; Larson, R. J. "Adaptation of Aquatic Microbial Communities to Quaternary Ammonia Compounds." *Appl. Environ. Microbiol.* **1986**, *51*, 356.

Walpole, R. E.; Myers, R. H. *Probability and Statistics for Engineers and Scientists*; MacMillan Pub. Co.: New York, 1972.

Wiggins, B. A.; Jones, S. H.; Alexander, M. "Explanations for the Acclimation Period Preceding the Mineralization of Organic Chemicals in Aquatic Environments." *Appl. Environ. Microbiol.* **1987**, *53*, 2617.

Characterization and Optimization of Treatment of Organic Wastes and Toxic Organic Compounds by a Lignolytic White Rot Fungus in Bench-Scale Bioreactors

Henry H. Tabak, John A. Glaser*
U.S. Environmental Protection Agency
Susan Strohofer, Margaret J. Kupferle,
Pasquale Scarpino, M. Wilson Tabor
University of Cincinnati

INTRODUCTION

The ability to degrade and detoxify organic and inorganic waste constituents requires two complementary features: the biochemical means (enzymes) to detoxify wastes and the capability of a single organism or a group of organisms to effect this required metabolism. An example of a single, highly competent organism is the wood-degrading fungus, *Phanerochaete chrysosporium*, which has the potential to degrade the aromatic components of toxic and hazardous waste. This capacity is based on its ability to degrade lignin, a persistent biogenic polymer, via the nonspecific extracellular enzymes, ligninases. This fungus has been utilized to treat liquid-phase wastes in a rotating biological contactor (RBC), the MyCoR process.

* U.S. Environmental Protection Agency, Risk Reduction Engineering Laboratory, Cincinnati OH 45268

This research program explores the use of the MyCoR process as a hazardous waste site clean-up technology, addressing issues of application and practicality to emphasize cost effectiveness and efficiency. Biochemical and mechanical parameters were optimized for biodegradation by the use of *P. chrysosporium* biofilms in bench-scale RBCs. The specific objectives were (1) to optimize the carbon source and other nutrient requirements in the primary growth phase and in the secondary enzyme synthesis/metabolic phase, (2) to optimize the operational parameters (*e.g.*, headspace oxygen, temperature, rotational speed of RBC discs), and (3) to investigate degradation kinetics and identify intermediate and end products of the metabolism of several representative chlorinated phenols. The variables were evaluated using a wood-processing pulp and paper mill (Kraft liquor) effluent and 2,4-dichlorophenol (DCP).

Background

Ligninolytic fungi, found throughout the northern hemisphere, are unique among microorganisms in that they can rapidly depolymerize and mineralize lignin, a complex, biorecalcitrant, biogenic, irregular, nonhydrolyzable, and environmentally persistent wood polymer of phenol propane units (Chen & Chang 1985; Kirk & Farrell 1987). Lignin degradation is part of their secondary metabolism, occurring at high rates only under conditions of nutrient limitation, and is cometabolic, i.e., a primary growth substrate such as cellulose or glucose is required. Lignin contains numerous substructures also found in common organic pollutants, phenolics, anisoles, aryl-*o*-ethers, and biphenyls. Its exceptional recalcitrance to attack by most microbes has led to a belief that any organism capable of mineralizing it must have highly nonspecific oxidizing systems that could be applicable to aromatic pollutants as well (Bumpus *et al.* 1985; Kirk & Chang 1981).

P. chrysosporium, a white rot Basidiomycete, has a superior ability to degrade lignin (Buswell & Odier 1987; Kelley & Reddy 1988; Kirk *et al.* 1986; Tien & Kirk 1983). Ligninases are peroxidases that utilize hydrogen peroxide to oxidize organic compounds. Although these enzymes resemble conventional peroxidases in some respects, their H_2O_2-oxidized states are electropositive (Hammel *et al.* 1986). They catalyze one-electron oxidations of a wide variety of lignin-related, polyalkoxylated model compounds to give carbon-centered, substrate-free radicals as initial transient products, and post-enzymatic reactions of these radicals account for the many end products found (Kersten *et al.* 1985; Schoemaker *et al.* 1985). In the case of lignins, carbon-carbon cleavage is the major reaction of interest and leads to fragmentation of the polymer, i.e., ligninolysis. This enzyme system is induced in response to specific nutrient deprivation in nature and has been simulated in the laboratory (Jeffries *et al.* 1981; Keyser *et al.* 1978). Studies of Fenn and Kirk (1981) established that changes in glutamate concentration played a major role in the onset of ligninolytic activity.

Indications that wood decay fungi might have desirable xenobiotic degrading capabilities came from studies of paper mill effluent treatment by white rot fungi (Fukuzumi *et al.* 1977). *P. chrysosporium* was found to decolorize this effluent, degrade its residual Kraft lignins, and remove its low molecular weight organics. These results were later applied to development of a successful bench-scale method for effluent treatment by *P. chrysosporium* (Huynh *et al.* 1985; Kirk & Chang 1981). The paper pulp treatment application of *P. chrysosporium* is based on the use of a liquid bioreactor configurated as a rotating biological contactor, the MyCoR process (Chang *et al.* 1985, 1987). The color-producing bodies and contaminating chlorinated compounds characteristic of these effluents are removed by this process (Glaser *et al.* 1990a,b, 1991; Lewandowski *et al.* 1990; Prouty 1990; Yin *et al.* 1989b).

Subsequent work showed that *P. chrysosporium,* because of the low specificity of lignin enzyme systems, has biodegradative capabilities that encompass a much broader range of organo-pollutants than those produced during wood pulping (Bumpus & Aust 1987; Hammel 1989). Recent studies have demonstrated that organo-pollutant degradation by *Phanerochaete* coincides temporarily with the production of lignin peroxidases in fungal cultures, and that these enzymes oxidize not only lignin-related compounds, but also a wide variety of environmentally significant aromatics and waste components. Recent literature lists studies that have shown that *P. chrysosporium* is able to degrade chlorophenols and pentachlorophenol (components of wood-treating process wastes), polycyclic aromatic hydrocarbons, polychlorinated biphenyls, polymeric dyes, pesticides and herbicides, explosives (munitions wastes), methoxybenzyl alcohols, chlorinated anilines, methoxylated benzenes, wood-preserving wastes, chloroorganics, and humic acids. Data presented in the recent literature are in fair agreement that the expression of the ligninolytic system may be necessary for degradation and that *P. chrysosporium* is unusually versatile at organo-pollutant destruction.

MATERIALS AND METHODS

Experimental Design

Eight operational parameters were selected for bench-scale RBC study under controlled experimental conditions representing physical and biochemical aspects of RBC design biotechnology. Those selected for bench-scale RBC studies with paper and pulp mill (Kraft liquor) effluent wastewaters were (1) atmosphere—air versus increased oxygen, (2) substrate concentrations—variable glucose, (3) nitrogen concentrations—ligninolytic versus nonligninolytic, (4) buffer, (5) RBC disk rotational speed, (6) reactor temperature, (7) growth in absence and presence of pulp effluent, and (8) treatment of dichlorophenol.

These experimental variables have been examined by other investigators in isolated laboratory studies, but not in a coordinated manner to provide the basis required for the extension of laboratory-scale observations through bench-scale verification to pilot- or full-scale application of this biodegradative process. Tabak and Cooke (1968) reported the facilitation of fungal growth by increased oxygen atmospheres, which was verified to influence positively the transition between primary and secondary metabolism in *P. chrysosporium* (Bar-Lev & Kirk 1981). Likewise, glucose had been shown to be utilized in both primary and secondary phases of metabolism (Chang *et al.* 1985, 1987) in the organism and to be minimally required for the decolorization of pulp mill effluents (Yin *et al.* 1989a). However, this required substrate created BOD problems in the scale-up from laboratory- to bench-scale (Kirk *et al.* 1978). Additionally, optimal growth and maintenance of the organisms as a degrader of xenobiotics had been shown to be dependent on the initial growth of the *P. chrysosporium* in nitrogen deficient media (Kirk *et al.* 1978). These operational variables and the studies reported by Huynh *et al.* (1985) as essentials for patenting of the fungal-RBC process (Chang *et al.* 1985, 1987) formed the basis of the experiments described herein to elucidate and to optimize the RBC process for effective biodegradation of hazardous wastes. Table 1 summarizes the design criteria for bench- and pilot-scale RBC units, although only the bench-scale work will be presented.

Materials

Chemicals. All chemicals used in experimentation were reagent grade or better. Gases (*e.g.*, 99.5% oxygen) were purchased from Wright Bros.

TABLE 1. Summary of design parameters for bench- and pilot-scale rotating biological contactor units. Bench-scale RBC units were built after the design of Chang *et al.* 1985, 1987), and pilot-scale RBC units were 77.5X scaled-up modified versions of this design.

Parameter	Bench	Pilot
Volume	2 L	155 L
Disk diameter	0.015 m	0.5 m
Disk surface area	0.28 m^2	32.4 m^2
Disk material	Plexiglas	HDPE
Disk submergence	40%	40%
Shaft RPM	1–10	1–10
Operation mode	Batch	Batch with recirculation

Internal standards for gas chromatography and gas chromatography/mass spectrometry analyses were provided by the EER laboratory at U.S. Environmental Protection Agency (EPA; Cincinnati OH).

Wastewater Source. Kraft liquor samples of 10 to 50 gal were obtained from Mead Paper Mill (Chillicothe OH) as needed to maintain a steady source for reactor treatment studies. These effluents contained approximately 80 percent of the total mill color effluent (Tyler & Fitzgerald 1972). The sample lots were from both hardwood and softwood processing; color values were determined to range from 1900 to 6200 platinum color unit intensity (PCU).

Methods

RBC Bioreactors. Bench-scale rotating biological contactors (RBC; Figure 1) were constructed from plexiglas based upon the design description in the patent of Chang et al. (1985, 1987) and by Huynh *et al.* (1985). Alterations to the design were made as follows: the reaction chamber, a half cylinder consisting of four baffled sections that can be

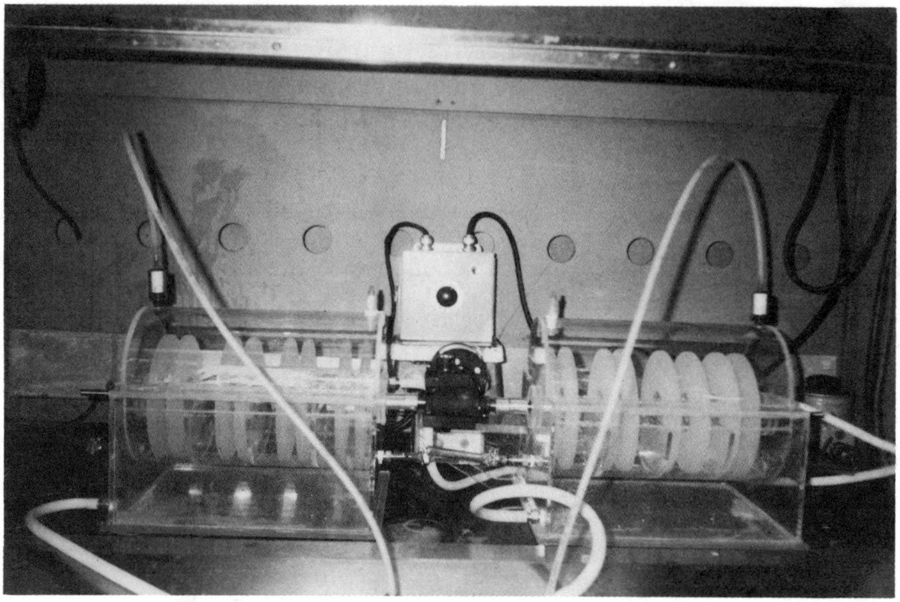

FIGURE 1. Bench-scale RBC units for studies of treatment of wastewaters by *P. chrysosporium*. Bench-scale RBC units were constructed after the design of Chang *et al.* (1985, 1987), but with a variable speed motor operating the disk shaft for two units, as shown.

plugged or used for flowthrough, had valve fittings on either end of the outside chamber to permit flow during continuous operation or to remain sealed for batch operations. Each section contained two plexiglas disks, 0.5 m in diameter by 8 mm thick, roughed by a lathe to facilitate fungal attachment. The eight disks per RBC provided a surface area of 0.28 m^2/RBC. Forty percent of the total disk area was submerged in 2-L mixtures of growth and/or treatment solutions for experimentation. A water jacket surrounded the RBC chamber to allow temperature control, and a half cylinder shaped lid enclosed the top of the disks to prevent contamination and evaporation and to maintain increased oxygen environments. Two round apertures on opposite sides of the lid were used for the air/oxygen inlet and gas sampling ports, respectively. Oxygen was supplied by cylinder and distributed to four individual RBC units through manifolds, containing valves to distribute the flow of oxygen to each reactor individually. The flow to each unit was adjusted via Gilmont (F-1100) steel ball flowmeters. Two

RBC units were driven by a Dayton Gearmotors (3M096) dual drive shaft motor mounted at the ends of respective RBCs. Motor speed was adjusted via a rheostat controller (Dart 125-50C).

White Rot Fungus. Culture of *P. chrysosporium* strain BKM F-1767 was obtained from Dr. Thomas W. Joyce of the North Carolina State University School of Forestry. Stationary cultures of fungus were maintained on basic media consisting of 3 percent agar, 1 percent malt, 1 percent glucose (Kirk *et al.* 1978), and minimal salts, pH 4.5, and grown at both room temperature and 34 C. Cultures were grown in Roux flasks and transferred to the RBC reactors. A concentration of spores equivalent to approximately 400 per mL at $A_{650} = 0.5$ was used per RBC (Chang *et al.* 1985).

Reactor Operation. The composition of growth and treatment solutions added to the RBC media mixtures was based on the work of Campbell (1983). Synthetic media were prepared aseptically with sterile deionized water. Each sample lot of Kraft liquor was refrigerated at a temperature of 4 C until use. Buffer was not added to any mixture containing Kraft liquor. The RBC units for Kraft liquor decolorization experiments were operated in both growth and treatment phases in a constant temperature room maintained at 33 to 35 C. The pH of the reactor media was maintained at 4.25 to 4.7 via daily adjustments to 4.4 to 4.5 with 1N HCl or NaOH as required.

Dichlorophenol (DCP) treatment medium was prepared by adding the required volume of stock solutions (2,000 or 200,000 ppm DCP) into the reactor media. Treatment batches of DCP at final concentrations of 200 and 100 ppm were prepared. Biodegradation of the DCP was assessed during experimental runs in the laboratory hood by duplicate sampling of each RBC unit; this was accomplished by dividing each of two RBC units into two individual 1-L reactors by plugging the central baffles of the units according to the method of Yin (1986). To maintain a temperature of 30 to 33 C in the RBC units, water was circulated constantly by a pump from a temperature-controlled waterbath through the water jackets of the RBCs.

RBC inoculation and development of fungal biofilms were usually conducted in accordance with literature reports (Chang *et al.* 1985). The initial inoculation media containing spores and synthetic growth media was grown for 2 days, following which the media was replaced with fresh growth media without spores. After 2 additional days of growth, nitrogen had been depleted and ligninolytic conditions were established (Kirk 1980). Biofilms grown in the presence of Kraft liquor remained under operation with the initial inoculation media until decolorization of that Kraft liquor occurred, generally after 5 days. This acclimated the biofilm to the Kraft liquor substrate, but is in contrast to work performed by Campbell (1983). Experiments to assess the loss of DCP continued for 1 week or until less than detectable limit (1 ppm) amounts were measured by analysis.

During experimental RBC runs, sample collections and pH adjustments were conducted daily. Sample removal and pH adjustments were conducted via a cannula and a syringe. In Kraft liquor decolorization experiments, approximately 30-mL reactor mixture samples were taken for measurement of absorbance, PCU, COD, and DOC. This sampling did not change the reaction mixture volume by more than 10 percent during the course of a given experiment. In DCP degradation experiments, 15-mL samples were withdrawn daily and analyzed for DCP only. To maintain sufficient volume in the RBC, reactor liquid matrix samples were filtered from the withdrawing syringe through a filter holder containing a 2.4-cm glass fiber filter (Whatman GF/C) before use of the sample in analytical procedures, except for selected samples for phenol and volatile organic analysis (VOA). Atmospheres were established by allowing free gas exchange with the ambient atmosphere for air-mediated experiments, whereas elevated oxygen concentrations were established by sealing the lid of the RBC with tape and supplying oxygen to the reactor. When oxygen was used as the atmosphere, the headspace of the reactors was flushed aggressively with oxygen for 5 to 15 min before the oxygen flow rate was adjusted to a constant rate of 20 mL/min. Wet weights of disk biofilms were measured upon completion of the reactor batch runs after draining the liquid phase from each RBC unit.

Final samples from the RBC units (at the end of batch runs) of 30 to 100 mL were filtered through glass fiber filters and analyzed for DOC, COD, TKN, ammonia, and nitrite. Absorbance and PCU measurements were taken for the final samples from Kraft liquor decolorization experiments. Once an experiment was completed, the biofilm was scraped off the disks, and the RBC was washed with soap and warm water and then sterilized by ethylene oxide gas.

Analytical Techniques. Dissolved oxygen was measured by a YSI Model 54A meter with a stirring BOD probe YSI Model 5720A. Ambient atmospheric oxygen concentration was measured with a Fisher Model 1200 Gas Partitioner. Spectroscopic measurements of UV/visible were taken with a Bechman Model 25 spectrophotometer. Measurements were made on filtered liquid samples. Wavelengths were chosen to monitor the reduction of substrate concentration throughout the batch treatment periods, which ranged from 2 days to 1 week. Because of the heterogeneity of the Kraft liquor, a single wavelength was considered inappropriate to measure rates of conversion; wavelengths of 580 nm (measuring the yellow-orange color region), 450 nm (for humic substance detection), and 280 nm (for aromatic structure and lignin fragment detection) were used (Alen & Hartus 1988; Janshedar *et al.* 1981). In Kraft liquors, a strong absorbance was observed at 220 nm, and a wide flattened peak was observed at 280 nm. Color was measured as platinum color units (PCU) according to the standard method developed by the paper and pulp mill industry (Gellman 1971).

Continuous liquid-liquid extraction (EPA Method 3250) was used to extract phenolic compounds, including DCP, for GC/MS analysis (EPA Method 8270). Quality assurance and quality control (QA/QC) measures for extractions included a daily blank plus a duplicate, a matrix spike, and a spike duplicate for every 10 samples. A 5-point calibration curve for the GC/MS was established and rechecked daily. Additionally, separatory funnel extractions (EPA Method 3510) were performed on samples from DCP experiments for quick turnaround of results. Extracts were analyzed via gas chromatography with flame ionization detection (EPA Method 8040). For these determinations, QA/QC measures before and after every series of injections included the use of a 5-point calibration curve of the DCP and two acid surrogates, 2-fluorophenol and phenol d-6. VOA analysis used EPA Method 1624. An internal standard of bromofluorobenzene was used for semiquantitation of identified compounds for select samples.

RESULTS

Growth. No discernible difference in growth or activity was observed in the biofilms inoculated with spores cultivated at room temperature compared to those at 34 C. Sufficient numbers of spores for the inoculation of two to six reactors were grown within 10 days at 34 C, whereas 2 weeks were required for growth of spores at room temperature. No significant differences were observed in the consecutive treatment performance of biofilms developed in the presence of pulp mill effluent compared with those developed on synthetic media. In both uses, an initial growth and acclimation period of 1 week was required after start-up. Under all conditions of growth and treatment, but not at all times, black and red pervasive coloration of the biofilm developed, and occasionally coloration was observed in the form of localized speckled areas. Such coloration is believed to result from one or more of the manganese-dependent ligninase enzymes. When environments of purified oxygen were established with the initial inoculation of a reactor, some of the fungal mass was observed to group into spherical clumps on the disks. This was not observed when biofilm growth was begun under air.

Decolorization. In these bench-scale studies with the pulp and paper mill effluent, visible color removal from Kraft liquor occurred in experiments in which oxygen, glucose, and rotational speed of the RBC disks were varied. Effects of air versus oxygen headspace were compared; the best decolorization kinetics were found using oxygen rather than air (Glaser *et al.* 1990a,b, 1991). The biomass developed under these two oxygen concentrations differed considerably. The nitrogen-rich biomass retained greater amounts of water than the nitrogen-deficient biomass. The color bodies in the Kraft bleach effluent were better sorbed by the nitrogen-rich biomass than by the nitrogen-deficient biomass. The biomass grown under oxygen was more stable mechanically and had greater decolorizing activity.

Nitrogen-Rich and Ligninolytic Conditions. Figure 2 illustrates that lignolytic conditions (nitrogen-deficient media) produce effective treatment of Kraft liquor waste. Experiments using nitrogen-rich media (conditions that do not induce the lignin-degrading secondary metabolic state in nonmutant fungal species) throughout the treatment resulted in a "nonactive" condition. Physical attachment of color bodies was noted on the biofilm for a period of 1 to 2 days, followed by their release back into the liquid matrix. Substrates monitored at the 280-nm wavelength also displayed this uptake/release phenomenon. This is to be contrasted with conditions of observed "inactivity" in which no initial uptake by the fungus is observed and no treatment occurs. In slower acting systems (i.e., under air), residual attached color from a previous batch is decolorized upon the addition of a new batch of effluent containing a fresh supply of glucose (i.e., H_2O_2).

Physical attachment or adsorption of substrate to the fungal biofilm is primary to initiating subsequent degradation by the fungus. The treatment removal was based on measurement of the liquid concentrations of color and individual organic substrates. For example, trichlorophenol concentrations of 18 µg/L in the pulp mill effluent were not detected after treatment even in the case where the desired enzymatic conditions for treatment had not been established. This loss

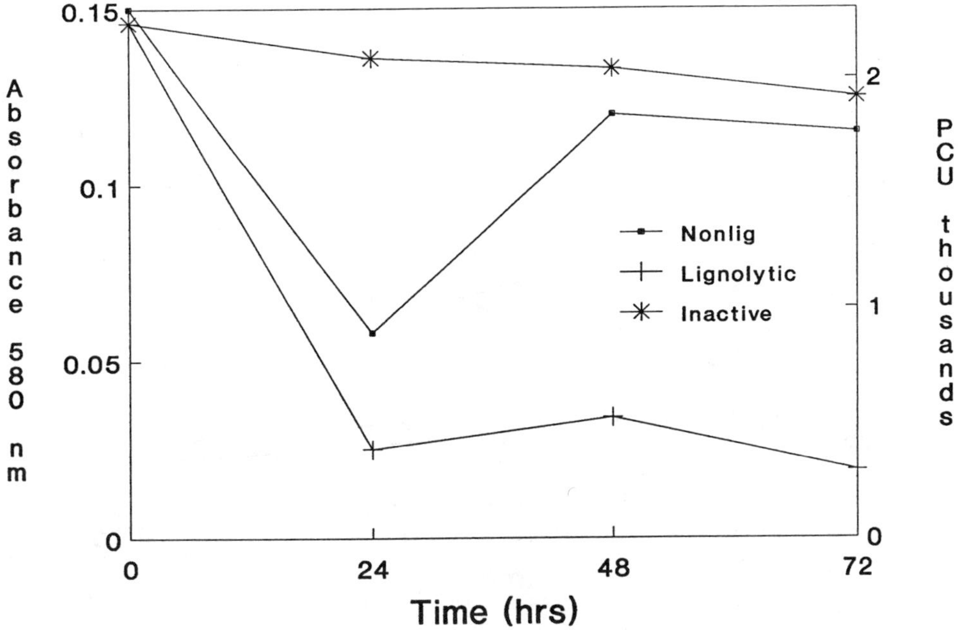

FIGURE 2. Effect of concentration of bioavailable nitrogen on treatment of Kraft liquor effluent by *P. chrysosporium.* Using bench-scale RBC units, the fungus was grown in media containing no added nitrogen (nonlig); low nitrogen, 2.2 mm NH_4Cl (lignolytic); and high nitrogen, 13.2 mm NH_4Cl (inactive) concentrations. Treatability was assessed by triplicate measurements of Kraft liquor effluent concentrations at each time point using both absorbance at 580 nm and platinum color units.

of phenol was attributed to the great retentive characteristics of the developed nonlignolytic biofilm.

Oxygen Concentration. The effect of 100 percent oxygen on concentration of Kraft liquor effluent is illustrated in Figure 3, which compares treatability data for concentrated and diluted waste in the presence of 100 percent oxygen and air. No difference in treatment was noted when cultures were placed initially under an increased oxygen atmosphere despite the smoother appearance of the biofilm. The cost of oxygen supports the use of air for early growth in pilot- and field-scale studies. Invariably with all other parameters held constant, more biomass developed under air than under increased oxygen. When

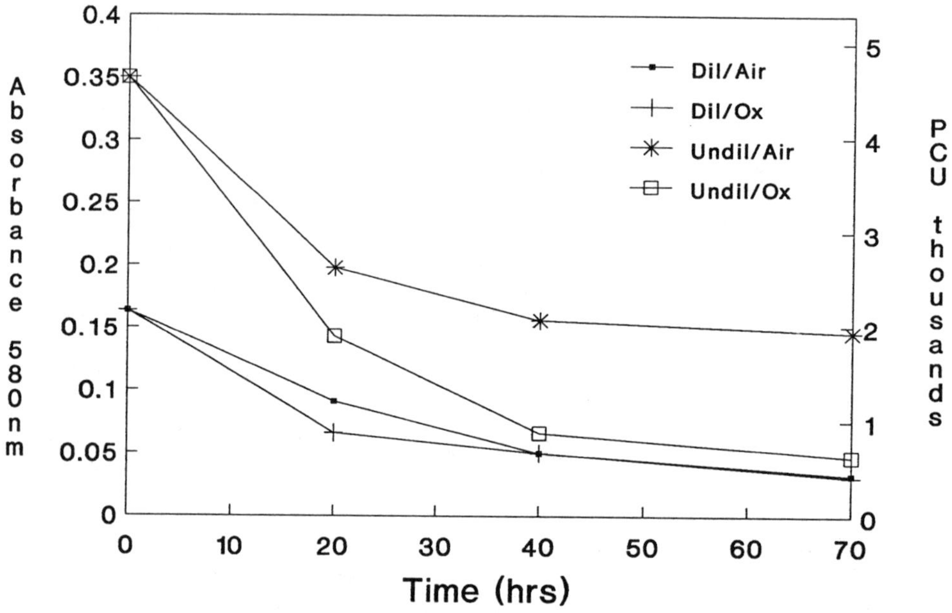

FIGURE 3. Effect of Kraft liquor effluent concentration and oxygen concentration on treatability by *P. chrysosporium*. Full strength, 4,000 PCU (undil), and diluted, ≤ 3,000 PCU (dil) Kraft liquor effluents were treated in bench-scale RBC units by the fungus grown and maintained in the presence of air (air) or 100 percent oxygen (ox). Treatability was assessed by triplicate measurements of Kraft liquor effluent concentrations at each time point using both absorbance at 580 nm and platinum color units.

treatment failed in a set of reactors under increased oxygen, the fungus visually appeared to grow to a weight comparable to that grown in the presence of air.

Initial acclimation to the pulp mill effluent was found to be slower in the presence of air than in the presence of higher oxygen concentrations. These reactors decolorized dilute effluents comparably. The oxygen-assisted reactor systems showed comparable treatment activity for both dilute (≤3,000 PCU) and concentrated (≥4,000 PCU) effluents. However, reactors with air atmospheres showed greatly diminished decolorization rates for concentrated pulp mill effluent compared with the rate observed for the more dilute effluents (Figure 3). In similar studies, Yin *et al.* (1989a,b) demonstrated that the kinetics of color removal by *P. chrysosporium* are concentration dependent with effluents containing ≤8,000 PCU. The effluent concentration used in our studies was 2,000 to 5,000 PCU; hence, oxygen concentration and its availability are determining factors for the biological treatment of these effluents in RBCs. These results suggest strong correlation between oxygen requirement and pulp mill effluent color concentration. Such relationships are integral to the commercial operation of a treatment system of this design.

Glucose. Bench-scale RBC studies evaluating the effect of glucose concentration on treatment efficiency generated data suggesting that glucose concentration at the levels studied did not affect the degree of treatment, but in conjunction with oxygen concentration, did affect the rate of treatment. Studies showed that following acclimation, a period of 1 day was considered significant time for effective treatment (decolorization) of the Kraft mill liquor waste. Accordingly, evaluation of data at 1 day indicates that treatment time can be considered a more central determinant of the treatment.

The data in this Table 2 indicate that at 100 percent oxygen, the amount of glucose (1 versus 0.5 percent) and the disk shaft rotational speed (1 versus 6.5 rpm) had little effect on treatment efficiency. Glucose is used *de novo* to produce hydrogen peroxide from molecular oxygen and is used by the ligninases in the oxidative depolymerization of lignin (Bar-Lev & Kirk 1981). The dual role for glucose in both growth and secondary metabolism is supported by the described results.

The amount of biofilm growth is not directly proportional to glucose concentration. Under increased oxygen concentrations with 0.5 percent glucose, treatment performance was as efficient as with 1 percent glucose (see Table 2). Under air, with only dilute effluent, the activities of the initial acclimation/treatment batch and the second consecutive batch were comparable at the two glucose concentrations. The treatment of later batches were accompanied with a decline in the rate of decolorizing activity at the 0.5 percent glucose level. However, the extent of decolorization ultimately was maintained by employing longer treatment periods. The decolorization rate was maintained at 1 percent glucose for dilute effluent feedstocks. An additional factor of carbon source is indicated here in terms of concentration, oxygen, and kinetics. A critical concentration between 0.1 and 0.2 percent glucose has been identified as necessary for decolorization with supplied oxygen (Yin *et al.* 1989a,b). Our results show that activity at 0.1 percent glucose proceeded at a decreased rate. Oxygen concentration in conjunction

TABLE 2. Effect of glucose concentration and disk shaft rotational speed on treatability of Kraft liquor by *P. chrysosporium*. Using bench-scale RBC units, fungal biofilms were grown and maintained under an atmosphere of 100 percent oxygen and in Kraft liquor containing 1 percent or 0.5 percent glucose, with RBC unit disks operated at indicated speeds. Treatability was assessed by triplicate measurements of Kraft liquor concentration at 8 and 17 h of treatment using both absorbance at 280 nm and platinum color units.

Reactor Conditions		Total Average Removal (%)			
		8 hr		17 hr	
Glucose %	Disk Speed (rpm)	280 nm	Color Units	280 nm	Color Units
1.0	1	10	39	19	57
1.0	6.5	13	41	23	59
0.5	6.5	11	33	19	55
0.5	1	5	27	20	51
		11	39	21	55

with Kraft pulp mill effluent feed strength and glucose were shown to have the greatest impact on treatment.

Buffers. No difference in growth or treatment extent was observed when potassium tartrate (KTar) was used in the place of 2,2-dimethyl succinate (DMS) buffer for biofilm development at 1 and 0.5 percent glucose concentrations with the reactor speed at 1 rpm. The DMS buffer has been previously used for the optimization of ligninase production (Fenn & Kirk 1979). Buffer choice is of critical economic importance for economic considerations for larger scale operations. In bench-scale studies, 2,2-DMS was shown to stabilize pH for an effective treatment and to stabilize pH more effectively than KTar. The effect of buffers on treatment of Kraft liquor, supplemented with 1 and 0.5 percent glucose, is shown in Figure 4.

Two inorganic buffers were tested as alternatives to DMS. One buffer was a Na_2HPO_4/KH_2PO_4 mixture adjusted to pH 4.5 with HCL

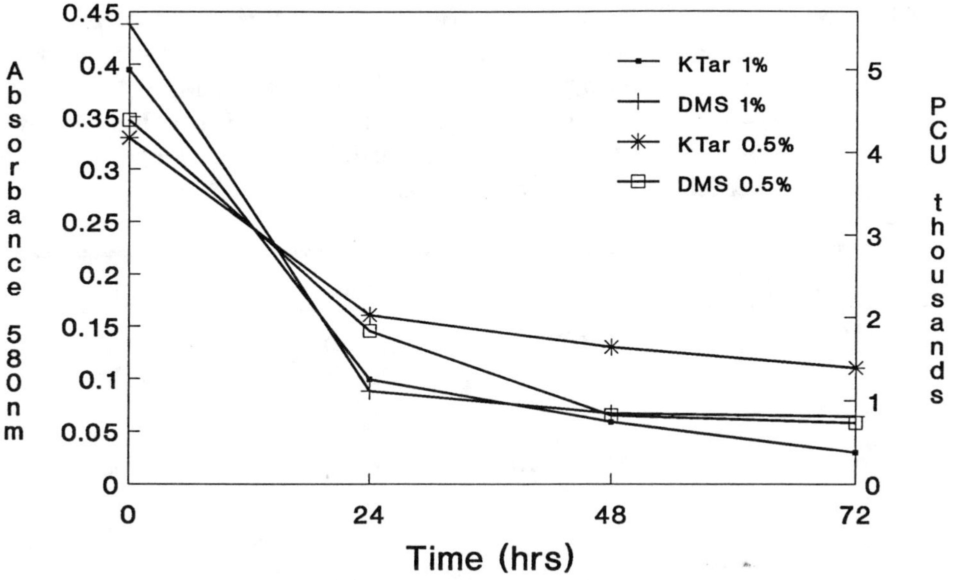

FIGURE 4. Effect of RBC units buffer on treatability of Kraft liquor by *P. chrysosporium*. Using bench-scale RBC units, fungal biofilms were grown and maintained in Kraft liquor containing 0.5 percent or 1.0 percent glucose in the presence of either 20 mm 2,2-dimethyl succinate (DMS) buffer (pH = 4.5) or 20 mm potassium tartrate (KTar) buffer (pH = 4.5). Treatability was assessed by triplicate measurements of Kraft liquor concentrations at each time point using both absorbance at 580 nm and platinum color units.

and the other was 0.05 M KH_2PO_4. The reactors were operated at 0.5 percent glucose and 4-week-old Kraft liquor at 1 rpm. The fungus was grown under air for the first 48 h (GR1) and then under oxygen for the second 48 h (GR2). The pH was measured at 0, 24, and 48 h for each batch of growth solution. The pH was adjusted with NaOH as necessary. The pH data are summarized in Table 3. Both buffers performed acceptably, although both reactors required a small addition of base to bring the pH to 4.5 after the first 24 h following introduction of fresh growth media.

TABLE 3. Effect of alternative buffers on pH maintenance of RBC unit media during treatment of Kraft liquor by *P. chrysosporium*. Using bench-scale RBC units, fungal biofilms were grown and maintained under an atmosphere of air using a Kraft liquor media containing 0.5 percent glucose and 0.05 M inorganic buffer, as indicated.

Buffer pH		Batch	Time
Na_2HPO_4	Growth	0 h	4.52
	1	24 h	4.00(c)
+ KH_2PO_4(a)		48 h	4.53
	Growth	0 h	4.50
	2	24 h	3.70(c)
		48 h	4.48
KH_2PO_4(b)	Growth	0 h	4.53
	1	24 h	4.15(c)
		48 h	4.28
	Growth	0 h	4.53
	2	24 h	3.50(c)
		48 h	4.43

(a) pH = 5.0, adjusted with HCl.
(b) pH = 4.53 with no pH adjustment.
(c) pH = 4.5, adjusted with NaOH.

Disk Rotation Speed. Development of biofilm at 1 rpm invariably resulted in bridging between at least two of the disks, whereas this never occurred at 6.5 rpm. In addition, it appeared that the films at 6.5 rpm were more evenly distributed with an overall smoother surface. No difference in treatment activity was observed, although dissolved oxygen concentrations at 6.5 rpm were generally double those at 1 rpm (Table 2). The influence of disk rotation rate may be more pronounced for larger scale operations.

Temperature. Bench-scale RBC studies indicated an improvement in growth and treatment at higher temperatures. A temperature range of 33 to 35 C was found to be optimum for growth and treatment, whereas temperature at 23 C provided a delay of growth cycle for 24 h.

Treatment of Organic Components of Wood-Processing Waste. In addition to monitoring the decolorization of Kraft liquor waste during treatment, samples were analyzed by GC/MS for residual organic compounds and chlorinated phenols. The GC/MS scans of waste sample before and after bench-scale fungal treatment showed a reduction of organics including chlorinated phenols (2,4-dichlorophenol and 2,4,6-trichlorophenol). Extraction surrogate, internal standard, and Kraft liquor effluent indigenous compounds and their concentration levels in Kraft liquor before and after treatment with *P. chrysosporium* are listed in Table 4.

Kraft Pulp Mill Effluent Variability. Differences in sampling lots of the Kraft pulp mill effluent and changes in the lot during the time of storage may explain the variability of treatment results. Among the five lots of effluent treated by our system, treatability varied distinctly. The sample lots can differ significantly. Their composition depends on daily mill processes and feedstock; pulp mill effluent for softwood is typically dark in color, whereas the hardwood wastewater is less concentrated. No trend can be observed to identify the time interval for greatest treatment activity; maximum treatment effects were observed for both early and late periods in the lifetime of the biofilms. An apparent toxicity developed within the effluent lots over time, normally encountered within 2 months after collection of the effluent supply. At this stage, fungal treatment became less effective and ultimately the fungus would not grow in the presence of that effluent lot. Biofilm became fragile and peeled off the disk. Toxins showing activity toward fungi have been previously noted in these effluents. Bacterial contamination is also suspected.

One batch of Kraft liquor was completely characterized, and the effect of holding time and aging prior to treatment was studied at four reactor operating conditions. Fresh biomass was grown up on the RBCs for each experiment. One-week-old, 4-week-old, and 8-week-old Kraft liquors were tested. Color units as well as absorption at 580 nm, 450 nm, and 280 nm were measured at 0, 24, and 48 h, and the removals are compared on a percentage basis in Table 5. Treatment was not affected in any of the reactors at 1 and 4 weeks, but at 8 weeks symptoms of toxicity to the fungal biomass developed: the fungus began sloughing from the disks after introduction of the 8-week-old Kraft liquor.

Treatability of Kraft liquor, held for 5.5 months, was assessed periodically during the holding time using the bench-scale RBC units.

TABLE 4. List of extraction surrogates, GC internal standards and Kraft liquor effluent indigenous compounds, and their concentration levels in Kraft liquor before and after treatment with *P. chrysosporium*.

Compound	Concentration (mg/L)	
	Before	After
1. Extraction Surrogate		
2-Fluorophenol	68.66	49.2
Phenol d-6	65.95	45.46
2,4,6-Tribromophenol	61.66	41.61
2. GC Internal Standard		
d4-1,4-Dichlorobenzene	40.00	40.00
d8-Naphthalene	40.00	40.00
d10-Acenaphthene	40.00	40.00
d12-Chrysene	40.00	40.00
d12-Perylene	40.00	40.00
3. Sample Compounds		
2,4-Dichlorophenol	1.51	–
Benzoic Acid	9.00	–
2,4,6-Trichlorophenol	4.10	–
Pentachlorophenol	0.92	–
Di-*n*-Butylphthalate	1.62	0.78
bis(2-Ethylhexyl)Phthalate	–	–
di-*n*-Octyl Phthalate	–	1.89

Notes:
1. Surrogate Compounds—spiked into Kraft liquor effluent before extraction.
2. GC Internal Standards—spiked into Kraft liquor effluent extract before GC/MS analysis.
3. Sample Compounds—indigenous to Kraft liquor effluent.

For each treatability run, the liquor was supplemented with 0.5 percent glucose and was maintained under an atmosphere of oxygen. Following 24 h of fungal treatment, Kraft liquor concentration was quantitated using the absorbance at 580 nm, and the platinum color units were derived using regression data. The results (data not shown) indicated that treatment efficiency decreased more than 70 percent during the time period. These results suggest that either microbial growth

TABLE 5. Effect of Kraft liquor holding time on treatability of Kraft liquor by *P. chrysosporium*. Using bench-scale RBC units, fungal biofilms were grown and maintained in Kraft liquor (aged 1, 4, and 8 weeks) under the experimental conditions indicated. Treatability was assessed by triplicate measurements of Kraft liquor concentrations at each time point using absorbance measurements at 580 nm, 450 nm, and 280 nm, and using platinum color units.

Reactor Age			% Removal at 580 nm	% Removal at 450 nm	% Removal at 280 nm	% Removal of C.U.
Reactor 1—1% glucose/1 rpm/oxygen						
1	week	24 h	23	18	4	21
	old	48 h	47	36	13	41
4	week	24 h	24	15	12	16
	old	48 h	40	35	20	35
8	week	24 h	18	14	6	17
	old	48 h	33	29	12	37
Reactor 2—1% glucose/1 rpm/air						
1	week	24 h	8	7	4.5	24
	old	48 h	21	17	12	31
4	week	24 h	22	12	16	14
	old	48 h	33	26	20	33
8	week	24 h	15	13	7	26
	old	48 h	30	22	13	38
Reactor 3—0.5% glucose/6.5 rpm/oxygen						
1	week	24 h	32	32	29	37
	old	48 h	60	51	32	61
4	week	24 h	45	29	17	24
	old	48 h	60	47	15	45
8	week	24 h	3	12	10	26
	old	48 h	21	22	15	33
Reactor 4—0.5% glucose/6.5 rpm air						
1	week	24 h	18	20	14	25
	old	48 h	35	30	17	46
4	week	24 h	19	13	11	7
	old	48 h	41	30	8	28
8	week	24 h	12	15	11	31
	old	48 h	33	25	16	37

in the Kraft liquor or production of toxins from these organisms are inhibiting the action of the *P. chrysosporium* biofilm.

2,4-Dichlorophenol Treatment in Bench-Scale RBC Reactors. Two concentrations of DCP were treated in replicate. These reactors were operated at 0.5 percent glucose and 6.5 rpm. The fungus was grown under air for the first 48 h and then under oxygen for the remainder of the study. The fungal biomass was grown on two consecutive 48-h growth solution batches prior to introduction of DCP and limitation of nitrogen for the treatment phase. DCP concentration in the reactor was measured with time, and these data are summarized in Table 6. The concentration of DCP was reduced from 17.1 mg/L to <1 mg/L within 48 h, and from 84.8 mg/L to <1 mg/L within 120 to 144 h.

Additional bench-scale studies with 2,4-DCP provided data on removals of the compound at three different initial concentrations (20, 50, and 100 mg/L target concentrations). No nitrogen source was supplied during treatment for these batches. The removal of 2,4-DCP at the three concentration levels versus time is shown in Figure 5. The final concentration of 2,4-DCP was less than 1 mg/L for all three initial concentrations. At the two lower concentrations, this level of reduction was obtained in 48 h or less. In the 100 mg/L reactor, 120 h were required. The biofilm was adversely affected at the 100 mg/L concentration; partial sloughing of the established growth was observed. A kinetics study was performed in which 2,4-DCP removal at 50 mg/L initial concentration was measured at 0, 2, 4, 8, 12, 24, 36, 48, and 72 h of incubation. The maximum removal took place between 0 and 24 h, with the sharpest drop between 12 and 24 h. Data between 0 and 24 h are plotted in Figure 6.

Studies were also performed with consecutive batch treatments of 2,4-DCP at 20 and 50 mg/L initial concentrations, with and without a nitrogen source (NH_4Cl) in the synthetic media. The effect of the addition of 35 mg/L nitrogen to the treatment batch reactors on the biodegradation rate of 2,4-DCP was measured in these consecutive batch treatment experiments. The consecutive bench-scale RBC batch treatment data for 2,4-DCP at 50 mg/L initial concentration in the reactor with and without added nitrogen showed comparable removals of the target compound after 24 h. DCP removal for up to three consecutive batch treatments at 20 and 50 mg/L initial concentration with and without added nitrogen showed reduction to levels of less than 1 mg/L within 48 h. For treatment of DCP at 100 mg/L initial concentrations, reductions to less than 1 mg/L were achieved within 120 h.

TABLE 6. Reduction in 2,4-dichlorophenol concentration by *P. chrysosporium* versus time. Using bench-scale RBC units, fungal biofilms were grown and maintained under an atmosphere of air and in media containing 0.5 percent glucose and 2,4-dichlorophenol (DCP), as indicated. Treatability was assessed by triplicate measurements of DCP concentration at each time point by EPA Method 3250.

Target	Batch	Time	mg/L DCP Reactor 1	mg/L DCP Reactor 2
20 mg/L DCP	1	0 h	17.1	17.1
		24 h	2.08	1.36
		48 h	< 1	< 1
		72 h	–	–
		96 h	< 0.1	< 0.1
	2	0 h	18.0	18.0
		24 h	0.52	1.5
		48 h	< 0.8	–
		72 h	0.52	–
100 mg/L DCP	1	0 h	84.8	84.8
		24 h	42.1	56.2
		48 h	31.7	43.3
		72 h	24.9	29.9
		96 h	5.55	17.7
		120 h	< 1	5.22
		144 h	–	0.16

Control bench-scale studies of 2,4-DCP treatability were conducted in duplicate as follows: Positive Control—fungus plus glucose/buffer/salts media plus 50 mg/L DCP; Negative Media Control—buffer/salts media plus 50 mg/L DCP; and Negative Complete Media Control—glucose/buffer/salts media plus 50 mg/L DCP. The results showed DCP treatability to a level of less than 1 mg/L in the positive control (experiment 1) within 24 h, and insignificant loss of DCP to the reactor or atmosphere in both negative controls (experiments 2 and 3).

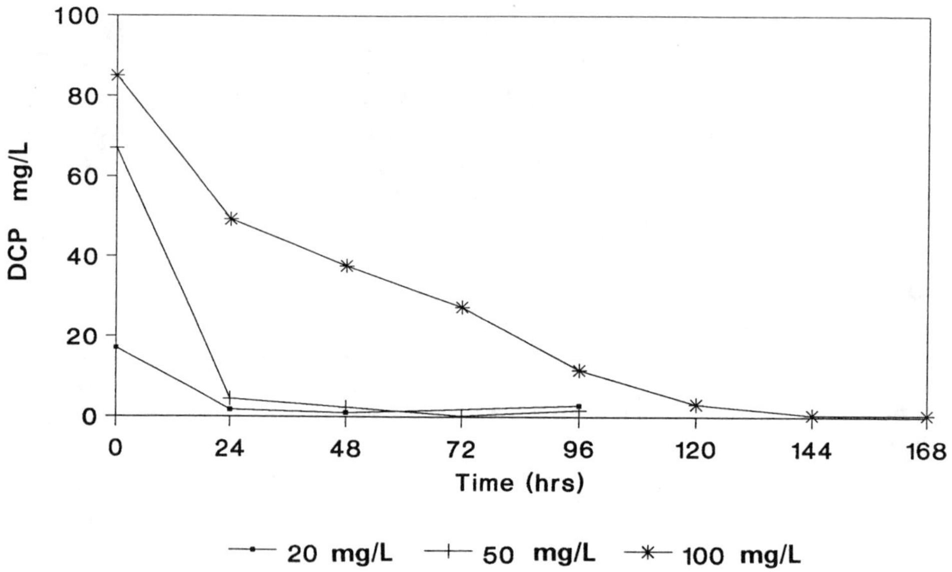

FIGURE 5. DCP removal by *P. chrysosporium* under conditions as described in Table 6. The initial concentrations of 2,4-DCP used in these studies were 20, 50, and 100 mg/L.

CONCLUSIONS

The following conclusions can be made based on the experimental results to date: (1) Substrate removal kinetics indicate that substrate concentration and oxygen requirements are codependent. (2) At 100 percent oxygen, 0.5 percent oxygen, 0.5 percent glucose, and 1 rpm are acceptable operating parameters for pilot plant studies. (3) 2,2-Dimethyl succinate is cost prohibitive for use at pilot scale; use of potassium tartrate requires supplemental pH control. (4) Both growth and treatment are temperature dependent. (5) Organic contaminants in wood processing wastes are effectively treated by the fungus. (6) 2,4-Dichlorophenol is effectively biodegraded by the fungus in bench-scale bioreactors.

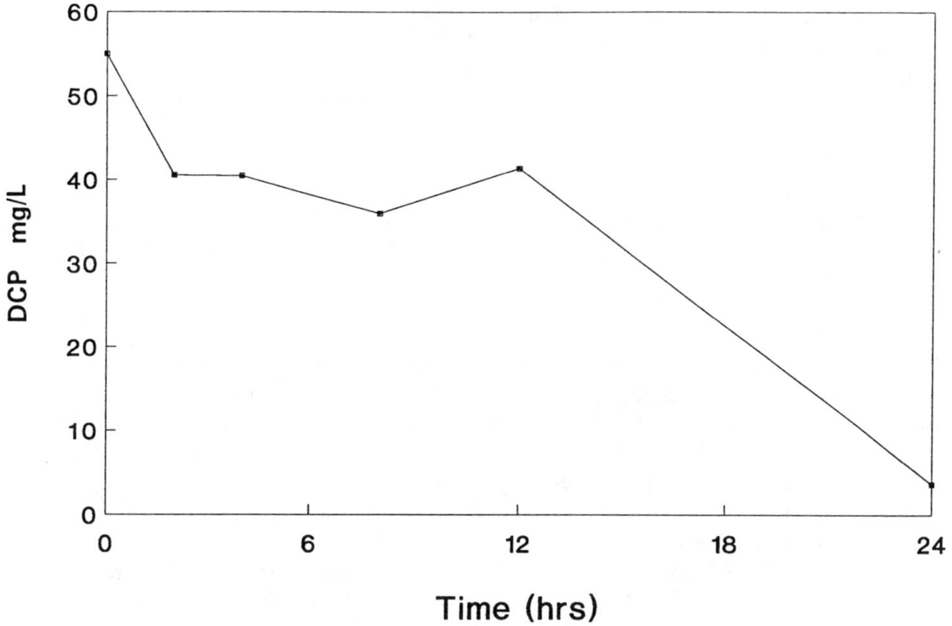

FIGURE 6. Kinetics of DCP removal by *P. chrysosporium* using nitrogen-supplemented growth media. Using bench-scale RBC units, fungal biofilm was grown and maintained under an atmosphere using Kraft liquor media containing 0.5 percent glucose and 2.2 mM NH_4Cl. Treatability was assessed at the time points indicated by triplicate measurements of DCP concentration by EPA Method 3250.

Acknowledgments

The authors wish to thank Ms. Suzanne Chase, Secretary in the U.S. Environmental Protection Agency's Risk Reduction Engineering Laboratory (Cincinnati OH) for her excellent and timely wordprocessing skills in preparing this manuscript.

REFERENCES

Alen, R.; Hartus, T. "UV Spectrophotometric Determination of Lignin From Alkaline Pulping Liquors." *Cell. Chem. Technol.* **1988,** *22,* 613.

Bar-Lev, S. S.; Kirk, T. K. "Effects of Molecular Oxygen on Lignin Degradation by *Phanerochaete chrysosporium.*" *Biochem. Biophys. Res. Comm.* **1981,** *99,* 126.

Bumpus, J. A.; Tien, M.; Wright, D.; Aust, S. D. "Oxidation of Persistent Environmental Pollutants by a White Rot Fungus." *Science* **1985**, *228*, 1434–1436.

Bumpus, J. A.; Aust, S. D. "Biodegradation of Environmental Pollutants by the White Rot Fungus *Phanerochaete chrysosporium*: Involvement of the Lignin Degrading System." *BioEssays* **1987**, *6*, 166–170.

Buswell, J. A.; Odier, E. "Lignin Biodegradation." *Crit. Revs. Biotechnol.* **1987**, *6*, 1–60.

Campbell, A. G. "A Bench-Scale Evaluation of the MyCoR Process for Decolorization of Bleach Plant Effluent Using the White-Rot Fungus, *Phanerochaete chrysosporium*." Ph.D. Thesis, North Carolina State University, Raleigh, NC, 1983.

Chang, H. M.; Joyce, T. W.; Kirk, T. K.; Huynh, V. B. "Process of Degrading Chloroorganics by the White Rot Fungi." U.S. Patent No. 4 554 075, 1985.

Chang, H. M.; Joyce, T. W.; Kirk, T. K. "Process of Treating Effluent From Pulp or Paper-Making Operation." U.S. Patent No. 4 655 926, 1987.

Chen, C. L.; Chang, H. M. 1985. "Chemistry of Lignin Biodegradation." In *Biosynthesis and Biodegradation of Wood Components*; T. Higuchi, Ed.; Academic Press, Inc.: New York, 1985; pp 535–556.

Fenn, P.; Kirk, T. K. "Ligninolytic System of *Phanerochaete chrysosporium*: Inhibition by o-Phthalate." *Arch. Microbiol.* **1979**, *123*, 307.

Fenn, P.; Kirk, T. K. "Relationship of Nitrogen to the Onset and Suppression of Ligninolytic Activity and Secondary Metabolism in *Phanerochaete chrysosporium*." *Arch. Microbiol.* **1981**, *130*, 59–65.

Fukuzumi, T.; Nishida, A.; Aoshima, K.; Minami, K. "Decolourization of Kraft Wate Liquor with White-Rot Fungi." *Mok. Gak.* **1977**, *23*, 290.

Gellman, I. "An Investigation of Improved Procedures for Measurement of Mill Effluent and Receiving Water Color." NCASI Technical Bulletin No. 253; National Council of the Paper Industry for Air and Stream Improvement, Inc.: New York, 1971.

Glaser, J. A.; Tabak, H. H.; Strohofer, S.; Kupferle, M.; Scarpino, P.; Tabor, M. W. "Bench-Scale Biodegradation Studies with Organic Pollutants Using a White Rot Fungus." In *Remedial Action, Treatment, and Disposal of Hazardous Waste—Proceedings*, 16th Annual EPA Research Symposium, Cincinnati, OH; 1990a, April 3–5 (in press).

Glaser, J. A.; Tabak, H. H.; Opatken, E. J.; Joyce, T. W.; Chang, H. M.; Strohofer, S.; Hummel, C. 1990b. "Use of White-Rot Fungus in a RBC." In *Abstracts of EPA Symposium* on Bioremediation of Hazardous Wastes: EPA's Biosystem Technology Development Program, Arlington, VA, 1990b; pp 5–8.

Glaser, J. A.; Tabak, H. H.; Opatken, E. J. "Treatment of Aqueous Streams of Hazardous Waste by a Wood-Degrading Fungus." In *Chemical and Biochemical Detoxification of Hazardous Waste II*; Glaser, J. A., Ed.; Lewis Publishers: Chelsea, MI, 1991.

Hammel, K. E. "Organopollutant Degradation by Lignolytic Fungi." *Enzyme Microbiol. Technol.* **1989**, *1*, 776–777.

Hammel, K. E.; Kalyanaraman, B.; Kirk, T. K. Oxidation of Polycyclic Aromatic Hydrocarbons and Dibenzo(p)dioxins by *Phanerochaete chrysosporium* Ligninase. *J. Biol. Chem.* **1986**, *261*, 16948–16952.

Huynh, V. B.; Chang, H. M.; Joyce, T. K. "Dechlorination of Chloroorganics by a White Rot Fungus." *Tappi J.* **1985**, *68*, 98.

Janshedar, H.; Brown, C.; Fiechter, A. "Determination of Lignin From Alkaline Pulping Liquors." *Anal. Chim. Acta.* **1981**, *130*, 81.

Jeffries, T. W.; Choi, S.; Kirk, T. K. "Nutritional Regulation of Lignin Degradation by *Phanerochaete chrysosporium*." *Appl. Environ. Microbiol.* **1981,** *42,* 290.

Kelley, R. L.; Reddy, C. A. 1988. "Glucose Oxidase of *Phanerochaete chrysosporium*." In *Methods in Enzymology*; Wood, W. A.; Kellogg, S. T., Eds.; Academic Press, Inc.: New York, 1988; Vol. 161, BIOMASS Part B, Lignin, Pectin, and Chitin, pp 307–315.

Kersten, P. J.; Tien, M.; Kalyanaraman, B.; Kirk, T. K. "The Ligninase of *Phanerochaete chrysosporium* Generates Cation Radicals from Methoxybenzenes." *J. Biol. Chem.* **1985,** *260,* 2609–2612.

Keyser, P.; Kirk, T. K.; Zeikus, J. G. "Ligninolytic Enzyme System of *Phanerochaete chrysosporium*: Synthesized in the Absence of Lignin in Response to Nitrogen Starvation." *J. Bacteriol.* **1978,** *135,* 790.

Kirk, T. K. "Studies on the Physiology of Lignin Metabolism by the White-Rot Fungi." In *Lignin Biodegradation: Microbiology, Chemistry, and Potential Application*; Kirk, T. K.; Higuchi, T.; Chang, H. M., Eds.; CRC Press, Inc.: Boca Raton, FL, 1980; Vol. 2, pp 51–64.

Kirk, T. K.; Chang, H. M. "Potential Applications of Bio-Ligninolytic Systems." *Enzyme Microb. Technolog.* **1981,** *3,* 189–196.

Kirk, T. K.; Farrell, R. L. "Enzymatic 'Combustion': The Microbial Degradation of Lignin." *Ann. Review Microbiol.* **1987,** *14,* 465–50.

Kirk, T. K.; Schultz, E.; Connors, W. J.; Lorenz, L. F.; Zeikus, J. G. "Influence of Culture Parameters on Lignin Metabolism by *Phanerochaete chrysosporium*." *Arch. Microbiol.* **1978,** *117,* 277.

Kirk, T. K.; Croan, S.; Tien, M.; Murtaugh, K. E.; Farrell, R. L. "Production of Multiple Ligninases by *Phanerochaete chrysosporium*: Effect of Selected Growth Conditions and Use of Mutant Strain." *Enz. Microb. Tech.* **1986,** *8,* 27–32.

Lewandowski, G. A.; Armenante, P. M.; Pak, D. "Reactor Design for Hazardous Waste Treatment Using White Rot Fungi." *Water Res.* **1990,** *24,* 75.

Prouty, A. L. "Bench-Scale Development and Evaluation of a Fungal Bioreactor for Color Removal from Bleach Effluents." *Appl. Microbiol. Biotechnol.* **1990,** *32,* 490–493.

Schoemaker, H. E.; Harvey, P. J.; Bowman, R. M.; Palmer, J. M. "On the Mechanism of Enzymatic Lignin Breakdown." *FEBS Lett.* **1985,** *183,* 7–12.

Tabak, H. H.; Cooke, W. B. "Effects of Gaseous Environments on the Growth and Metabolism of Fungi." *Botanical Review* **1968,** *34,* 126.

Tien, M.; Kirk, T. K. "Lignin-Degrading Enzyme from the Hymenomycete *Phanerochaete chrysosporium* Burds." *Science* **1983,** *221,* 661–662.

Tyler, M. A.; Fitzgerald, A. D. "A Review of Color Reduction Technology in Pulp and Paper Mill Effluent." In 58th Annual Meeting of the Paper and Pulp Industry, Technical Section CPPA, 1972.

Yin, C. "Kinetics of Fungal Decolorization in the MyCoR Process and Biotreatment of its Effluent." M.S. Thesis, North Carolina State University, Raleigh, NC, 1986.

Yin, C. F.; Joyce, C. W.; Chang, H. M. "Role of Glucose in Fungal Decolorization of Wood Pulp Bleaching Effluents." *J. Biotechnol.* **1989a,** *8,* 67–76.

Yin, C.; Joyce, T. W.; Chang, H. M. "Kinetics of Bleach Plant Effluent Decolorization by *Phanerochaete chrysosporium*." *J. Biotechnol.* **1989b,** *10,* 67–76.

Nutrient-Enhanced Bioremediation of Oil-Contaminated Shoreline: The Valdez Experience

John A. Glaser[*]
U.S. Environmental Protection Agency

INTRODUCTION

On March 24, 1989, the collision of the supertanker Exxon Valdez with a submerged reef in Prince William Sound AK, released 41.6 million L (11 million gal) of Prudhoe Bay crude oil. The oil spread with time to contaminate an estimated 565 km (350 miles) of shoreline. Beach gravel, rock surfaces, and faces of vertical cliffs were coated with the oil as it came ashore. The major deposition of oil occurred in the intertidal zone. Weathering of the oil resulted in a loss of 15 to 20 percent by volatilization. Little loss by sedimentation was observed because the concentration of particulate matter, which could become entrained in the oil and cause subsequent gravity separation, was low in Prince William Sound at the time of the spill. Aliphatic hydrocarbons of 12 carbon atoms and less and low molecular weight aromatic hydrocarbons (benzene, toluene, xylene, and some naphthalenes) were lost through volatilization mechanisms. High molecular weight waxes and asphaltenes comprised 40 to 50 percent of the oil composition found on beaches.

The degradation of oil components by biological mechanisms has been intensively studied during the last 20 years (Atlas 1984). The general outline of biodegradation pathways for aliphatic and aromatic

[*] U.S. Environmental Protection Agency, 26 West Martin Luther King Drive, Cincinnati OH 45268

hydrocarbons has been formulated and continues to be developed in greater detail (Rehm & Reiff 1981). Consequently, the microbial decomposition of oil in aquatic environments is well understood to include descriptions of biodegradation kinetics (Atlas 1975; Gibbs 1975; Gibbs *et al.* 1975); temperature effects for biodegradation can be described by an Arrhenius relationship. Even cold-water environments have been shown to support the biodegradation of oil components (Atlas 1975; Haines & Atlas 1983; Walker & Colwell 1974).

A panel of experts was assembled to assist the U.S. Environmental Protection Agency (EPA) in determining the best treatment strategy to accelerate the natural biodegradation process in Prince William Sound. A bioremediation research plan was recommended with the following major objectives:

- Examine the extent to which natural biodegradation of oil on the contaminated beaches is occurring
- Determine if nutrient addition will enhance natural biodegradation of contaminated beaches in the field
- Develop methodology for full-scale application of nutrients to contaminated beaches

The bioremediation research plan was supported by a cooperative agreement between the EPA and Exxon Company.

Criteria for Fertilizer Selection

The selection of nutrient formulations for field evaluation was critical to this project (Lee & Levy 1986). Optimally, fertilizer formulations would slowly release nitrogen and phosphorus nutrients without causing eutrophication or holding nutrients in contact with surface microbial communities over extended time periods (Atlas & Bartha 1972, Jacobson *et al.* 1974). Nutrient formulations were selected for the characteristics of durability, ease of application, long-term nutrient release, and commercial availability. Three types of fertilizer were selected:

- Solid, slow-release fertilizer. Nutrients are released slowly by dissolution from a point source, with tidal action and rainfall effects used to distribute the nutrients over the beach surface (Hauck 1972; Olivieri *et al.* 1976).

- Liquid oleophilic fertilizer. Nutrients partition to the oil covering the rock and gravel surfaces to promote the growth of oil-degrading microbial populations in close proximity to the oil contamination (Atlas & Bartha 1972; Sirvins & Angeles 1986; Sveum; Tramier & Sirvins 1985).

- Water-based nutrient solutions. Inorganic nitrogen and phosphorus nutrient components are dissolved in seawater and distributed to the contaminated shoreline via fixed sprinkler systems.

These requirements aided in the selection of several candidate commercial nutrient formulations. Subsequently the nutrient release characteristics of each formulation were determined through static bench tests, the enhancement of oil biodegradation was established through analysis of oxygen uptake respirometry, and oil composition changes were analyzed by gas chromatography (Figures 1–3).

IBDU Briquettes. This fertilizer formulation is manufactured in the form of briquettes containing isobutylidene diurea (IBDU), a chemical that hydrolyzes into isobutyl aldehyde and urea as the briquette dissolves with time. Hydrolysis is temperature dependent, with slower

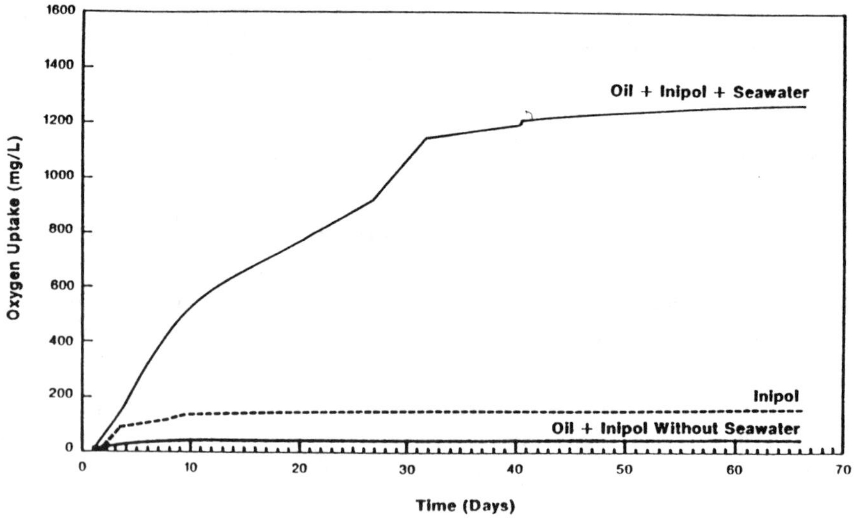

FIGURE 1. Oxygen uptake for Alaskan slope crude.

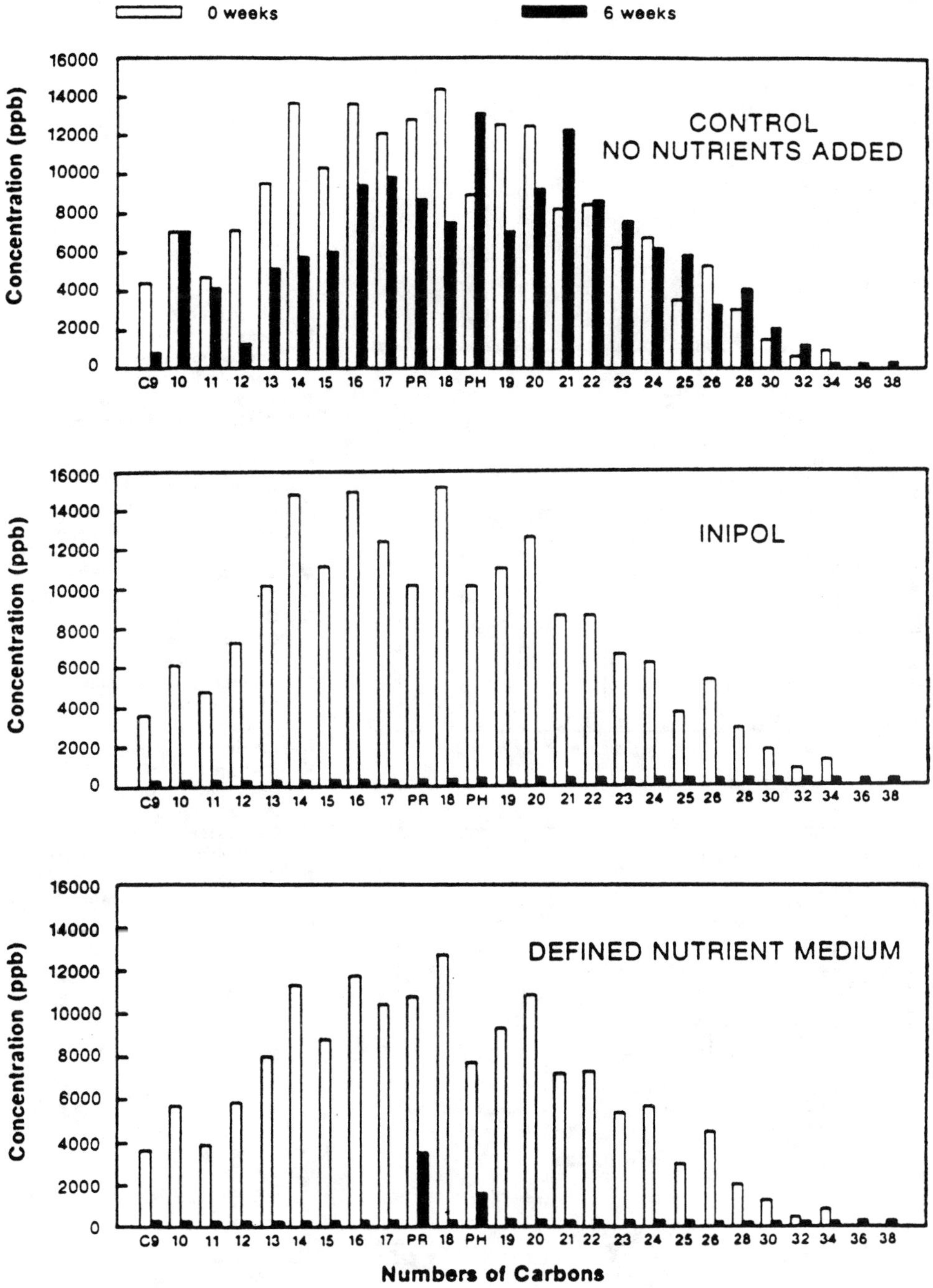

FIGURE 2. Concentration profiles of alkanes.

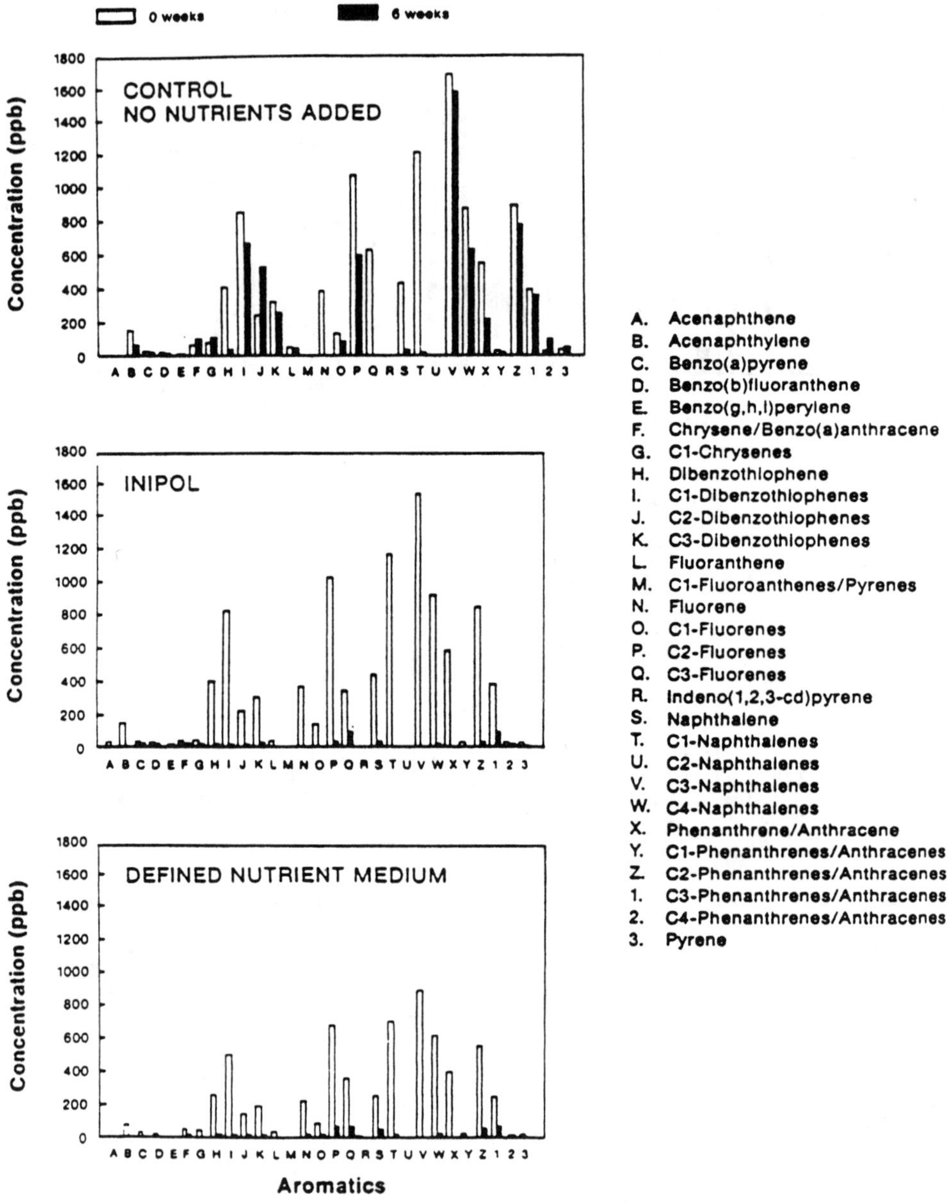

FIGURE 3. Concentration profiles of aromatics.

rates of hydrolysis at lower temperatures. The source of phosphorus is a citric acid soluble phosphate fertilizer. Each briquette weighs approximately 17 g and has a specific gravity of 1.5 to 1.8. The N:P:K ratio is 14:3:3.

Granular Fertilizer. This fertilizer formulation consists of inorganic nutrient sources (ammonium nitrate, calcium phosphate, and ammonium phosphate) contained in a vegetable oil coating (polymerized by reaction with a cyclic diene). The coating gives the fertilizer its slow-release characteristic. The N:P:K ratio is 28:8:0. The granules have a specific gravity of 1.8.

Oleophilic Fertilizer. Inipol EAP 22 is an oil-in-water microemulsion of nutrients. Oleic acid and surfactants promote the partitioning of the nutrient mixture to the oil-phase contamination. Inipol EAP 22 is a clear liquid with a specific gravity of 0.996 and a pour point of 11 C. The N:P:K ratio is 7.3:2.8:0. A microemulsion is to be understood in terms of three pseudo-phases or domains—aqueous, organic, and interfacial—each having a different composition (Freiberg & Bothorel 1987). The chemical components of each for Inipol EAP 22 are given in Table 1.

TABLE 1. Chemical composition of Inipol EAP 22.

Ingredient	Chemical Formula	Purpose
Oleic acid	$CH_3\ (CH_2)_7\ CH{=}CH(CH_2)_7\ COOH$	Hydrophobic phase
Lauryl phosphate	$C_{12}H_{25}PO_4$	Phosphorous source and surfactant
2-Butoxyl-1-ethanol	$HO{-}C_2H_4{-}O{-}C_4H_9$	Cosurfactant, emulsion stabilizer
Urea	$NH_2{-}CO{-}NH_2$	Nitrogen source
Water	H_2O	Solvent

Water-Soluble Fertilizer. A mixture of nutrients was formulated in concentrated solution and delivered through an aspirated chamber to the treatment plot in diluted form. The concentrated solution consisted of 7.7 kg (17 lb) ammonium nitrate and 3.2 kg (7 lb) triple superphosphate commercial grade so that the nitrogen delivered to the beach was in the ratio of 10:1 carbon:nitrogen.

SITE SELECTION AND CHARACTERISTICS

Selection of the test sites was based on the following criteria:

- Typical shoreline of Prince William Sound; i.e., mixed sand and gravel and cobblestone beaches
- Sufficient area with fairly uniform distribution of sand, gravel, and cobble for the test plots
- Protected embayment with adequate staging areas and sufficient size to support several test and control plots
- Uniform oil contamination

Two test sites on Knight Island were selected for the field demonstration project, Snug Harbor and Passage Cove. Snug Harbor was selected to serve as an oiled beach that approximated the degree of contamination remaining after a heavily oiled beach had been physically washed. The Passage Cove site was selected as the main reference beach for the large-scale application of fertilizers and as a means to evaluate a sprayer system for liquid fertilizer application. This shoreline had previously been physically washed by Exxon operations.

Snug Harbor Project Site

Site Characterization. Snug Harbor is located on the southeastern side of Knight Island. The area is surrounded by mountains reaching an elevation of approximately 700 m (2,000 ft) with steep vertical ascents. Major sources of freshwater runoff are from precipitation and snowmelt, which is typical of islands in Prince William Sound.

Snug Harbor was chosen for its long expanse of shoreline consisting of several beach compositions. Table 2 identifies the plot dimensions and types of treatment used at Snug Harbor. Each plot (Figure 4) was divided into 21 blocks, 7 across and 3 deep, with each row of blocks occupying a different tidal wash area: high, intermediate, and low.

TABLE 2. Description of fertilizer treatment plots at Snug Harbor.

Beach Type	Fertilizer Treatment	Length (m)	Depth (m)
Sand/gravel (control)	None	21	12
Sand/gravel	Oleophilic fertilizer	21	12
Sand/gravel	IBDU briquettes	35	12
Cobble	IBDU briquettes	28	12
Cobble	Oleophilic fertilizer	28	12
Cobble (control)	None	21	8

Fertilizer Application. Field application of the briquettes was complicated by their buoyancy and redistribution by tidal action. The solution to these difficulties was answered by packaging the briquettes in mesh bags that were tethered to a buried iron bar in the beach. Granular fertilizer forms were broadcast for application.

Large mesh bags composed of herring seine and filled with briquettes were placed on the beach to provide complete exposure of the beach material to the nutrients leaching from the bags. Each bag contained 15 kg (33 lb) of briquettes. Application of the briquette bags occurred on June 11, 1989. The total quantity of briquettes applied to the 35 m × 12 m plot was 363 kg (800 lb), representing approximately 45 kg (100 lb) nitrogen and 11 kg (24 lb) phosphorus (as P_2O_5). The bags were tethered to 1-m sections of steel rod that were buried 15 cm (6 in) below the surface of the beach.

On June 20 and 21, 1989, the bags were repositioned since those located at the topmost row were not submerged consistently at high tide. Preliminary field data indicated that the nutrients were being channeled vertically down the beach. Four more bags were added to the beach, resulting in 417 kg (920 lb) of fertilizer or 59 kg (130 lb) of nitrogen.

Tidal fluctuations affected the amount of time each zone was under water, which in turn affected nutrient dissolution and transport. For example, in the sand and gravel plot treated with the fertilizer briquettes, the top row of fertilizer bags was placed at a relative tidal height of 3.9 m (13 ft). The top row of bags in this plot was only under water approximately one-fourth of the days in June. Consequently, precipitation was the primary factor controlling the dissolution and transport of the nutrients in this zone.

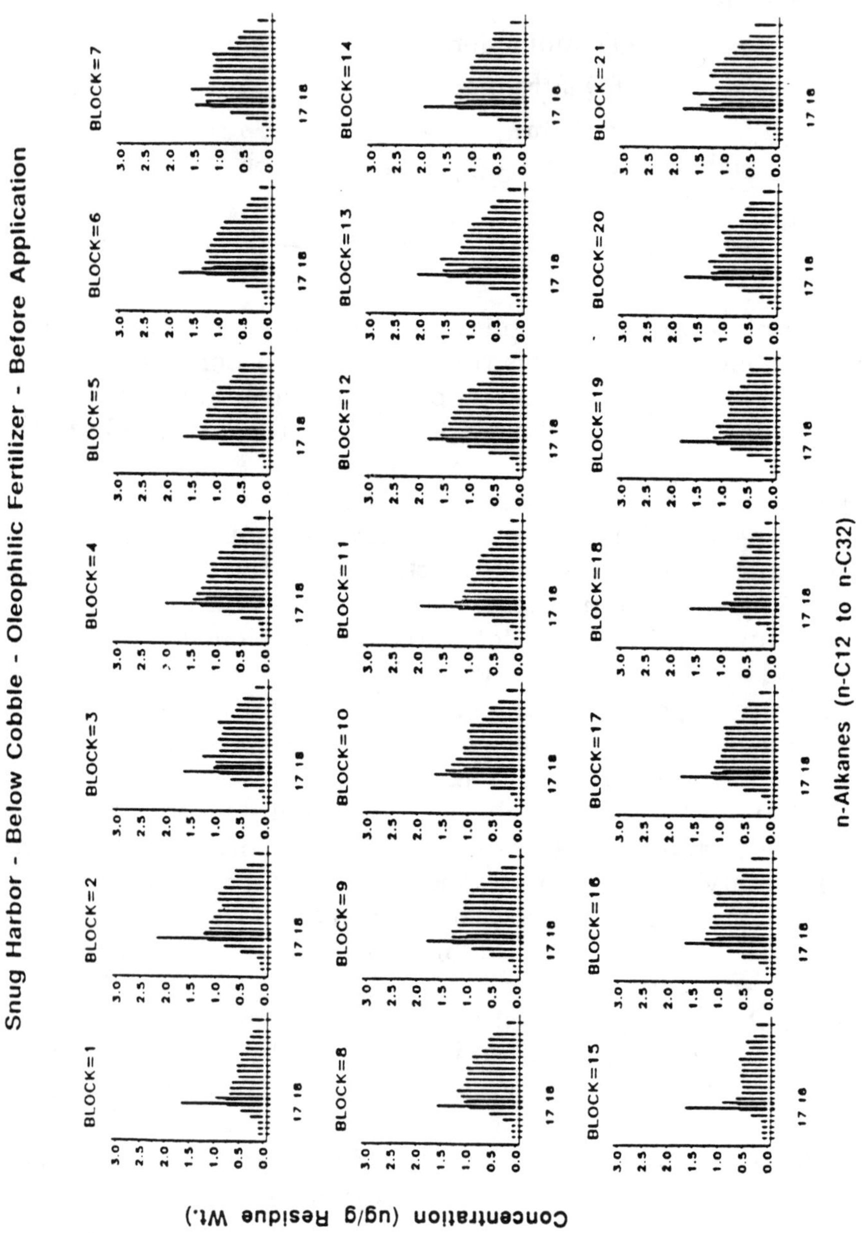

FIGURE 4. Zero time GC data profiles.

Oleophilic fertilizer (Inipol EAP 22) was first applied to Otter Beach in Snug Harbor (mixed sand and gravel) on June 8, 1989. A total of 38 kg (83 lbs) was applied, which represented approximately 5 percent of the estimated weight of the oil on the treated beach. A second application of 40 kg (87 lb) of Inipol was made on June 17, 1989, to the Otter Beach plot based on recommendations from Elf Aquitaine representatives.

An application of 53 kg (16 lb) of Inipol EPA 22 at Seal Beach in Snug Harbor (cobble) occurred on June 9, and a second application of 49 kg (107 lb) on June 18.

Inipol was applied to the plots in the evening as the tide was ebbing. Application was initiated at the top of the beach, an hour after the tide was past the lowest zone in the plot. A 15-L (4-gal) backpack sprayer was used to apply the liquid Inipol. The pour point (11 C) of the product became a major obstacle in its field use. These problems were resolved by warming the product to a temperature higher than the pour point to ensure uniform application and to prevent clogging of the spray nozzle.

The weather during the first applications was rainy and cool. During the second application, both days were clear and sunny with temperatures around 15 C. Examination of the plots the day after the second application revealed a sheen on the surface of the treated rocks. The sheen lasted for two days, during which time wave action was minimal. This sheen was not observed with the first application.

Passage Cove Project Site

Site Characterization. Passage Cove is located on the northwestern side of Knight Island. This site was originally heavily contaminated with oil and was subjected to physical washing by Exxon. Even after physical washing, considerable amounts of oil remained at this site, distributed on the surface of rocks and in beach material below the rocks. Pools of oil and mousse-like material were minimal on the surface. Contamination extended about 50 cm (19.5 in) below the beach surface. All of the beach areas at this site used for application testing consisted of cobblestones set on a mixed sand and gravel base. Table 3 lists beach types, plot dimensions, and fertilizer treatments at Passage Cove.

Fertilizer Application. A treatment of slow-release granules (Sierra Chemical Co.) and Inipol EAP 22 were applied to selected shoreline in Passage Cove. The Inipol was applied at the same rate as on the Snug

TABLE 3. Description of fertilizer treatment plots at Passage Cove.

Beach Type	Fertilizer Treatment	Length (m)	Depth (m)
Cobble and sand/ gravel (control)	None	28	21
Cobble and sand/ gravel	Inipol and water soluble	35	21
Cobble and sand/ gravel	Water-soluble nutrient sprinkler system	28	21
Sand/gravel and patchy cobble	Inipol and slow-release granules	21	7

Harbor plots. The granular fertilizer was applied using a commercial broadcast fertilizer spreader at a rate of approximately 0.14 g/m^2 (0.0033 lb/ft^2). The total application of nitrogen and phosphorus by slow-release granules in Passage Cove was approximately 182 kg (400 lb) and 18 kg (40 lb), respectively. The granules stuck to the oil on the rock surfaces and were not easily displaced from the beach or redistributed by tidal action.

A separate beach in Passage Cove was used to evaluate the effectiveness of application of nitrogen and phosphorus by spray irrigation. Inorganic salts of nitrogen and phosphorus were dissolved in seawater and sprayed onto the beach daily. The spray irrigation system used sprinkler heads typical of lawn sprinklers. The fertilizer solution was pumped by a gasoline-driven well pump to four sprinkler heads set on each side of the plot. Typical applications were about 1 cm (0.4 in) of water per day. Application rates were established to supply 7 mg/L of nitrogen and 4 mg/L of phosphorus to pore water in the saturated beach material to a depth of 2 m.

FIELD TEST RESULTS

Snug Harbor

Visual Observations. Test beaches at Snug Harbor were moderately contaminated. Visually, the cobble plots had a thin coating of dry, sticky, black oil covering rock surfaces and gravel areas under the cobble. Oil did not penetrate more than a few centimeters below the gravel surface. In mixed sand and gravel plots, oil was well distributed over exposed surface areas and commonly found 20 to 30 cm below the surface. In many areas of the test plots, small patches of thick oil and mousse (emulsified oil) could be found. This material was very viscous and mixed with extensive debris.

Approximately 8 to 10 days following oleophilic fertilizer application to the cobble beach plot, a reduction in the amount of oil on rock surfaces was visually apparent; it was particularly evident from aerial observations where the contrast with oiled areas surrounding the plot was dramatic. A clean rectangle was etched onto the beach surface. The contrast was also impressive at ground level, where a precise demarcation between fertilizer-treated and untreated areas was clearly visible.

Close examination of the treated cobble plot showed that much of the oil on the surface of the rocks was gone. There were still considerable amounts of the oil under rocks and in the mixed gravel below the rocks. The remaining oil was not dry and dull as was the case with oil in other areas of the beach, but appeared softened, more liquid, and sticky to the touch. It had little tendency to come off the rocks. At the time of these observations, oil slicks or oily materials were observed leaving the beach during tidal flushing.

The mixed sand and gravel beach treated with oleophilic fertilizer exhibited visually reduced amounts of oil after 8 to 10 days. Differences between treated and untreated plots, however, were not as dramatic as on the similarly treated cobble beach. Loss of subsurface oil in treated areas was visually apparent.

All other plots looked as oiled as they did at the beginning of the field study. There were essentially no visual indications of oil removal on plots treated with slow-release fertilizer briquettes.

Over the next two to three weeks, the cleaned rectangle on the cobble beach remained clearly visible. Oil below the rocks remained, but was less apparent.

Six to eight weeks after fertilizer application, the contrast between the treated and untreated areas on the cobble beach narrowed. This resulted from reoiling from subsurface material concurrent with the slow removal of oil on the beach material surrounding the plot. It was evident, however, that the total amount of oil on the treated plots had decreased substantially relative to the control plots. The corresponding mixed sand and gravel plot was also reoiled, but to a lesser extent. All of the other plots still had observable oil contamination, but generally less than that seen at the beginning of the study.

Toward the end of the summer season, the area used for the nutrient application study became steadily cleaner, including most of the area surrounding the test plots. This was attributed to several storms and more frequent rainfall. A heavily contaminated area to the south, which was never treated, remained heavily contaminated by all visual criteria.

Passage Cove

Visual Observations. Original contamination in Passage Cove was heavy. Following complete physical washing, oil was well distributed over most of the surface of the cobble and gravel. The oil appeared as a tacky film, black, dry, and dull. It was spread in a thin layer over the beach material. Relatively few patches of pooled oil or mousse were present, but the oil in such patches was thick and viscous. Oil was also found at depth in the beach, generally 30 to 40 cm below the surface.

Within approximately two weeks following application of Inipol and slow-release granules, it became apparent that the treated beach was considerably cleaner than the control plots. In contrast to the observations at Snug Harbor, not only did the rock surfaces look cleaner, but the oil under the rocks and on the gravel below was also disappearing. In another two weeks, oil could be found only in isolated patches and below 10 cm in the subsurface. At no time were oil slicks or oily material seen leaving the beach area.

Unexpectedly, the beach treated with fertilizer solution from the sprinkler system appeared as clean as the Inipol/granule-treated plot. The only difference was that it lagged behind the Inipol/granule-treated beach by approximately 10 to 14 days. By the end of August, both beaches looked equally clean. In contrast, the control plot appeared much as it did at the start of the field study. Oil remained in the subsurface in all plots.

Changes in Oil Composition. Data analysis for oil residue weight and chemistry has been completed (Appendix 1). Over 1,100 samples have been analyzed, and the resulting information is being incorporated into the database. Six different approaches for analyzing trends in the data are being used, involving analysis through time of the following:

- Oil residue weights (methylene chloride extractable material)

- Ratios of C_{17}/pristane and C_{18}/phytane
- Gas chromatographic profiles of aliphatic hydrocarbons
- Total concentration of aliphatic hydrocarbons
- Average individual aliphatic hydrocarbon concentrations
- Relationship of degradation extent to oil residue weight

The residue weight of oil taken at each sampling was recorded as a gravimetric measure of the petroleum remaining on the beach. This measurement has limited value for several reasons. The gravimetric concentrate will contain both the hydrocarbons from the petroleum and any metabolic products. Since the residue is a function of the surface area and volumetric loading of the sample, results from such poorly sorted (wide size distribution) beach materials as found in Snug Harbor will be highly variable. If any mineral debris was entrained in the residue despite extraction by methylene chloride, the resulting gravimetric data would be highly skewed. The Snug Harbor treatment plots were not boomed to exclude any new floating oil, so reoiling may have occurred, although it was not recognized in the Snug Harbor treatments. The Passage Cove plots were protected by a boom stretching across the mouth of the cove.

Clear compositional changes in the deposited oil did occur on Snug Harbor plots over the four weeks after oleophilic nutrient treatment (Figures 4 and 5). Almost every block in the plot showed significant disappearance of alkane fractions. By contrast, changes observed with the IBDU briquettes were not as obvious as with the Inipol treated plot; however, still significant changes in the oil composition with this treatment could be seen. The data from all plots were analyzed to determine global rates of oil removal. Results of the global rate analysis were difficult to interpret because of the high variability in the data. Global rates for the treatment plots appeared to be similar to those for the control plots for reasons discussed above. One example of the data generated for the global rate estimate is summarized in Figure 6. This graph depicts the change in the concentration of a selected range of hydrocarbons, n-C_{18} to n-C_{27}, over a 6-week period.

The data are also being analyzed to confirm the strong possibility of spatially distinct rates among the treatment plots. If this analysis has any merit, it may enable a reduction of variability in the rate estimates since the variability will be spatially correlated. Comparison of

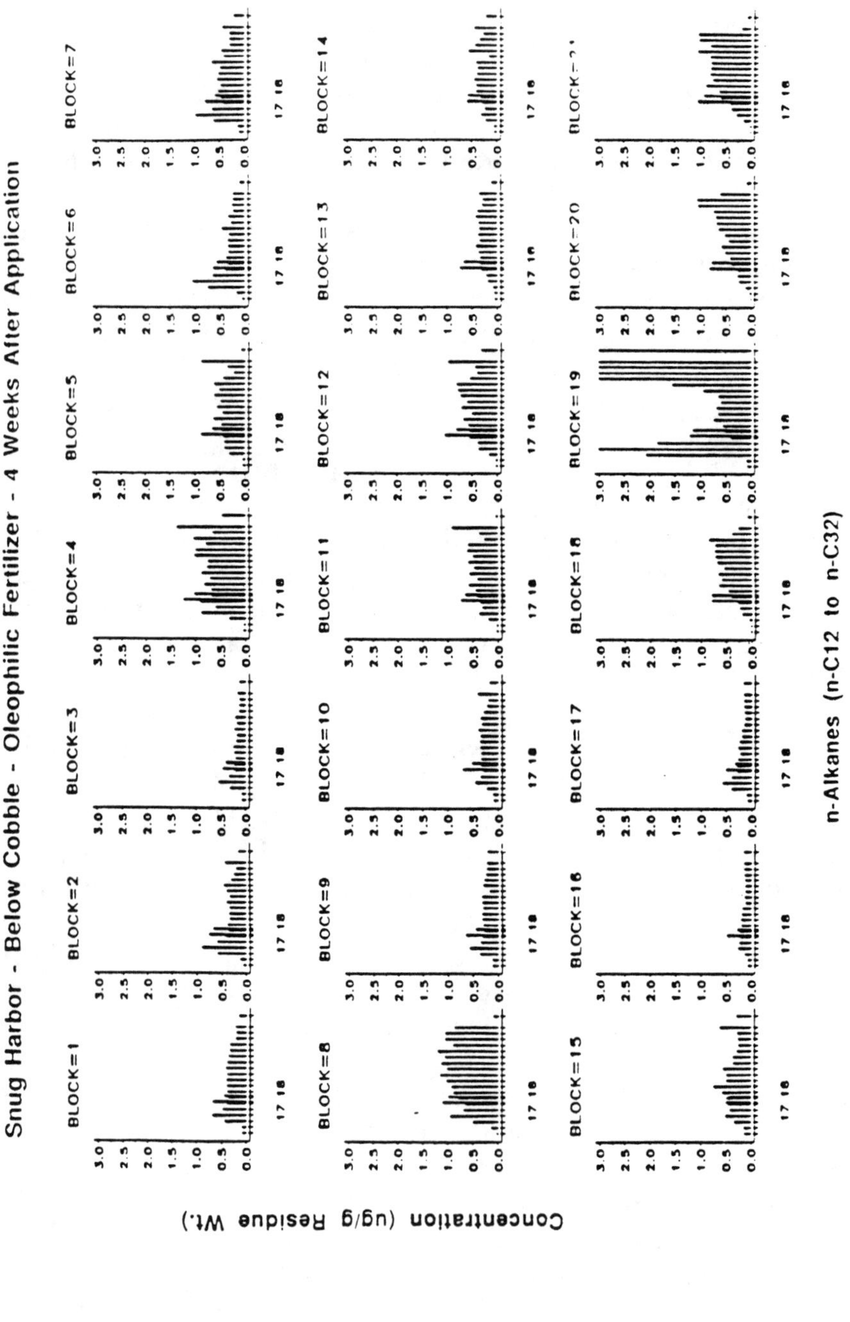

FIGURE 5. Four-week treatment GC data profiles.

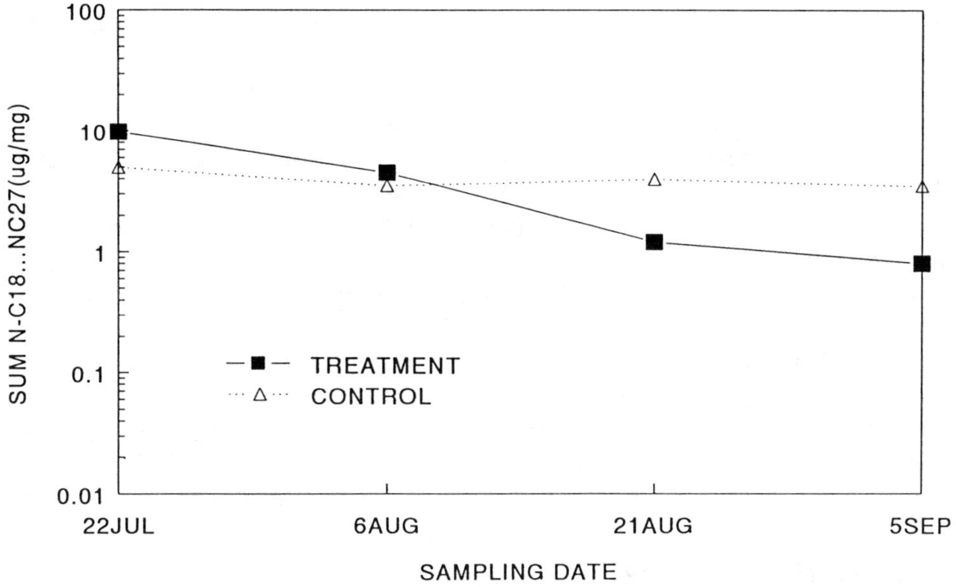

FIGURE 6. Application of soluble nutrients.

ranges of biodegradation rates may then permit the development of a clear statistical measure of treatment effectiveness.

The n-C_{17}/pristane and n-C_{18}/phytane ratios, which have been used to indicate the rate and extent of biodegradation, were not very useful for that purpose with this data set. The branched chain hydrocarbons were observed to be depleted at rates corresponding to those for straight chain isomers. The fact that these compounds degraded at comparable rates indicates that the organisms in Prince William Sound waters were already active degraders of branched chain hydrocarbons, perhaps because of terpene inputs from the conifer forests indigenous to the region.

CONCLUSIONS

One analysis of the field data based on global rate estimation has recently been completed, but the information contained in this report does not include the results based on the spatial rate difference evaluation studies. Conclusions drawn are based on trends selected from the project data set.

Visual observations suggest that enhanced biodegradation occurred on the beaches treated with Inipol, slow-release briquettes, and dissolved solutions of inorganic nutrients. The visual characteristics associated with treatment were very noticeable on the Inipol plot at Snug Harbor and the sprinkler plot at Passage Cove. Analysis of oil from control plots showed that changes in oil composition were substantial and progressed steadily through time. Apparently natural biodegradation of the oil has occurred at a surprisingly rapid rate.

Samples of oil from fertilizer-treated beaches, particularly from cobble surfaces, taken at about the time when the oil was visually disappearing, showed substantial changes in hydrocarbon composition, indicating extensive biodegradation. This suggests that biodegradation was effecting the removal of oil through direct decomposition and possibly through the production of biochemical products (bioemulsifiers) known to be produced by bacteria as they consume oil and hydrocarbons as sources of food (Okami 1986).

Acknowledgments

The collaboration of a group of EPA scientists has led to the success of this project. Each is identified with their areas of contribution; Dr. P. Hap Pritchard (project coordination), Dr. Albert Venosa (nutrient evaluation/application), Dr. John Haines (nutrient application), Dr. John Rogers (microbial analysis), Dr. Fran Kremer (field site selection/coordination), Drs. James Clark and Richard Coffin (ecological assessment), Dr. Larry Claxton (genotoxicity), and Daniel Heggem (quality assurance). The special contributions of the following individuals require recognition: John Baker, Claude R. Cripe, Steve McCutcheon, Richard Valentinetti, Dennis Miller, Henry Tabak, and Steve Safferman.

This project was a joint effort supported cooperatively by Exxon and the EPA. Drs. Steve Hinton, Roger Prince, and Russell Chianelli of Exxon provided valuable information and interaction. Drs. Eric Bretthauer and John Skinner, director and deputy director of the EPA Office of Research and Development, respectively, offered guidance, encouragement, and wisdom throughout the project.

REFERENCES

Atlas, R. M. *Appl. Microbiol.* **1975**, *30*, 396.
Atlas, R. M. *Petroleum Microbiology*; Macmillan: New York, 1984.
Atlas, R. M.; Bartha, R. *Biotechnol. Bioeng.* **1972**, *14*, 297.
Atlas, R.M.; Bartha, R. *Environ. Sci. Technol.* **1973**, *7*, 538.

Friberg, S.; Bothorel, P. *Microemulsions: Structure and Dynamics*; CRC Press: Baton Rouge, FL, 1987.
Gibbs, C. F. *Proc. R. Soc. Lond. Ser. B.* **1975**, *188*, 61.
Gibbs, C. F.; Pugh, K. B.; Andrews, A. R. *Proc. R. Soc. Land. Ser. B.* **1975**, *188*, 83.
Haines, J. R.; Atlas, R. M. *Oil Petrochem. Pollut.* **1983**, *1*, 85.
Hauck, R. D. In *Organic Chemicals in the Soil Environment*; Goring, C.A.I.; Hamaker, J. W., Eds.; Marcel Dekker: New York, 1972; Vol. 2, pp 633–690.
Jacobson, A.; McLaughlin, M.; Cook, F. D.; Westlake, D. W. *Appl. Microbiol.* **1974**, *27*, 166.
Lee, K.; Levy, E. "Biodegradation of Petroleum in the Marine Environment: Limiting Factors and Methods of Enhancement"; Canadian Technical Report of Fisheries and Aquatic Sciences, No. 1442; 1986.
Okami, Y. *Microb. Ecol.* **1986**, *12*, 65–78.
Olivieri, R.; Bacchin, P.; Robertiello, A.; Oddio, N.; Deggen, L.; Tondo, A. *Appl. Environ. Microbial.* **1976**, *31*, 629.
Rehm, H. J.; Reiff, I. *Adv. Biochem. Eng.* **1981**, *19*, 115–215.
Sirvins, A.; Angeles, M. In *Strategies and Advanced Techniques for Marine Pollution Studies: Mediterranean Sea*; Gram., C. S.; Dou, H. J., Eds.; Springer-Verlag: Berlin, 1986; pp 357–404.
Sveum, P. "Accidentally Spilled Gas-Oil in a Shoreline Sediment on Spitzbergen: Natural Fate and Enhancement of Biodegradation"; internal report N-7034; Sintef, Applied Chemistry Div.: Trandheim, Norway [Undated]; pp 1–16.
Tramier, B.; Sirvins, A. "Enhanced Oil Biodegradation: A New Operational Tool to Control Oil Spills." *Proceedings*, 1985 Oil Spill Conference, U.S. Coast Guard, Amer. Petrol. Inst., Environ. Prot. Agency, Los Angeles, 1985; pp 115–119.
Walker, J. D.; Colwell, R. R. *Microb. Ecol.* **1973**, *1*, 63.

Appendix 1. Analytical Procedures for Oil Chemistry

Beach samples of either mixed sand and gravel or cobble were preserved by freezing. The samples were thawed and mixed thoroughly, without any pH adjustment prior to initiating oil analysis. A weighed 100-g subsample was removed and mixed thoroughly with 300 mL of methanol in a separatory funnel. The slurry was shaken for 5 min, and the methanol was decanted into a 2-L separatory funnel. The samples were similarly re-extracted twice with 300 mL HPLC-grade methylene chloride. The three organic fractions were combined and back-extracted with 100 mL of 3 percent aqueous sodium chloride. The phases were separated, and the aqueous portion was extracted with 50 mL of fresh methylene chloride. This aqueous extraction in methylene chloride was added to the combined organic fraction.

The combined organic fractions were reduced in volume within a 1-L, round-bottom flask fitted with a three-ball Snyder column. The volume of solvent was reduced until the color was approximately the color of dilute weathered oil (*ca.* 7.5 mg/mL methylene chloride). The final volume of the extract was measured with a syringe having an appropriate graduated cylinder, and an aliquot was transferred to a gas chromatographic (GC) autosampler vial.

All of the cobblestones were extracted using the same procedure (methanol, followed by methylene chloride), except that shaking was replaced by gentle swirling to remove oil from the rock surfaces.

GC analysis was accomplished with an instrument capable of reproducible temperature programming with a flame ionization detector and a reliable autosampler. The GC conditions were:

Column: DB-5, 30 m × 0.25 mm, film thickness 0.25 µm
Initial Temperature: 45 C, 5-min hold
Temperature Rate: 3.5 C/min
Final Temperature: 280 C, 60-min analysis
Injector: splitless, 1-min valve closure
Injector Temperature: 285 C
Injection: 2.0 µL
Detector: FID, 350 C

Those samples that demonstrated significant evidence of biodegradation were fractionated to allow separate determination of aliphatics and aromatics. Extracts selected for fractionation were solvent-exchanged to hexane. A 50-µL volume of hexamethylbenzene (80 ng/µL) and 25 µL of *n*-decyclohexane (1 µg/µL) was added to each sample extract prior to fractionation. The fractionation was accomplished using a 60/200 mesh silica gel activated at 210 C for 24 h. The aliphatic fraction was eluted with 30 mL of hexane, and the aromatic fraction was eluted with 45 mL of hexane/benzene (1:1). The aliphatic and aromatic fractions were analyzed using the GC methods described above.

Subsamples of the final concentrated extract were subjected to mass spectral analysis. The analytical procedure is given in the Fucus oil analysis protocols.

Subsamples (5 to 15 mL) of the final concentrated extract were also removed, filtered through sodium sulfate, and placed in tared watch glasses. After passive evaporation of the solvent, the oil residue weight was determined.

Changes in oil composition were determined by comparing the total weight of all alkanes appearing on the chromatograph, normalized to the total residue weight of oil, on a sample-by-sample basis.

Transformation of Pesticides in Groundwater under Anaerobic Conditions

*Gregory G. Wilber**, *Gene F. Parkin*
University of Iowa

INTRODUCTION

Concern about the presence of pesticides in groundwater supplies has been growing in recent years. A number of water quality surveys, both nationwide and specific to the Midwest, have reported finding many of the most commonly used pesticides in both raw groundwaters and in finished drinking waters (Cohen *et al.* 1986; Hallberg 1987). Furthermore, evidence exists that the percentage of wells with detectable amounts of pesticides is increasing (Hallberg *et al.* 1987). Nationally, the two most commonly used pesticides are the acetanilide herbicide alachlor and the triazine herbicide atrazine, which account for 25 percent of total pesticide use by weight. These pesticides are also two of the most frequently detected pesticides in Midwestern groundwater supplies. A survey by Hallberg (1987) reported that atrazine and alachlor have been detected in a number of wells in the Midwest, in typical concentration ranges of 0.0001 to 0.01 mg/L, apparently as a result of routine agricultural practices. In addition, a number of sites with much higher concentrations, greater than 0.5 mg/L, have been found for which the contamination can be attributed to a quasi-point

* Department of Civil and Environmental Engineering, University of Iowa, Iowa City, IA 52242 (319) 335-5646

source, such as a spill at a farm chemical supply dealership (Hallberg 1987).

It is known that the ingestion of drinking water contaminated with pesticides can be dangerous to human health (NAS 1977). Although nationwide drinking water standards for these compounds have not been established, the EPA has issued proposed Maximum Contaminant Levels (MCL) for alachlor and atrazine of 0.002 and 0.003 mg/L, respectively (Parmelee 1988).

The discovery of sites with groundwater contaminated by these pesticides and concern about their effects on health has led researchers to seek new methods to use in remediating these sites. Among the promising techniques currently under investigation is *in situ* bioremediation, in which the native microbial population is stimulated to increase in number and activity and to metabolize the pollutants present (Lee *et al.* 1988). The potential advantages of this technique include the potentially complete mineralization of the harmful pollutants to inorganic compounds, as well as its potentially significant cost savings over other remediation alternatives (Lee *et al.* 1988). It is known that the addition of exogenous carbon substrates to an aquifer may result in rapid depletion of dissolved oxygen, resulting in anaerobic electron acceptor conditions (Bouwer & Cobb 1987). The electron acceptor condition will then be determined by that electron acceptor present in the media which is most thermodynamically favored and thus utilized by the microbial population for deriving energy from an electron donor (Stumm & Morgan 1981). Bouwer and Cobb (1987) report the sequential utilization of nitrate, sulfate, and carbon dioxide as electron acceptors in a simulated groundwater column that was fed acetate, indicating the potential for multiple electron acceptor conditions within a single *in situ* biodegradation scheme. Furthermore, it is known that the electron acceptor condition under which bioremediation is performed can affect the success and rate of the biodegradation of xenobiotic compounds, including pesticides (Berry *et al.* 1987; Bouwer & McCarty 1985; Kuhn & Suflita 1989).

Recent work (Lynch 1990) has shown that alachlor, and to a lesser degree, atrazine, can undergo microbially mediated degradation under four different electron acceptor conditions: aerobic respiration, nitrate reduction, sulfate reduction, and methanogenesis. Others (Pothuluri *et al.* 1990) have also studied the effects of redox conditions and reported that alachlor was more readily degraded under aerobic than anaerobic conditions in soil. The addition of extraneous carbon sources to soil samples was found to enhance the rate of alachlor biotransformation. Novick *et al.* (1986) performed mineralization studies on alachlor in soil-water suspensions and lake water. They found that only very small percentages of ring labeled ^{14}C were recoverable as $^{14}CO_2$ after 30 days of incubation. Novick and Alexander (1985) reported degradation of alachlor in sewage and lake water, but could isolate no organisms that mineralized alachlor and found that no ^{14}C was incorporated into the biomass. It was concluded that the degradation of alachlor was likely co-metabolic, meaning the microorganisms converted the alachlor to organic products but did not obtain energy from the reaction or use its carbon for biosynthesis (Novick & Alexander 1985). Microbial metabolism of atrazine has also been investigated previously. For example, Behki and Khan (1986) reported on the biodegradation of atrazine by a *Pseudomonas* strain. Erickson and Lee (1989) also reviewed a number of studies and concluded that degradation of atrazine is favored under aerobic conditions in soil.

With these facts in mind, a research project was initiated to investigate the fate of atrazine and alachlor under conditions similar to those found in Iowa groundwater and under the following three anaerobic electron acceptor conditions: nitrate reduction, sulfate reduction, and methanogenesis. Among the primary objectives of this research were the following:

- To determine the effect of electron acceptor condition on the biotransformation of atrazine and alachlor

- To obtain preliminary information on the effects of influent pesticide concentration on cultures grown under each of the three anaerobic electron acceptor conditions

- To investigate the relationship between microbial utilization of a primary substrate (acetate), growth of biomass, and secondary substrate (pesticide) biotransformation

METHODS AND PROCEDURES

Experimental Systems

Column Reactors. Three attached-growth, mixed-culture reactors were developed and grown under conditions of nitrate reduction, sulfate reduction, and methanogenesis, respectively, and their preliminary characteristics with respect to pesticide degradation were determined

(Lynch 1990). The nitrate-reducing culture was originally seeded with sediment from the Iowa River, while the sulfate-reducing and methanogenic cultures were seeded with sediments from an anaerobic farm pond. The presence of the desired electron acceptor conditions were confirmed by monitoring the utilization of nitrate and the production of gas in the nitrate-reducing column, loss of sulfate and appearance of sulfide in the sulfate-reducing column, and the production of methane in the methanogenic column.

Each experimental system consisted of a plexiglass column 100 cm in length and 5 cm in diameter, filled with 3-mm-diameter glass beads. The reactors were housed in the dark in an incubator maintained at 16 C. The inorganic, simulated groundwater medium was stripped of oxygen by a flow of N_2 and CO_2 gas (in a 95:5% ratio) and was fed down-flow through each column with a peristaltic (Masterflex) pump. Effluent samples were collected after the pump, which required the use of a small segment of silicone tubing. All other tubing was either Teflon® or glass. Organic constituents (acetate, atrazine, and alachlor) were introduced at the top of each column via a syringe pump and a Teflon needle. At the 2-, 10-, 21-, 51-, and 91-cm intervals down each column were sample ports fitted with gas-tight Mininert (Supelco) valves. Samples were taken from these ports by plugging the flow at the bottom of each column and connecting the pump to a Teflon line run from the sample port of interest.

Influent pesticide concentrations were varied and transformation of the pesticides was monitored in the effluents both over time and at each port down each column once steady state for each experimental period was reached. The initial experimental conditions (Experimental Period 1) included a flow rate of 0.5 L/day, resulting in a hydraulic retention time of 34 h for each column. Influent pesticide concentrations for this period were 0.1 mg/L. During this period, an attempt was made to measure the amount of biomass present along each column to correlate biotransformation activity with the amount of biomass present. In Experimental Period 2, flow rates remained the same, but influent pesticide concentrations were tripled to 0.3 mg/L alachlor and 0.325 mg/L atrazine. At the end of this period, pesticides were removed from the influent for a period of 1 week to observe washout and desorption characteristics. During Experimental Period 3, conditions were returned to those of the first experimental period in an effort to see if the effluent quality returned to that observed previously. For all three periods, the carbon and energy source added was acetate, at 20 mg/L to the denitrifying column and 50 mg/L to the sulfate-reducing and methanogenic columns. In each of the systems, the preferred electron acceptor was provided in sufficient excess to ensure that it would never be the limiting substrate for microbial growth.

Abiotic Reaction with Bisulfide. Previous work has reported the abiotic reaction of alachlor with bisulfide ion (Lynch 1990). An experiment was performed to better quantify the rate of this reaction and estimate its relative importance in anaerobic systems. A solution containing the same inorganic media fed to the sulfate-reducing system was stripped of oxygen with nitrogen gas, dosed with a phosphate buffer at pH 7.1 (the pKa of hydrogen sulfide), 0.3 mg/L alachlor and atrazine, and 17 mg/L total sulfide. With the pH buffered at the pKa of H_2S, half of the total sulfide was in the form of bisulfide ion. The solution was then quickly distributed among a series of 120-mL serum bottles with no headspace, preventing any volatilization of the hydrogen sulfide. Pesticide and sulfide concentrations were measured over time, and from this data a pseudo first order rate constant could be calculated.

Influent Media and Organics. The inorganic feed solution was designed to resemble a typical groundwater (Freeze & Cherry 1979). The media for all three columns included the following (mg/L): K_2HPO_4 (2.1), $NaHCO_3$ (3.3), K_2CO_3 (27.6), $MgCl_2 \bullet 6H_2O$ (16.3), $CaCl_2 \bullet 2H_2O$ (11.8), $FeCl_2 \bullet 4H_2O$ (0.7), $CoCl_2 \bullet 6H_2O$ (0.05), KI (0.05), $MnCl_2 \bullet 4H_2O$ (0.06), $ZnCl_2$ (0.007), $Na_2MoO_4 \bullet 2H_2O$ (0.01), H_3BO_3 (0.007), and $NiCl_2 \bullet 2H_2O$ (0.01). In addition, the medium for the denitrifying column contained 0.008 mg/L $CuCl_2 \bullet 2H_2O$, 4.5 mg/L $MgSO_4 \bullet 7H_2O$, and 75.2 mg/L $Ca(NO_3)_2 \bullet 4H_2O$, and that for the sulfate-reducing column contained 208.5 mg/L $MgSO_4 \bullet 7H_2O$ and 0.4 mg/L NH_4HCO_3. The methanogenic column was fed no sulfate, but rather received 0.4 mg/L NH_4HCO_3 and 1.05 mg/L $Na_2S \bullet 9H_2O$.

Reagents and Analytical Methods. Aqueous stock solutions of each pesticide were prepared from pure, analytical grade chemicals obtained from Supelco (Bellefonte, PA) for atrazine or The Monsanto Company (St. Louis, MO) for alachlor. All other compounds used for the feed solution or for the analyses were analytical grade or better (Fisher Scientific). Pesticide concentrations in the aqueous phase were measured by a solid-phase extraction (SPE) method using PrepSep C-18 (octadecyl silane) columns from Fisher Scientific. Aqueous samples (100 mL) were extracted with a SPE column and then eluted with 2 mL ethyl acetate and analyzed by GC (Hewlett Packard 5890A, with an electron capture detector and a fused silica capillary column [J & W] with DB-5 stationary phase). The extraction method had an efficiency of approximately 96 percent. Analysis of replicate samples demonstrated a reproducibility for the method of ±6 percent. Nitrate and sulfate were measured by ion chromatography (Dionex 2000i). Acetate was measured by GC using a packed column (Supelco) and a flame ionization detector (Hewlett Packard 5890A).

Biomass in each of the columns was determined in the following manner. Near the end of Experimental Period 1, several beads were removed from each of the five sample ports along each column. Each set of beads was placed in a dilute Tris buffer, and the samples were sonicated to disrupt the attached cells and solubilize the cellular proteins. The liquid samples were then assayed for total protein using the Pierce assay (Smith *et al.* 1985; Pierce 1989). Liquid samples from each sample port were also assayed in a similar manner to quantify the amount of suspended biomass in the liquid fractions along each column.

RESULTS AND DISCUSSION

Column Data

Alachlor and atrazine concentrations from the effluent of each column were monitored over time during each experimental period. Results from these analyses appear in Figures 1, 2, and 3. The amount of variability in each reactor's effluent pesticide concentrations can be attributed in part to slight amounts of variation in the influent concentrations, which result from fluctuations in either the media pump or syringe pump operation, resulting in an approximate variation of ±6 percent. With this in mind, a number of differences in the amount of biotransformation among the three electron acceptor conditions can be observed.

Experimental Period 1. Beginning with Experimental Period 1, the largest percentage removals of alachlor occurred in the sulfate-reducing system, which showed approximately 67 percent removal at steady state (Figure 1). The methanogenic system showed about 55 percent removal at steady state, and the denitrifying system transformed about 47 percent. Disappearance of atrazine occurred to a much lesser degree. In the denitrifying, sulfate-reducing, and methanogenic systems, removals observed were approximately 11, 6, and 15 percent, respectively.

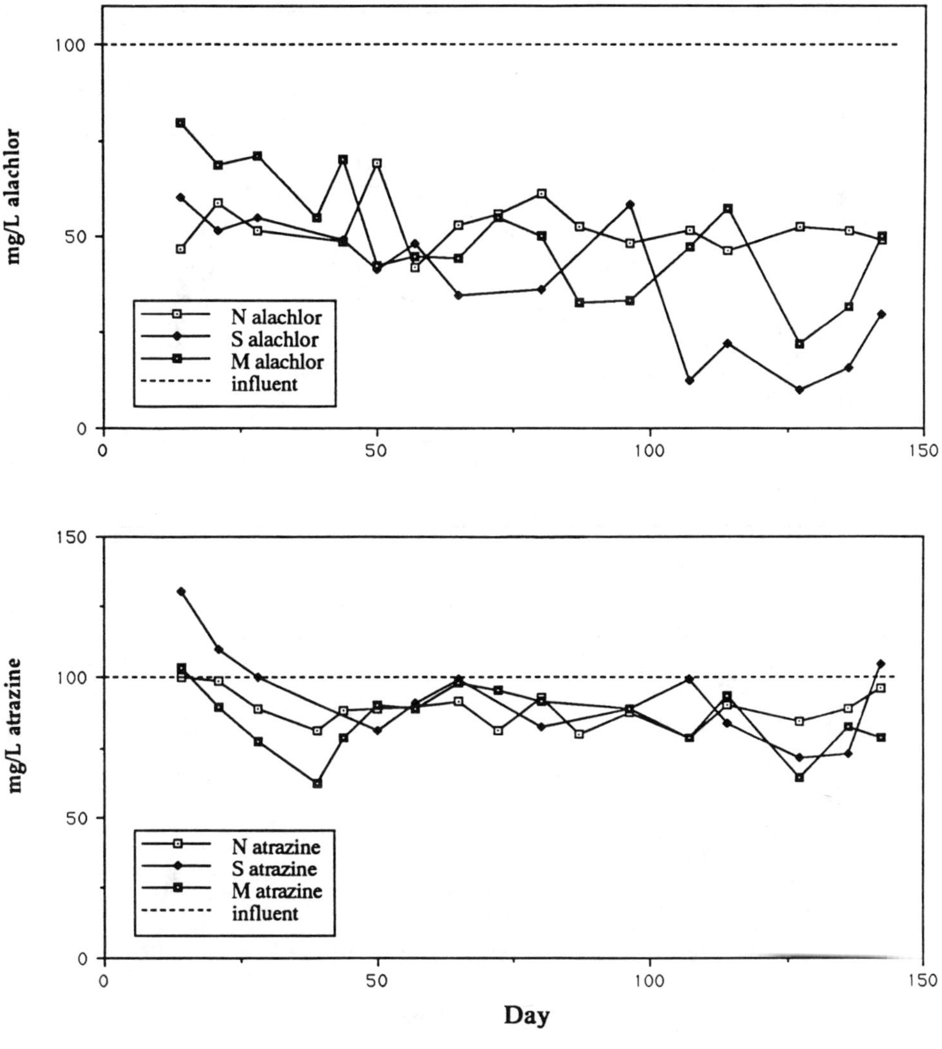

FIGURE 1. Pesticide effluent data for Experimental Period 1 for the three columns (N, nitrate-reducing; S, sulfate-reducing; and M, methanogenic).

Also during this period, an attempt was made to identify any products of the pesticide transformations (Monsanto 1990). No transformation products were conclusively identified in the effluents from the denitrifying column. However, in the sulfate-reducing and methanogenic column effluents, what appears to be the reductive dechlorination product of alachlor, "nochloroalachlor," on which the

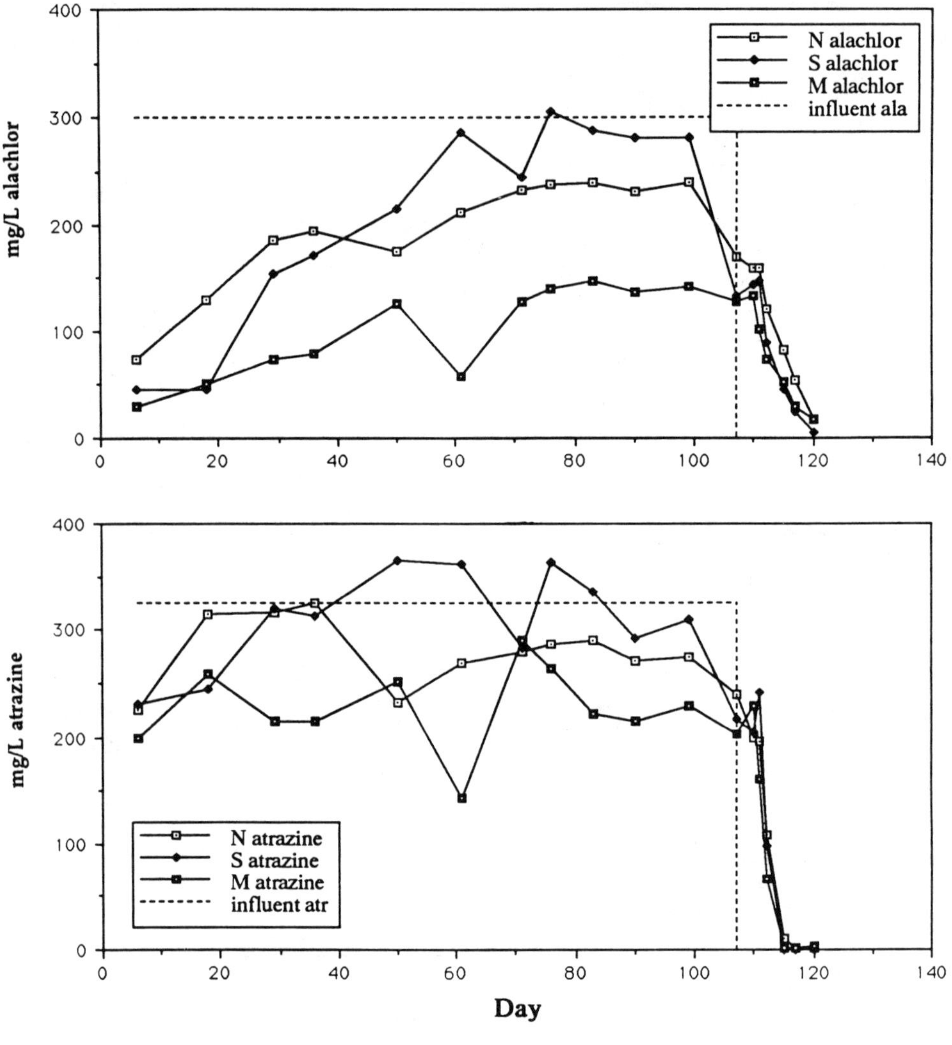

FIGURE 2. Pesticide effluent data for Experimental Period 2 for the three columns (N, nitrate-reducing; S, sulfate-reducing; and M, methanogenic).

chlorine atom has been replaced with a hydrogen, was identified. This is also consistent with the results reported by Lynch (1990).

Experimental Period 2. It would be expected that the relative differences among the three reactors would be the same following a threefold increase in pesticide influent concentration, with all other parameters remaining the same. However, this was not observed. During

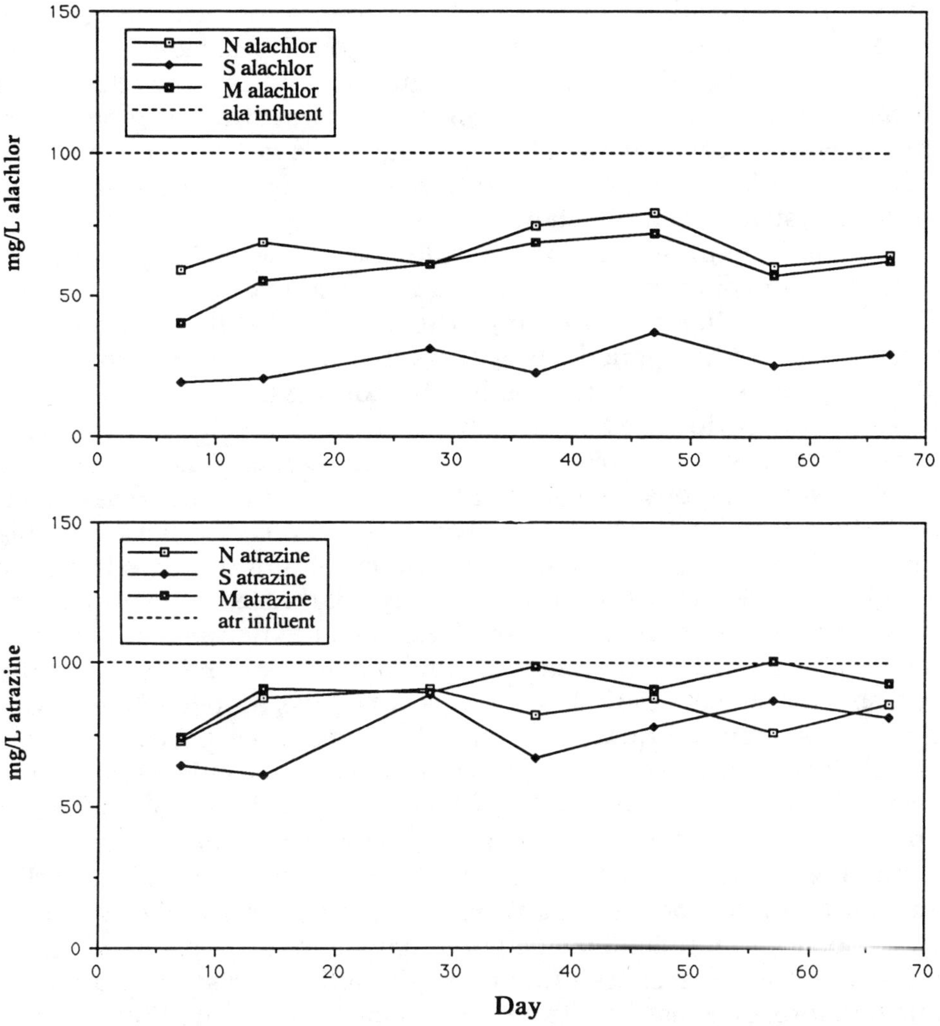

FIGURE 3. Pesticide effluent data for Experimental Period 3 for the three columns (N, nitrate-reducing; S, sulfate-reducing; and M, methanogenic).

Experimental Period 2, a number of unanticipated results were obtained, as shown in Figure 2. The first is the long time required to reach a new steady state, particularly in the sulfate-reducing column. Pesticide effluent values rose gradually in all three columns over a period of about 60 days, completely unlike the immediate step-function increase of the influent concentration. It is likely that a number of factors contributed to this occurrence. It is possible that some adsorption onto tubing and plexiglass surfaces in each reactor took place. The relatively long washout time once pesticides had been removed from the influent (Day 107, Figure 2) implies some retardation of alachlor movement through the column. However, adsorption experiments with all of the materials that come in contact with the alachlor ruled out the likelihood that adsorption is a predominant phenomenon in these systems (data not shown).

The most dramatic change resulting from the tripling of the pesticide influent concentrations occurred in the sulfate-reducing column. Figure 2 shows that this culture gradually lost all ability to transform alachlor during this period. It appears that the higher concentrations of atrazine or alachlor, or the combination of these, present during this experimental period are toxic or inhibitory to all or some of the microorganisms present in this reactor. Others (Pratt *et al.* 1988) have reported toxic responses by several types of bacteria to atrazine in concentrations similar to those used here. Additional evidence for this hypothesis is that acetate utilization in this column decreased rapidly after the pesticide influent concentration was increased. Acetate utilization continued at approximately half the rate of Experimental Period 1, until both pesticides were removed from the influent on Day 107 of the second experimental period. Once this was done, acetate effluent concentrations returned quickly to what they had been previously (Figure 4). This rapid recovery implies that much of the culture may have been inhibited, but had not been killed. Thus, if the inhibition to normal cellular metabolism occurred over a long enough period, an increasing fraction of the culture would become inactive, and the effluent concentrations of pesticide would continue to gradually rise. This slow decrease of microbial activity may help to explain why the increase in alachlor concentrations in the effluent rose so gradually. Furthermore, decreased utilization of acetate also resulted in a corresponding decrease in the production of sulfide. Thus, even the abiotic transformation of alachlor from sulfide was greatly slowed. (This reaction is discussed in more detail below.) Experiments are currently under way to determine the cause of this apparent toxic response and whether atrazine, alachlor, or a combination of the two is responsible for these effects.

The other two systems responded differently than the sulfate-reducing system during Experimental Period 2. The denitrifying system showed no apparent effects with respect to acetate utilization. However, the percentage of alachlor transformed was considerably less than during Experimental Period 1 (21% compared to 47%). Acetate utilization in the methanogenic column also remained unchanged, and

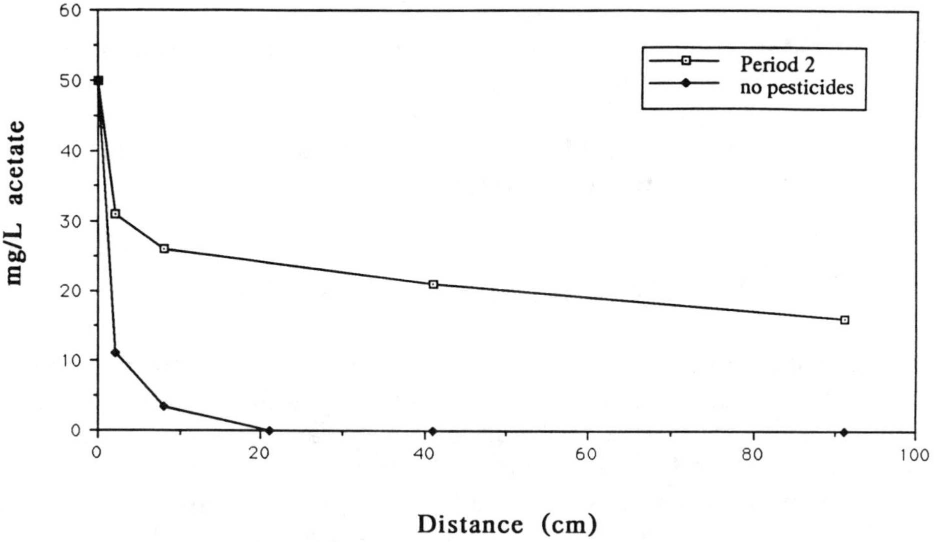

FIGURE 4. Comparison of acetate utilization along the sulfate-reducing column during Experimental Period 2 (influent pesticide = 0.3 mg/L) and after pesticides were removed from the influent.

its percent removals of alachlor were very similar in both experimental periods (55% and 53%).

With respect to atrazine transformation during Experimental Period 2, percent removals for the three reactors were as follows. The sulfate-reducing column again showed very little activity, with only about 3 percent removal, whereas the denitrifying and methanogenic columns showed 14 and 25 percent removals, respectively. Note that compared to Experimental Period 1 this represents a significant increase in percent removal in the methanogenic column, whereas the percent removal is unchanged in the denitrifying system.

Experimental Period 3. The primary observation made from Experimental Period 3 (Figure 3), in which the pesticide influent concentrations were returned to the original levels of 0.1 mg/L, is that the denitrifying column did not return to its former level of alachlor degradation, with an observed percent transformed of 28 compared to 47 percent during Experimental Period 1. Atrazine removals were also slightly less, approximately 5 percent transformed during this period compared to 11 percent previously. The sulfate-reducing column recovered from the apparent toxic effects of Experimental Period 2 very quickly, returning almost immediately to similar, or even better, levels of removal than those observed during Experimental Period 1. About 76 percent of the alachlor and 22 percent of the atrazine were transformed in this column during this period. The methanogenic column also returned fairly closely to its previous levels of removal: 41 percent for alachlor and 6 percent for atrazine.

Pesticide, Acetate, and Biomass Profiles

During each experimental period, once the effluent data from each column appeared to have reached a quasi-steady state, acetate and pesticide concentrations were determined at each sampling port down each column. At the end of Experimental Period 1, biomass protein concentrations down the length of each column were also estimated. Results from the protein and acetate analyses for Experimental Period 1 appear in Figure 5. It is apparent that in all three columns, the greatest amount of biomass present in each column is located in the first 20 cm. Most of the acetate in all three columns has been used within the first 20 cm as well. It has not yet been determined why measurable biomass protein is present beyond the 20-cm mark in the sulfate-reducing and methanogenic columns, even though no acetate is present beyond this point. It is thought that this may represent protein from dead bacteria which have sloughed off from the upper, more densely populated areas. Profiles resulting from the pesticide analyses for Experimental Periods 1 and 2, which appear in Figures 6 and 7, support this hypothesis. Clearly, a majority of the pesticide transformation also occurs in the first several centimeters of each column, and there is little degradation below the areas of greatest biomass concentration. This implies that if the measured biomass near the middle and bottom of the sulfate-reducing and methanogenic columns is alive, it is not actively degrading pesticide to the same degree as that in the first 20 cm.

The information from these substrate and biomass profiles, however, is not sufficient to draw any definite conclusions about the nature of the pesticide biotransformation reactions. It does appear as though the reaction may be co-metabolic, in that pesticide transformation occurs primarily in those regions in each column containing significant amounts of the primary substrate, acetate.

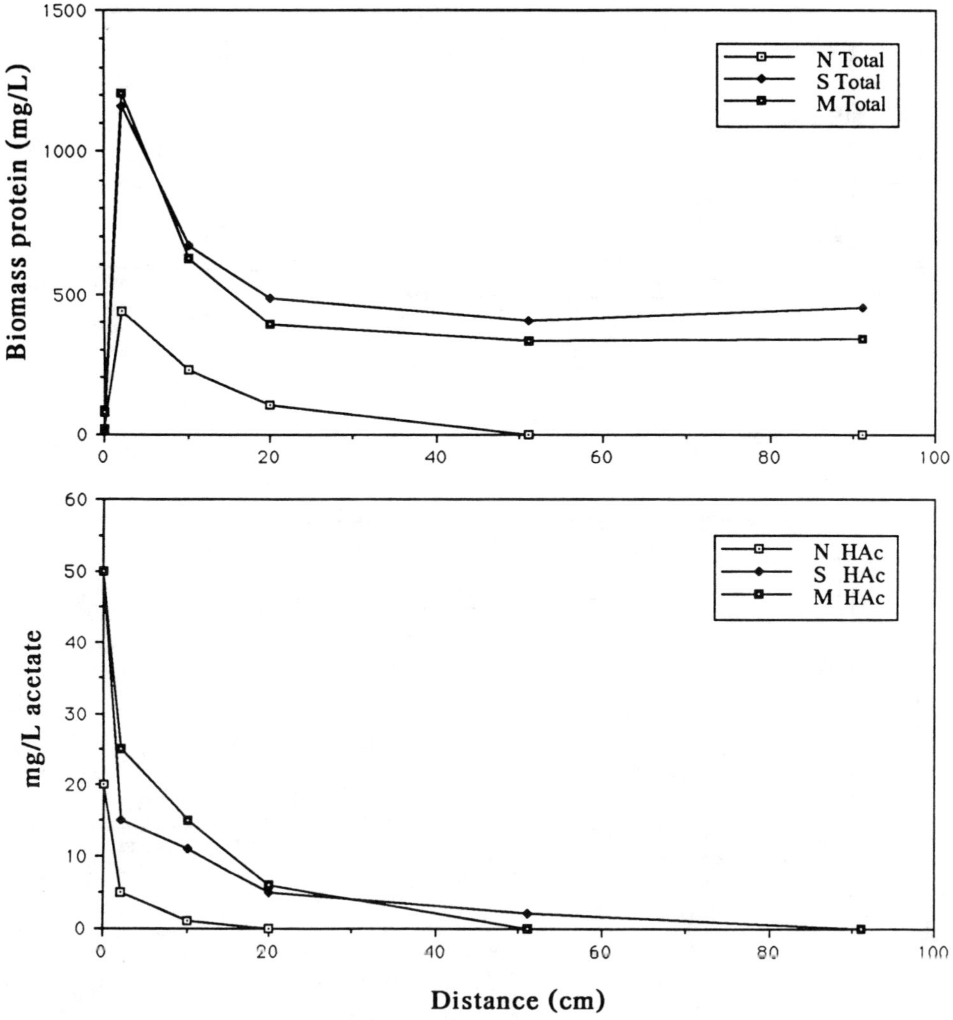

FIGURE 5. Profiles along each column of biomass protein concentration, including both attached and suspended fractions and typical acetate utilization in each column during Experimental Period 1 (N, nitrate-reducing; S, sulfate-reducing; and M, methanogenic).

Abiotic Kinetic Experiment Results

Results from the abiotic sulfide reaction kinetic experiment appear in Figure 8. The data have been linearized by treating them as first order degradation data. The bisulfide concentration was constant at 8.5 mg/L

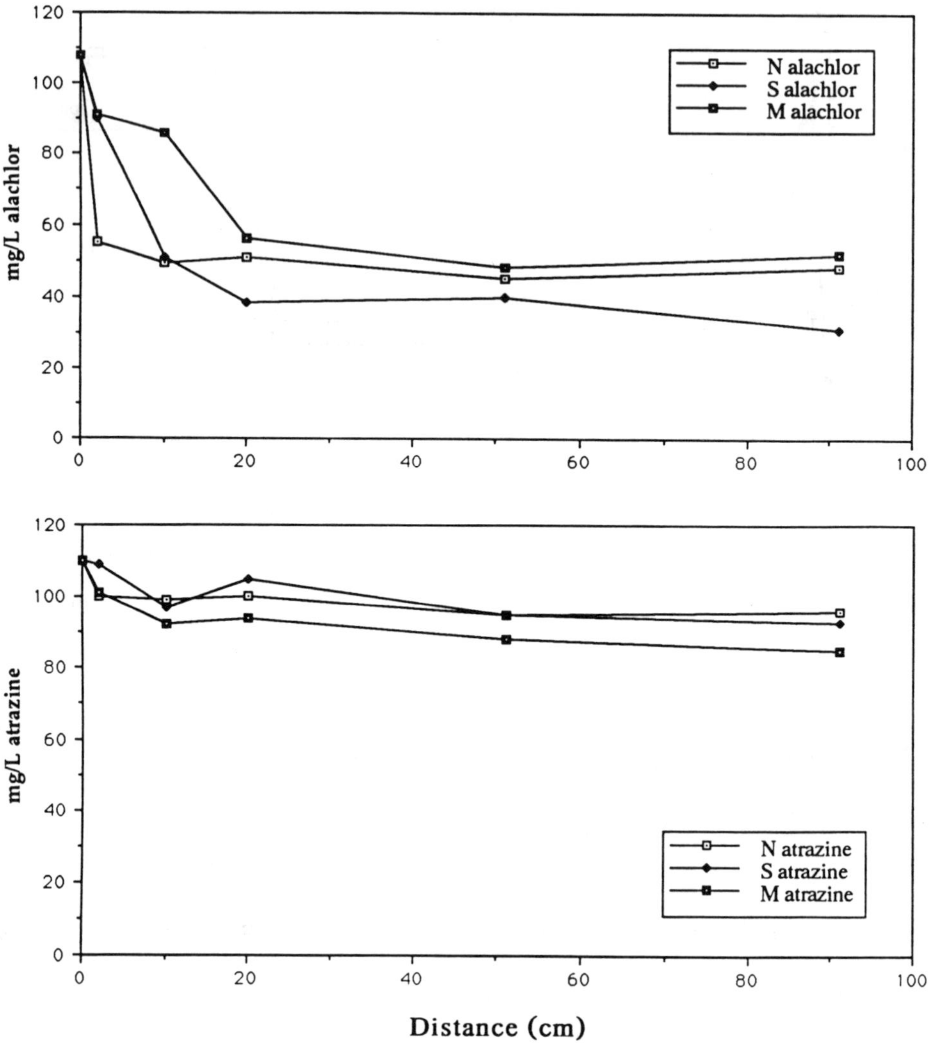

FIGURE 6. Pesticide transformation profiles along each column, Experimental Period 1 (N, nitrate-reducing; S, sulfate-reducing; and M, methanogenic).

throughout the experiment. Thus, from the graph, a second order overall reaction rate of 0.013/h is shown, which can then be converted to a pseudo first order rate constant by dividing by the bisulfide concentration. The resulting rate constant is 0.0015/h mg/L [HS^-]. Thus, for example, if the average bisulfide concentration in the sulfate-reducing column is about 2.5 mg/L and pH ≈ 7.1, and the retention time is 34 h,

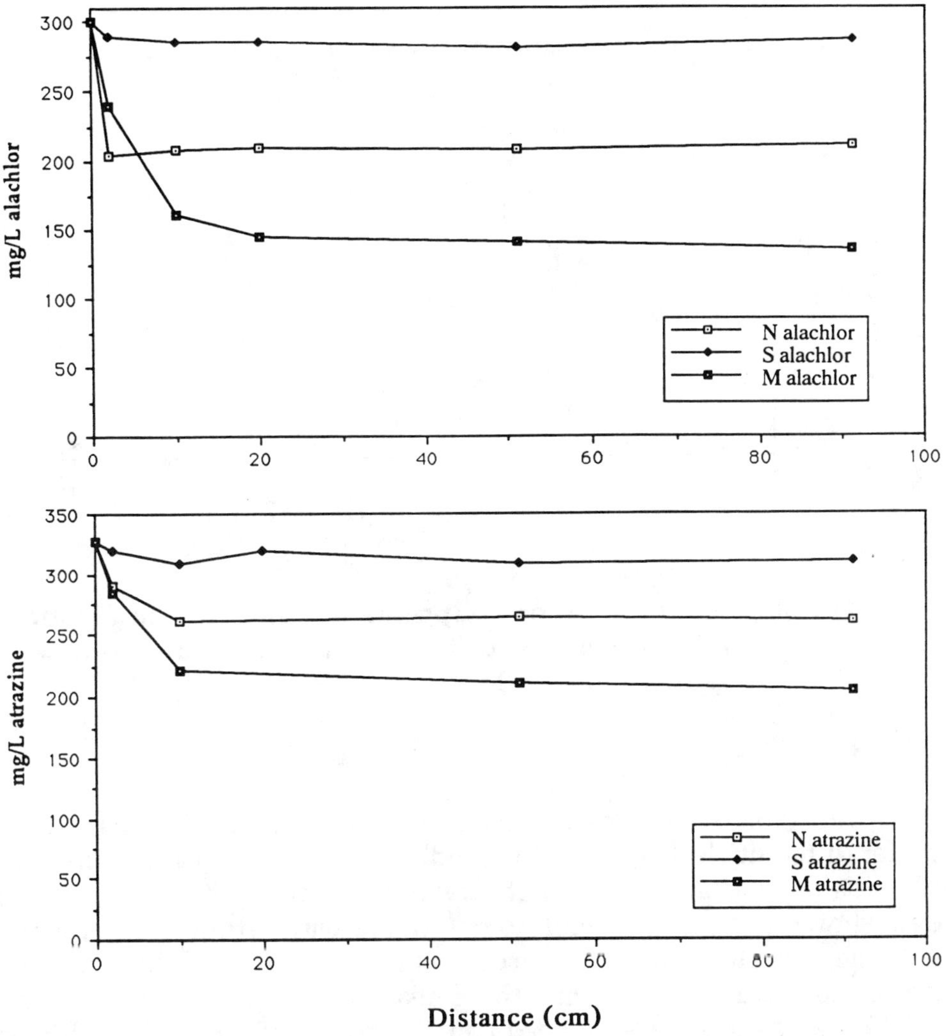

FIGURE 7. Pesticide transformation profiles along each column, Experimental Period 2 (N, nitrate-reducing; S, sulfate-reducing; and M, methanogenic).

then it can be approximated that, of a 0.1 mg/L influent of alachlor, about 0.012 mg/L will be transformed abiotically.

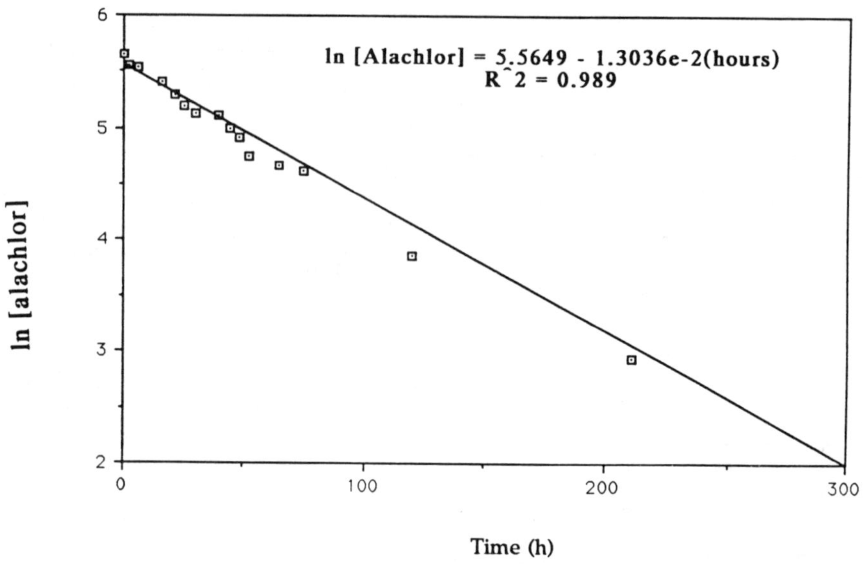

FIGURE 8. Plot of alachlor disappearance with time in abiotic reaction with bisulfide. Total dissolved sulfide = 17 mg/L; bisulfide = 8.5 mg/L; pH = 7.

SUMMARY

The results of these experiments show that microorganisms grown under different electron acceptor conditions will demonstrate different patterns of biotransformation of atrazine and alachlor. Based on the data shown here, it is difficult to determine with certainty whether or not the transformations observed are the result of co-metabolic processes. Other work with these reactors, using ^{14}C ring-labeled alachlor and similar acetate feed concentrations, reported that little or no mineralization of alachlor took place in these systems (Lynch 1990). The results shown here also suggest that biotransformation of alachlor must occur simultaneously with active metabolism of acetate, the primary carbon and energy source. More work is needed for definite conclusions about this aspect of the transformation reactions.

When influent concentrations of these pesticides exceeded 0.1 mg/L, an inhibitory or toxic effect on normal microbial metabolism seemed to occur in the sulfate-reducing culture, resulting in decreased acetate and pesticide removal. Once the pesticides were removed from the influent, this culture recovered quickly. The denitrifying and methanogenic systems showed no signs of toxicity at either influent concentration. At levels at or below 0.1 mg/L, the sulfate-reducing culture had the greatest capacity for transforming alachlor, in part because the bisulfide formed during normal metabolism is involved in an abiotic transformation of alachlor as well. A pseudo first order rate constant for this reaction was determined to be 0.0015/h mg/L [HS^-].

Within each column, measurement of biomass confirms that the greatest degree of pesticide biotransformation is occurring in the regions of greatest biomass concentration. Reductive dechlorination appears to be among the biotransformation mechanisms at work in the sulfate-reducing and methanogenic columns.

Acknowledgments

This research was funded by a grant from the Iowa State Water Resources Research Institute, Ames, IA.

REFERENCES

Behki, R. M.; Khan, S. U. "Degradation of Atrazine by *Pseudomonas*: N-dealkylation and Dehalogenation of Atrazine and Its Metabolites." *Journal of Agricultural and Food Chemistry* **1986**, *29*, 1132.

Berry, D. F.; Francis, A. J.; Bollag, J.-M. "Microbial Metabolism of Homocyclic and Heterocyclic Aromatic Compounds under Anaerobic Conditions." *Microbiological Reviews* **1987**, *51*(1), 43–59.

Bouwer, E. J.; Cobb, G. D. "Modeling of Biological Processes in the Subsurface." *Water Science and Technology* **1987**, *19*, 769–779.

Bouwer, E. J.; McCarty, P. L. "Utilization Rates of Trace Halogenated Organic Compounds in Acetate-Grown Biofilms." *Biotechnology and Bioengineering* **1985**, *27*, 1564–1571.

Cohen, S. Z.; Eiden, C.; and Lorber, M. N. "Monitoring Groundwater for Pesticides." In *Evaluation of pesticides in groundwater*; W. Y. Gamer *et al.*, Eds.; ACS Symposium Series 315; American Chemical Society: Washington, DC; 1986; pp 170–196.

Erickson, L. E.; Lee, K. H. "Degradation of Atrazine and Related S-Triazines." *Critical Reviews in Environmental Control* **1989**, *19*(1), 1–14.

Freeze, R. A.; Cherry, J. A. *Groundwater*; Prentice-Hall: Englewood Cliffs, NJ, 1979.

Hallberg, G. R. "Agricultural Chemicals in Ground Water. Extent and Implications." *American Journal of Alternative Agriculture* **1987**, *2*(1), 3–15.

Hallberg, G. R.; Libra, R. D.; Long, K. R.; Splinter, R. C. "Pesticides, Groundwater, and Rural Drinking Water Quality in Iowa." In *Pesticides and Groundwater: A Health Concern for the Midwest*; The Freshwater Foundation: Navarre, MN, 1987; pp 83–104.

Kuhn, E. P.; Suflita, J. M. "Dehalogenation of Pesticides by Anaerobic Microorganisms in Soil and Groundwater—A Review." In *Reactions and Movement of Organic Chemicals in Soil,* Soil Science Society of America and American Society of Agronomy, SSSA Special Publication No. 22, 1989; pp 111–180.

Lee, M. D.; Thomas, J. M; Borden, R. C.; Bedient, P. B.; Ward, C. H.; Wilson, J. T. "Biorestoration of Aquifers Contaminated with Organic Compounds." *CRC Critical Reviews in Environmental Control* **1988**, *18*(1), 29–89.

Lynch, N. L. 1990. "Transformation of Pesticides and Halogenated Hydrocarbons in the Subsurface Environment." PhD. Dissertation, Department of Civil and Environmental Engineering, The University of Iowa, May 1990.

Monsanto Company 1990. Unpublished data from analyses of samples from the reactors described in this paper. St. Louis, MO.

NAS (National Academy of Science). *Drinking Water and Health;* NAS: Washington, DC, 1977.

Novick, N. J.; Alexander, M. "Cometabolism of Low Concentrations of Propachlor, Alachlor, and Cycloate in Sewage and Lake Water." *Applied and Environmental Microbiology* **1985**, *49*(4), 737–743.

Novick, N. J.; Mukherjee, R.; Alexander, M. "Metabolism of Alachlor and Propachlor in Suspensions of Pretreated Soils and in Samples from Ground Water Aquifers." *Journal of Agricultural and Food Chemistry* **1986**, *34*, 721–725.

Parmelee, M. A. "Update." *American Water Works Ass'n. Journal* **1988**, *80*(7).

Pierce Chemical Company. *BCA Protein Assay Reagent Instructions Manual;* Rockford, IL, 1989.

Pothuluri, J. V.; Moorman, T. B.; Obenhuber, D. C.; Wauchope, R. D. "Aerobic and Anaerobic Degradation of Alachlor in Samples from a Surface-to-Groundwater Profile." *Journal of Environmental Quality* **1990**, *19*, 525–530.

Pratt, J. R.; Bowers, N. J.; Niederlehner, B. R.; Cairns, J. "Effects of Atrazine on Freshwater Microbial Communities." *Arch. of Environmental Contamination and Toxicology* **1988**, *17*, 449–457.

Smith, P. K.; Krohn, R. I.; Hermanson, G. T. "Measurement of Protein Using Bicinchoninic Acid." *Analytical Chemistry* **1985**, *150*, 76–85.

Stumm, W.; and Morgan, J. J. *Aquatic Chemistry,* Second Edition; Wiley: New York, 1981.

Combined Biotreatment of Hazardous Leachate

*Sanjoy K. Bhattacharya**, *Reddy M. Nandipati*
Tulane University

The objective of this research was to determine the treatability of hazardous leachate using different combined bioprocesses based on the premise that a combined anaerobic-aerobic biological treatment process should be able to biodegrade most of the organic compounds present in hazardous leachates. Some organic compounds such as chlorinated aliphatic hydrocarbons are efficiently biodegradable anaerobically (Bouwer & McCarty 1983). On the other hand, homocyclic and heterocyclic aromatic volatile compounds are not easily biodegradable anaerobically presumably because of their ring structure. Some researchers (Schink 1985) reported no indication of methanogenic degradation for a number of compounds, including toluene and benzene; whereas, some other researchers (Vogel & Grbić-Galić 1986) have shown that oxygen (^{18}O labeled) from water is incorporated in the anaerobic fermentative transformation of toluene and benzene.

A review of literature shows that, depending on the source of the hazardous leachate, the composition of hazardous compounds in the leachate could vary significantly. An earlier work (Bhattacharya *et al.* 1990) on anaerobic treatment of hazardous leachate has shown that 90 percent or higher removals of acetone, methyl ethyl ketone, methyl isobutyl ketone, and trichloroethylene were achieved by a pilot-scale, upflow, anaerobic filter. Removals of chlorobenzene and toluene were less than 15 percent. Such low removals of these compounds are

* Department of Civil Engineering, 206 Civil Engineering Building, Tulane University, New Orleans LA 70118

expected in anaerobic systems. When the effluent from the anaerobic filter was further treated using a conventional activated sludge system, the total removal of chlorobenzene and toluene exceeded 90 percent. Extraction of organics from sludge samples showed that adsorption was about 5 percent or less for all compounds. These data indicated the effectiveness of combined anaerobic-aerobic treatment using a pilot-scale anaerobic filter followed by a conventional activated sludge system (Bhattacharya *et al.* 1989). In this ongoing study, such combined systems are further evaluated for specific groups of compounds such as nitrophenols which are expected to be present in hazardous leachates. This technical note includes some preliminary results from this study.

Methodology

Based on the information available from literature, a synthetic leachate was prepared using acetic acid as the primary carbon-source (2 g/L) and spiking selected volatile and semivolatile compounds (chlorinated hydrocarbons, phenols, etc.) and heavy metals to the feed.

Preliminary serum bottle studies were performed for anaerobic toxicity assay (ATA). Each serum bottle (150 mL) had a culture volume of 50 mL. Three process trains are used for this combined biotreatment study (Figure 1). In the first process, acidogenic bacteria are expected to biodegrade some of the toxic organics. Since sulfate could be present in significant amounts in some leachates, studying the role of sulfate reducing bacteria is very important. Process trains 2 and 3 will be used for this purpose. Acetic acid and other volatile acids were measured using a GC. The concentrations of the toxicants were measured using a GC or an HPLC.

Results

The preliminary results obtained from serum bottle studies and process train 2 are presented in this technical note. Figure 2 shows the results from a typical serum bottle study; in this case, the toxicant was 4-nitrophenol. Triplicate serum bottles for controls and each test concentration were maintained. Methane production was adversely affected when 4-nitrophenol was spiked. However, with 20 and 30 ppm of 4-nitrophenol, the methane production went back to normal after 1 to 2 days (Figure 1). With 40 ppm, however, it took 4 to 5 days to recover. Similar studies were performed with 2-nitrophenol,

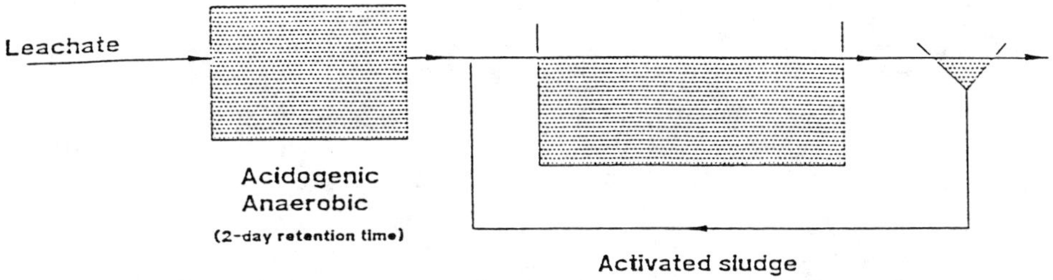

Process Train 1: Acidogenic (Anaerobic) System Followed by Activated Sludge

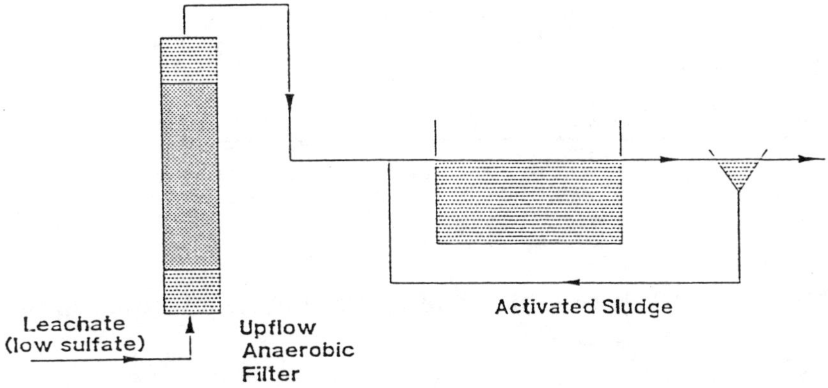

Process Train 2: Upflow Anaerobic Filter Followed by Activated Sludge

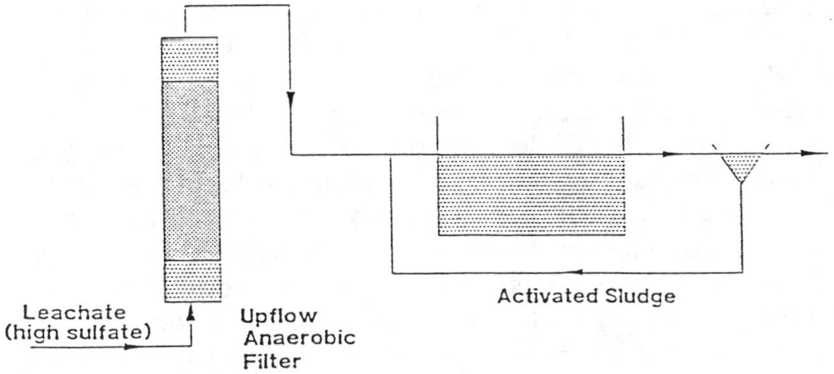

Process Train 3: Upflow Anaerobic Filter (with sulfate-reducing bacteria) Followed by Activated Sludge

FIGURE 1. Process trains for combined anaerobic-aerobic treatment.

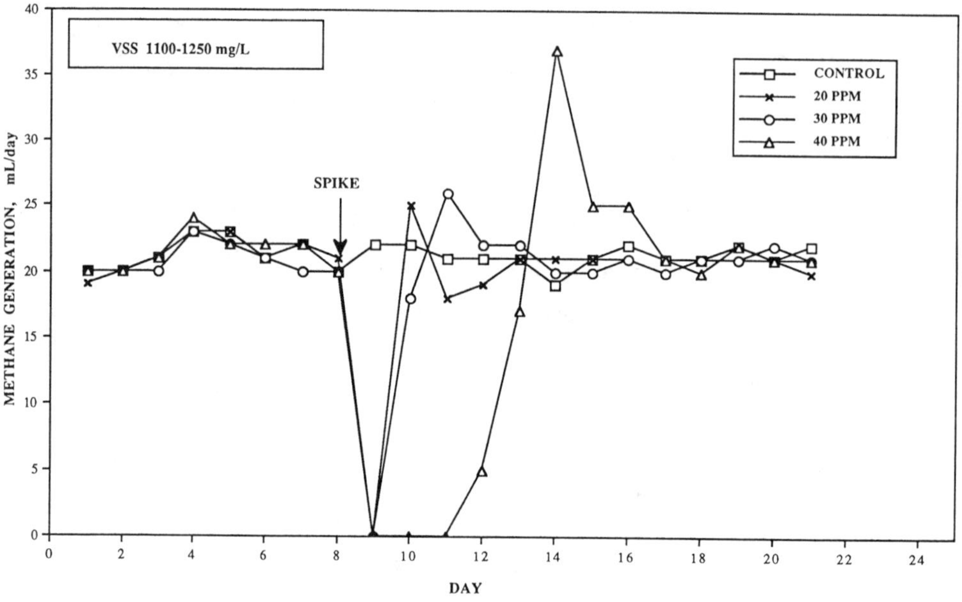

FIGURE 2. Effects of 4-nitrophenol on methane formation in serum bottles.

2-4 dinitrophenol, 2,4,6-trichlorophenol, chlorobenzene, ethyl benzene, dibutyl phthalate, nitrobenzene, and others.

The information developed from serum bottles was used to select the toxicant concentrations for the column studies (Process trains 2 and 3 in Figure 1). The columns (12-L volume and 2-day hydraulic retention time) were seeded with methanogenic enrichment culture, and the toxicants were spiked after the systems reached steady state. Figure 3 shows the results from one such study with 20 mg/L of 4-nitrophenol. The graph shows the influent and effluent volatile acids (mainly acetic acid) concentration. With toxic effects of 4-nitrophenol, volatile acids accumulated in the system. That means, the methanogens were initially unable to use acetic acid as a carbon source in presence of 4-nitrophenol. The toxicant was withdrawn after 8 days, the acetic acid effluent concentration started going down indicating recovery. On day 17, with a respike of 4-nitrophenol, acetic acid accumulated until day 27. At that point, 4-nitrophenol was withdrawn again and the system recovered. However, little evidence of acclimation was observed.

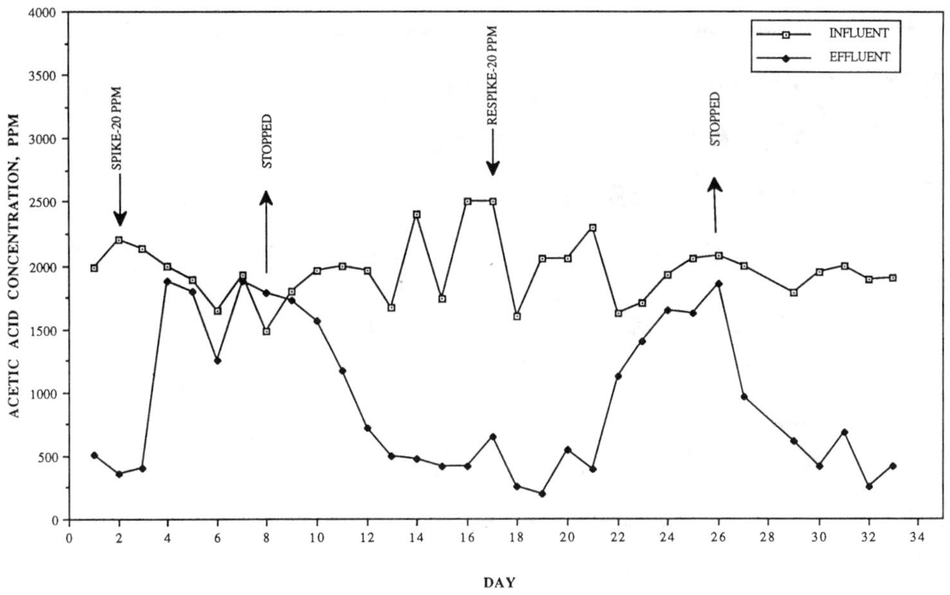

FIGURE 3. Effects of 4-nitrophenol on acetic acid utilization in anaerobic column.

The column study is continuing with individual compounds. Eventually all the compounds (i.e., the synthetic leachate) will be added simultaneously as feed to these columns. The effluents will be further treated in a conventional activated sludge system. The first phase of the study will be completed in February 1991 and the whole project will be completed in June 1992.

REFERENCES

Bhattacharya, S. K.; Angara, R.V.R. Presented at the 15th Annual EPA Research Symposium, Cincinnati OH, 1989.

Bhattacharya, S. K.; Dobbs, R. A.; Angara, R.V.R. Presented at the 16th Annual EPA Research Symposium, Cincinnati OH, 1990.

Bouwer, E. J.; McCarty, P. L. *Environ. Microbiol.* **1983**, *45*, 1286.

Schink, B. *FEMS Microbiol. Ecol.* **1985**, *31*, 69.

Vogel, T. M.; Grbić-Galić, D. *Appl. Environ. Microbiol.* **1986**, *52*, 200.

Slurry-Based Biotreatment of Contaminants Sorbed onto Soil Constituents

William V. Black, Robert C. Ahlert, David S. Kosson*
Rutgers University
with
John E. Brugger
U.S. Environmental Protection Agency

INTRODUCTION

A biotreatment process using slurried soil fines was developed (Ahlert *et al.* 1990). The slurry-based system aided mass transfer by facilitating greater contact between contaminated soil and the microorganisms. Obtaining the slurry involved a pretreatment process in which the whole soil is fractionated, and the most contaminated fraction was obtained as a suspension of soil fines for subsequent biodegradation experiments. Separation of soil fines reduced the mass of material requiring further treatment. Biological treatment involved the use of a mixed aerobic culture in a slurry-based system operated under aerobic conditions in a batchwise manner. Reducing volatile losses of low molecular weight compounds in an aerobic system and quantifying these losses are two of the greatest challenges associated with this system. The goal of these experiments was to improve the extent of biodegradation, minimize losses from volatility, and demonstrate the feasibility of the physical pretreatment process as a way of reducing the amount of soil requiring further treatment.

A number of options are available for treating soils contaminated with aromatic compounds including chemical extraction, incineration, and biotreatment. Of these, biotreatment is often the least expensive. The potential for reduced costs and permanent remediation of soils contaminated with these materials underscores the need to

* Chemical and Biochemical Engineering, Rutgers University, P.O. Box 909, Piscataway NJ 08855

evaluate this method of soil treatment. Many studies have been carried out to investigate the biodegradability and transformation pathways of individual polycyclic aromatic hydrocarbons (PAH). A number of researchers have shown compounds typically found in distillate bottoms (aromatics, polyaryls, and PAHs) to be biodegradable under aerobic conditions (Bauer & Capone 1988; Shiaris 1989; Shocken & Gibson 1984). Complete mineralization of four-ringed PAHs, and transformation of compounds having greater than five rings, has been demonstrated (Heitkamp 1988; Heitkamp & Cerniglia 1989). Moreover, during the past decade, a number of bioreactor configurations have been used successfully to treat soils contaminated with PAH and polyaryl wastes from a variety of spill, disposal, and production sites. Mass transfer of these compounds from soil to microorganisms is usually the most significant barrier to biodegradation processes.

Polyaryl compounds and PAHs are found in atmospheric particulates, fresh and brackish waters, and soils and sediments. In addition to BTX production, other sources of aromatic hydrocarbons include combustion residues, petroleum products and refinery wastes, coal gasification and liquefaction effluents, and wastes from creosote production and wood treatment facilities. BTX and 16 PAHs are considered priority pollutants by the EPA. Mammalian enzymes are thought to metabolize many of these compounds into carcinogenic substances including epoxides and *trans*-dihydrodiols (Cerniglia 1984). Their presence in soil and potential mobility as fine particulates and/or into groundwater are matters of great concern. Biotransformation of polyaryls and PAHs can dramatically increase the solubility of these compounds and thereby affect partitioning. For this reason, achieving mineralization or at least detoxification of soils contaminated with polyaryls and PAHs is necessary to reduce the potential hazards posed by these compounds. The EPA has determined that the levels of distillate bottoms contaminants at this site must be significantly reduced before it can be properly closed.

Soil surrounding a chemical manufacturing facility was contaminated in a number of locations with organic wastes. The location examined in this study was contaminated with distillate bottom residues from a benzene, toluene, and xylenes (BTX) production process, which consisted of cracking naptha in the presence of sulphuric acid. Distillate bottoms and sludges from the BTX production process were impounded in two lagoons for several decades. The impoundments exceed 40,000 m^2 in extent and contains an estimated 75,000 m^3 of residues. The contents of the lagoons have separated into several distinct layers that include, in bottom-to-top sequence, a solid organic layer, a layer of viscous organic matter, an oil-like layer, and a floating aqueous layer. Possible contaminants include, but are not limited to, single-ring aromatic species, bi- and polyaryl compounds, PAH, as well as sulfonated derivatives of these compounds.

MATERIALS AND METHODS

Reactor Configuration. Fermentations were carried out in the apparatus shown in Figure 1. The apparatus was operated in a batch mode (put and draw operation). Air was sparged into the bottom of the 2.3-L reactor vessel (working volume, 2.0 L) with the effluent air passing through a chilled (4 C) condensor trap. The trap was equipped with a siphon tube that continuously returned condensate to the bioreactor. Initially, the reactors were charged with 50 percent soil slurry (1.0 L), 33 percent mixed culture inoculum (0.67 L), and 17 percent mineral salts medium (0.33 L). At each feeding, the reactors were drained to a volume of 1.4 L and fed 0.6 L of fresh soil slurry.

Inoculum. The mixed culture used in all biological experiments was aerobic activated sludge obtained from a secondary treatment basin at the Somerset-Raritan Valley Sewage Authority in Bound Brook NJ.

Mineral Salts Medium. The mineral salts medium contained the following compounds: ammonium sulfate, 3,000 mg/L; magnesium sulfate, 600 mg/L; ferric chloride, 3 mg/L; manganese sulfate, 60 mg/L; yeast extract, 300 mg/L; potassium phosphate (dibasic), 1,000 mg/L; and potassium phosphate (monobasic) to pH 7.

Chemical Analyses

COD. Overall contaminant levels were evaluated using chemical oxygen demand (COD) assays (Black 1990). This procedure involved oxidation with potassium dichromate in the presence of boiling sulfuric acid. After refluxing for 2 h, the solution was cooled and titrated with ferrous ammonium sulfate.

Toluene Extraction. The extraction/pH neutralization step of soil pretreatment resulted in the coating of Teflon® beads with a tarlike substance. (See "Soil Characterization and Fractionation.") Ten of these tar-coated Teflon beads were mixed with 10.0 mL of toluene and the solution passed through a 0.2-µm filter and was analyzed by GC/MS.

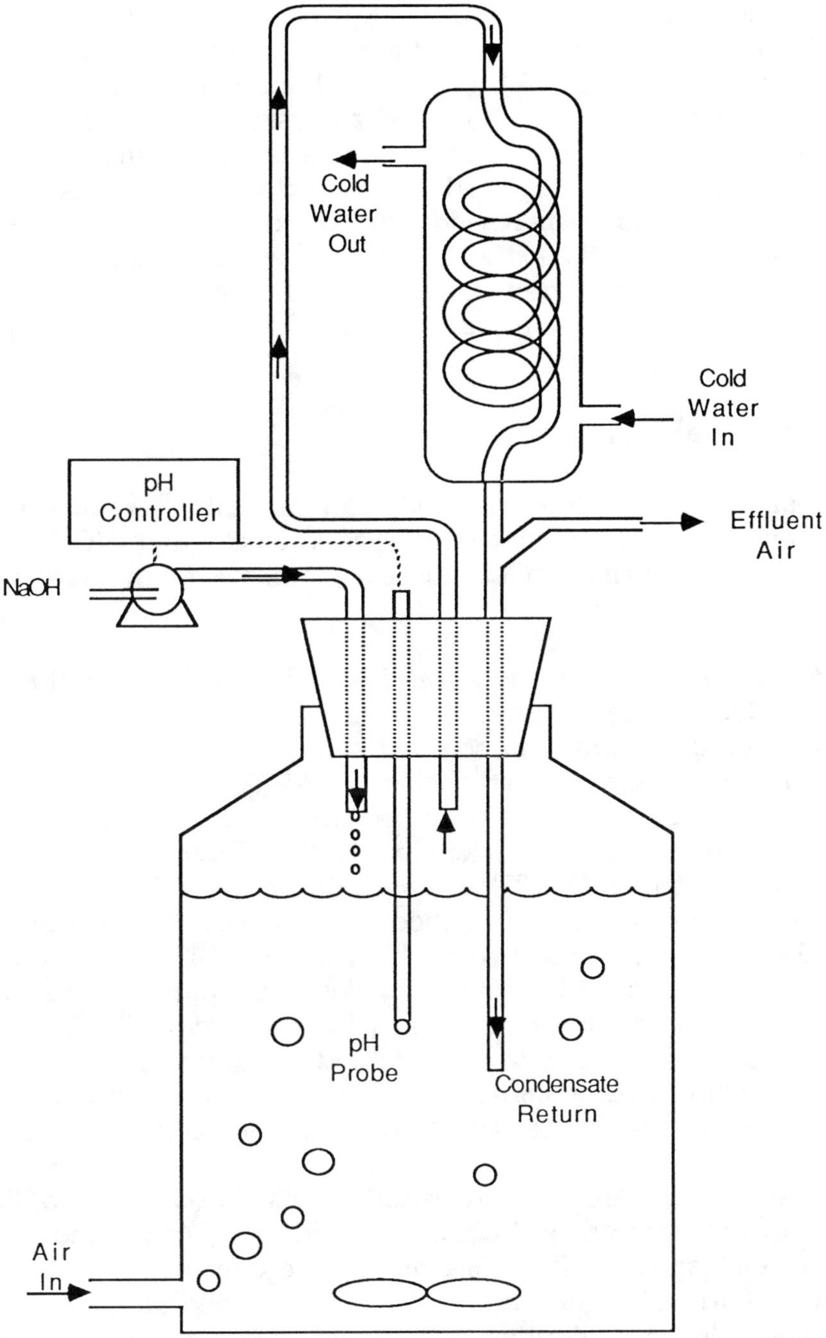

FIGURE 1. Bioreactor apparatus.

Cyclohexane Extraction. The slurry of soil fines contaminated with distillate bottoms was extracted by shaking 15.0 mL of slurry with 3.0 mL of cyclohexane for at least 24 h, filtering through a 0.2-µm filter, and analyzing the extract by GC. For HPLC analysis, 25.0 mL of slurry was extracted with 5.0 mL of cyclohexane for at least 24 h. After the extraction mixture was then centrifuged and the cyclohexane layer separated and filtered; the solids were resuspended, dried at 105 C, and weighed. The mass of solids in the extraction mixture was then used to correct high-performance liquid chromatography (HPLC) measurements of contaminant concentrations to a soil mass basis.

Instrumental Analyses

Soil contaminants were identified and quantified using several forms of instrumental analysis including HPLC, gas chromatography (GC), aqueous total organic carbon (TOC), and gas chromatography/mass spectrometry (GC/MS).

Gas Chromatography. U.S. EPA Test Methods 602 (for aromatics) and 610 (for polycyclic aromatic hydrocarbons) were suitable for GC analysis of cyclohexane extracts of soil and soil slurries. The GC used in these analyses was a Hewlett-Packard HP-5880 with an HP-5880A Series GC Terminal integrator equipped with either a flame ionization detector (FID) or a photo-ionization detector (PID) and electrolytic conductivity detector (ELCD) arranged in series. The PID/ELCD combination was used in Test Method 602. The FID was used for Test Method 610. The column used for Test Method 602 (aromatic compounds) was a 1.83-m, 2.16-mm i.d. stainless steel column packed with Supelcoport 100/120 mesh coated with 5 percent SP-1200 and 1.75 percent Bentone-34. EPA Test Method 610 (polynuclear aromatic hydrocarbons) utilizes a 1.8-m, 2-mm i.d. glass column packed with Chromosorb W-AW-DCMS (100/120 mesh) coated with 3 percent OV-17.

Polynuclear aromatic hydrocarbon standards (Supelco PAH Kit 610-S, catalogue number 4-8755) contained 200 µg/mL of individual PAHs in methanol. A PAH locator mixture (Supelco PAH Locator Mixture 610-M, catalogue number 4-8743) contained 16 PAHs in 1:1 methanol:methylene chloride. PAHs included acenaphthene, acenaphthylene, anthracene, benzo(a)anthracene, benzo(a)pyrene, benzo(b)-fluoranthene, benzo(ghi)perylene, benzo(k)fluoranthene, chrysene, dibenzo(a,h)anthracene, fluoranthene, fluorene, indeno(1,2,3-cd)pyrene, naphthalene, phenanthrene, and pyrene.

High-Performance Liquid Chromatography. Cyclohexane extracts were passed through a 0.2-µm filter before HPLC analysis. The program used for most HPLC analyses is given in Table 1. Step 2 involved a linear composition gradient that started with the composition of phase 1 (5.5:4.5 water:acetonitrile) and finished with pure acetonitrile. Injection volume was 10 µL and the column was operated at 35 C. The liquid chromatograph was a Perkin-Elmer Series 4 Liquid Chromatograph equipped with a Perkin-Elmer LC-95 UV/Visible Spectrophotometer Detector, a Perkin-Elmer Lab Computer Integrator Model LCI-100, and a Perkin-Elmer 7500 Professional Computer. The HPLC column was a Supelco LC-PAH 25 cm long, 6.35 mm o.d., 4.6 mm i.d., with 5-µm particle support, octadecylsilyl phase (Supelco 5-8229). The PAH standards were the same as those used for GC analysis.

TABLE 1. HPLC solvent program.

Step	Time (min)	Flow (mL/min)	Water (%)	Acetonitrile (%)
Equilibration	5.0	1.5	55	45
1	1.0	1.5	55	45
2	18.0	1.5	0	100
3	6.0	1.5	0	100

Total Organic Carbon Analysis. Each (aqueous) sample was passed through a 0.45-µm filter and injected into an OI Instruments Model 700 Total Organic Carbon Analyzer. Injection volume was between 0.1 and 1.0 mL. The injection size was chosen to keep the detector response in a linear output range (100 and 1,000 mV).

Gas Chromatography/Mass Spectrometry. This analysis was used to identify some of the compounds present in the toluene extract. The oven temperature program is given in Table 2. Detector and injector temperatures were 280 C and 275 C, respectively. Injection volumes of 5 µL were introduced manually. The GC/MS utilized a Hewlett-Packard HP-5890 Gas Chromatograph equipped with HP-5970 Series

Mass Selective Detector. The gas chromatograph has automatic database search capability and operates in selected ion (SIM) or scan modes. A 30-m capillary column was used for all analyses. Activated carbon traps used in effluent gas sampling are desorbed and analyzed by GC/MS. The tubes are 20.3 cm long and packed with 1.00 g of Carbotrap 20/40 mesh activated carbon. The temperature program is summarized in Table 3.

TABLE 2. GC/MS temperature program.

Step	Time (min)	Temperature (C)	Ramp (C/min)
1	3	80	–
2	22	80–300	10
3	15	300	–

TABLE 3. GC/MS temperature program for effluent gas analysis.

Step	Time (min)	Temperature (C)	Ramp (C/min)
1	7	50	–
2	23	80–225	10
3	15	225	–

Biological Methods. Two variables were manipulated in an effort to improve the biodegradation process, feeding frequency and air flow rate. Two feeding frequencies have been examined, 1 week and 2 weeks. Air flow was set at two different rates, 1.5 and 0.125 L/min. Air flow was reduced in an attempt to minimize the effects of air stripping while not depriving the system of the oxygen needed for biological processes.

Volatilization Controls. Two studies were carried out to determine the effects of air stripping on contaminant reductions. The volatilization studies were carried out under the same conditions as the active bioreactors, except that 2 percent mercuric chloride was added and the activated sludge inoculum was replaced with deionized water. These studies used reactors that were equipped with activated carbon traps and were operated in parallel with the active bioreactors. At the higher air flow rate (1.5 L/min), sampling time was 5 min. At the reduced air flow rate (0.125 L/min), sampling time was 15 min.

SOIL CHARACTERIZATION AND FRACTIONATION

Whole Soil Properties

Many of the compounds found in the lagoon containing acidified distillate bottoms have a slight solubility in water and/or an affinity for some of the constituents of soil and have migrated into and through the soil immediately adjacent to and underneath the lagoon. In January 1988, soil samples (15 L each) were taken from three locations at the north, south, and east edges of the lagoon. An uncontaminated soil sample was taken from adjacent land. Contaminated soil used in these biological studies came from the south end of the lagoon. Physical properties of this sample, such as color, consistency, and odor, fell somewhere between those of the north and east lagoon soil samples. This sample contained a great deal of excess water; approximately 1 L of water was decanted from the surface of the south sample before it was homogenized.

The contaminated and uncontaminated soil samples were homogenized by coning and quartering. Soil samples were forced through a 5-mm mesh sieve to remove rocks, sticks, roots, and other large objects, and the resulting cone of sieved soil was quartered through the apex. The quarters were passed through the sieve twice more, and the soil was then stored in 0.94-L mason jars and refrigerated to minimize volatilization losses. Homogenization of the soil by this method not only assured a uniform distribution of soil particles and moisture, but also served to eliminate densely packed clumps. After homogenization, pH, moisture, and COD were measured (Table 4).

GC analysis of soil extracts (EPA Test Method 602) identified several volatile and semivolatile compounds present in whole soil (Table 5). Cyclohexane extractions followed by HPLC analysis indicated the presence of numerous high molecular weight compounds. Identification of some of these compounds was done by matching retention times with standards. Only BTX and naphthalene were identified by this method. Subsequent GC/MS analysis identified the two largest peaks in the solvent extract as 1,1-*bis*(*p*-tolyl) ethane and 1,2-dimethyl-4-(phenylmethyl) benzene. These names were then

TABLE 4. Whole soil properties—contaminated and uncontaminated soils.

pH	3.6–3.8
Moisture	16%
COD (contaminated soil)	89.6 g O_2/kg
COD (clean background soil)	19.0 g O_2/kg

TABLE 5. Volatile and semivolatile contaminants in soil.

Contaminant	Conc. (mg/kg soil)
Benzene	2,200
Toluene	3,000
p-Xylene	600
o-Xylene	800
m-Xylene	1,400
Ethyl Benzene	600
1,2-Dichlorobenzene	6,400
Naphthalene	21,800

assigned to the two largest peaks in the HPLC chromatogram. The structures of these and other contaminants are shown in Figure 2.

Soil Pretreatment and Fractionation

Homogenized contaminated soil (100 g) was extracted and neutralized with 350 mL of a sodium hydroxide (0.05 M) and sodium chloride (0.05 M) solution. The composition of this solution was determined to be that best suited to neutralize soil acidity and disperse clay particles, reducing the pore space available for trapping contaminants.

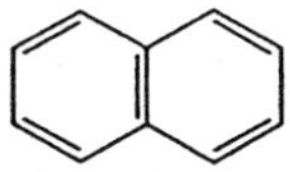

Naphthalene
($C_{10}H_8$, MW=128)

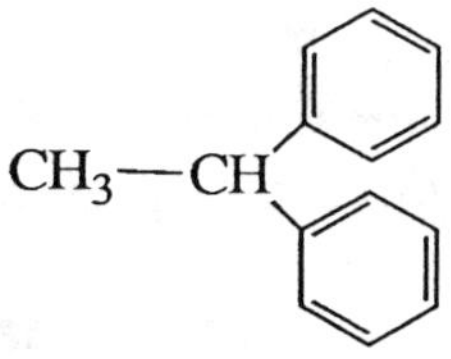

Benzene, 1,1'-ethylidenebis
($C_{14}H_{14}$, MW=182)

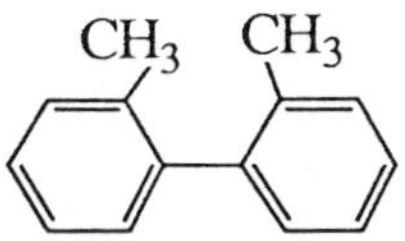

1,1'-Biphenyl, 2,2'-Dimethyl-
($C_{14}H_{14}$, MW=182)

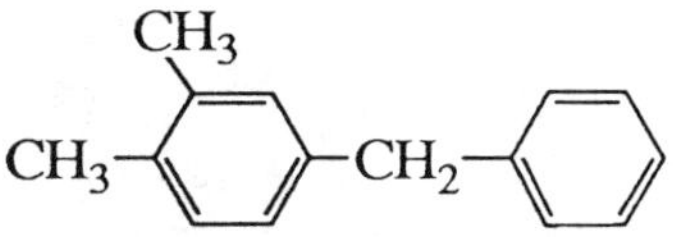

Benzene, 1,2-dimethyl-
4-(phenylmethyl)-
($C_{15}H_{16}$, MW=196)

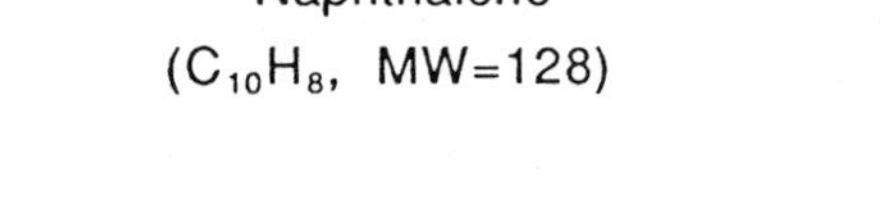

1,1-Di-(p-toyl)ethane
($C_{16}H_{18}$, MW=210)

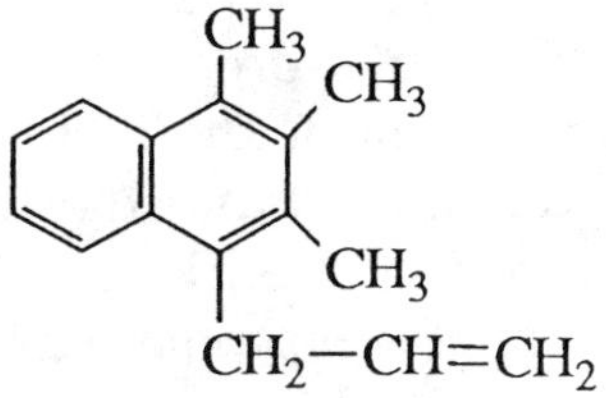

Naphthalene, 1,2,3-trimethyl-
4-propenyl
($C_{16}H_{18}$, MW=210)

FIGURE 2. Names and structures of most abundant distillate bottoms contaminants.

Sodium chloride/sodium hydroxide extraction of soil contaminated with still bottoms created three distinct fractions: coarse sand, an aqueous suspension of soil fines, and an organic (tar-like) fraction. COD analysis was performed on each of these three fractions. This method of analysis was chosen for its ability to evaluate overall levels of contamination associated with the soil fractions without the use of complex extraction procedures. After separation, the coarse sand (30 percent by weight of the whole soil) contained less than 3 percent of the whole soil COD, greatly reducing the mass requiring further treatment (Table 6). The organic fraction was tar-like in consistency, appearance, and odor. This material was dissolved in an organic solvent (toluene or cyclohexane) and analyzed by GC/MS.

TABLE 6. Contaminant distribution (COD) within soil fractions.

Soil Fraction	Mass Fraction	COD (g O_2/kg of fraction) (Dry Basis)	COD (g O_2/kg of whole soil) (Dry Basis)	% COD (of whole soil)
Whole Soil	–		89.6[(a)]	100
Water	0.16 ± 0.02	0	0	0
Large Particles	0.29 ± 0.02	7.9	2.7	2.9
Soil Fines	0.55 ± 0.02	127	83.1	88.3
Tar-Like	0.003 ± 0.0005	2330	8.3	8.8
Slurry (Volumetric Basis)	–	7.9 g O_2/L	–	–
Uncontaminated Soil	–		19.0	100

(a) Determined independently using whole soil; it is within 5 percent of total from individual fractions.

Up to 8.8 percent of whole soil COD was removed by adhesion of the tar-like organic fraction to nonpolar surfaces. These nonpolar surfaces include the Teflon lid of the glass bottle in which the extraction was carried out and the 6.35-mm-diameter Teflon beads. These beads were added specifically to remove the organic material by increasing the available nonpolar surface area. The tar coating on the Teflon beads was usually lumpy in nature, except when the bottle was rotated to ensure a uniform coating of tar on the bottle as well as the beads. If too many beads are added, the tar would not remain on the beads, but formed small (>1-mm-diameter) spheres. These spheres had a tendency to float in the frothy surface layer or (if sufficiently large) to settle and mix with the coarse sand.

RESULTS

Two semibatch bioreactors were started and fed a 3:2:1 mixture of a slurry, deionized water, and mineral salts medium at regular intervals. Feeding involved sampling the reactor contents immediately before feeding (after the 1-week biodegradation period) and immediately after feeding. In the case of the bioreactor that was fed every 2 weeks, samples were taken immediately after feeding, after 1 week of biodegradation (at the midpoint between feedings), and after the 2-week biodegradation period (immediately before the reactors next feeding).

Reactors used in volatilization studies were not fed, but were sampled at regular, shorter intervals. Samples taken from the bioreactors were analyzed for COD and extracted with cyclohexane for analysis with HPLC; the aqueous phase was assayed for dissolved TOC.

Each sample analyzed by HPLC generated a chromatogram containing approximately 20 peaks, with each peak integrated and quantified as a dimensionless area. The peak areas were corrected for soil mass to account for variations in slurry density. Since standards are unavailable for the types of polyaryl compounds examined in these studies, compound reductions were evaluated on a relative basis, i.e., as percent reductions of peak areas measured at the time the reactors were fed. Previous experiments (Black 1990) using serial dilutions of the cyclohexane extracts found that dilutions of at least 90 percent (in most cases, at least 99 percent) were necessary to bring compound concentrations below HPLC detection limits. Extrapolation based on the concentrations of known aromatic compounds indicates that HPLC detection limits for these polyaryl compounds are less than 1 mg/L of extract, implying polyaryl soil concentrations of less than 10 mg/kg.

Compound reductions in the active bioreactors and mercuric chloride poisoned reactor (volatilization study) at an air flow rate of 1.5 L/min are shown in Table 7. Reductions of 100 percent indicate compound removals to levels below HPLC detection limits. Significant reductions of low molecular weight compounds, such as benzene, xylenes, ethyl benzene, naphthalene, etc., can be attributed to volatilization in reactors operated at the high air flow rate. All but two of the medium and high molecular weight compounds showed reductions of less than 30 percent in the volatilization study. The larger reductions for compounds with retention times of 11.80, 13.20 (dimethyl-4-phenylmethyl benzene), 13.76, 15.10 (1,1-*bis*(*p*-tolyl) ethane), 15.79, 16.58, and 19.33 min in the biodegration experiments indicates an active microbial population is utilizing the distillate bottoms as a primary source of carbon and energy.

GC/MS analysis of the off-gas from the first volatilization control study, at an air flow rate of 1.5 L/min, did not detect the presence of any compounds with a molecular weight greater than naphthalene. Compounds identified included BTX, ethyl benzene, ethylmethyl benzenes, and dichlorobenzenes. Since 9 of the 12 compounds analyzed by HPLC are heavier than naphthalene, it appears that losses of these compounds resulted either from slow volatilization or from biodegradation processes unhindered by the presence of mercury.

A volatilization study carried out in a reactor operated in parallel with the active bioreactors suggested a lower air flow rate could

TABLE 7. Compound reductions (%) at high air flow rate (1.5 L/min).

Compound Retention Time (min)	Compound Identity (Tentative)	Volatilization Study	Reactor 1 Feeding Interval 1 Week	Reactor 2 Feeding Interval 1 Week	Reactor 2 Feeding Interval 2 Weeks
3.49	Benzene	100.0	100.0	100.0	100.0
6.46	Toluene	6.8	58.9	100.0	100.0
8.23	Xylenes[a]	100.0	100.0	100.0	100.0
10.43		70.8	55.6	61.8	79.5
11.80		19.7	100.0	100.0	100.0
13.20	(b)	22.0	100.0	100.0	100.0
13.76		65.3	100.0	97.8	100.0
14.33		51.7	49.5	71.5	87.0
15.10	(c)	30.0	77.5	100.0	100.0
15.79		n.v.l.	52.4	100.0	100.0
16.58		n.v.l.	24.6	67.2	82.5
19.33		n.v.l.	11.2	32.9	50.6

n.v.l. = No volatile losses of this compound were detected.
(a) Compounds eluting at this retention time include xylenes, ethyl benzene, and naphthalene.
(b) Compound eluting at this retention time has been tentatively identified as 1,2-dimethyl-4-phenylmethyl benzene.
(c) Compound eluting at this retention time has been tentatively identified as 1,1-*bis*(*p*-tolyl) ethane.

result in less volatilization; air flow was reduced by 92 percent, from 1.5 L/min to 0.125 L/min. Losses of low molecular weight compounds were reduced (see Table 8). Many volatile species such as benzene and toluene were still detectable after 2 weeks. At the lower air flow rate (0.125 L/min), losses attributed to volatilization still accounted for almost all of the reductions in toluene and xylene concentrations. On the other hand, biodegradation processes appeared to account for at least half the removal of benzene, ethyl benzene, and ethyl benzene from the bioreactor. For medium and high molecular weight compounds with HPLC retention times of 13.20 (dimethyl-4-phenylmethylbenzene), 13.76, and 15.10 (1,1-*bis*(*p*-tolyl) ethane), reductions in the active bioreactors were three to five times greater than those in the poisoned bioreactors. Reductions of other compounds (retention times of 11.80, 15.79, 16.58, and 19.33 min) can be attributed entirely to biodegradation processes.

Analysis by COD did not detect losses caused by volatility at either flow rate. COD analysis of all of the active bioreactors showed contaminant reductions ranging from 0.5 to 1.5 g O_2/L. Acid

TABLE 8. Compound reductions (%) at low air flow rate (0.125 L/min).

Compound Retention Time (min)	Compound Identity (Tentative)	Volatilization Study		Biodegradation Study
		1 Week	2 Weeks	
3.49	Benzene	36.8	46.4	100.0
6.46	Toluene	35.8	36.3	16.5
8.23	Xylenes[(a)]	93.6	95.8	99.0
10.43		34.6	47.6	67.8
11.80		n.v.l.	n.v.l.	37.8
13.20	(b)	18.0	28.6	79.6
13.76		8.1	17.1	100.0
14.33		17.2	23.4	31.0
15.10	(c)	10.7	14.2	56.2
15.79		n.v.l.	n.v.l.	16.5
16.58		n.v.l.	n.v.l.	26.3
19.33		n.v.l.	n.v.l.	18.4

n.v.l. = No volatile losses of this compound were detected.
(a) Compounds eluting at this retention time include xylenes, ethyl benzene, and naphthalene.
(b) Compound eluting at this retention time has been tentatively identified as 1,2-dimethyl-4-phenylmethyl benzene.
(c) Compound eluting at this retention time has been tentatively identified as 1,1-*bis*(*p*-tolyl) ethane.

production in the reactors was sufficiently high to reduce the pH from 7.2 to 6.8 during the course of 1 week, despite the use of a phosphate buffer. The aqueous phase of the mercury poisoned reactor was nearly clear, whereas that of the active bioreactors was a brownish yellow, indicating the presence of high concentrations of dissolved organic compounds.

CONCLUSIONS

Soil pretreatment is an effective way to reduce the quantity of material requiring further treatment. It also prepares the soil for subsequent biotreatment by adjusting pH and increasing the surface area for mass transfer. These experiments have demonstrated that an acclimated microbial population capable of utilizing still bottoms (i.e., aromatic and polyaryl species) as its sole source of energy can be established. The population has been maintained for a period of 11 months, indicating high potential for long-term applications. Reducing the air flow rate reduced volatility losses by at least 50 percent while not significantly compromising biodegradation processes. This implies that additional adjustments may further reduce volatility losses while maintaining a viable microbial population.

REFERENCES

Ahlert, R. C.; Black, W.; Bruger, J.; Kosson, D.; Süenter, J. In *Remedial Action, Treatment, and Disposal of Hazardous Waste*; Proceedings of the Sixteenth Annual RREL Hazardous Waste Research Symposium; U.S. Environmental Protection Agency: Washington, DC, 1990; pp 536–547.

Bauer, J. E.; Capone, D. G. *Applied and Environmental Microbiology* **1988**, *54*, 1649–1655.

Black, W. V. Master's Thesis, Rutgers, The State University of New Jersey, Jan. 1990.

Cerniglia, C. E. "Microbial Transformation of Aromatic Hydrocarbons." In *Petroleum Microbiology;* Atlas, A. M., Ed.; Macmillan: New York, 1984.

Heitkamp, M. A. *Applied and Environmental Microbiology* **1988**, *54*, 1612–1614.

Heitkamp, M. A.; Cerniglia, C. E. *Applied and Environmental Microbiology* **1989**, *55*, 1968–1973.

Shiaris, M. P. *Applied and Environmental Microbiology* **1989**, *55*, 1391–1399.

Shocken, M. J.; Gibson, D. T. *Applied and Environmental Microbiology* **1984**, *48*, 10–16.

Aerobic Biodegradation of Low Levels of Dissolved Organic Compounds in Groundwater: Research and Field Data

Frank Lenzo[*], *David G. Ward, Jr.*
Hydro Group, Inc.

The use of microbes to oxidize organic compounds dissolved in wastewater is a proven and well accepted unit process in industrial wastewater treatment. Until the early 1980s, however, biodegradation of dissolved organic compounds in groundwater was seldom applied. This paper will present laboratory-scale testing conducted in 1985 for

[*] Hydro Group, Inc., 97 Chimney Rock Road, Bridgewater NJ 08807

an aerobic biodegradation process specifically targeted at low levels of biodegradable compounds dissolved in groundwater. The process has been used to treat waters containing petroleum hydrocarbons, ring compounds (such as benzene, toluene, and xylenes), and ketones. The treatment system relies on acclimated microbes surviving in the decay phase of the typical growth curve. A subsequent field demonstration project begun in January of 1988 will also be discussed.

The ultimate purpose of the laboratory and field studies was to develop a transportable, low-cost treatment system capable of treating low levels (less than 40 mg/L) of organic compounds commonly found in groundwater supplies. The primary objective of the laboratory-scale study was to determine, in a controlled environment, whether sustained biodegradation was possible at low concentrations of a single organic compound. The field-scale study was aimed at determining how well the process translated from the laboratory to the real world.

The treatment system tested is a submerged fixed film aerobic biodegradation process. The system operates in a countercurrent flow scheme (water downflow, air upflow) and utilizes a structured media to provide the necessary surface area for biofilm development. The operating system is engineered to allow sustained biodegradation of dissolved organic compounds in the decay phase of the standard microbial growth curve.

Prior to treatment of trace levels of the organic compound(s) of concern, a suitable biomass must be developed. Biomass development in the packed reactor can take as long as 8 weeks. Once the biomass has been developed treatment can begin.

Bench-Scale Study

The bench-scale bioreactor used was a 10-cm-diameter by 1.8-m-high clear acrylic column containing 1.5 m depth of PVC structured media (Johnson 1985). The media utilized had a specific surface area of 225 m^2/m^3. The column operated in the downflow mode while air flowed countercurrently up from a distribution system at the bottom of the column.

Startup of the system incorporated an initial charge of acclimated biomass from an activated sludge process, to seed biofilm growth on the packing surface. A mixture of suspended biomass along with a feed of 50 to 100 mg/L of benzene and nutrients was recirculated through the column for 10 weeks. During this time, biomass development was enhanced using an immersion water heater within the closed loop system to raise the average water temperature from 13 to 24 C. At the same time operational and process details associated with the pilot system were resolved.

During flow-through operation a variable rate feed pump was used to pump water to the bioreactor. The organic compound being tested was fed, along with nutrients, into the water influent line prior to entering the bioreactor. The nutrient feed rate selected was based on achieving a 100:5:1 carbon:nitrogen:phosphorous ratio.

Figure 1 summarizes the results of the bench-scale test conducted for benzene treatment. Benzene served as the sole substrate during the test, its concentration varied between 1.2 and 7.6 mg/L. Influent and effluent water samples were collected to measure the benzene concentrations and to monitor treatment system efficiency. The test was begun at a water flow rate of 9.4 L/h. The system was operated at this rate for 5 weeks; during this period of time various airflow rates were tested to determine the system's response. The airflow rate was varied between 0.2 and 1.0 L/min. Removal rates of greater than 95 percent were seen consistently, although some variations existed (mostly from mechanical problems with the treatment system). It was determined that an air-to-water ratio of less than 1:1 was sufficient to provide the necessary oxygen to the microbial population and to assure adequate mixing action through the media.

In subsequent weeks water flow rates of 13.1, 18.8, and 28.1 L/h were tested. Removal efficiencies of the column averaged between 97 and 98 percent during this period of time. The benzene feed concentration of 1.4 to 7.4 mg/L was treated to less than detectable levels (0.005 mg/L) more than half of the time. During the last 2 weeks of this test period, the nutrient feed was stopped while the water flow rate was maintained at 28.1 L/h. The system efficiency remained at or near 98 percent during this period.

A flow rate of 37.5 L/h was then operated for 2 weeks with an average removal efficiency of 92 percent. Nutrients were not added to the system during this period of time. At the end of this 2-week period, the effluent levels began to rise and biofilm breakdown was observed. Nutrients were again added to the benzene feed, and the flow was decreased to 9.4 L/h to prevent further biofilm loss.

The second organic compound tested was methylethylketone (MEK). Figure 2 is a summary of the test results with MEK as the sole carbon source. Prior to sampling to evaluate system performance, the biological population was fed MEK for approximately 2 weeks to allow the population to acclimate itself. The column was operated at an average water flow rate of 8.8 L/h during the first 8 weeks of

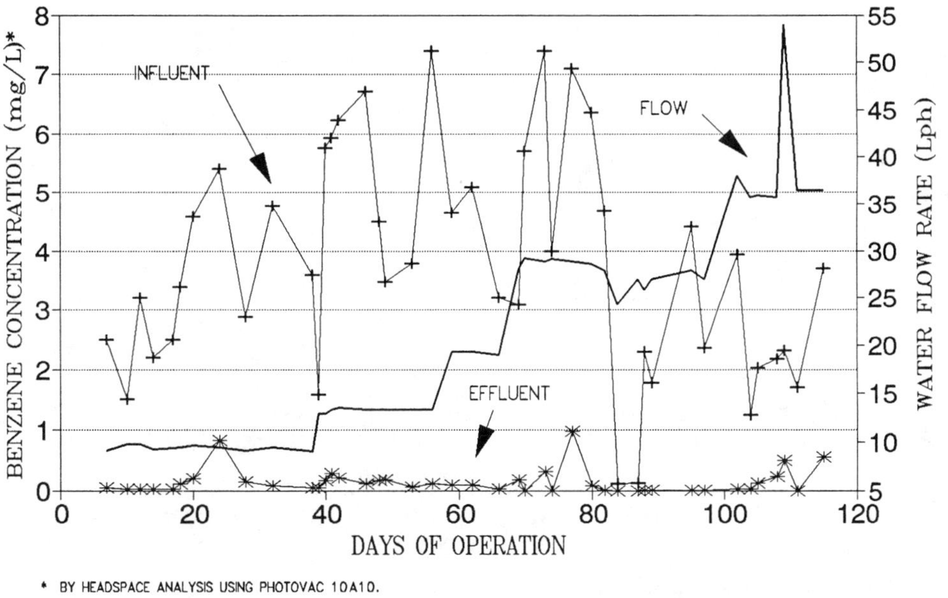

FIGURE 1. Bench-scale benzene treatment. Asterisk indicates headspace analysis using PHOTOVAC 10A10.

operation. The flow was increased to 16.9 L/h during the last 3 weeks of testing. During this test 1 to 10 mg/L of MEK (plus nutrients) was fed. The treatment system consistently exhibited MEK removal efficiencies in excess of 99 percent. The MEK test was run for approximately 11 weeks before the test was terminated.

Field Demonstration

To field test the treatment process a field-scale process demonstration unit (PDU) was developed. The PDU is 1.1 m in diameter and 1.8 m high with a 1.2-m depth of structured media identical to that used in the bench-scale study. The system was designed to handle water flow rates between 4 and 19 L/min and airflow rates of 5 to 25 L/min. The treatment system consisted of the PDU, a bag filter to collect any solids carried over from the PDU, and an activated carbon filter that served as a polishing step prior to discharge to the sanitary sewer.

The site chosen for the demonstration test was a contaminated well located in central New Jersey. A leaking underground fuel oil storage tank had caused limited contamination of the aquifer with

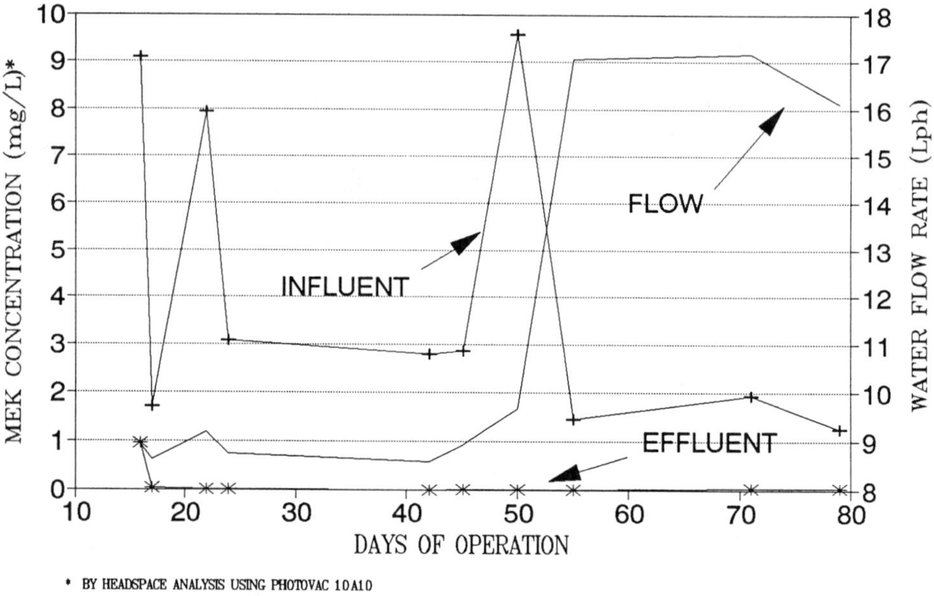

FIGURE 2. Bench-scale MEK treatment. Asterisk indicates a headspace analysis using PHOTOVAC 10A10.

benzene, toluene, xylenes (BTX), and petroleum hydrocarbons (PHC). During the field test, benzene was monitored as an indicator compound. Figure 3 summarizes the treatment efficiency of the bioreactor, as well as the influent concentration for benzene over the 13-month test period.

The PDU successfully treated the BTX and PHC present in the groundwater. Benzene removal efficiencies averaged greater than 94 percent over the test period. After a prolonged shutdown (see Figure 3) the system was easily restarted with high treatment efficiencies achieved within several weeks of startup. The water flow rate ranged between 4.7 and 19.9 L/min over the test period; the airflow rate ranged from 7 to 19 L/min. The nutrient feed was maintained throughout operation.

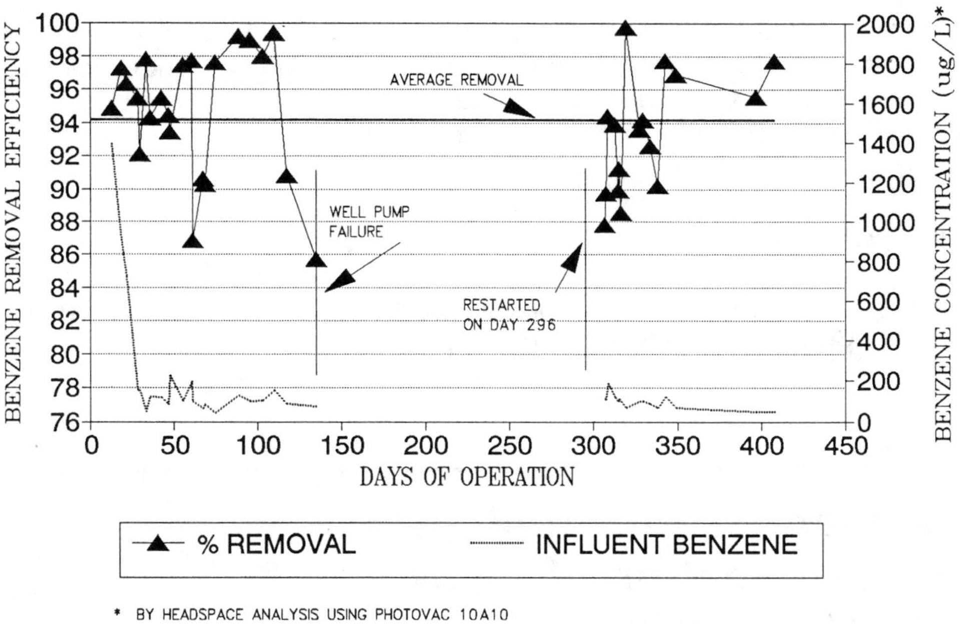

FIGURE 3. Field demonstration system efficiency. Asterisk indicates headspace analysis using PHOTOVAC 10A10.

Summary

The bench-scale study and subsequent field test described above demonstrate the ability of a properly designed biological reactor to treat low levels of biodegradable organic compounds dissolved in water. The structured packing provides a mechanically stable media for biofilm development and allows for air to be bubbled up through the packing, thus maintaining adequate dissolved oxygen levels for aerobic biodegradation of the organic compounds. The biological population, once raised and acclimated to the organic compounds of concern, will continue to use the organics as a carbon source even when levels of available carbon have driven the microbes into the decay phase of their growth cycle. A supplementary nutrient feed is necessary to maintain the microbial population in the decay stage. Occasional regrowth stages may be required to replenish biomass lost during the decay stage of operation.

This process is presently being operated, or has been tested, at more than six sites across the United States. Future work will be geared toward a broader range of compounds and will focus on a better definition of the limitations of the system. As more and more real life operating data are collected, both the process and the equipment may be refined to enhance system performance.

REFERENCE

Johnson, F. T. "Summary of Low Concentration Biological Reactor Research." Hydro Group, Inc., unpublished internal report, 1985.

Sequential Anaerobic-Aerobic Biodegradation of PCBs in the River Model

*P. J. Anid**, *L. Nies*, *T. M. Vogel*
The University of Michigan

Laboratory studies have shown that PCBs can be partially dechlorinated under anaerobic conditions (Quensen *et al.* 1988) and aerobically oxidized by mixed and pure cultures of microorganisms (Bedard *et al.* 1986, 1987; Bopp 1986). Generally, anaerobic microbes can reduce the chlorine number through reductive dehalogenation, and aerobic microbes can oxidize lightly chlorinated PCBs. The complete degradation process might, therefore, require sequential anaerobic dechlorination and aerobic mineralization.

Anaerobic dechlorination and aerobic degradation have not been shown to occur in sequence within the same natural system. However, illustration of the sequential anaerobic-aerobic degradation of PCBs in batch reactors using river sediment is under way and might provide evidence for the potential of *in situ* degradation. This research

* Department of Civil and Environmental Engineering, Environmental and Water Resources Engineering, The University of Michigan, Ann Arbor MI 48109-2125

concentrates on the methodology of inducing anaerobic dechlorination and switching from anaerobic to aerobic conditions. For example, recent studies have shown that dechlorination is dependent on the availability of an adequate carbon source (Nies & Vogel 1990). Unfortunately, an *in situ* method using both anaerobic and aerobic processes has not yet been attempted.

Determining whether sequential anaerobic-aerobic *in situ* bioremediation techniques are likely to succeed depends on results from bench-scale experiments that simulate natural conditions and help optimize design parameters. Using both anaerobic and aerobic processes, the objectives of this research were: (1) to demonstrate that a sequential anaerobic-aerobic biodegradation scheme is possible in batch incubations by reductively dechlorinating PCBs in sediments where anaerobic conditions had been induced, and then degrading many of the less chlorinated products under aerobic conditions; (2) to model a Hudson River bottom sediment to determine whether PCBs can be mineralized under sequential anaerobic-aerobic conditions using substrate addition that might enhance anaerobic dechlorination; and (3) to examine the design parameters that determine the applicability of the river model findings to *in situ* conditions.

MATERIAL AND METHODS

Batch Incubations. Initial anaerobic batch incubations of Hudson River sediments were started in order to begin the first phase of the anaerobic-aerobic biodegradation sequence. The batches (100-mL serum bottles) containing the Hudson River sediments (30 g) were spiked with Aroclor 1242 (300 mg/kg) using very little acetone (2.5 µL) as a solvent and inoculated with a 2-mL sediment slurry from an actively dechlorinating methanol-fed enrichment (derived from Hudson River sediment). The solubility of pure Aroclor 1242 in water did not increase significantly even in 5 percent solvent solutions (data not shown). Therefore, it is doubtful that methanol significantly affects solubility or bioavailability of PCBs in sediments. High levels of methanol (6,670 mg/kg of sediment) have been found to enhance reductive dechlorination (Nies & Vogel 1990). After 20 weeks, initial batches showed an increase in the rates and extent of dechlorination relative to controls. An aerobic biphenyl degrading Hudson River isolate (S3) was selected to test for aerobic degradation of the PCB congeners remaining after anaerobic dechlorination. The isolate was added to the incubations after purging the headspace of the serum bottles with oxygen. After 96 h, live and autoclaved samples were extracted.

River Model. The river model (1.8 × 0.6 × 0.6 m) (Figure 1) was filled with 0.8 ton of Hudson River sediment which was originally received lightly contaminated with approximately 5 mg/kg of mono-, di-, and trichlorobiphenyl congeners. The congener distribution in the sediments indicated that substantial dechlorination had occurred. To demonstrate the feasibility of the sequential anaerobic-aerobic process, PCB was added to the dechlorinated sediment. The additional Aroclor 1242 (neat from vial, then washed with methanol) was added and mixed with the sediment to achieve a final concentration of approximately 300 mg/kg on a wet basis. Aroclor was mixed into the sediments in individual buckets that afterwards were poured into the glass reactor. Anaerobic conditions were induced by adding methanol (MeOH) in three increments of 6,670 mg MeOH/kg sediment/week, along with a mineral media (Table 1), through eight injection wells connected to a multichannel pump (Figure 1). To simulate the river flow, a stream of water on top of the sediment was continuously recirculated. The flow rate (0.15 L/sec) was low enough so that no resuspension of sediment occurred. The reactor was kept at room temperature (22 C). To monitor dechlorination, 24 sediment samples were collected each month, using a special corer inserted through eight top access ports, and analyzed. Samples were taken from the bottom, middle, and top of the model.

In parallel, river model batch incubations were prepared by adding 30 g of sediment from the reactor (after the Aroclor 1242 was added) to 100-mL serum bottles. Methanol (6,670 mg/kg sediment) and mineral media were added once to all bottles (samples and autoclaved controls). None of these batches was inoculated with an enrichment culture. Triplicate bottles were kept at 12, 25, and 30 C, respectively. Sterile control bottles were prepared identically and autoclaved twice at a 24-h interval. All bottles were sealed with Teflon®-coated stoppers and aluminum crimp caps. Gas production by the batch incubations was measured by inserting the needle of a wetted glass-barreled syringe through the stopper of the serum bottle and measuring gas volume. Sample extraction, PCB analysis, and Aroclor peak identification are described elsewhere (Nies & Vogel 1990).

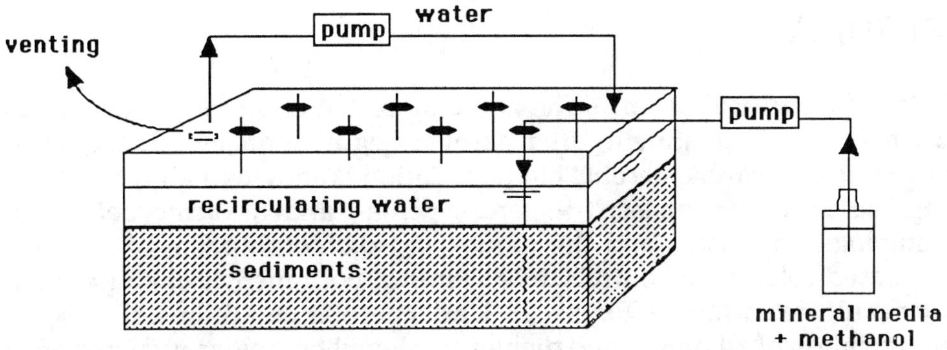

FIGURE 1. Schematic representation of the physical river model and sediments. Water is pumped across top of sediment simulating river flow, and nutrients are injected into sediments at eight locations.

TABLE 1. Mineral media composition.

Mineral	mg/L
$(NH_4)_2HPO_4$	80
NH_4Cl	1000
K_2HPO_4	200
$NaCl$	10
$CaCl_2$	10
$MgCl_2$	50
$CoCl_2 \bullet 6H_2O$	1.5
$CuCl_2 \bullet 2H_2O$	0.2
$Na_2MoO_4 \bullet H_2O$	0.23
$ZnCl_2$	0.19
$NiSO_4 \bullet 6H_2O$	0.2
$FeSO_4 \bullet 7H_2O$	1.0
$AlCl_3 \bullet 6H_2O$	0.4
H_3BO_3	0.38

RESULTS

Initial Batch Incubations. The homolog distribution for Aroclor 1242, after anaerobic treatment, after aerobic degradation by S3, and after sequential anaerobic-aerobic biodegradation is shown in Figure 2. During the anaerobic phase, tri-, tetra-, penta-, and hexachlorobiphenyl homologs decreased, and the mono- and dichlorobiphenyl homologs increased relative to controls, as a result of both methanol and enrichment culture additions (Figure 2). Under aerobic conditions, significant degradation of all mono- and dichlorobiphenyl homologs occurred after 96 h. The total PCB concentration in the sediment was reduced 130 mg/kg by the sequential anaerobic-aerobic biodegradation treatment in initial batch incubations. Loss to volatilization was controlled by keeping the serum bottles sealed and injecting oxygen through the septa. The di-*ortho* substituted congeners, 2, 2′- and 2, 6-, were more resistant to aerobic degradation (only 18% degraded) than other dechlorinated biphenyls. The recalcitrance of 2, 2′- and 2, 6- results in an accumulation of dichlorobiphenyl homologs after sequential anaerobic-aerobic biodegradation (Figure 2). No trichlorobiphenyls were significantly degraded aerobically by the S3 bacteria.

River Model. Methanol addition to the river model quickly initiated anaerobic (methanogenic) conditions within the reactor, as evidenced by the methane bubbling through sediments. However, considerably less gas was observed whenever the methanol had been depleted. After 24 weeks of operation, a shift of the PCB congener pattern from the highly chlorinated congeners to the less chlorinated congeners was observed, indicating that the dechlorination process was under way (Figure 3). The extent of dechlorination was measured by comparing the removal of chlorine from PCBs relative to that in batch controls. In general, an increase in the mono- and dichlorobiphenyl concentrations was observed relative to autoclaved controls. In addition, a decrease in tri-, tetra-, penta-, and hexachlorobiphenyl concentrations relative to controls was also observed. The individual congeners that underwent significant dechlorination were 2,3,3′,4′- and 2,3,4,4′-; 2,3,4′-, 2,3,3′-; and 2,3,4-, and 2,2′,5-, which were first to be dechlorinated as indicated by the relatively rapid reduction in their peak heights relative to controls. At the same time, congeners 2,4′-, 2,3-, 2,2′-, and 2,6-dichlorobiphenyls were the major products of dechlorination after 24 weeks of operation.

No dechlorination was observed in river model batches incubated at 12 C after 16 weeks, whereas incubations at 30 C and at room temperature (22 C) exhibited similar and significant dechlorination just

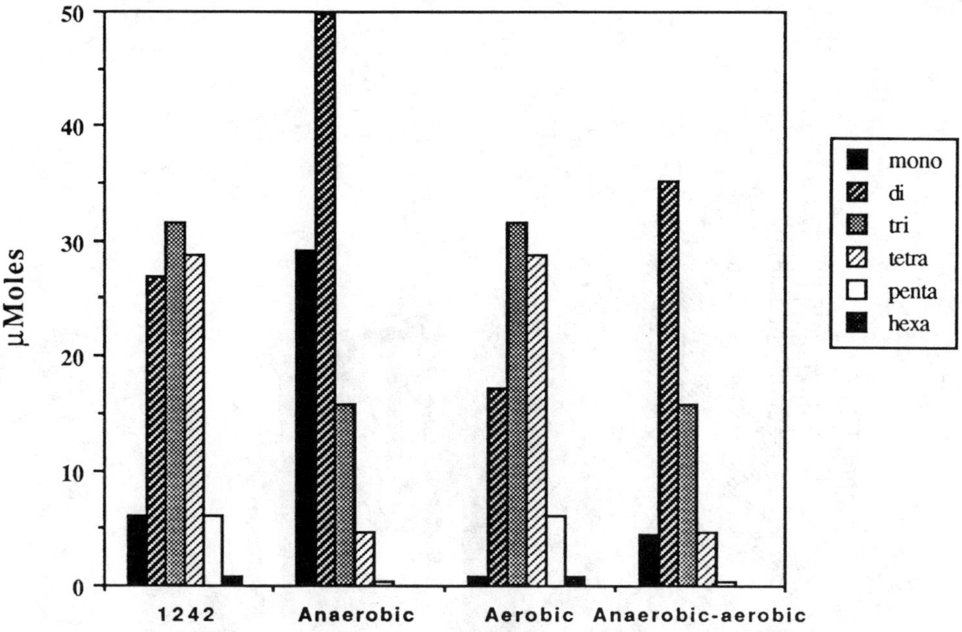

FIGURE 2. Distribution of PCB homologs (normalized to 100 µM initially) in Aroclor 1242, after anaerobic dechlorination, aerobic biodegration, and sequential anaerobic-aerobic biodegradation in batch incubations.

as in the Hudson River model sediment in both the extent and pattern of dechlorination. Gas production was monitored in the batch incubations. In general, the volume of gas measured was lowest in the 12 C incubations and highest in batches kept at 22 and at 30 C.

DISCUSSION

In some PCB-contaminated sediments, anaerobic microbial communities seem to mediate the dechlorination reaction. However, in nature, dechlorination is probably a slow process. The sediments used in this study were respiked with PCBs in the laboratory after extensive dechlorination had occurred in the Hudson River. The bioavailability of PCBs added in the laboratory compared to PCBs found in contaminated sediments is an issue that must be considered when using laboratory data for environmental predictions. In other experiments using nondechlorinated sediments (not of Hudson River origin) with high levels of

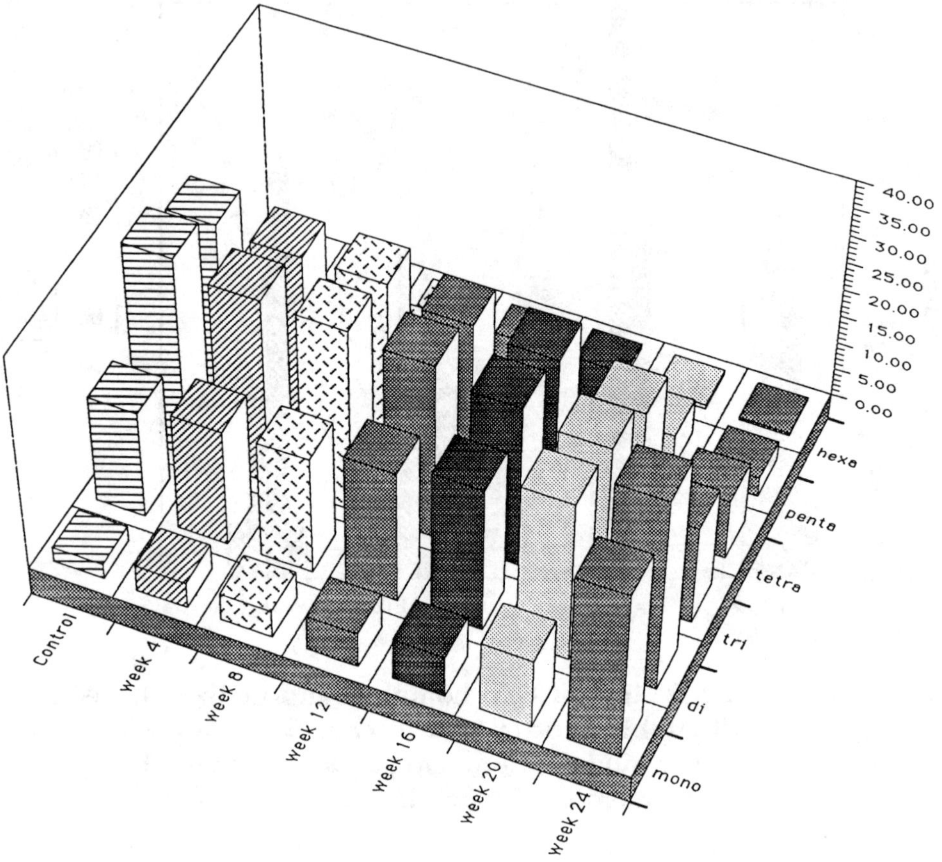

FIGURE 3. Distribution of PCB congeners (mole percent) classified by homolog (mono-, di-, trichlorobyphenyl) as a function of time in the physical river model (middle location) during the anaerobic phase.

background endogenous PCBs, dechlorination was achieved in the laboratory, although at a slower rate (experiment still in progress). Although bioavailability might explain the difference in the dechlorinating capacity of two sediments, no experimental evidence yet defines the required conditions for the successful dechlorination of endogenous PCB.

This study is the first attempt to initiate dechlorination through substrate addition while simulating *in situ* conditions in a lab-scale river model. Methanol injection initiated anaerobic conditions and, hence, dechlorination without the addition of a specific reducing agent. In addition, the similarity of the results between the river model batch incubations (microbial enrichment not added) and the river model sediments suggests that it might be possible to transpose small-scale work to *in situ* conditions. However, in both cases, the observed dechlorination rates are slower than what is observed in the initial batch incubations that were inoculated with a slurry from an actively dechlorinating methanol-fed enrichment. One possible explanation is that the dechlorinating enrichment culture added to the batches might accelerate the onset of methanogenesis and hence dechlorination, or that this enrichment addition provides an increase in the number of competent PCB-dechlorinating microbes.

The addition of a dechlorinating enrichment for the purpose of inducing and enhancing anaerobic dechlorination might be permitted only in the laboratory because of regulatory difficulties of *in situ* application of microbes. Therefore, the demonstration of dechlorination in the river model without inoculation indicates the potential of *in situ* technology. Sequencing the anaerobic dechlorination of the highly chlorinated congeners through substrate addition with the aerobic degradation of the reductively transformed PCB congeners is promising. To date, the two processes have been coupled with PCB-spiked sediments in the initial batch incubations. Initial results have shown an accumulation of some di-*ortho* substituted congeners that resisted aerobic oxidation. Although the accumulation of those congeners has also been reported in other laboratory studies (Quensen *et al.* 1988), longer aerobic treatment as well as the induction of more competent indigenous aerobic microorganisms, or more extensive dechlorination during the anaerobic phase, might improve the performance of the sequenced process. Nevertheless, with the use of an easily isolated aerobic organism with only "average" PCB degrading ability, like the S3 organism, it was possible to significantly reduce PCB levels in the sediments. The transition from the anaerobic to the aerobic phase *in situ* once anaerobic dechlorination has neared completion would require the stimulation of the indigenous aerobic microorganism (S3-like organisms), parallel to inducing aerobic conditions within the sediments. The anaerobic sediment would need to be oxygenated by adding pure oxygen or hydrogen peroxide through the injection wells.

A disadvantage to using pure oxygen, however, is related to operational problems associated with the delivery and chemistry of oxygen in the sediments. Often, subsurface *in situ* remediation schemes use hydrogen peroxide as a source of oxygen, because of its high aqueous solubility. Therefore, batch incubations with dechlorinated sediments are presently being tested to evaluate the efficiency of hydrogen peroxide as an oxygenation agent. However, in order to hasten the distribution of aerobic organisms adapted to PCB degradation throughout the sediments, the upper aerobic layer of the water will be pumped into the sediments. This might be helpful in introducing PCB degraders into the bulk of the sediments. It is still unclear how recirculating the upper aerobic layer can be achieved with minimum disturbance of the sediments. The next aerobic phase might give some insight on the effect of sediment recirculating on the physical stability of the sediments in the river model.

A conceptual *in situ* model can be formulated using the results of the river model. A possible scenario would consist of a barge equipped with injection capabilities. Those injectors would deliver the substrate (*e.g.*, methanol) or nutrient incrementally into the anaerobic sediments. The barge would move along the contaminated section of the river and repeat injection operations. Sediments would be monitored to measure the extent of dechlorination so that once appreciable amounts of dechlorinated products have accumulated, the same injection procedure will be repeated although hydrogen peroxide would replace methanol. Sequencing anaerobic and aerobic conditions might ultimately lead to a complete detoxification of the contaminated sediments. The river model constitutes an avant garde approach that provides insight into the feasibility of the process under real river and estuarine conditions.

REFERENCES

Bedard, D. L.; Unterman, R.; Bopp, L.; Brennan, M. J.; Harbel, M. L.; Johnson, C. *Appl. Environ. Microbiol.* **1986**, *51*, 761–768.

Bedard, D. L.; Wagner, R. E.; Brennan, M. J.; Haberl, M. L.; Brown, J. F., Jr. *Appl. Environ. Microbiol.* **1987**, *53*, 1103–1112.

Bopp, L.H.J. *Ind. Microbiol.* **1986**, *1*, 23–29.

Nies, L.; Vogel, T. M. *Appl. Environ. Microbiol.* **1990**, *56*, 2612–2617.

Quensen, J. F., III; Tiedje, J. M.; Boyd, S. A. *Science.* **1988**, *242*, 752–754.

Methanotrophic Removal of TCE in an Aquifer Simulator

D. D. Hicks
CH2M Hill
R. P. Egg[*], *D. L. Reddell, C. G. Coble*
Texas A&M University

Trichloroethylene (TCE) is resistant to aerobic bioremediation, and anaerobic methods can form vinyl chloride, a known carcinogen. TCE can be oxidized by methanotrophic bacteria. Cometabolism of TCE by methanotrophs results from methane monooxygenase (MMO) activity. MMO is a nonspecific enzyme produced to oxidize methane to methanol. Methanotrophic TCE removal has been accomplished in serum vials (Little *et al.* 1988) and sand columns (Mayer *et al.* 1988). This paper presents results of a study to determine the effects of groundwater velocity on TCE degradation by methanotrophs in a model aquifer. Groundwater velocity will affect the hydraulic retention time (HRT) and, therefore, the rate of substrate utilization within a unit volume of aquifer. The objectives of this study were to stimulate a naturally occurring population of methanotrophic bacteria in a model aquifer, and determine changes in TCE treatment efficiency resulting from changes in flow velocity.

Materials and Methods

A model aquifer (1 m × 1 m × 3 m) was constructed of stainless steel (Figure 1). A nonsterile medium brick sand from a local concrete supplier was used as fill material. Constant head chambers at the inlet and outlet provided uniform flow. A 15-cm clear water well, encompassing the cross section of the model aquifer, was located 0.75 m from the aquifer inlet. Water was circulated from the clear water well through two gas transfer tanks, connected in parallel, where methane

[*] Agricultural Engineering Department, Texas A&M University, College Station TX 77843

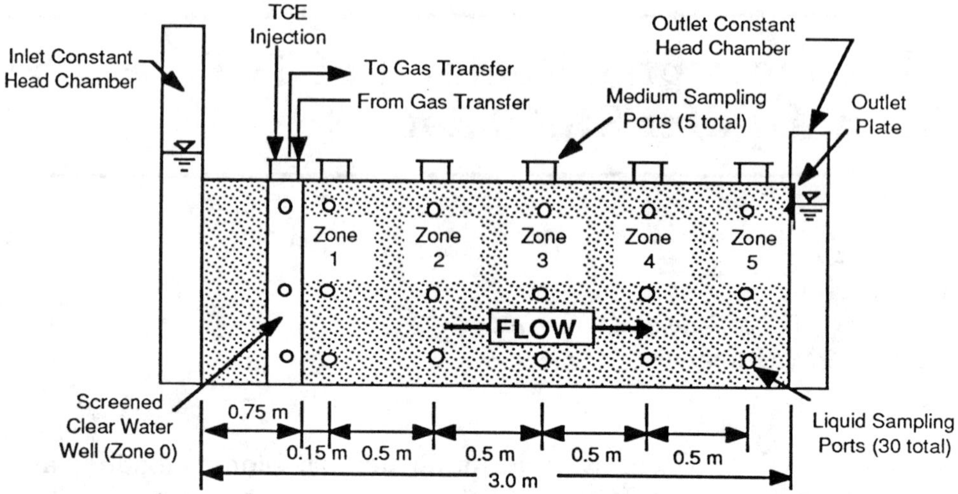

FIGURE 1. Aquifer simulator showing locations of sampling zones.

and oxygen were dissolved into the water. TCE was metered into the clear water well by gravity through a rotameter. TCE was dissolved in 10 mL of methanol (cosolvent), which was then diluted to 2.0 L with distilled water. The TCE solution was stored in a collapsible Teflon® bag. An activated carbon column removed residual chlorine from the tap water source. All tests were conducted at 23 C.

Liquid samples from the aquifer were analyzed for dissolved oxygen, methane, and TCE. Samples were withdrawn through sampling tubes located in the clear water well and five downstream sampling zones (Figure 1). Each sampling zone contained five sampling tubes spaced over the aquifer cross section. Equal volumes were withdrawn from each tube in a sampling zone and mixed to form composite samples for that zone. Sample volumes for dissolved oxygen, methane, and TCE were 15, 35, and 35 mL, respectively. Two samples were taken each day for three days for each test. Dissolved oxygen was analyzed using a hand-held probe (Presto-Tec). Dissolved methane was analyzed with a Fisher model 1200 gas partitioner equipped with a thermal conductivity detector, a Hewlett-Packard model 3390A integrator, and two packed columns (80/100 Poropak and 60/80 Molecular Sieve; Supelco) using a headspace equilibrium technique (Kampbell *et al.* 1989). TCE was analyzed with the headspace equilibrium technique on a gas chromatograph (Perkin Elmer model 8500) equipped with an electron capture detector and a packed column (1% SP-1000 on 60/80 Carbopak B; Supelco).

After stimulation of a methanotrophic population in the aquifer, five tests were conducted using different Darcy velocities (flowrate/gross cross-sectional area) and TCE concentrations. Methanotrophic stimulation and Test #1 were performed at a Darcy velocity of 1.25 m/day (2 day HRT, 19 L/h flow rate). Tests #2 and #3 were at 2.50 m/day (1 day HRT, 38 L/h flow rate). The velocity was then returned to 1.25 m/day for Tests #4 and #5. Test #5 was a zero methane control. Table 1 shows initial concentrations of dissolved gases, methanol, and TCE, as well as flow velocities for all tests. The system was operated as an unconfined aquifer.

TABLE 1. Conditions for all aquifer simulator tests.

Test	Darcy Velocity (m/day)	TCE Concentration[a] (µg/L)	Methane Concentration[a] (mg/L)	Oxygen Concentration[a] (mg/L)	Methanol Concentration[a] (mg/L)
#1	1.25	43.2	4.27	11.18	2.87
#2	2.50	42.6	2.18	10.55	2.88
#3	2.50	41.2	2.80	5.79	2.88
#4	1.25	138-395	3.95	11.23	2.25
#5	1.25	125.6	0.00	13.14	2.87

(a) Average concentration in clear water well.

Results and Discussion

A naturally occurring methanotrophic population was stimulated with methane and oxygen. After about a month, concentration profiles for methane and oxygen were consistent (Figure 2). Oxygen and methane were used rapidly between the clear water well and the first sampling zone. An unexpected observation was the increase in methane downstream from the first sampling zone. This occurred in nearly all test runs and began where the dissolved oxygen concentration was below about 2 mg/L. Methanogenesis may have occurred in areas of low oxygen. The oxygen demand for complete oxidation of the methane and methanol would have resulted in anaerobic conditions in the aquifer. However, no anaerobic degradation products of TCE (*cis-* or *trans-*DCE) were ever detected.

Figure 3 shows the TCE profile along the length of the aquifer simulator for Tests #1, #2, and #3. Test #1 resulted in rapid TCE removal between the clear water well and the first sampling zone.

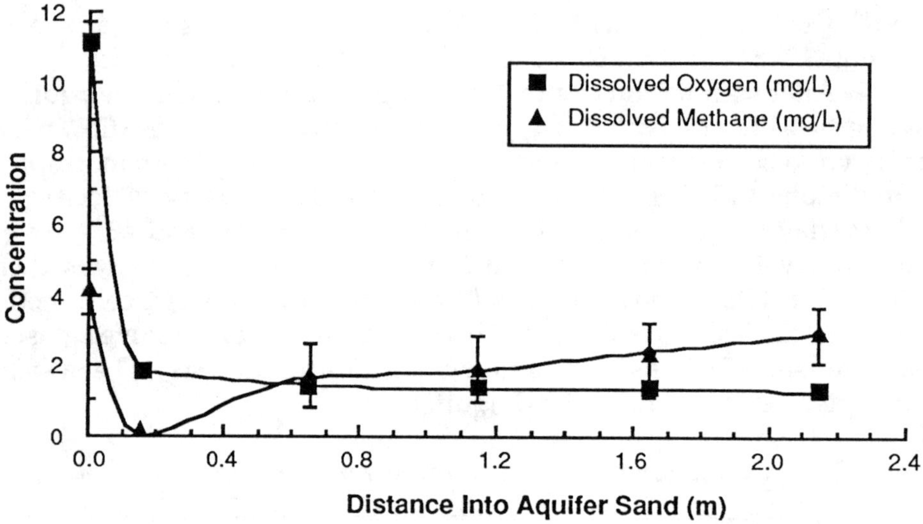

FIGURE 2. Oxygen and methane concentration profiles in the model aquifer after one month of methanotrophic stimulation.

Average TCE removal was 43.5 percent over the length of the aquifer. Test #2 showed essentially no TCE removal. When the Darcy velocity for Test #2 was doubled to 2.50 m/day, the flow from the methane gas transfer tank remained constant at 3.5 L/min. This dilution factor resulted in an initial dissolved methane concentration of about 2.0 mg/L. The methane gas transfer tank flow rate was increased to 6.0 L/min for Test #3, and after two weeks, TCE removal efficiency averaged 69 percent.

The average initial methane concentration for Test #3 was 2.8 mg/L. Possibly the aquifer system had not stabilized when data for Test #2 was gathered; however, the speed at which the system recovered after the methane tank flow rate was increased suggests poor TCE removal may have resulted from low methane levels during Test #2. The effect of methane concentration on TCE removal is not well defined. High methane levels can inhibit TCE degradation while low levels have had mixed effects depending on the cultures used (Roberts *et al.* 1989).

For Test #4 the methane gas transfer tank flow rate was reset to 3.5 L/min. The TCE concentration in the clear water well was measured at 395 µg/L, resulting from a calculation error when filling the

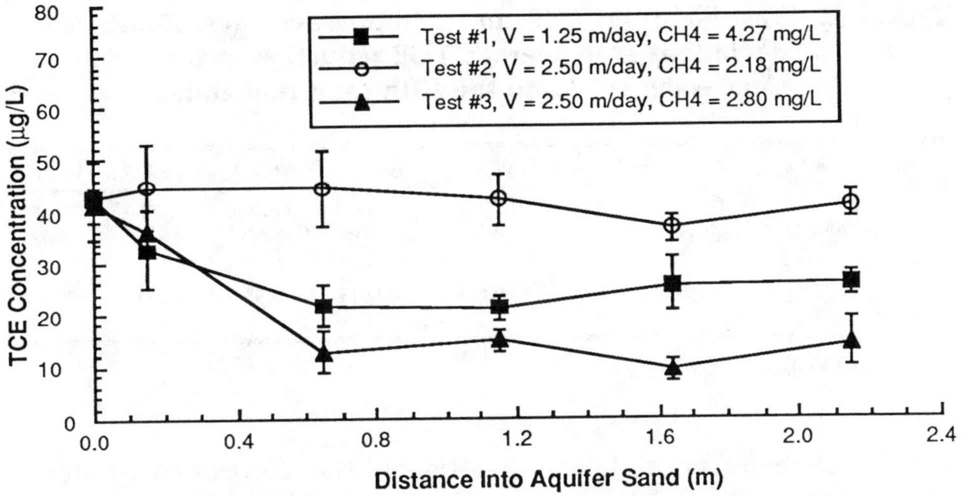

FIGURE 3. Comparison of TCE concentration profiles from velocity fluctuations. V in the legend is the Darcy velocity and CH_4 is the clear water well dissolved methane concentration.

TCE storage bag. After one week, TCE reduction averaged about 98 percent. Over the next six weeks, TCE reduction ranged from 64 to 99 percent (Table 2). Throughout Test #4, dissolved methane was completely utilized between the clear water well and the first sampling zone. Several problems occurred with the TCE feed system during Test #4 causing fluctuations in the TCE feed rate. Because of the weekly variation in the average TCE concentration, the data in Table 2 are probably indicative of transient rather than steady state conditions.

The aquifer simulator was next flooded with water to test for leaks; none were found. Methane was then removed for Test #5. After two weeks, no TCE removal occurred (data not shown).

The reason for the high TCE removal efficiency in Test #4 is not known. Possibly the high velocity during Test #3 carried methane farther into the medium, stimulating methanotrophic bacterial growth and MMO activity in a larger volume of the aquifer. When the flow velocity was decreased, the MMO was still active and the HRT in the effective treatment zone increased, resulting in the higher TCE removal rates.

The effect of the cosolvent on TCE degradation could not be evaluated directly. The initial methanol concentration was held constant for most tests (Table 1) although some variation occurred during

TABLE 2. Test #4 initial TCE concentration averages, standard deviations, and average TCE reductions between the clear water well and the fifth sampling zone.

Time into Test #	Week 0	Week 1	Week 2	Week 3	Week 4	Week 5	Week 6
Initial TCE Concentration (μg/L)	41.21 ± 3.14	395.01 ± 54.79	187.00 ± 28.93	192.16 ± 0.86	317.27 ± 6.97	192.78 ± 13.15	137.89 ± 8.09
Average TCE Reduction (%)	65.71	97.93	98.55	63.61	83.55	86.23	92.60

(a) Week 0 denotes Test #3.

Test #4. Oldenhuis *et al.* (1989) investigated the effect of co-substrates on TCE reduction by *Methylosinus trichosporium* OB3b. TCE degradation occurred with formate or methane, but not methanol; cell growth occurred only with methane. Their data and results of Test #5 (no TCE removal without methane), suggest that methanol had little effect on the changes in TCE degradation observed.

Summary

The effects of fluctuating groundwater velocities on TCE biodegradation were tested in an aquifer simulator. A natural population of methanotrophic bacteria was stimulated by water enriched with methane and oxygen. Increasing the Darcy velocity from 1.25 to 2.50 m/day resulted in increased TCE removal from 44 to 69 percent, respectively. When the Darcy velocity was retumed to 1.25 m/day, TCE removal exceeded 98 percent.

REFERENCES

Kampbell, D. H.; Wilson, J. T.; Vandegrift, S. A. *Intern. J. Environ. Anal. Chem.* **1989**, *36*, 249–257.

Little, C. D.; Palumbo, A. V.; Herbes, S. E.; Lindstrom, M. E.; Tyndall, R. L.; Gilmer, P. J. *Appl. Envir. Microb.* **1988**, *54*, 951–956.

Mayer, K. P.; Grbić-Galić, D.; Semprini, L.; McCarty, P. L. *Wat. Sci. Tech.* **1988**, *20*(11/12), 175–178.

Oldenhuis, R.; Vink, R.L.J.M.; Janssen, D. B.; Witholt, B. *Appl. Environ. Microbiol.* **1989**, *55*, 2819–2826.

Roberts, P. V.; Semprini, L.; Hopkins, G. D.; Grbić-Galić, D.; McCarty, P. L.; Reinhard, M. *In-situ Aquifer Restoration of Chlorinated Aliphatics by Methanotrophic Bacteria*; U.S. Environmental Protection Agency: Washington, DC, 1989; EPA/600/2-89/033.

Production-Scale Trials on the Decontamination of Oil-Polluted Soil in a Rotating Bioreactor at Field Capacity

*Ger P. M. van den Munckhof**, *Martin F. X. Veul*
Witteveen+Bos, Consulting Engineers

FRAMEWORK

As part of the development of biotechnological methods for cleaning soil, Witteveen+Bos Consulting Engineers, Deventer, has carried out production scale trials on the decontamination of oil-polluted soil in a rotating bioreactor at field capacity. The aim was to develop a method for quick cleaning of polluted soil under controlled conditions at field capacity.

The Microbiology Department of the Wageningen Agricultural University and Broerius Soil Sanitation Ltd., Voorthuizen, also took part in the study, which was partially financed by the Netherlands Ministries of Housing, Physical Planning and Environment (VROM) and of Economic Affairs (EZ).

Feasibility Study

A feasibility study was carried out at the Microbiology Department of the Wageningen Agricultural University from July through November

* Witteveen+Bos, Consulting Engineers, P.O. Box 233, 7400 AE Deventer, The Netherlands

of 1988. Optimum environmental conditions for biological decomposition were determined in incubation experiments with soil in which oil degradation was monitored by respiration measurements and oil analyses.

Mesophilic decomposition of oil was shown to be most efficient at a temperature of 30 to 35 C, a moisture content of approximately 10 percent by weight, and in the presence of 150 mg N nitrogenous fertilizer per kg of dry soil (oil concentration of 5,000 mg/kg ds). Under these conditions, on a laboratory scale, oil is completely decomposed (to <100 mg/kg ds) within 2 weeks.

The Rotating Bioreactor

The reactor used in the production scale trials was a modified DANO-composting installation for household garbage operated by the Municipality of Soest and Baarn (Figure 1). Broerius Soil Sanitation Ltd. renovated and adapted the installation for soil treatment. The reactor is 25 m long and 3.5 m in diameter. As the reactor rotates, the soil is mixed and homogenized so that biomass and oil substrate come into close contact. Temperature and oxygen levels are maintained with the aid of a warm-air blower, and soil moisture content is adjusted with a sprinkler installed in the bioreactor. Nutrients are added to the soil on the conveyor before entering the reactor. The soil treatment process is shown schematically in Figure 2. Witteveen+Bos has obtained the patent for this treatment process.

PRODUCTION SCALE TRIALS

Batch Experiments

Five loads (batches) of oil-polluted soil, each weighing 50 tonnes (wet weight), were treated in the reactor as follows: one batch of petrol-polluted soil, three batches of diesel-polluted soil, and one batch of soil polluted with both diesel and lubrication oil. The oil concentrations in the soils varied from 1,000 to 6,000 mg/kg ds. Four of the five batches consisted of fine to loamy fine sand, and one batch consisted of fairly coarse sand.

Each soil was incubated in the reactor under optimal environmental conditions as determined in the laboratory feasibility study. This was only partly successful because the soil temperature in the

FIGURE 1. The bioreactor.

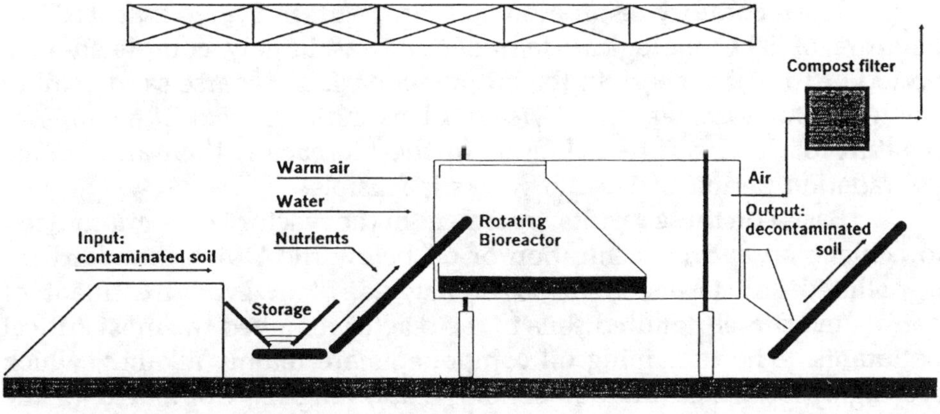

FIGURE 2. Biological treatment process in the rotating bioreactor.

reactor did not rise above 20 to 22 C. In addition, the structure of the relatively loamy soil batches partly deteriorated. Some soil particles tended to stick to one another around small particles of rock, thus fo1 ming soil balls sometimes several centimetres in diameter.

Soil and air samples taken from the reactor were analyzed at regular intervals to monitor and manage the microbial degradation process. To take the soil samples, three sampling flaps were made, at a quarter, half, and three-quarters of the length of the reactor (sampling points 1, 2 and 3, respectively). In addition to the chemical analyses, such as moisture content, humus content, pH, oil concentration, volatile aromatics, and PAH, counts were done of oil-degrading microflora, for which the most probable number (MPN) method was used. The air samples were analyzed for oil and volatile aromatics to establish a mass balance (degradation versus volatilization).

For comparison with the measurements in the reactor, respiration measurements were also done on soil samples from each batch in the Microbiology Laboratory at Wageningen Agricultural University. The results of the measurements on oil biodegradation in the five batches are summarized in Table 1 and presented in Figures 3, 4, and 5.

In the rotating bioreactor at a soil temperature of about 20 C, the end concentration of the oil biodegradation was reached on average within a period of approximately 2.5 weeks. Biodegradation was most rapid in the first week. The end concentration of oil varied from <50 to 250 mg/kg ds.

The maximum level of oil-degrading microflora in the reactor was reached within 1 to 3 days. The number of microorganisms grew from MPN/10^6 to MPN/10^8 per gram of soil.

The laboratory respiration measurements indicated that at a temperature of 30 C biological degradation was largely completed in a period of 1 to 1.5 weeks. In the humus-poor, fairly coarse sandy soil in batch 1, it took longer, 2 to 3 weeks. This indicates that, if a soil temperature of 30 C can be achieved in the bioreactor, then an average degradation period of 1 to 1.5 weeks is feasible.

Based on these results, treatment in the reactor is not guaranteed to achieve an end concentration of oil below the Dutch Standard for unpolluted soil (A-level) of 50 mg/kg ds. However, treatment of petrol- and diesel-polluted soils in the reactor removed the most critical pollutants. The remaining oil components are mainly alkanes, which are highly branched and/or of long chain length. Compared to the original oil product, these components are less volatile and less soluble, do not cause odor nuisance, and are probably less toxic.

Thus, treatment in the bioreactor is environmentally beneficial.

Petrol-polluted soils should not be treated in the rotating bioreactor without provisions to reduce emission of volatile aromatics into the air. There are no significant emissions of aromatics from diesel-polluted soils. However, sometimes at the beginning of the

TABLE 1. Results of the batch experiments.

Batch no	Soil type	Oil type	Oil-concentration (mg/kg ds) start	Oil-concentration (mg/kg ds) end	Degradation time (weeks)	Point in time max. level microflora (days)	Results of respiration measurements (% of total respiration)	Volatilization (weight %)
1.	poor coarse sand	diesel oil	6,000	<100	2.5-3.0	3	80% after 2 weeks	
2.	fine sand	petrol	920	<50	0.5			>30
3.	loamy fine sand	diesel oil/ lubrication oil	730	<100 (GC) 180 (IR)	2.5-3.0	3	90% after 1 week	30
4	fine sand	diesel oil	1,000	140 (GC/IR)	2.5	2-3	85% within 1½ week	0.5
5.	loamy fine sand	diesel oil	1,000 1,500	250 (GC/IR)	1.5	1-2	90% within 1 week	0

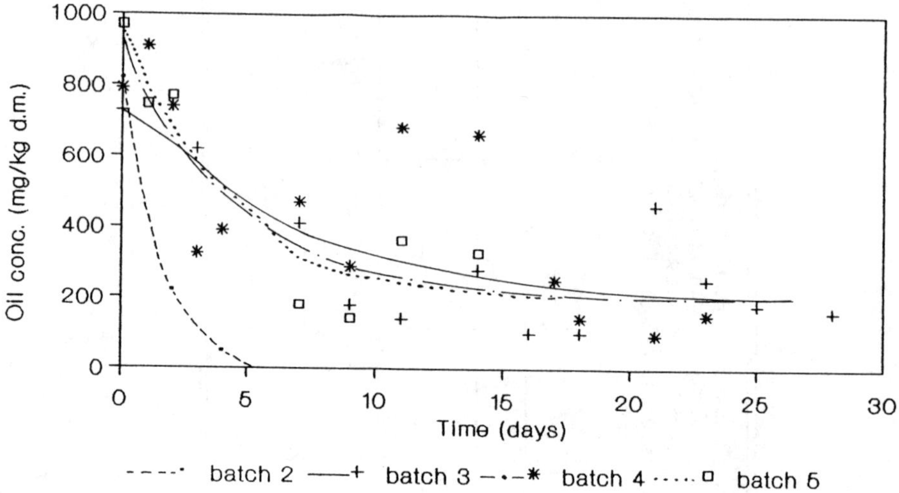

FIGURE 3. Decrease in oil concentration in the batch experiments 2, 3, 4, and 5.

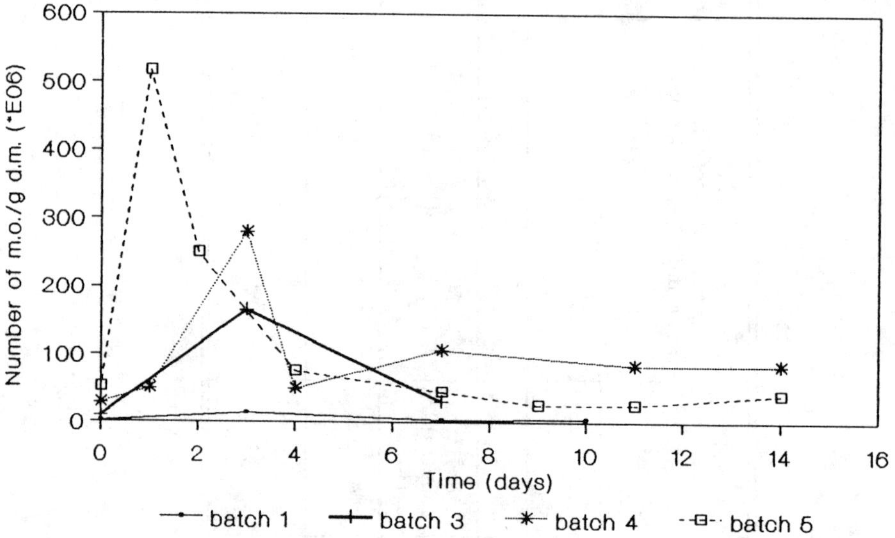

FIGURE 4. Development of oil-degrading microflora in batch experiments 1, 3, 4, and 5.

incubation, hydrocarbon emissions of up to a few hundred mg per m^3 may occur.

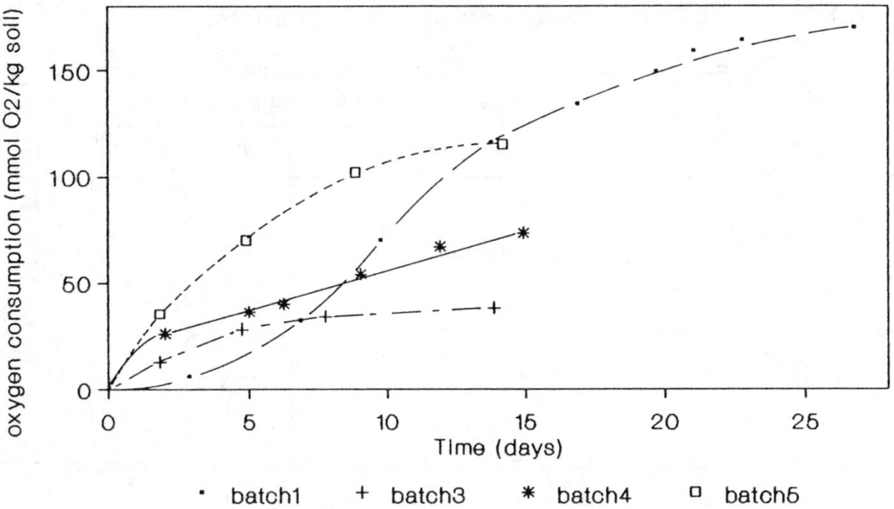

FIGURE 5. Oxygen consumption assessed by laboratory respiration measurements on soil samples from batch experiments 1, 3, 4, and 5.

Semicontinuous Experiments

In the semicontinuous experiments, 3.5 tonnes of decontaminated soil were removed daily from the end of the reactor and the same amount of contaminated soil placed in the front section. The total quantity of soil in the reactor remained at 50 tonnes, and the average retention time was 14 days.

Two semicontinuous experiments were carried out, one for 1 week (experiment 1) and the other for 2 weeks (experiment 2). Each day soil samples were taken at all sampling points, analyzed for oil concentration, and MPN counts of the oil-degrading microflora made. The results are summarized in Table 2.

The soil entering the reactor had an average oil concentration of about 1,000 mg/kg ds. At the defined sampling points 1, 2, and 3 and in the soil leaving the reactor, the oil concentration was the same, namely 200 to 350 mg/kg ds. Thus, it may be concluded that biodegradation to 200 to 350 mg/kg ds occurred quickly (within 3 to 4 days). The highest microbial activity occurred in the front section of the reactor (up to sp 1). At sp 1, counts of oil-degrading microflora were MPN 10^8 to 10^9 per gram of soil. At sp 2 and sp 3, the number of oil-degrading microorganisms had declined to the level in the soil entering the reactor (MPN 10^7 per gram of soil).

TABLE 2. Results of semicontinuous experiments.

Parameter	Sampling point				
	input	sp 1	sp 2	sp 3	output
* Mean oil concentration (IR) (mg/kg ds)					
- expt. 1	980	215	370	350	350
- expt. 2	995	220	205	190	230
* Mean of MPN-counts ($.10^6$/g. soil)					
- expt. 1	25	183	33	16	16
- expt. 2	26	131	46	4	

The results for the semicontinuous experiments are in line with those of the batch experiments. The highest microbial activity and the greatest breakdown occurred in the first 3 to 4 days. The end concentration was slightly above the Dutch A-level for unpolluted soil. On the basis of the available data, preference cannot be given to either semicontinuous or batch incubation.

PROGRESS AND PERSPECTIVE

The production scale trials indicate that, at a soil temperature of approximately 22 C, oil can be decomposed within 1 week to an end concentration varying from <50 to 350 mg/kg ds. In this process, the highest microbial activity occurs in the first 3 to 4 days.

For the present Dutch situation, a soil-cleaning technique is only fully applicable if the end concentration of a contaminant is below the A-levels defined in the Interim Act on Soil Sanitation. For mineral oil in a standard soil (containing 10% organic matter), this A-level is 50 mg/kg ds.

On the basis of the results of these experiments, soil treatment in the bioreactor cannot be guaranteed to achieve the required A-level for mineral oil. But the same is also true for land farming by which oil concentration is reduced to between 500 and 1,000 mg/kg ds after 2 years.

We must therefore conclude that, on the basis of current Dutch policy, there are no practical applications for the biotechnological cleaning of oil-polluted soil. Further research is recommended. In comparison to other techniques, biotechnological treatment is environmentally friendly and energy-saving, and the end product is a living, fertile soil. Further research should be directed not only to microbiological but also environmental hygiene aspects.

Microbiological Aspects

Further laboratory and pilot scale studies would determine the optimal process parameters to increase biological availability and rate of oil breakdown and to reduce the end concentration. Attention should be given to the effects of introducing selected microorganisms and chemical pretreatment. With a degradation time of up to 1 week, the bioreactor treatment is competitively priced with such other techniques as thermal decontamination.

Environmental Hygiene Aspects

The Dutch A-level for mineral oil was established on the basis of physical and chemical properties and the toxicity of oil products. However, during the biological breakdown process the oil changes in composition. This means that environmental hygiene characteristics, such as ecotoxicity and leaching, may have changed to such an extent that it would be appropriate to reconsider the risks to public health and the environment.

Compared with the oil-polluted soil, the components of biologically treated soil are less volatile, less soluble, and probably less toxic. Thus the question arises, to what extent is the A-level for mineral oil applicable to biologically treated soil?

There is draft legislation in the Netherlands (the "Concept Bouwstoffenbesluit" of 1989) that allows for the use of decontaminated soil with an end concentration above the A-level. The "Concept Bouwstoffenbesluit" states that decontaminated soil from which no environmental hygiene problem is expected may be used without restriction. To evaluate more precisely the environmental hygiene quality of oil-polluted soil that is biologically decontaminated (and of soil in general), it is important that appropriate ecotoxicological tests (bioassays) and leaching tests be developed.

A Biological Treatment System for the Purification of Waste Gases Containing Xenobiotic Compounds

R.M.M. Diks[*], *S.P.P. Ottengraf*
Eindhoven University of Technology

In the last decade, the biological treatment of waste gases has become an important alternative to many physical and physicochemical exhaust gas purification techniques in many different branches of industry. Its application generally appears to be reliable and effective at relatively low cost. The application of a biological treatment system is primarily based upon the biodegradability of the compounds by microorganisms, which use the organic compounds as their sole source of energy (catabolism) and carbon (anabolism) (Figure 1). The degradability of a compound often reflects its origin; biogenic compounds are easily biodegradable, whereas anthropogenic (i.e. engineered) compounds sometimes possess such unnatural structures (xenobiotics) that biological degradation is very difficult (recalcitrant compounds) or even impossible (persistent compounds). Through intensive microbiological research in recent years, progress has been made in isolating, selecting, or constructing new strains or mixed cultures of microorganisms (mainly bacteria) that are capable of degrading some recalcitrant compounds with growth rates of the same order of magnitude as those encountered in the degradation of biogenic compounds. Thus, the application of a biological system for the elimination of such xenobiotics offers real prospects at this time.

Three groups of biological waste gas purification systems are known; these can be distinguished by the behaviour of the liquid phase (which is either continuously moving or stationary in the contact apparatus) and of the microorganisms (which are either freely dispersed in the aqueous phase or immobilized on a carrier material). This paper provides a short overview of some process engineering aspects of the

[*] Department of Chemical Engineering, Eindhoven University of Technology, P.O. Box 513, 5600 MB, The Netherlands

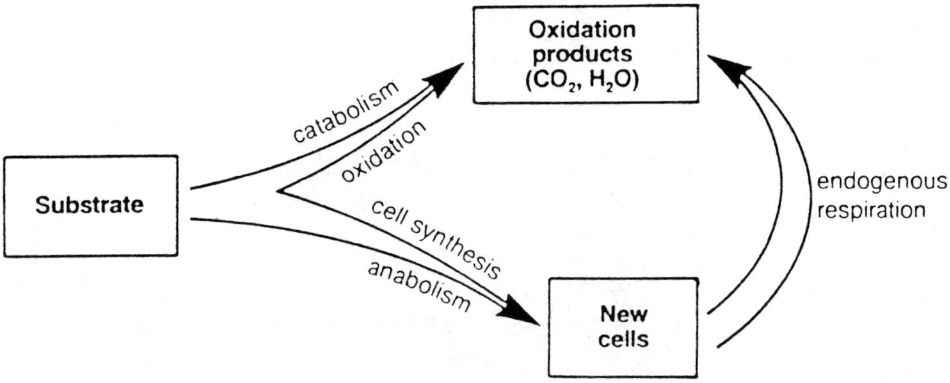

FIGURE 1. Substrate elimination by microbial oxidation.

two systems using immobilized microbial cultures. Experimental results concerning the removal of dichloromethane in a trickling filter will be discussed.

TABLE 1. Distinctions between different biological waste gas purification systems.

Microbial flora	Aqueous phase	
	Mobile	Stationary
Dispersed	Bioscrubber	–
Immobilized	Trickling filter	Biofilter

Biofilters

Biofilters reflect an easier mode of operation than other biological purification systems. In a biofilter (Figure 2) the waste gas is forced to rise through a simply structured packed bed of materials, in which a suitable microbial population develops naturally in the course of time, or by inocculation of the material with suitable microbial strains. As inorganic nutrients (phosphate, nitrogen, sulfur) must also be supplied by the carrier material, mostly natural materials (*e.g.*, compost, peat) mixed with wooden chips, lava, or polystyrene particles are applied. The latter materials are added to improve the structure of

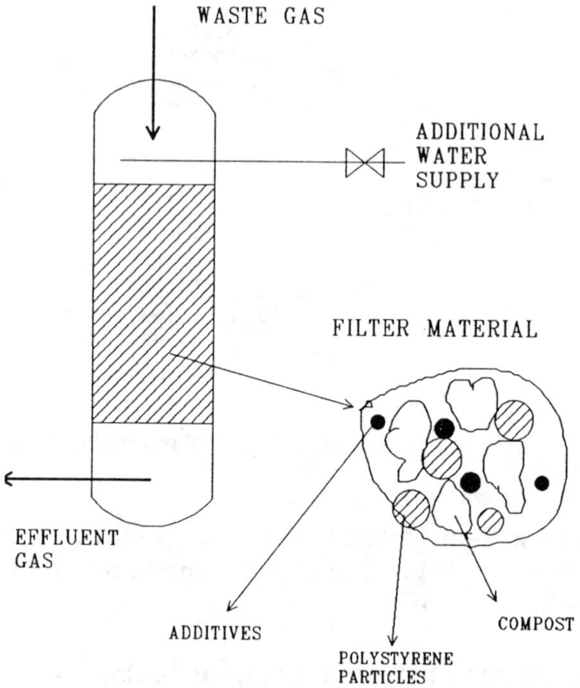

FIGURE 2. Experimental setup of a biofilter system. The filter material consists of compost mixed with additives.

the filter bed, hence decreasing the pressure drop and preventing aging phenomena. Acidification of the filter bed can be prevented by the addition of neutralizers.

It is generally assumed that the constituent particles of the packing material are surrounded by a wet biolayer. The volatile compounds and oxygen present in the waste gas are transferred from the gas phase into this biofilm, where the microbial degradation takes place. The packing material normally consists of small particles (d < 10 mm), hence a high specific surface area (300 to 1,000/mL), and good mass transfer is established. The effective operation of a biofilter depends mainly upon the water activity in the carrier material and the temperature and the structure of the packed bed.

As a result of experimental investigations of different volatile organic compounds, a model has been developed that describes the biological process inside the biofilter (Ottengraf & Oever, 1983) . Full-scale biofilters have been designed and installed in different branches of industry, both to remove odorous, easily biodegradable compounds (*e.g.*, food processing industry) and to prevent the emission of toxic xenobiotics (*e.g.*, chemical industry, lackeries, dye industry).

Biological Trickling Filters

Although applicable in most cases, a biofilter generally cannot be used if acidic metabolites are produced in large amounts during the degradation of the substrates present in the waste gas. Therefore, the concentration of these acid-producing compounds must be so low that neutralization can be achieved inside the packing material for a fairly long period of time (years). This situation is encountered in the degradation of, for example, halogenated hydrocarbons, ammonia, and hydrogen sulfide. For example, the degradation of dichloromethane yields hydrochloric acid:

$$CH_2Cl_2 + O_2 \xrightarrow{\textit{Hyphomicrobium sp.}} CO_2 + 2\,HCl$$

A second problem may arise if neutralization products, which can inhibit the biological activity, accumulate in the liquid phase within the packing material at inhibitory levels (Ottengraf & Diks, 1990). Both of these problems can easily be solved through the application of a biological trickling filter (Figure 3). In this system the gas flow is forced to rise through a bed of inert packing elements (rings, saddles) on which the microbial flora is immobilized. A water phase is recirculated continuously through this system to transfer the contaminants from the gas phase to the biolayer. This process also removes the acids produced in the filter bed. The pH of the liquid is controlled by adding an alkaline solution. By continuously refreshing the liquid phase, the concentration of the product of neutralization (NaCl) is kept below inhibitory levels (<200 mM NaCl in the case of eliminating dichloromethane; Janssen *et al.* 1987).

Experimental Results of the Removal of Dichloromethane in a Biological Trickling Filter. Dichloromethane is used in industry in huge amounts as a solute, degreasing agent, and brine. As this compound is suspected to have carcinogenic properties, the gaseous emission of dichloromethane, which amounted to some 4,000 tons/year in 1984 in The Netherlands (Bremmer *et al.* 1988), should be reduced. Therefore, experiments investigated the removal of dichloromethane from waste gases in a biological trickling filter (Diks & Ottengraf 1991). The biological elimination of dichloromethane in the trickling filter by

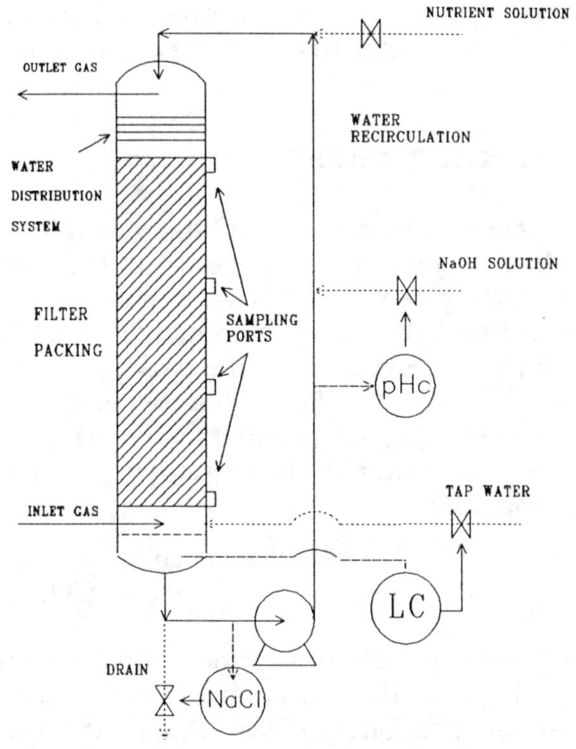

FIGURE 3. Experimental setup of a biological trickling filter system. The packing material consists of conventional packing elements.

Hyphomicrobium GJ21 (Janssen *et al.* 1987) was studied experimentally and theoretically at gas-phase concentrations in the range of 100 to 10,000 mg/m^3. Important results are discussed below.

A stable elimination performance was achieved for 2 years of operation at 20 C as shown in Figure 4. In this graph the elimination capacity, or the amount of substrate degraded per unit of reactor volume and time, is plotted as a function of the gas inlet concentration. The laboratory-scale trickling filter concerned here (height 3 m, diameter 0.4 m) was packed with 1.2-cm (0.5-in.) ceramic novalox saddles. At gas concentrations below 8 g/m^3/h, the elimination capacity is a function of the inlet gas concentration, whereas above this value the elimination capacity does not increase above a maximum of about 157 g/m^3/h.

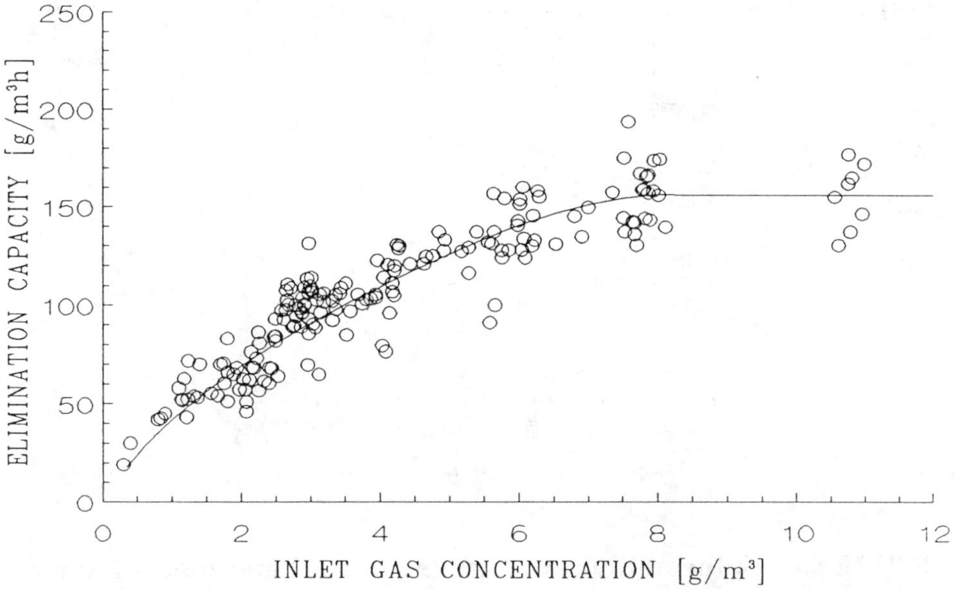

FIGURE 4. The dichloromethane elimination capacity vs. the inlet gas concentration: u_1 = 3.6 m/h; u_g = 160 m/h; maximum EC = 157 g/m³/h; C_{lcr} = 69 g/m³; data (o); theoretical model (solid line).

In the trickling filter, cocurrent flow as well as countercurrent flow of the gas and liquid phases was applied. From calculations using the model equations, relatively small performance differences appeared to exist between the two modes of operation. This is illustrated by the calculation of some theoretical axial gas and liquid concentration profiles in Figure 5a and 5b. Although considerable concentration gradients exist, an equal effluent gas concentration is reached. The experimental data presented in Figure 6 show that the elimination capacities for the two modes of operation are comparable.

Based on these results, a simplified model can be developed in which no relevant axial concentration gradients are assumed to exist in the liquid phase. If plug flow conditions of the gas phase are assumed, the dimensionless differential mass balance for the gas phase is given by

$$\left(\frac{dC_g^*}{d\xi}\right) = N_{og} \bullet (C_g^* - C_l^*) \tag{1}$$

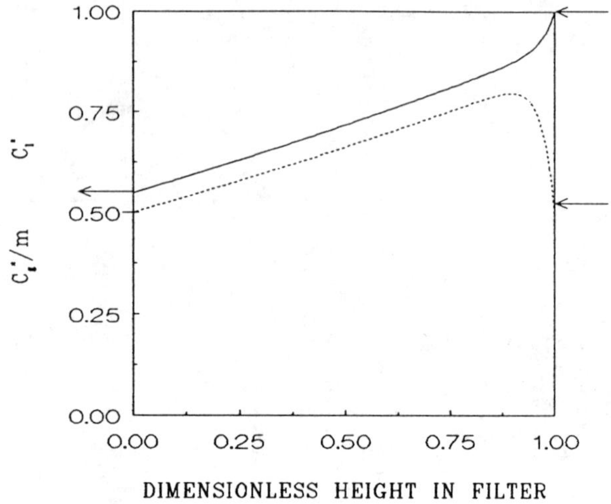

FIGURE 5a. A theoretical axial gas (solid line) and liquid (dotted line) concentration profile for cocurrent flow: $N_r = 0.7$; $N_{og} = 6$; $E = 3.8$; $C^*_{lcr} = 1.6$.

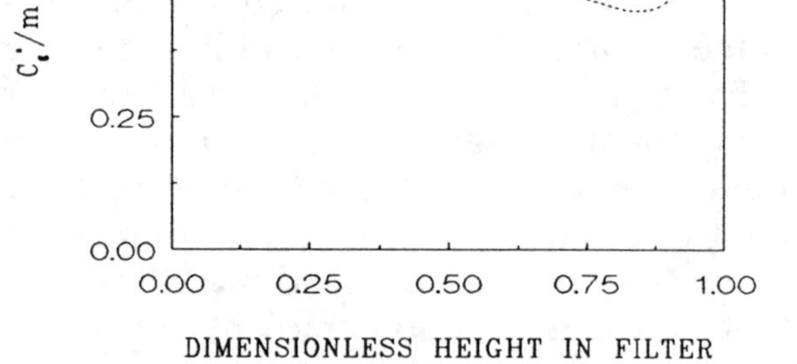

FIGURE 5b. A theoretical axial gas (solid line) and liquid (dotted line) concentration profile for countercurrent flow.

The overall dimensionless reactor balance is given by

$$1 - C_g^* (1) = N_r \bullet \eta \tag{2}$$

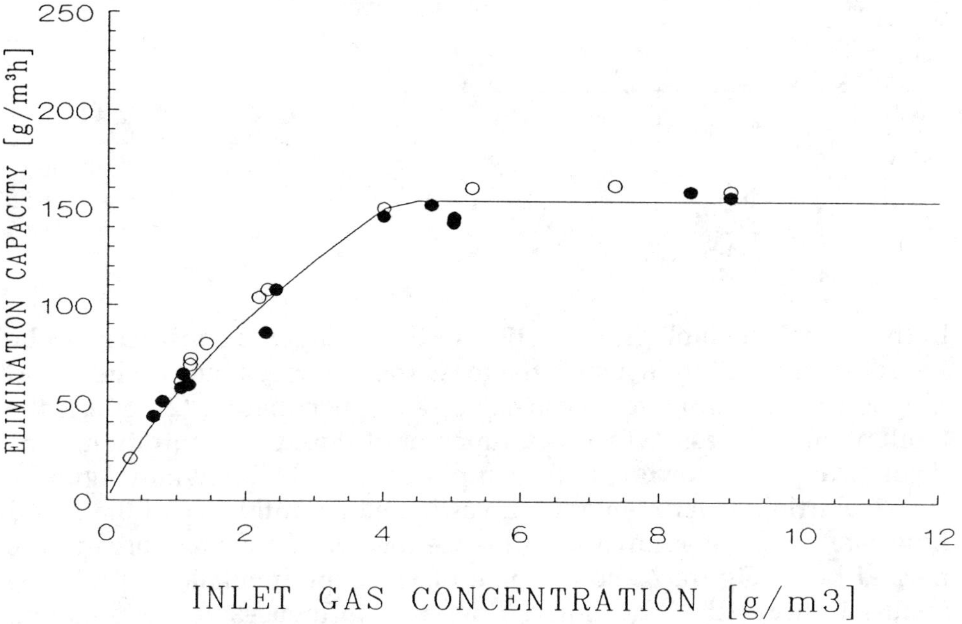

FIGURE 6. A trickling filter performance curve at cocurrent (o) and countercurrent flow (o): Theoretical model (solid line); u_1 = 10.8 m/h; u_g = 200 m/h; EC_{max} = 155 g/m^3/h; C_{lcr} = 29 g/m^3.

The boundary conditions belonging to this set of equations are

$$\xi = 0 \qquad C_g^* = 1 \tag{3a}$$

$$C_l^*(0) = C_l^*(1) \tag{3b}$$

The boundary condition according to equation (3b) accounts for the recirculation of the liquid phase and denotes that no substrate removal is assumed in the liquid phase outside the filter bed.

After solving this set of equations, the elimination capacity is given by

$$EC = EC_{max} \bullet \left\{ -K_1^* + \left[\left(K_1^*\right)^2 + K_2^* \right]^{1/2} \right\} \tag{4}$$

where

$$K_1^* = \frac{N_r}{2\, C_{lcr}^* \, (1 - \exp^{-Nog})}$$

$$K_2^* = \frac{N_r}{C_{lcr}^*}$$

In the model the biological reaction in the biolayer is considered to be a zero order reaction, whereas the mass transfer resistance in the liquid film near the biofilm interface is neglected. η in equation (2) reflects the biofilm effectiveness, which is a function of liquid concentration. The theoretical path according to the simplified model is shown in Figure 4.

Further experimentation investigated the influence of the liquid flow rate, which is known to influence the specific contact area in the packed bed. Figure 7 shows some of the experimental results. The results indicate that the liquid flow rate influences the elimination performance to a greater extent than can be expected from wetting correlations (Onda *et al.* 1968). This can be explained by the existence of an axial pH gradient in the filter bed from HCl production, which causes the microbial activity to decrease along the column from top to bottom. The pH gradient depends upon the liquid flow rate and the concentration of buffering components in the liquid phase (ammonium, phosphates, carbonates). By taking these into account, the influence of the liquid flow rate could be described quite well (Figure 7).

An investigation of the influence of water temperature revealed that the elimination rates at 20 and 30 C were comparable for inlet concentrations less than 8 g/m^3, as shown in Figure 8, although the biological reaction rate is known to depend upon the liquid temperature. From batch experiments using suspended microorganisms, it followed that the influence of the temperature between 20 and 30 C could be described by an activation energy of about 40 to 50 kJ/M, which is a common value for biological reactions. However, from the maximum elimination capacities in the trickling filter at both temperatures, an activation energy of only 10 to 20 kJ/M was calculated. This decreased temperature effect may result from the immobilized state of the microorganisms (Hauk 1987). Apart from the reaction rate, both the gas/liquid partition coefficient and the biofilm diffusion coefficient are dependent upon temperature. Taking these phenomena into account, the effect of temperature on trickling filter performance can be predicted quite well. The positive effect of a higher rate of degradation is

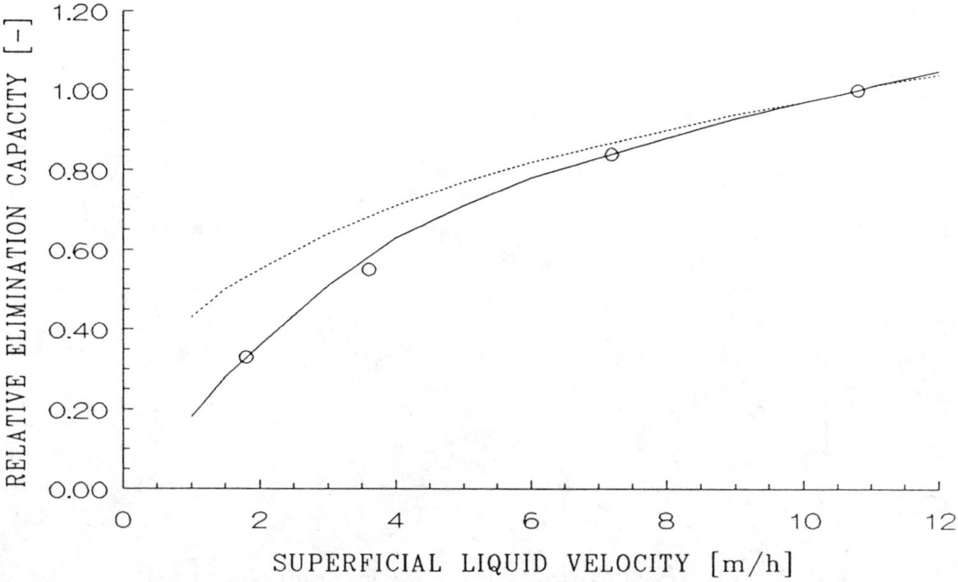

FIGURE 7. The influence of the liquid velocity on the relative EC according to the wetted area (dotted line), corrected for the axial pH gradient (solid line). Data (o); relative EC_{max} at u_1 = 10.8 m/h ≡ unity.

reduced by increasing the gas/liquid partition coefficient at higher temperatures, resulting in a lower concentration in the liquid phase at a constant gas inlet concentration. Experimental data as well as the curves predicted by the model are given in Figure 8.

From the present investigation, it can be concluded that stable elimination of dichloromethane is possible in a biological trickling filter. This process is well described with a simplified analytical model (Diks & Ottengraf 1991).

NOMENCLATURE

a_w	:	Specific wetted area (m^2/m^3)
C_{go}	:	inlet gas phase concentration (g/m^3)
C_g^*	:	C_g/C_{go} dimensionless gas concentration
C_{lcr}	:	critical liquid concentration (g/m^3)
C_{lcr}^*	:	$(m\ C_{lcr})/C_{go}$ dimensionless critical concentration

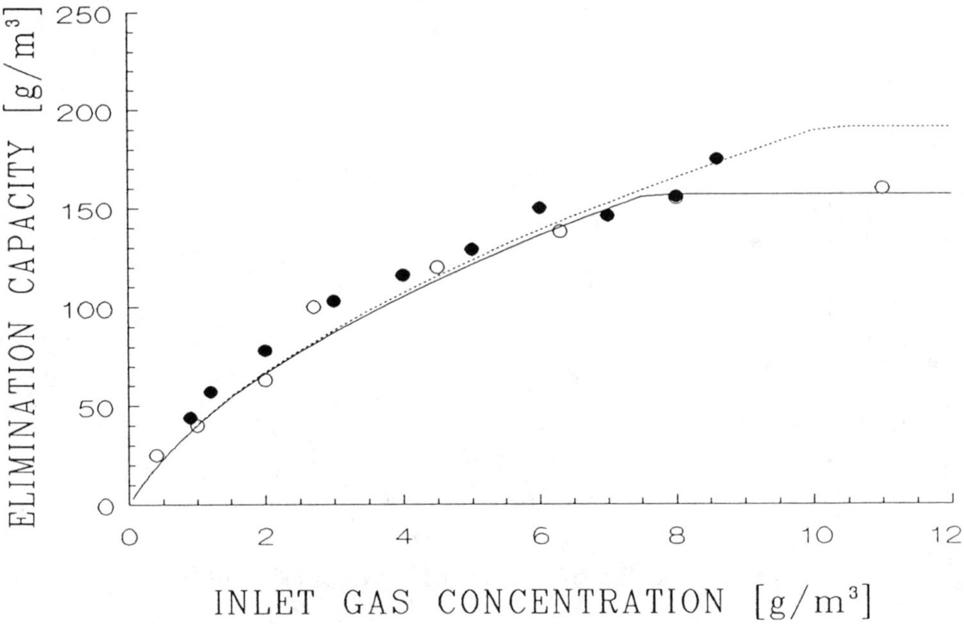

FIGURE 8. The experimental and theoretical performance of a countercurrent trickling filter at 20 C (o,——) and 30 C (o, ···) at u_1 = 3.6 m/h and u_g = 160 m/h.

C_l^*	:	$(m\ C_l)/C_{go}$ dimensionless liquid concentration
E	:	$(m\ u_g)/u_l$ extraction factor
E_{act}	:	activation energy for the biological reaction (kJ/m)
EC	:	elimination capacity, or the amount of substrate degraded per unit of reactor volume and time (g/m³/h])
EC_{max}	:	maximum elimination capacity (g/m³/h)
H	:	height of the packed bed (m)
k_{og}	:	overall gas phase mass transfer coefficient (m/h)
K_1^*	:	dimensionless constant given by equation (4)
K_2^*	:	dimensionless constant given by equation (4)
m	:	C_g/C_l gas liquid distribution coefficient (m^3_{liquid} / m^3_{gas})
N_{og}	:	$(k_{og}\ a_w\ H)/u_g$ number of gas phase transfer units
N_r	:	$(EC_{max}\ H)/(u_g\ C_{go})$ number of reaction units

T	:	absolute temperature (K)
u_g	:	superficial gas velocity (m/h)
u_l	:	superficial liquid velocity (m/h)

Greek symbols:

η	:	biofilm effectiveness factor
ξ	:	dimensionless height in the filter bed

REFERENCES

Bremmer, H., et al. Report of the National Institute of Public Health and Environmental Protection in The Netherlands (RIVM), No. 738608002, March 1988.

Diks, R.M.M.; Ottengraf, S.P.P. *Bioproc. Eng.* **1991**, *3*, Part I and Part II, 390–399.

Hauk, G. Thesis, Munchen Technical University, 1987.

Janssen, D.; Kuijk, L.; Witholt, B. Presented at Int. Meet. on Biological Treatment of Waste Gases, DECHEMA, Heidelberg, Germany, 24–26 March 1987.

Onda, K., *et al. J. Chem. Eng. Japan* **1968**, *1*, 56–62.

Ottengraf, S.P.P.; van den Oever, A.H.C. *Biotechnology and Bioengineering*; Wiley: New York, 1983; pp 3089–3102.

Ottengraf, S.P.P.; Diks, R.M.M. *Chim. Oggi* **1990**, *8*, 41–45.

Bioremediation of Contaminated Land: Bench Scale to Field Applications

M. Talaat Balba[*], *Anthony C. Ying, Thomas G. McNeice*
TreaTek® Incorporated

Despite a number of brief reports in the literature, there are only a few examples of detailed data from full-scale bioremediation projects. Most of the results presented have referred to pilot-scale field trials. There are probably two major reasons for this. Contaminated sites are often remarkably heterogeneous in nature so that the initial starting data vary from very low to very high concentrations over a relatively small area. In addition, large volumes, generally on the order of thousands of cubic yards of soil, are involved. Under such circumstances, it is very difficult to obtain statistically meaningful data without analyzing a massive number of samples which would be prohibitively expensive.

The following projects present some examples of the use of biotechnology in the remediation of contaminated soil and groundwater.

Case Study #1: Bioremediation of Heavy Petroleum Hydrocarbon Contaminated Soil

Numerous locomotive maintenance yards are operated by major railroad companies across the country. In recent years, an increasing number of these railroad yards are being identified as having soils contaminated with various concentrations of petroleum hydrocarbons. The source of the contamination is refueling operations and general locomotive servicing where hydrocarbon-type products such as diesel fuel and heavy motor oil are routinely used.

The current available means for cleaning up these oil-contaminated soils is off-site disposal at a secured land disposal facility.

[*] TreaTek Inc., 2801 Long Road, Grand Island NY 14072

In addition to the significant cost of this approach, the inherent liabilities associated with waste transportation and land disposal make this a very unattractive remedial option.

In early 1988, TreaTek®, Inc., was commissioned to develop and design a cost-effective biological treatment program specifically for this type of heavy engine oil contamination. This project was partially funded by the Alternative Technology Section of California's Department of Health Services under the California Hazardous Waste Reduction Grant Program.

Laboratory Program. A multi-step laboratory treatability program was carried out using actual oil-contaminated soil samples collected from an operating railyard. The soil contained from 5,000 ppm to 60,000 ppm petroleum hydrocarbons. The program was designed to focus primarily on microbial and chemical additives for enhancing the natural degradation of oil in the soil.

The oil extracted from the railyard soil was characterized and found to contain mostly linear and branched alkanes in the C-22 plus range. The oil consists mainly of heavy engine oil with a boiling point of more than 200 C and an asphaltene content of 0.1 percent.

The experimental program also involved several other preliminary tasks including microbial assessment and inocula optimization. These initial steps were then followed by two actual bench-scale treatment operations. The first involved 130 model ecosystems (or microcosms), each containing 500 g of contaminated soil. These microcosms were treated with multiple combinations of different microbial and chemical reagents. Oil degradation in these microcosms was measured and compared over a 16-week period. The second set of treatment simulations involved a much larger quantity of soil (approximately 9 kg each). These scaled-up systems enabled actual soil aeration testing and provided a means to measure potential oil removal via volatilization or leachate generation.

The laboratory data produced at 15 C showed that up to 94 percent of the total petroleum hydrocarbons in soil were degraded in less than 16 weeks.

Rigorous aeration of the soil in the simulation units showed minor volatilization of the contaminant at high concentration (soil with >3% oil), while no volatilization was observed with soil below 1 percent oil.

The project is now in the field demonstration phase; to date, more than 150 yd^3 of heavily contaminated soil (>10% oil) from the site has been excavated and placed on two enclosed rectangular beds (40 ×

4 m each), which are resting on concrete liners to prevent leaching into subsurface soils. The beds drain into an adjacent sump from which leachate can be collected and analyzed. The soil is kept aerated and mixed by using an agricultural spading machine. Nutrients and microbial inoculation of the soil are being applied to the soil beds by spraying, followed by soil mixing. Biodegradation parameters such as pH, nitrogen, phosphorous, and microbial population count are regularly monitored and adjusted when required. The beds are being sampled once every 4 weeks for total petroleum hydrocarbons (TPH) analysis using a sampling grid consisting of 13 samples from each bed. Routine TPH analyses are being carried out using EPA Method 418.1, and gas chromatographic analysis using a flame ionization detector is performed only intermittently to confirm biodegradation.

In spite of the complex nature of the contaminant and its extremely high starting concentration in the soil (>100,000 mg/kg), the treatment has been successful in substantially reducing the contamination level. More than 85 percent biodegradation has been achieved, so far within less than 28 weeks (Figure 1). This field data shows that hydrocarbon-containing soils from locomotive maintenance yards can be successfully remediated using bioremediation techniques.

Case Study #2: Bioremediation of an Oil-Polluted Refinery Site

This study was designed to assess the feasibility of bioremediation within an oil-polluted refinery site. The site occupies several hectares and is situated near a major river. Oil contamination penetrates to depths in excess of 6 m, and therefore excavation and removal of the polluted area would prove very expensive and complicated by fluctuations in the groundwater level. Furthermore, certain surface layers of the refinery soil contain contaminated clay lattices not amenable to *in situ* processes. Therefore, a dual treatment process was envisaged, incorporating microbial remediation of heavy contaminated excavated clay soil combined with bioaugmented *in situ* techniques for deeper layers.

During the field trials, two areas of contamination were selected. Site 1 underwent *in situ* treatment and consisted of approximately 1,600 m^3 of oil-contaminated land (0–8 m depth). Site 2 consisted of approximately 200 m^3 of more heavily contaminated subsurface soil layers (0–2 m depth).

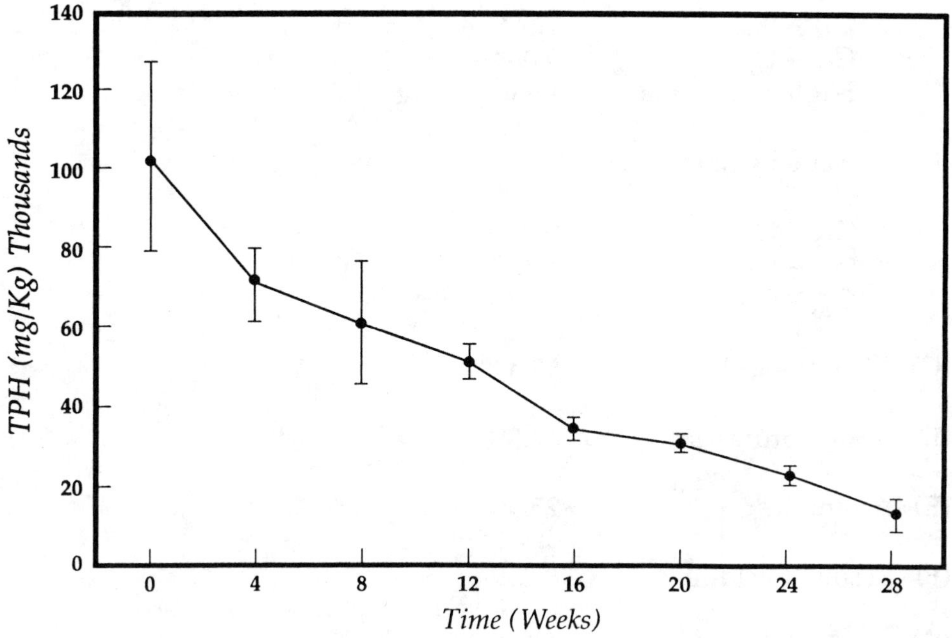

FIGURE 1. Bioremediation of heavy petroleum oil in soil at a railroad maintenance yard (figures represent the means of 13 samples).

Laboratory Program. The laboratory program consisted of several tasks including the following:

- Soil, oil, and groundwater characterization
- Microbial isolation and screening of oil-degrading microorganisms
- Surfactant screening for oil dispersants
- Soil microcosm optimization

Detailed analysis of the oil extracted from the refinery soil showed that it had the following composition:

(A) Alkane:

$C_5 - C_{10}$	Insignificant
$C_{11} - C_{13}$	0.22%

$C_{14} - C_{25}$	1.59%
$C_{26} - C_{27}$	0.04%
Higher Carbons	Insignificant

(B) Branched Chain Alkane

$C_{11} - C_{13}$	0.08%
$C_{14} - C_{25}$	1.10%
$C_{26} - C_{27}$	0.03%

(C) Naphthenes	59.44%
(D) Total Saturates	62.5%
(E) Aromatics	25.0%
(F) Resins/Residue	6.25%
(G) Asphaltines	6.25%

Axenic cultures capable of oil degradation were successfully isolated from the soil and screened for their oil degradation capabilities and survival in the contaminated soil. The results in Table 1 show that not only did each species tested survive, but also proliferated within the soil, clearly indicating their tolerance to the soil contaminants.

Oil dispersants and surfactants were also screened and tested for their ability to mobilize oil contaminants in soil. More than 50 compounds were examined and the most promising of these agents were subjected to permeability tests using methods described elsewhere. Surface active compounds were further examined using soil column leach tests with and without recirculation of the leachate. These investigations focused attention on the ability of the various surfactants to effect oil movement from contaminated soil under *in situ* conditions and also monitored their influence on soil permeability. One compound, Cyanamer P70 (Cyanamid), was selected because it removed a relatively high proportion (47 percent) of the oil contaminant while maintaining soil permeability even when applied at high concentration (>1,000 mg/L). Another advantage of this compound was its descalent properties which alleviated the possibility of metal precipitation during groundwater extraction, aeration, and infiltration. Metal precipitation of specially iron and manganese is a common problem in *in situ*

TABLE 1. Characteristics of oil-degrading microorganisms isolated from the refinery soil.

Designated Code	Microorganisms Species	Growth of Oil Fraction[a]			Degradation of Refinery Oil (%)[b]	Survival of Cells in Contaminated Soil (%)[c]
		Gas Oil	Hexadecane	Asphaltene		
ETB001/2	Pseudomonas	+	+	---	81.0	55.9
ETB004	Rhodococcus	+	+	---	66.0	70.8
ETB004/2	Acinetobacter	+	+	---	56.0	17.2
ETB005/2	Pseudomonas	+	+	---	61.0	51.7
ETB007	Mycobacterium	+	+	+	93.0	ND
ETB015/3	Arthrobacter	+	+	---	82.0	55.6
Control	None	---	---	---	0	---

(a) Growth was monitored qualitatively (increase in absorbance of >0.5 compared with controls at 620 nm) in liquid culture after 10 days incubation (25 C, 200 rpm). Medium consisted of basal salts containing the relevant oil fraction (0.2% w/v).

(b) Refinery oil was determined quantitatively using infrared spectroscopy.

(c) Survival (viable cell counts as a percentage of the initial values) after 10 days incubation in oil-contaminated soil.

ND = not determined.

remediation and can result in severe soil clogging problems, which affect the success of the remediation process.

Toxicity of this surfactant was also examined toward three common soil microorganisms: *Arthrobacter*, *Rhodococcus*, and *pseudomonas*, plus the oil degrading species isolated from the refinery site (Table 1). Each species examined was found to tolerate up to 1,000 mg/L surfactants in liquid culture, and the presence of 2,000 mg/L did not affect growth on solid media.

The laboratory program also included soil microcosm tests to optimize microbial treatment of contaminated soil; a series of soil microcosm tests were carried out. More than 80 different treatment combinations were tested with the objective of optimizing the nutrient status of the soil and mobilizing oil contaminants in order to enhance degradation by microbial inocula. The results showed that nutrient addition and moisture optimization alone resulted in approximately 79 percent oil reduction within 23 weeks, while the use of nutrient and microbial inocula plus the surfactants resulted in 95 percent oil reduction during the same period (Figure 2). These successful laboratory data led to field trials which were performed over a 6-month period to treat a portion of the refinery site; the area examined represented a volume of approximately 2,000 m^3 of soil. The field trials involved two separate trials; one concerned of the soil, the other *in situ* treatment.

***In Situ* Treatment.** The treatment of groundwater and subsurface soil over an area measuring 10 × 20 m and to a depth of 8 m was achieved over a 15-week period. Treatment was effected using groundwater extraction and infiltration systems. *In situ* biodegradation was enhanced by applying the surface-active agent, nutrients, aeration, and inoculation with key microorganisms. Groundwater was extracted from the central well or from 10 well points surrounding the treatment zone. Two vacuum pumps, operating at approximately 30 m^3/h, forced liquid from an 8-m-deep central well to an aerated 50-m^3-capacity reactor tank and then piped back into gravel infiltration trenches at the perimeters of the treatment area (Figure 3).

Tracer dye tests were performed using a series of well points around the *in situ* treatment area in order to follow the nature of flow during treatment. Approximately 5 percent of flow was diverted to an on-site wastewater treatment facility to enhance groundwater movement and to maintain an effective zone of influence.

Organic and inorganic nutrients were applied to maintain an optimum C:N:P ratio. Microorganisms were used as inoculants on four occasions during the trial period.

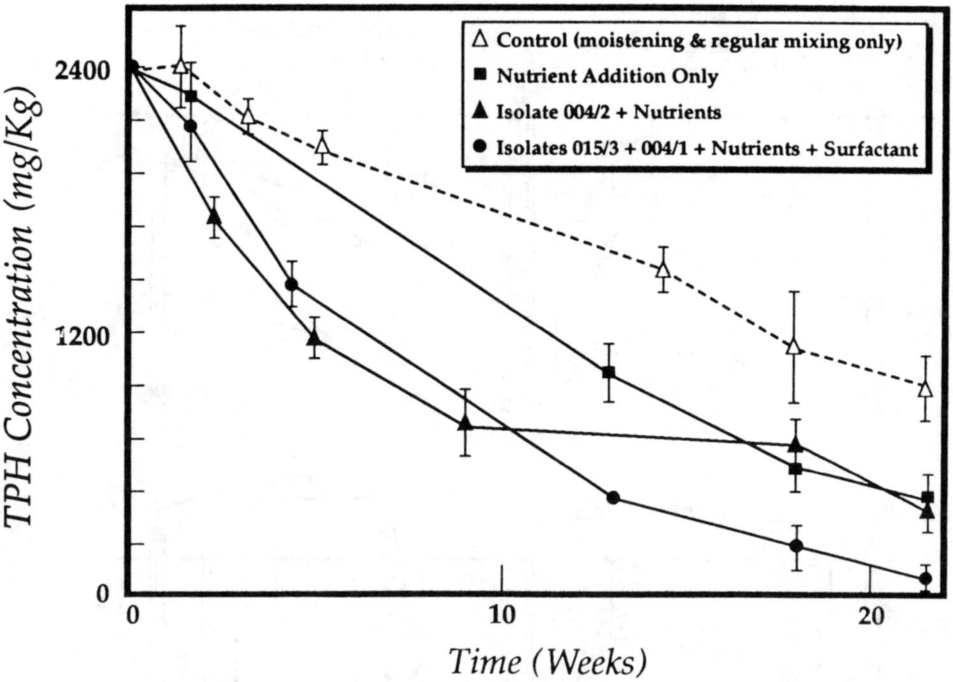

FIGURE 2. Effects of biological treatment on oil concentrations in soil microcosm test (mean values of four replicates).

Soil and groundwater samples (1 kg) were taken before and after treatment over a 15-week period. A 20-point sampling grid was used, with samples being taken at five depths per point (100 samples per sample time). Liquid samples were taken within the treatment area and at five points surrounding the site in order to monitor possible interference and changes in groundwater characteristics.

Liquid entering the *in situ* treatment area was found to contain an oxygen concentration of 8.4 mg/L but only 2.4 mg/L in extracted leachate. This provided some indication of the rate of oxygen consumption within the zone of treatment during the remediation program and confirmed biological degradation.

Results of oil analyses from the *in situ* treatment of the site are shown in Figure 4, including the mean oil concentrations at various depths. Clearly there were significant reductions at every depth tested, and the overall mean oil hydrocarbon concentration was reduced from 185 ± 49 mg/kg to 26 ± 6 mg/kg (86% reduction) within 15 weeks. Leachate analysis on the surrounding groundwater area resulted in no detectable oil hydrocarbons.

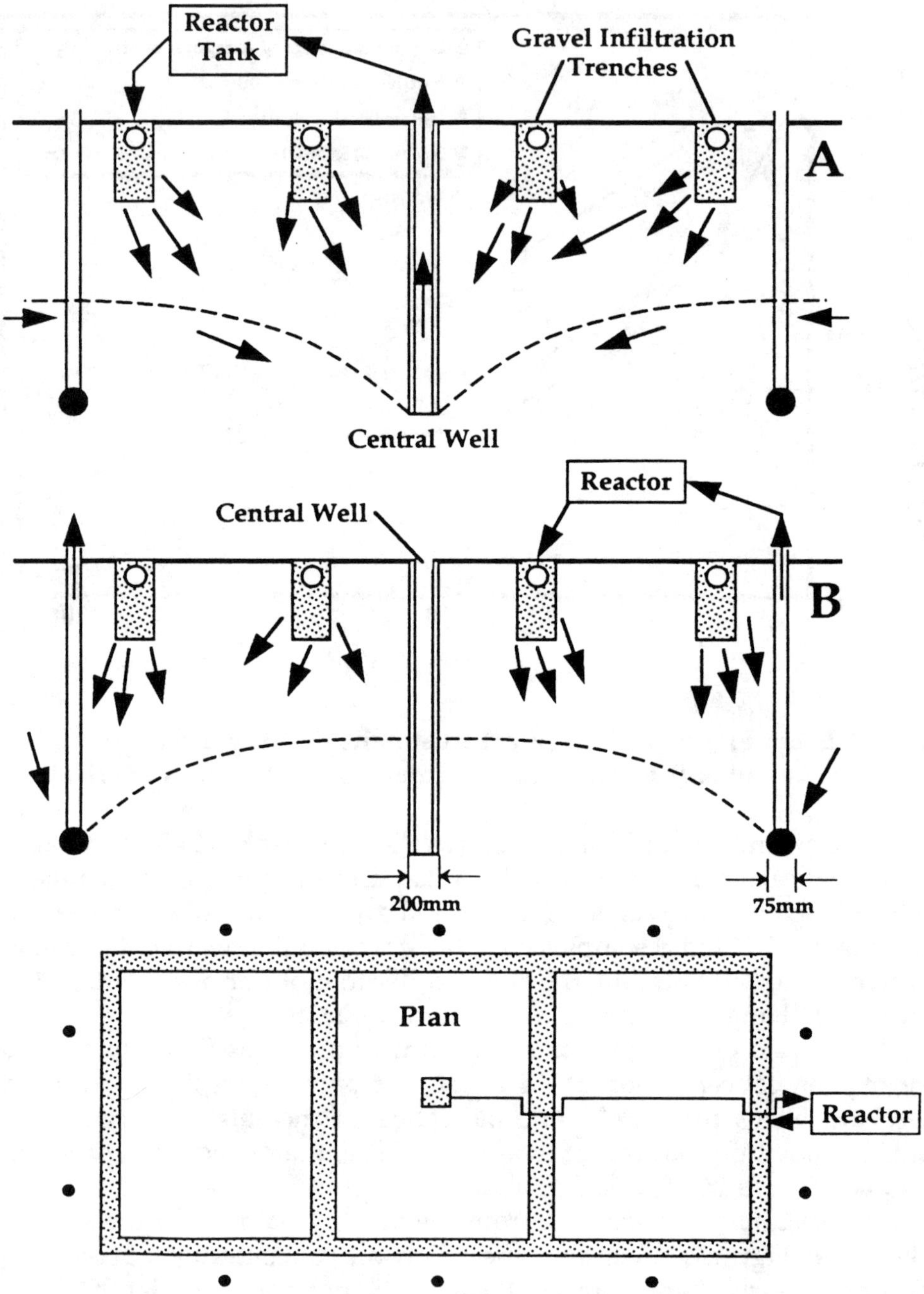

FIGURE 3. ***In situ*** **treatment operation modes: (A) During normal pumping out through the central well. (B) When the system was reversed and liquid was pumped out using the perimeter wells.**

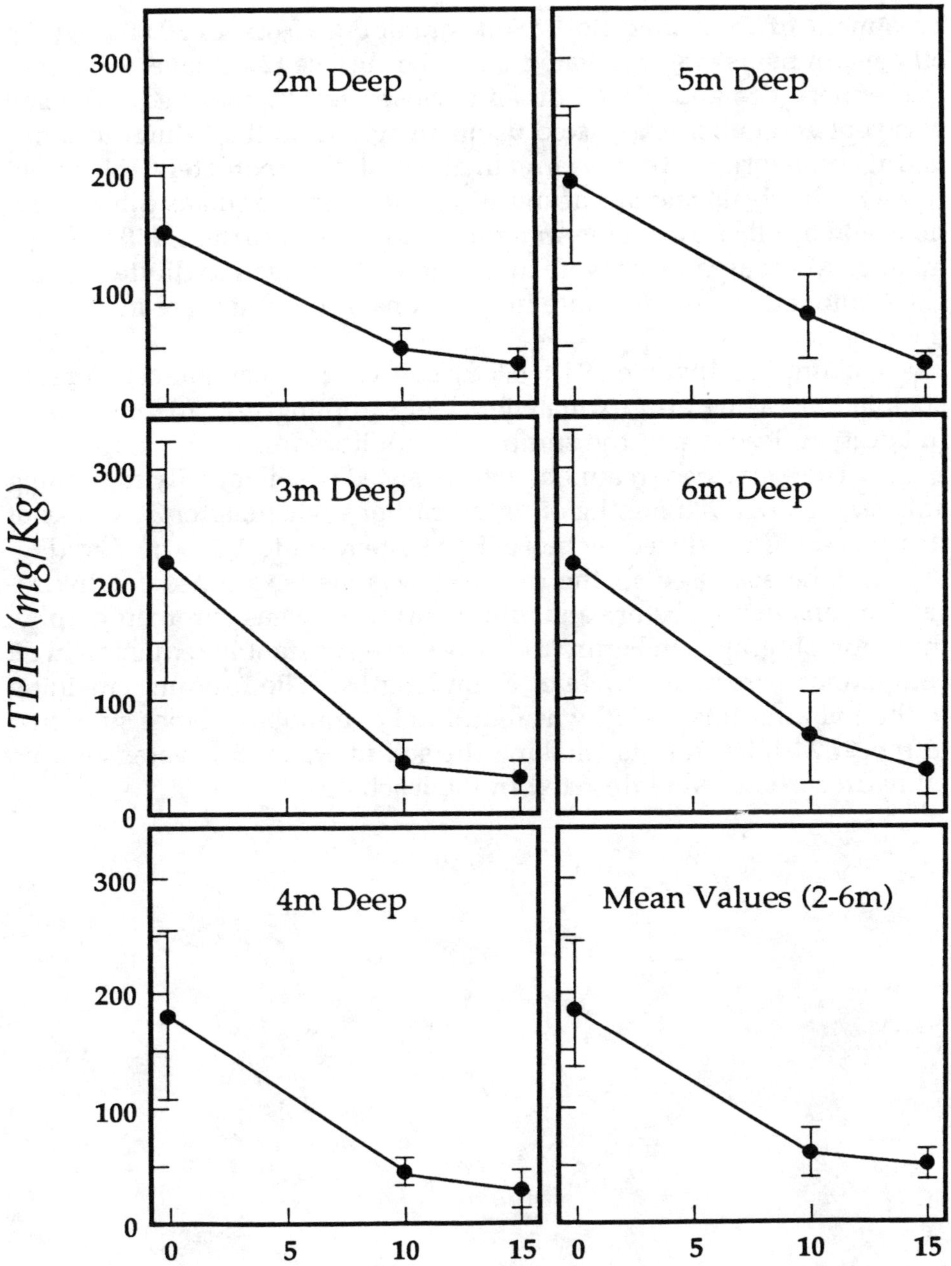

FIGURE 4. ***In situ*** **treatment of refinery land. Figures indicate the mean concentrations of oil hydrocarbons at various depths (20 samples per depth).**

Treatment of Excavated Soil. Soil containing 10,000 to 30,000 mg/kg oil contaminants was excavated from the surface (2 m) layer of part of the refinery site and placed on an enclosed rectangular bed. The soil was kept aerated; it was mixed using an agricultural spading machine, and its moisture content was maintained at approximately 15 percent (w/w). Nutrients and surfactant were added to the soil as dilute solutions and applied by tractor-spraying onto the soil surface, followed by mixing. Microorganisms were inoculated into the soil as dilute suspensions; altogether, five separate inoculations of the soil were made over the trial period.

Sampling involved 99 × 0.5 kg soil samples being taken at regular intervals using approximately a 2-m sampling grid (0.4 m depth). In addition, leachate in the sump was monitored.

The excavated treatment had consisted of soil initially containing an average of 12,980 mg/kg oil hydrocarbons which, after 34 weeks of treatment, were reduced by more than 90 percent to 1,273 mg/kg (Figure 5). The soil used in the trial test was also examined for hydrocarbon constituents before and after treatment by gas chromatography; the chromatographs in Figure 6 clearly show substantial reduction in oil compounds across the range of chain lengths. The residue remaining in the soil after this period was found to be immobile. Since subjected to repeat 24-h leach tests (shaking the soil in water, 5:1 water:soil), no oil hydrocarbons were detected in the leachate.

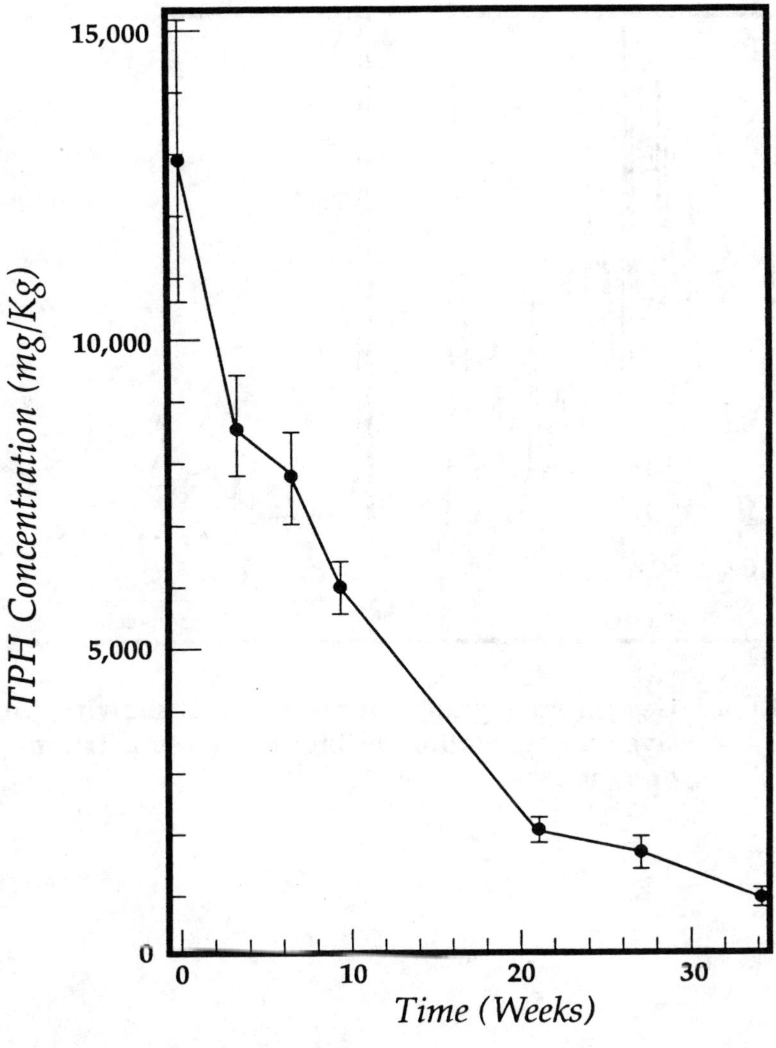

FIGURE 5. ***Ex situ*** **treatment of refinery soil. Figures indicate the means of 100 samples.**

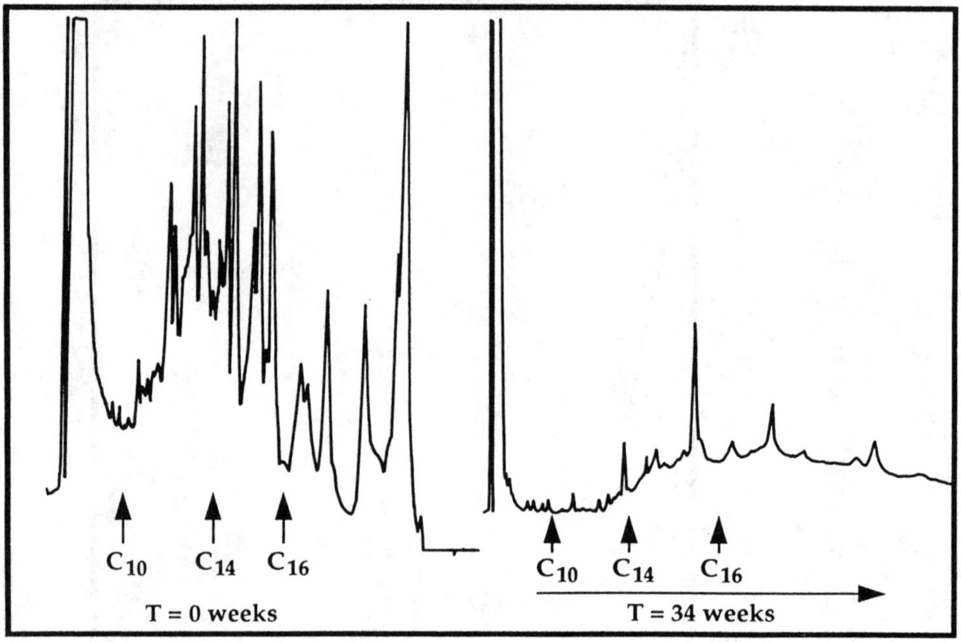

FIGURE 6. Gas chromatographs of refinery soil showing progressive oil degradation during the bioremediation program.

Biological Treatment of Hanford Groundwater: Development of an *Ex Situ* Treatment Process

Thomas M. Brouns, Sydney S. Koegler, James K. Fredrickson, Stuart P. Luttrell, Kathy A. Borgeson*
Pacific Northwest Laboratory

Liquid wastes containing radioactive, hazardous, and regulated chemicals have been generated throughout the 40 years of operations at the U.S. Department of Energy's (DOE) Hanford Site in southeastern Washington state. Some of these wastes were discharged to the soil column, and many of the waste components, including nitrate (NO_3^-) and carbon tetrachloride (CCl_4), have been detected in the groundwater of the unconfined aquifer at Hanford.

The use of facultative anaerobic microorganisms is a promising technology for the *ex situ* and *in situ* removal of nitrates and organics from contaminated aqueous streams. This paper describes an *ex situ* biological process being developed at Pacific Northwest Laboratory (PNL) for removal of NO_3^- and CCl_4 from contaminated groundwater, and approaches for *in situ* denitrification and CCl_4 biodegradation.

Ex Situ Treatment: Design and Testing

To develop an *ex situ* biological treatment system for the destruction of NO_3^- and CCl_4 contaminants, laboratory-, bench-, and pilot-scale testing has been performed. This paper summarizes the results of pilot-scale testing and provides a comparison of the results to previously reported bench-scale data. Laboratory tests were conducted to select microbial cultures; identify physical, chemical, and biological requirements; confirm the degradation of CCl_4; and determine the potential for toxicity

* Pacific Northwest Laboratory, P.O. Box 999, Richland WA 99352

Pacific Northwest Laboratory is operated by Battelle Memorial Institute for the U.S. Department of Energy under Contract DE-AC06-76RLO 1830.

of contaminant and cocontaminant concentrations to the microorganisms. The bacterial consortium used for the *ex situ* process was obtained from groundwater samples from the Hanford unconfined aquifer and shown to be capable of dissimilatory NO_3^- reduction and CCl_4 degradation (Hansen 1990; Koegler *et al.* 1989). In these laboratory-scale tests, biodenitrification in Hanford groundwater samples was most efficient at pH 8, 29 C, and with acetate as the primary carbon source (Koegler *et al.* 1989). In addition, CCl_4 degradation tests using ^{14}C-labeled CCl_4 confirmed that CCl_4 was being degraded to CO_2 under the same conditions (Hansen 1990; Koegler *et al.* 1989). The capability of denitrifying bacteria to mineralize CCl_4 is consistent with reports from other researchers (Bower & McCarty 1983; Bower & Wright 1988; Egli *et al.* 1988). Carbon tetrachloride, cyanide, and hexavalent chromium, at the concentrations these contaminants are found in Hanford groundwater, did not significantly inhibit the metabolism or viability of the microorganisms (Brouns *et al.* 1990; Hansen 1990).

Bench-scale tests were conducted to determine the degradation rate of NO_3^- and CCl_4 and to develop reaction rate information that could be used for engineering scale-up. As anticipated, the rate of NO_3^- degradation was dependent on both the acetate and volatile suspended solids (VSS) concentrations (Brouns *et al.* 1990; Koegler *et al.* 1989). Like denitrification, CCl_4 destruction appeared to depend on the metabolism of the primary carbon source, acetate (Brouns *et al.* 1990). This dependence and linkage to acetate metabolism was confirmed with additional laboratory-scale tests using ^{14}C-labeled CCl_4 (Hansen 1990).

A pilot-scale biological treatment system has been constructed for use in evaluating the applicability of biodenitrification for Hanford groundwater remediation. The system consists of two groundwater feed tanks; a continuous stirred-tank bioreactor (CSTR); a fluidized-bed bioreactor (FBR); a biomass settling tank or clarifier; and pressure, temperature, off-gas, and pH monitoring and control equipment. Two 1,800-L, fiberglass-reinforced plastic feed tanks supply nitrate-bearing wastewater or groundwater to the bioreactors. With a nominal design feed flow rate of 240 L/day, each feed tank is capable of supplying groundwater to the bioreactors for 7 days. Low-flow metering pumps also supply a concentrated acetic acid solution to the bioreactors as the primary carbon source. Carbon tetrachloride was added directly to the CSTR via a syringe pump to study CCl_4 degradation with simulated groundwater.

The 50-L CSTR is equipped with internal baffles and flat-blade impellers. Thermocouples and a submersible pH electrode are included for monitoring and automatic control of pH and temperature. The reactor effluent, which contains suspended biomass, is fed to a settling vessel (clarifier) to recover the biomass for recycle and to clarify the denitrified simulated groundwater. The FBR is constructed of glass pipe, with a 1.5-L fluidized-bed section (D = 3.8 cm, L = 130 cm) and a de-entrainment section (expansion zone) equipped with an overflow clarifier. Based on samples collected during testing, most of the denitrification occurs in the fluidized-bed section. The overall liquid volume of the FBR is 15.5 L. Anthracite coal provides the support medium for microorganism attachment.

In 1989, the system was used in pilot-scale tests to demonstrate degradation of NO_3^- and CCl_4. The CSTR was initially inoculated with 4 L of the bacterial consortium (bacterial characterization of the consortium is described in Hansen, 1990). The system was operated in a batch mode through several growth cycles, then fed continuously with simulated groundwater at 1.0 L/h (2-day residence time). The composition of the simulated groundwater is described in Brouns *et al.* (1990). The feed flow rate was increased to 3.0 L/h and then to 6.0 L/h (8-h residence time). The system was operated for a total of 160 days to obtain steady-state denitrification data. Degradation of CCl_4 was tested under conditions similar to those used during the denitrification test (8-h liquid residence time, recycle ratio of 0.4, and 30 C), with the exception that a CCl_4 solution (3.5 wt% CCl_4 in methanol) was injected directly into the bioreactor using a low-flow syringe pump. The quantity of CCl_4 injected produced a feed concentration of 0. 14 mg/L in the simulated groundwater. All liquid influent and effluent streams in the process were sampled and analyzed for CCl_4 during the test using standard purge and trap techniques followed by gas chromatography with electron capture and flame ionization detectors. The off-gas flow rate and CCl_4 concentration were also measured during the test. Steady state was achieved as determined by steady flow rates and NO_3^- and VSS concentrations.

At the conclusion of the CCl_4 test, a control test in which the groundwater and nutrient feed streams were maintained, but biomass was eliminated, was performed to complete a CCl_4 material balance, and to confirm that biological degradation mechanisms were responsible for the observed disappearance of CCl_4. Greater than 99 percent of the CCl_4 was recovered and accounted for in this test.

Denitrification was studied in the FBR by inoculating the reactor with approximately 2 L of the bacterial consortium (8 g VSS) and 0.3 L (270 g) of anthracite coal. To promote initial growth of the microorganisms on the coal particles, the system was operated in a batch mode using total recycle. Following this start-up period, the FBR was operated continuously at feed rates of 6 and 12 L/h to obtain denitrification rates. The reactor operated at steady state, as determined by steady flow rates and NO_3^- and VSS concentrations.

Results and Discussion

Results of CSTR and FBR pilot-scale tests are summarized in Table 1. At the pilot-scale, both the CSTR and the FBR reduced concentrations of NO_3^- in simulated groundwater to levels well below the drinking water standard of 44 mg/L. The CSTR also demonstrated continuous degradation of CCl_4 over a 2.5-month test period.

Although specific degradation rates do not adequately represent true reaction rates except for zero-order reactions, specific rates are useful for comparison when limited data are available. The specific NO_3^- and CCl_4 removal rates from pilot-scale testing are generally within the range of specific degradation rates obtained previously from bench-scale tests. Specific rates for NO_3^- degradation ranged from 35 to 350 mg NO_3^-/g VSS/h in batch tests with initial NO_3^- concentrations of 44 to 180 mg/L (Brouns *et al.* 1990). Specific rates for CCl_4 degradation ranged from 0.0097 to 0.083 mg CCl_4/g VSS/h in batch tests with initial CCl_4 concentrations of 0.1 to 2.0 mg/L (Brouns *et al.* 1990; Stuart 1990). Although first-order rate constants for substrate (acetate) metabolism were derived from bench-scale data and correlate well with previously reported kinetics for denitrifying cultures (Stuart 1990), the high degree of variability in specific NO_3^- and CCl_4 degradation rates from bench-scale tests could not be explained by differences in acetate, NO_3^-, or VSS concentrations (Brouns *et al.* 1990; Stuart 1990). However, the specific removal rates for both NO_3^- and CCl_4 are generally consistent with the lower values of the range obtained from bench-scale tests. Lower values would be expected because both pilot-scale reactors operated at extremely low effluent NO_3^- (NO_3^--limited) conditions. The volumetric denitrification rates for the FBR are significantly greater than those obtained for the CSTR and most likely can be attributed to the higher VSS concentrations obtained using an attachment medium.

TABLE 1. Summary of results of pilot-scale testing.

Test System	NO_3^- Concentration (mg/L) Feed	NO_3^- Concentration (mg/L) Effluent	CCl_4 Concentration (mg/L) Feed	CCl_4 Concentration (mg/L) Effluent	Residence Time (h)	Sludge Age (day)	VSS (g/L)	NO_3^- or CCl_4 Removal Rate
CSTR	384	<1	0.0	0.0	8.3	≈100	1.0	44 mg NO_3^-/g VSS/h 49 mg NO_3^-/L/h
CSTR	387	<1	0.142	0.006	8.3	≈100	1.2	40 mg NO_3^-/g VSS/h 47 mg NO_3^-/L/h 14 µg CCl_4/g VSS/h
FBR	400	<1	0.0	0.0	2.6	≈12	17	8.8 mg NO_3^-/g VSS/h 150 mg NO_3^-/L/h
FBR	400	<1	0.0	0.0	1.3	≈19	23	13 mg NO_3^-/g VSS/h 305 mg NO_3^-/L/h

Summary

Bioremediation is a promising technology for the simultaneous removal of nitrates, via denitrification, and CCl_4, via biodegradation, from contaminated groundwaters. Microorganisms native to the Hanford Site have been shown to denitrify and degrade CCl_4 when supplied with an electron donor such as acetate. To apply this degradative capability, a pump-and-treat *ex situ* process that uses the native microorganisms to destroy both NO_3^- and CCl_4 has been developed. The pilot-scale treatment process was demonstrated with a simulated groundwater feed. Concentrations of NO_3^- and CCl_4 were reduced to levels below the drinking water standards of 44 mg/L and 0.005 mg/L, respectively. In the future, the process will be evaluated using actual groundwater from the Hanford Site.

Based on the promising results of pilot-scale testing, an *in situ* process for treating contaminated groundwater is also being developed. The approach will be to stimulate the indigenous microorganisms to promote the transformation of NO_3^- and the degradation of CCl_4. *In situ* treatment offers the potential benefits of lower operating costs, a more complete removal of sorbed contaminants, and lower potential exposure of workers to hazardous chemicals. A demonstration site at Hanford for *in situ* biological treatment has been selected, and hydrological, chemical, and biological characterization of the site has been initiated. Methods for supplying nutrients to the subsurface are currently being developed and will be tested in the laboratory prior to field experimentation.

REFERENCES

Bower, E. J.; McCarty, P. L. "Transformation of Halogenated Organic Compounds under Denitrification Conditions." *Applied and Environmental Microbiology* **1983**, *45*, 1295–1299.

Bower, E. J.; Wright, J. P. "Transformation of Trace Halogenated Aliphatics in Anoxic Biofilm Columns." *Journal of Contaminant Hydrology* **1988**, *2*, 155–169.

Brouns, T. M.; Koegler, S. S.; Heath, W. O.; Fredrickson, J. K.; Stensel, H. D.; Johnstone, D. L.; Donaldson, T. L. *Development of a Biological Treatment System for Hanford Groundwater Remediation: FY 1989 Status Report*; PNL-7290, Pacific Northwest Laboratory: Richland, WA, 1990.

Egli, C.; Tschan, T.; Scholtz, R.; Cook, A. M.; Leisinger, T. "Transformation of Tetrachloromethane to Dichloromethane and Carbon Dioxide by *Acetobacterium woodii*." *Applied and Environmental Microbiology* **1988**, *45*, 2819–2824.

Hansen, E. J. "The Transformation of Tetrachloromethane under Denitrifying Conditions by a Subsurface Bacterial Consortia and Its Isolates." M.S. Thesis, Washington State University, 1990.

Koegler, S. S.; Brouns, T. M.; Heath, W. O.; Hicks, R. J. *Biodenitrification of Hanford Groundwater and Process Effluents: FY 1988 Status Report*; PNL-6917, Pacific Northwest Laboratory: Richland, WA, 1989.

Stuart, S. L. "Biodegradation of Carbon Tetrachloride by Two Mixed Denitrifying Cultures." M.S. Thesis, University of Washington, 1990.

Construction and Characterization of Heavy Metal Resistant, PCB-Degrading *Alcaligenes* sp. Strains

*L. Diels**, *D. Springael, S. Kreps, M. Mergeay*
Laboratory of Genetics and Biotechnology

The use of specialized bacterial strains, which are able to degrade organic pollutants such as polychlorinated biphenyls (PCBs) and pesticides, is becoming increasingly important in the decontamination of polluted soils, sludges, and groundwaters. Most of the contaminations are not due to single compounds but to multiple compounds. Therefore, either bacterial cocktails are added or new constructions are made in one bacterium (Mokross *et al.* 1990).

However, about 50 percent of polluted sites also contain inorganic pollutants such as heavy metals. These heavy metals can inhibit the activity of native organisms or the introduced organic compound degraders. Moreover, sometimes the heavy metals themselves are important contaminants that render the soil, even after degradation of the organics, unusable. To overcome such problems, the construction of bacteria that are able to degrade organic pollutants and, at the same time, resist multiple heavy metals could be environmentally important. *Alcaligenes* sp. are known to be resistant to heavy metals (Diels & Mergeay 1990) or to degrade chloroaromatic compounds (Don & Pemberton 1985). Therefore, heavy metal resistant PCB-degrading *Alcaligenes sp.* strains were constructed. The new strains were obtained by mating *Alcaligenes* sp. A5 (harbouring plasmid pSS50, involved in 4-chlorobiphenyl degradation) with various *A. eutrophus* var. *metallotolerans* strains (harbouring plasmids conferring multiple resistance to heavy metals).

The construction and characterization of these new strains are presented here along with initial results of applications.

* Laboratory of Genetics and Biotechnology, S.C.K.-C.E.N., Boeretang 200, B-2400 Mol, Belgium

Materials and Methods

Strains and Plasmids. Strains and plasmids used in this study are presented in Table 1. *Alcaligenes* sp. A5 harbours a 53-kb, IncP1, broad host range conjugative plasmid pSS50 involved in degradation of 4-chlorobiphenyl. This strain is able to metabolize 4-chlorobiphenyl to 4-chlorobenzoate (Shields *et al.* 1985). *Alcaligenes eutrophus* CH34 contains the plasmids pMOL28 (163 kb), specifying nickel, mercury, chromate, cobalt, and thallium resistance; and pMOL30 (240 kb), specifying zinc, cadmium, cobalt, mercury, copper, lead, and thallium resistance. *A. eutrophus* DS185 contains pMOL85 (240 kb), specifying zinc, cadmium, cobalt and copper resistance. Metals such as cadmium, zinc, nickel, and copper can be removed from solution by these strains under certain conditions (Diels 1990). The strains induce a precipitation and crystallization of metal carbonates and hydroxides. This precipitation is due to a steep pH gradient at the cell surface and to the metabolism of some carbon sources (*e.g.*, lactate).

Different heavy metal resistance genes could be cloned, namely the Cd-Co-Zn genes (*czc*) (Nies *et al.* 1987), the Hg genes (Diels *et al.* 1985), and the Co, Ni, Cr genes (*cnr, chr*) (Nies *et al.* 1989; Siddiqui *et al.* 1989). Gene fusions have been made with the Pb, Cu, and Tl resistance genes.

Construction of Heavy Metal Resistant, PCB-Degrading *Alcaligenes* sp. Strains. Heavy metal resistant, PCB-degrading strains were constructed using the broad host range and hermaphroditic characteristics of pSS50 (IncP1).

After matings between *Alcaligenes* sp. A5 and the different *A. eutrophus* var. *metallotolerans* strains, PCB-degrading, heavy metal resistant strains were selected based on minimal medium supplemented with heavy metals (1 mM Ni or 2 mM Zn) and biphenyl as the sole source of carbon.

PCB Degradation Experiments. Cells were grown on minimal medium containing biphenyl as sole source of carbon and energy to an optical density at 660 nm of 1.0 and washed twice with minimal medium 284 with or without heavy metals (Zn, 1 mM and Ni, 1 mM). Afterward, 1 mL cell suspensions (OD = 1.0) were incubated in 20-mL screw-cap glass vials containing Aroclor 1242 or a defined PCB congener mixture (final concentration on 10 ppm PCBs). The cells were incubated for 72 h or 48 h at 30 C in a gyratory shaker. Following incubation, each assay was stopped by the addition of perchloric acid (final

TABLE 1. Tested strains and plasmids.

Strain	Plasmids	Plasmid Markers	Characteristics	Reference
Alcaligenes sp. A5	pSS50 (53 Kb)	Involved in 4-CBP degradation	Nic^-, Chr, Cob^-, Cad^-, Mer?, Plu?, Cup^-, Zin^-, Bif^+	Shields *et al.*, 1985
A. eutrophus var. *metallotolerans*				
CH34	pMOL28 (165 kb)	Nic^+, Chr^+, $CobA^+$, Tha^+, Mer^+ (Tn4378)	Bif^-	Mergeay *et al.*, 1985
	pMOL30 (240 kb)	Cad^+, $CobB^+$, Zin^+, Tha^+, Plu^+, Cup^+, Mer^+ (Tn4380)		
DS185	pMOL90	–	Bif^-, Cad^-, Cup^-	Diels & Mergeay 1990
	pMOL85 (250 kb)	Zin^+		
	pMOL80 (4 kb)	–		
SV661	pMOL284	Zin^+	Bif^-	Diels & Mergeay 1990
	pMOL304	–		

Notes:
4-CBP = 4-chlorobiphenyl.
Bif = biphenyl.
Tested concentration of heavy metals (= Minimum inhibitory concentration in *A. eutrophus* var. *metallotolerans*): Zin:Zinc (2 mM); Nic:Nickel (1 mM); Chr:Chromate (0,6 mM); Cad:Cadmium (0,8 mM); Cob:Cobalt (5 mM); Mer:Mercury (0.5 mM); Plu:Lead (8 mM); Cup:Copper (2 mM); Plu?:lead resistance not tested.

concentration, 0.7%). The PCBs were twice extracted by one volume of *n*-hexane and the phases separated by centrifugation when necessary (Bedard *et al.* 1986). PCB analysis was done with a GC-MS QMD-1000 on a column of WCOT fused silica.

Results and Discussion

The matings lead to heavy metal resistant *Alcaligenes* sp. A5 transconjugants (A5.4, A5.9) (retrotransfer of the poorly self-transferable, heavy metal resistant megaplasmids by pSS50) and to PCB-degrading *A. eutrophus* var. *metallotolerans* CH34 and SV661 transconjugants by transfer of pSS50 itself (AE707, AE860). The properties concerning heavy metal resistance of the constructed strains are reported in Table 2, while the plasmid patterns are shown in Figure 1. A5.9 (Zin^+, Cad^+, Cup^+, Bif^+) is an A5 derivative that received pMOL85 from DS185. Interestingly, A5.9 exhibited cadmium and copper resistance, at 30 C, which is expressed in DS185 only at 37 C, confirming previous results (Diels & Mergeay 1990) that showed that pMOL85 carries these two resistances in addition to Zn. A5.4 is an A5 derivative that inherited pMOL28 from CH34. However, Co resistance is not expressed in the new host.

All transconjugants show increased pSS50 plasmids. The pSS50 derivatives of A5.4 and A5.9 probably arose after cointegrate formation between pSS50 and the heavy metal resistant megaplasmids, and subsequent resolution in strain A5. Thus, although pMOL85 was shown to be a self-transferable plasmid (Diels & Mergeay 1990), it seems to be transferred to A5 by means of pSS50.

AE707 and AE860, which are CH34 and SV661 transconjugants, respectively, and which received pSS50 from A5, harbour pSS50 derivatives with greatly enlarged sizes (100 kb). The cause of these rearrangements are being examined.

Growth curves based on minimal medium with biphenyl as carbon source (2,000 mg/L), in the absence and in the presence of heavy metals (Figures 2, 3, and 4), show a higher growth rate and shorter lag phase for CH34 transconjugants in comparison with the parent strain of A5 and A5 heavy metal-resistant transconjugants. The heavy metal-resistant A5 derivatives show the same growth pattern as that of CH34 and DS185 with gluconate as a carbon source and in presence of Ni or Zn (data not shown).

Aroclor 1242 was degraded in a resting cell experiment with and without Ni and Zn ions. Table 3 presents results on degradation of

TABLE 2. Characteristics of the heavy metal resistant, PCB-degrading *Alcaligenes* sp. constructs.

Strain	Origin (recipient × donor)	Selection	Plasmids	Heavy Metal Resistance Properties	Biphenyl-Degrading Property
AE707	CH34 × A5	Bif, Zin	pSS50(++), pMOL28, pMOL30	Nic^+, Chr^+, Cob^+, Mer? Cad^+, Zin^+, Plu^+, Cup^+	+
A5.4	A5 × CH34	Bif, Nic	pSS50(+), pMOL28	Nic^+, Chr^+, Cob^+Mer?	+
A5.9	A5 × DS185	Bif, Zin	pSS50(+), pMOL85	Cad^+, Cob^+, Zin^+, Cup^+	+
AE860	SV661 × A5	Bif, Zin	pSS50(++), pMOL284, pMOL304	Zin^+	+
AE712	CH34 × A5	Bif, Zin	pSS50, pMOL28, pMOL30	Nic^+, Chr^+, Cob^+, Mer^+ Cad^+, Zin^+, Plu^+, Cup^+	+

Notes:
pSS50(++) = enlarged pSS50 derivative (100 kb).
pSS50 (+) = enlarged pSS50 derivative (67 kb).
Bif = biphenyl.
For tested concentrations of heavy metals, see Table 1.

Lane 1 : A5 (pSS50)

Lane 2 : CH34 (pMOL28,pMOL30)

Lane 3 : A5.4 (pSS50(+),pMOL28)

Lane 4 : A5.9 (pSS50,pMOL85)

Lane 5 : AE707 (pSS50(++),pMOL28,pMOL30)

Lane 6 : AE712 (pSS50,pMOL28,pMOL30)

Lane 7 : A5 (pSS50)

Lane 8 : DS185 (pMOL85)

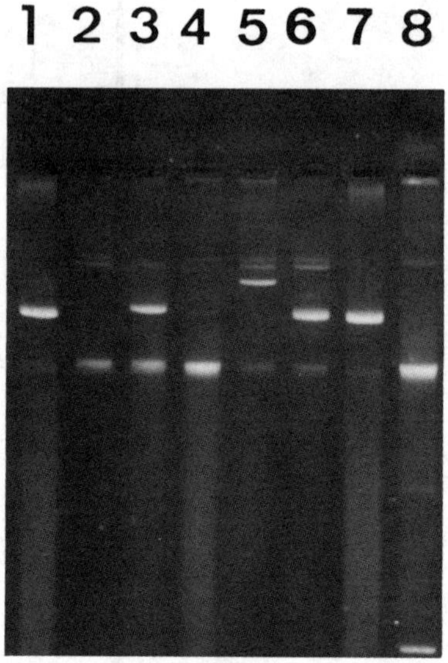

FIGURE 1. Plasmid patterns of the heavy metal resistant, PCB-degrading *Alcaligenes* sp. strains.

di- and trihalogenated biphenyls. The results indicate a good degradation of di- and trichlorinated biphenyls in the presence of Zn. The Ni concentration used (1.0 mM) was, however, still toxic for AE707. Nonetheless, a good degradation of the chlorinated biphenyls could be obtained in the presence of Ni ions with A5.4. However, higher chlorinated biphenyls could not be degraded by A5.1.

The results show the remarkable properties of the genus *Alcaligenes* to express genes of environmental concern. Also a 2.4-D (2.4-dichlorophenyl acetic acid) degrading, heavy metal resistant *Alcaligenes* has been constructed using plasmid pJP4 which codes for 2,4-D catabolism. The construction presented will be a second example of combining of organic catabolism and inorganic resistance in one strain. Further strains will be developed with bacteria having a broader spectrum of PCB degradation. Such strains will be used in bioreactors (slurry reactors) to treat soils contaminated with PCBs or other organic pollutants and for heavy metal reduction. With the use of CH34 in a slurry reactor (at lab scale), between 70 and 90 percent of Cd could be removed from a sandy garden soil. Combination of this feature and degradation of inorganics in one strain will lead to advantageous treatments of soil for removal of inorganics and organics in one time.

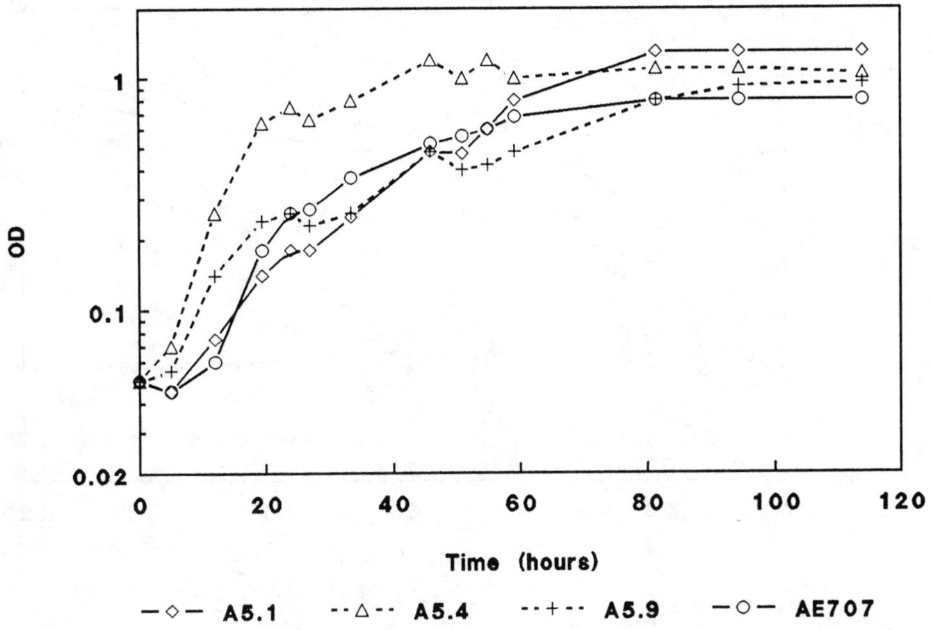

FIGURE 2. **Growth of the constructed strains based on minimal medium with biphenyl as a carbon source in the absence of heavy metals.**

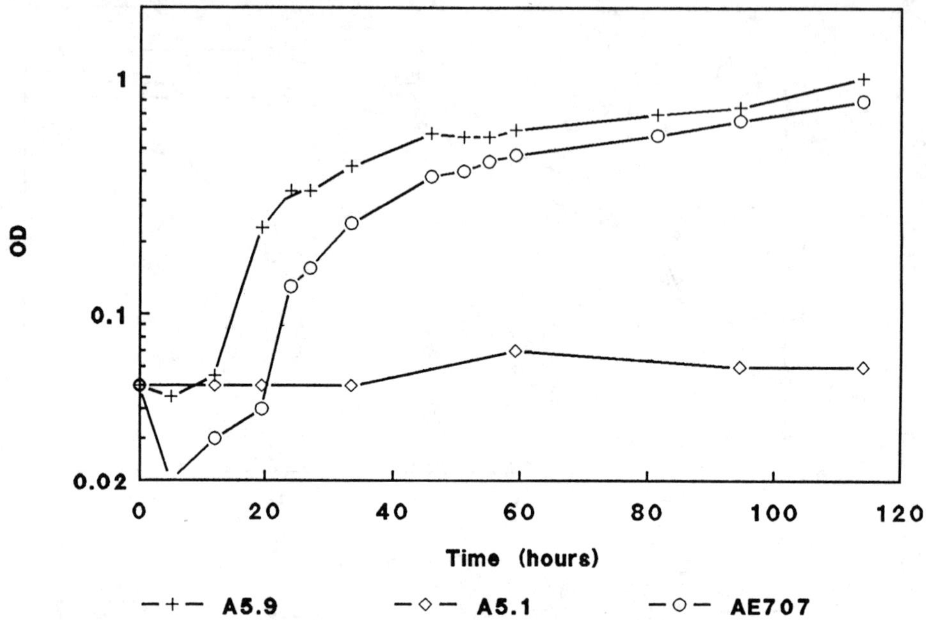

FIGURE 3. **Growth of the constructed strains based on minimal medium with biphenyl as a carbon source and 2 mM zinc.**

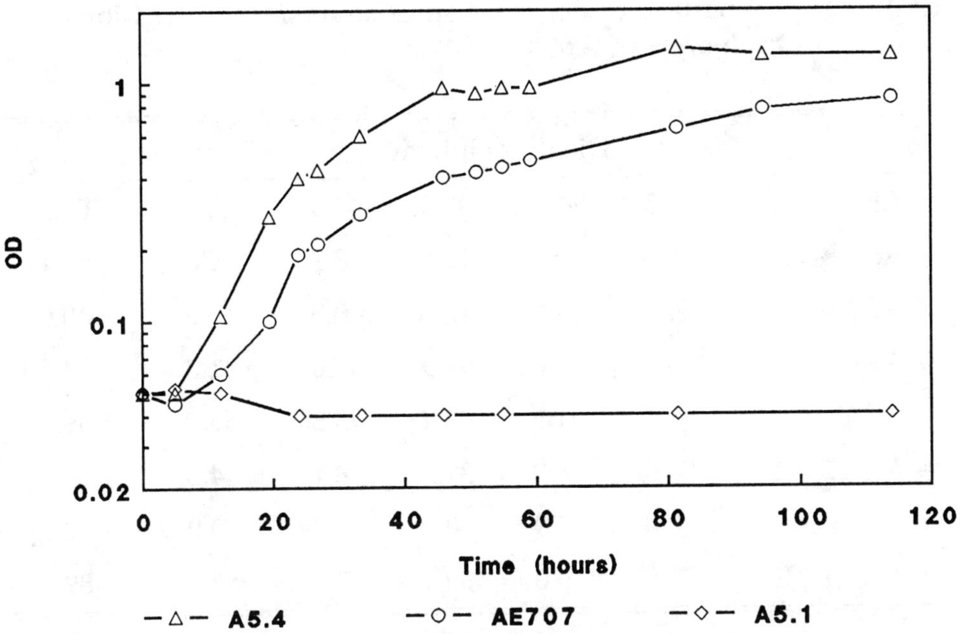

FIGURE 4. **Growth of the constructed strains based on minimal medium with biphenyl as a carbon source and 1 mM nickel.**

TABLE 3. Percentage of degradation of some di- and trichlorobiphenyls of Aroclor 1242.

Strain	Dichlorobiphenyl				Trichlorobiphenyl	
+ A5.1	92.5	91.5	92.3	42.0	28.6	60.3
+ A5.1 + Zn	0.0	12.2	16.5	8.7	4.8	1.1
+ A5.1 + Ni	40.0	14.7	17.3	0.0	0.0	0.0
+ AE707	100.0	97.6	97.9	97.8	81.2	84.4
+ AE707 + Zn	95.0	91.5	93.2	93.8	56.7	69.9
+ AE707 + Ni	50.0	26.9	31.2	8.1	4.8	10.3
+ CH34	0.0	4.9	0.0	0.0	0.0	0.0
+ A5.4 + Ni	72.5	100.0	99.4	97.8	88.6	89.8

Notes:
The percentage of degradation of the different di- and trichlorobiphenyls are presented.
Higher chlorinated biphenyls are not degraded by the strain A5.1.

REFERENCES

Bedard, D. L. *Appl. Environm. Microbiol.* **1986**, *51*, 761–768.

Diels, L. "Accumulation and Precipitation of Cd and Zn Ions by *Alcaligenes eutrophus* Strains." *Biohydrometallurgy* **1990**, 369–377.

Diels, L.; Mergeay, M. "DNA Probe Mediated Detection of Resistant Bacteria from Soils Highly Polluted by Heavy Metals." *Appl. Environm. Microbiol.* **1990**, *56*, 1485–1491.

Diels, L.; Faelen, M.; Mergeay, M.; Nies, D. "Mercury Transposons from Plasmids Governing Multiple Resistance to Heavy Metals in *Alcaligenes eutrophus* CH34." *Arch. Int. Physiol. Biochim.* **1985**, *93*, B27–B28.

Don, R. H.; Pemberton, J. M. "Genetic and Physical Map of the 2,4-Dichlorophenoxyacetic acid—Degradative Plasmid pJP4." *J. Bacteriol.* **1985**, *161*, 466–468.

Mergeay, M.; Nies, D.; Schlegel, H. G.; Gerits, J.; Charles, P.; Van Gijsegem, F. *J. Bacteriol.* **1985**, *162*, 328–334.

Mokross, H.; Schmidt, E.; Reineke, W. "Degradation of 3-Chlorobiphenyl by In Vivo Constructed Hybrid Pseudomonads." *FEMS Microbiol. Letters* **1990**, *71*, 179–186.

Nies, D.; Mergeay, M.; Friedrich, B.; Schlegel, H. G. "Cloning of Plasmid Genes Encoding Resistance to Cadmium, Zinc and Cobalt in *Alcaligenes eutrophus* CH34." *J. Bacteriol.* **1987**, *169*, 4865–4868.

Nies, A.; Nies, D.; Silver, S. "Cloning and Expression of Plasmid Genes Encoding Resistance to Chromate and Cobalt in *Alcaligenes eutrophus*." *J. Bacteriol.* **1989**, *171*, 5065–5070.
Shields, M. S.; Hooper, S. W.; Sayler, G. S. "Plasmid Mediated Mineralization of 4-Chlorobiphenyls." *J. Bacteriol.* **1985**, *163*, 882–889.
Siddiqui, R. A.; Benthin, K.; Schlegel, H. G. "Cloning of pMOL2 8-Encoded Nickel Resistance Genes and Expression of the Genes in *Alcaligenes eutrophus* and Pseudononas spp." *J. Bacteriol.* **1989**, *171*, 5071–5078.

Use of Bacterial Stress Promoters to Induce Biodegradation under Conditions of Environmental Stress

C. Deane Little[*], *Cresson D. Fraley, Michael P. McCann, Abdul Matin*
Stanford University

Trichloroethylene (TCE) is a suspected carcinogen and a major groundwater contaminant in the United States. In recent years, a variety of aerobic microorganisms have been shown to epoxidate TCE when grown on alternate carbon substrates. These organisms include methane-, ammonia- and propane-oxidizing bacteria (Arciero *et al.* 1989; Little *et al.* 1988) and a variety of bacteria that catabolize aromatic substrates such as toluene and phenol (Nelson *et al.* 1986; Wackett & Gibson 1988). When epoxidated, TCE breaks down spontaneously in water to form harmless or easily metabolized products, ultimately resulting in complete conversion to CO_2. The use of such microorganisms, or recombinant strains expressing their genes, may be a potentially important technique in the bioremediation of TCE-contaminated sites. At least one major research project has focused on *in situ* biodegradation of TCE and other chlorinated alkenes by stimulating native methane-oxidizing bacteria in a subsurface environment. Only about 20 percent of the TCE was degraded, although greater success was obtained with less chlorinated compounds (Roberts *et al.* 1989).

* Department of Microbiology and Immunology, Stanford University, Stanford CA 94305

A major obstacle to *in situ* bioremediation efforts is the low metabolic activity of environmental bacterial populations (Morita 1988). This limited activity is probably a result of a variety of stress conditions inherent to the environment, including carbon, nitrogen, and phosphorus starvation (Matin *et al.* 1989). In general, bioremediation efforts have relied on nutrient amendments to counter these limitations, but such efforts may result in other complications, including oxygen limitation, growth of competing microbial populations, and clogging of pore spaces with biomass. Ideally, a microbial population would possess moderately high biodegradation activity under natural environmental conditions.

To address these problems we have constructed recombinant *E. coli* strains that express the toluene monooxygenase (TMO) gene under control of the powerful GroEL stress promoter. The TMO gene, isolated from *Pseudomonas mendocina,* has been previously shown to biodegrade TCE when expressed under control of the Trp-Lac (TAC) promoter in growing *E. coli* (Winter *et al.* 1989). Although it is known as a heat shock protein, GroEL is an important biomolecule in a variety of cellular stress responses besides heat shock. These include carbon, nitrogen, and phosphorus starvation, oxidative shock, and chemical toxicity (Matin *et al.* 1989). GroEL appears to play a role in renaturing damaged proteins in stressed cells and may be present at levels as high as 3 to 4 percent of total cell protein under some conditions (Matin *et al.* 1989).

Materials and Methods

Construction of Plasmid Vectors. Plasmid pMY402 was obtained as a gift from Dr. Burt Ensley of AMGEN, Inc. This multicopy plasmid is a derivative of RSF 1010 and contains the TMO gene under control of the TAC promoter. Plasmid pAMC101 is a derivative of PBR322, obtained by replacing 1.3 kB of pBR322 DNA with a 4.5-kB fragment from pMY402 encoding the TMO genes. Both fragments were excised by sequential cutting with *Eco*RI and *Ppu*MI and recovery from low-melt gels. Plasmid pAMC103 was constructed by cutting pAMC101 with *Eco*RI and *Bam*HI in a polylinker upstream of the TMO genes. A 300-bp *Eco*RI-*Sau*3A fragment encoding the GroEL promoter was cloned into this site. The GroEL fragment was obtained from plasmid pOF39, a gift from Dr. Costa Georgopoulos of the University of Utah.

Growth of *E. coli* Strains. Cultures of *E. coli* strain AMS66 (K12: lac UV169, recA 56, srl::Tn10) were transformed with plasmids pMY402 and pAMC103. The resulting strains were grown in a Tris-buffered mineral salts medium with 0.2 percent glucose as an energy source. This medium allowed specific limitation of glucose, nitrogen, or phosphorus from the medium (Groat *et al.* 1986). Cultures were grown in 250-mL Erlenmeyer flasks at 37 C on a rotary shaker. After obtaining an optical density of 1.2, cultures were centrifuged at 5,000 x G for 10 min in a prewarmed rotor. Cells were resuspended in mineral salts medium minus nitrogen, phosphorus, and glucose. Aliquots of these suspensions were transferred to mineral media containing either all nutrients or lacking glucose, nitrogen, or phosphorus. Isopropylthiogalactoside (IPTG, 1 mM) was added to AMS66:pMY402 cultures to induce the TAC promoter. After 30 min, phenol (120 to 140 µM) was added to all suspensions and monitored for biodegradation using an assay adapted from Martin (Folsom *et al.* 1990). Phenol concentrations were monitored every 15 to 20 min for 2 h.

Results

E. coli strains AMS103 (AMS66:pAMC103, PGroEL:TMO) and AMS66:pMY402 (Ptac:TMO) both degraded phenol at significant rates when grown on a mineral salts medium with added glucose, nitrogen, and phosphorus (MS + GNP) (see Figures 1 and 2). Under these conditions, strain AMS103 degraded phenol at an initial rate of 8.5 nmol/min/mg protein; strain AMS66:pMY402 degraded phenol at a rate of 3.5 nmol/min/mg protein. Phenol concentrations were reduced from 120 to 140 µM to 40 µM in about an hour; phenol biodegradation below the 40-µM level did not occur in these cultures for unknown reasons. The rate differences between cultures probably reflect differences in plasmid number per cell rather than real differences between promoter activities.

Under TAC promoter control (pMY402), no significant phenol degradation occurred in AMS66 cultures starved for glucose, nitrogen, or phosphorus (Figure 2). In phosphorus or nitrogen-starved AMS66:pMY402 cultures, phenol degradation rates were 0.6 nmol/min/mg protein; in glucose-starved cultures rates were 0.1 nmol/min/mg protein. No phenol disappearance was observed in sterile controls (data not shown). The phenol degradation rates observed in starved AMS66:pMY402 cultures are approximately one-tenth of the rate observed in similar unstarved cultures (see Figure 2 and above).

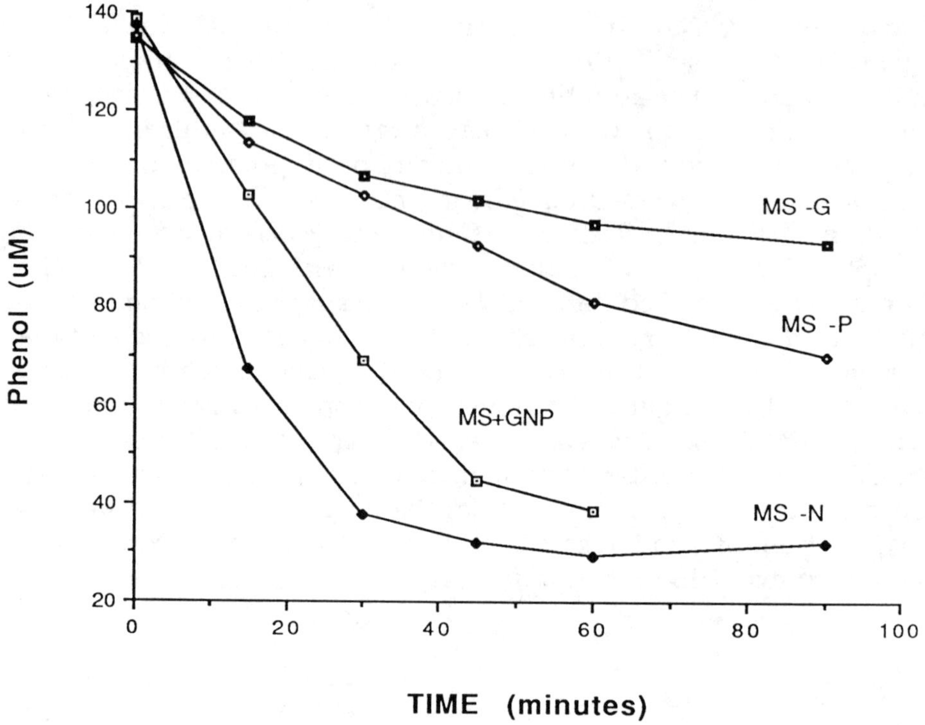

FIGURE 1. Effect of glucose, nitrogen, or phosphorus starvation on phenol biodegradation in AMS103 (PGroEL:TMO) cultures. Significant biodegradation occurred under all starvation conditions.

In contrast, when TMO expression is under control of the GroEL promoter (AMS103), significant phenol biodegradation occurred under such starvation conditions (Figure 1). In AMS103 cultures phenol degradation rates in glucose-, nitrogen-, and phosphorus-starved cultures were 3.0, 9.6, and 3.6 nmol/min/mg protein, respectively. Although the rates observed in glucose and phosphorus-starved cultures are lower than in unstarved PGroEL:TMO cultures (8.5 nmol/min/mg protein), they are nearly as high as the rate observed for unstarved cultures under TAC promoter control.

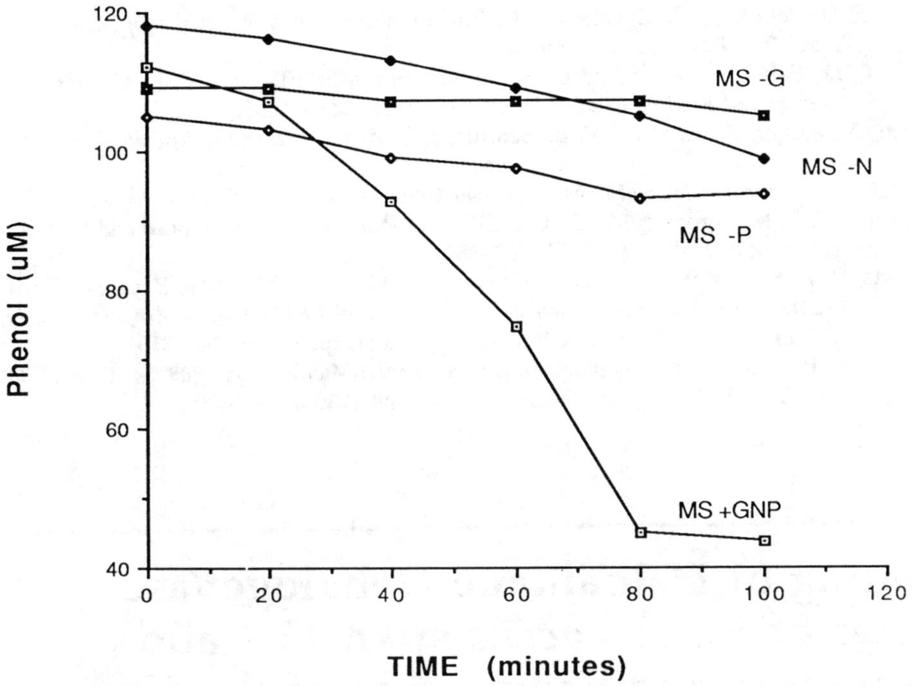

FIGURE 2. Effect of glucose, nitrogen, or phosphorus starvation on phenol biodegradation in AMS66:pMY402 (Ptac:TMO) cultures. Degradation rates were significantly lower under starvation conditions.

Summary

In general, phenol degradation rates in the presence of nutrient starvation were greater when the TMO gene was under *groEL* promoter control than under TAC promoter control. Thus, it appears feasible to use stress promoters such as GroEL to obtain significant expression of selected biodegradative activity in the metabolically sluggish bacterial populations encountered in nature.

REFERENCES

Arciero, D.; Vanelli, T.; Logan, M.; Hooper, A. B. *Biochemical and Biophysical Research Communications* **1989**, *159*, 640–643.

Folsom, B. R.; Chapman, P. J.; Pritchard, P. H. *Applied and Environmental Microbiology* **1990**, *56*, 1279–1285.

Groat, R. G.; Schultz, J. E.; Zychlinsky, E.; Bockman, A.; Matin, A. *Journal of Bacteriology* **1986**, *168*, 486–493.
Little, C. D.; Palumbo, A. V.; Herbes, S. E.; Lidstrom, M. E.; Tyndall, R. L.; Gilmer, P. J. *Applied and Environmental Microbiology* **1988**, *54*, 951–956.
Matin, A.; Auger, E. A.; Blum, P. B.; Schultz, J. E. *Annual Review of Microbiology* **1989**, *43*, 293–316.
Morita, R. Y. *Canadian Journal of Microbiology* **1988**, *34*, 436–441.
Nelson, M.J.K.; Montgomery, S. O.; O'Neill, E. J.; Pritchard, P. H. *Applied and Environmental Microbiology* **1986**, *52*, 383–384.
Roberts, P. V.; Semprini, L.; Hopkins, G. D.; Grbic-Galic, D.; McCarty, P. L.; Reinhard, M.. In-Situ *Aquifer Restoration of Chlorinated Aliphatics by Methanotrophic Bacteria*; U.S. Environmental Protection Agency Technical Report No. 310, 1989.
Wackett, L. P.; Gibson, D. T. *Applied and Environmental Microbiology* **1988**, *54*, 1703–1708.
Winter, R. B.; Yen, K.-M.; Ensley, B. D. *Biotechnology* **1989**, *7*, 282–285.

Cloning of Haloalkane Dehalogenase Genes of *Rhodococcus* sp. m15-3 and Expression and Sequencing of the Genes

Toshio Omori[*], *Toshiaki Kimura, and Tohru Kodama*
The University of Tokyo

Chlorinated paraffin is resistant to microbial attack, causes environmental pollution problems, and seems to be persistent in the natural environment. Its recalcitrance may result from its many chlorine atoms on both terminal and internal carbons of *n*-paraffin.

A little is known about the microbial degradation of polyhalogenated alkanes (Omori *et al.* 1987). Polyhalogenated alkanes seem to be totally degraded *via* halogenated aliphatic acids, as shown in Figure 1 (Yokota *et al.* 1986). The terminal chlorine of polychlorinated alkanes is dehalogenated by oxygenase or hydrolase (Keuning *et al.* 1985), chlorine at the second position by hydrolase, and chlorine at the third position by enzymes in β-oxidation system. Our observations indicated that three types of enzymatic reaction were involved in the removal of halogens from

[*] Department of Agricultural Chemistry, Faculty of Agriculture, The University of Tokyo, Bunkyo-ky, Tokyo 113, Japan

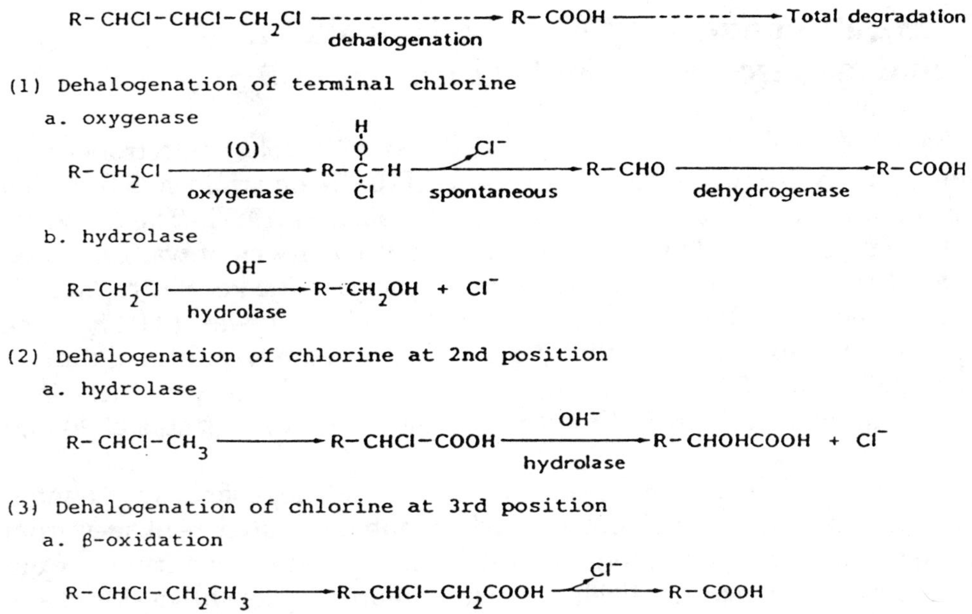

FIGURE 1. Modes of microbial dehalogenation of halogenated aliphatic compounds.

polyhalogenated alkanes in addition to the spontaneous dehalogenation of 4-chlorobutyric acid and 5-chlorovaleric acid.

The chloride release from 1-chlorobutane (1-CB) under anaerobic conditions by 1-CB-utilizing bacteria suggested the presence of halidohydrolase. Lacking the ability to dehalogenate 2-halogenated aliphatic acid, this halidohydrolase is different from the specific enzymes involved in the dehalogenation of 2-halogenated aliphatic acids. Organisms possessing the dehalogenase that acts on haloalkanes seem to play an important role in the degradation of environmental pollutants. It is interesting to investigate the physiological role of the enzymes that act on xenobiotics and to elucidate how the bacteria could acquire those enzymes.

To elucidate the physiological role of dehalogenation, haloalkane halidohydrolase was purified from *Rhodococcus* sp. ml5-3 capable of using 1-CB as the sole carbon and energy source, and the gene specifying the enzyme was sequenced.

Purification and Properties of Haloalkane Dehalogenase from *Rhodococcus* sp. ml5-3

A haloalkane dehalogenase was purified about 400-fold to electrophoretic homogeneity from cell extracts of the 1-CB-utilizing strain ml5-3, which was identified as *Rhodococcus* sp. ml5-3 (Yokota *et al.* 1987). The *Km* value of the enzyme for 1-CB was 0.18 mM. Its molecular weight was estimated to be 36,000 by sodium dodecyl sulfate-polyacrylamide gel electrophoresis and 33,000 by gel filtration. The isoelectric point was pH 4.5. The optimum pH for enzyme activity was found to be 9.4, and the optimum temperature was 30 to 35 C. The enzyme was stable for 1 h at temperatures ranging from 4 to 30 C, but was progressively less stable at 40 and 50 C.

A broad substrate specificity of the dehalogenase was demonstrated. The enzyme dehalohgenated monohalogenated *n*-alkanes from bromoethane to 1-chlorododecane, dihalogenated *n*-alkanes from 1,2-dibromoethane to 1,9-dichlorononane, chloroalkanes with a side chain, and 2-chloroalkanes. The order of reactivity was determined by using chloro-, bromo-, and iodobutane and was as follows: bromo- > chloro- > iodo-. The enzyme was active with some chloroalcohols and weakly active against some haloacids.

Cloning of the Total DNA Fragment from *Rhodococcus* ml5-3 Carrying Genes Required for Growth on 1-Chlorobutane

An *Sau*3AI gene bank of genomic DNA from *Rhodococcus* sp. ml5-3 was constructed in *Escherichia coli* DC682 by using the expression vector plasmid pTTQ18, which has *tac* promoter and *lacI*^{q}. A genomic library of total DNA of *Rhodococcus* sp. ml5-3 was used to transform cells of *E. coli* DC682. Transformants were washed and transferred onto 1-CB plate. 1-CB-positive transformants were screened for plasmid content by a rapid alkaline lysis procedure. Out of 120,000 transformants, two 1-CB-positive clones were found. Restriction digest analysis demonstrated that the recombinant plasmids, designated pHD18 and pHD28, contained 8.5-kilobase (kb) and 4.6-kb *Sau*3AI inserts, respectively.

Subcloning and Localization of the Genes *dehA* and *dehB* Encoding Haloalkane Dehalogenase

From plasmids pHDl 8 and pHD28, subclones were prepared by using restriction enzyme *Xba*I. The positive subclones contained plasmid pHD118, which had a 2.0-kb-*Xba*I insert derived from pHD18, and pHD218, which had a 1.6-kb-*Xba*I insert derived from pHD28. To further localize the *dehA* gene on plasmid pHD118, a series of deletion plasmids were made by exonuclease III. The expression of the dehalogenase genes in several recombinants was examined in both resting cell and cell-free extract reactions (Table 1). A host cell of *E. coli* DC682, induced by IPTG, showed the haloalkane dehalogenase activity in both reactions, indicating that the expression of the haloalkane dehalogenase was dependent on the *tac* or *lac* promoter of the vector pTTQ18 or pUC118 and pUC119, respectively. On the other hand, *E. coli* JM109 and MV1184, which were induced by IPTG, showed the activity only in the cell-free extract reactions. The activity levels were similar to those found in 1-CB-induced cells of *Rhodococcus* sp. ml5-3. This difference might result from the difference of the permeability of the substrate to the cell membranes of the strains.

Southern Blot Analysis

Total DNA isolated from *Rhodococcus* sp. ml5-3 or *E. coli* DC682 was digested with *Xba*I. DNA fragments thus obtained were run by agarose gel electrophoresis, transferred to a nylon membrane filter, and probed with ^{32}P-labeled 1.6-kb- and 2.0-kb-*Xba*I fragments. Southern blot experiments showed that the cloned fragment was derived from ml5-3 and not present in the genomes of the other strain.

Nucleotide Sequencing of the Cloned DNA Fragment

The 2.0-kb-*Xba*I fragment of pHD118 was inserted into pUC118 and pUC119. A series of unidirectionally deleted mutants with exonuclease III and mung bean nuclease was constructed. By using the dideoxy-chain termination method, about 2,000 base pair (bp), including the haloalkane dehalogenase-coding region, were sequenced. A 408-bp open reading frame, starting at a codon of ATG and terminating at a TGA codon, was found in the nucleotide sequence of the *Xba*I fragment of pHD118. Deletion analysis of the *dehA* qene suggested that the gene was involved in the

TABLE 1. Expression of the *dehA* gene.

Strain	IPTG[b]	Chloride Released[a] (μg/ml) Resting Cell	Cell-free Extracts
E. coli DC682 (pHD18)	–	0	1
E. coli DC682 (pHD18)	+	30	45
E. coli DC682 (pTTQ18)	–	1	0
E. coli DC682 (pTTQ18)	+	1	1
E. coli JM109 (pHD18)	–	0	1
E. coli JM109 (pHD18)	+	3	38
E. coli JM109 (pTTQ18)	–	0	0
E. coli JM109 (pTTQ18)	+	1	0

(a) Cells were grown in LB broth containing ampicilline 50 μg/ml and no NaCl at 37 C. The reaction was carried out by incubating at 30 C with 10 mM of 1-chlorobutane for 2 hours. Chloride liberation was followed colorimetrically.
(b) One millimolar of IPTG was added.

dehalogenation, and the *dehA* gene product showed very similar substrate specificity to that of the purified enzyme of *Rhodococcus* sp. ml5-3. Although the cells induced by 1-CB gave a major dehalogenase peak through the purification of the enzyme, the NH_2-terminal amino acid sequence of the purified haloalkane dehalogenase was not found in the sequenced gene. A *dehA* gene that was normally cryptic might be expressed by *tac* promoter of the vector. The physiological function of the *dehA* gene product has not yet been determined.

The 1.6-kb-*Xba*I fragment of pHD218 was inserted into pUC118 and pUC119. A series of the deletion mutants was constructed as described above. The nucleotide sequence corresponding to the NH_2-terminal amino acid sequence (Ser-Glu-Ile-Gly-Thr-Gly-Phe-Pro-Phe-Asp-Pro-His-Tyr-Val-Glu-Val-Leu-Gly-Glu-Arg) of the purified haloalkane dehalogenase was found in this fragment. Sequencing of the gene is now under investigation.

REFERENCES

Keuning, S.; Janssen, D. B.; Witholt, B. *J. Bacteriol.* **1985**, *163*, 635–639.
Omori, T.; Kimura, T.; Kodama, T. *Appl. Microbiol. Biotechnol.* **1987**, *25*, 553–557.
Yokota, T.; Fuse, H.; Omori, T.; Minoda, Y. *Agric. Biol. Chem.* **1986**, *50*, 453–460.
Yokota, T.; Omori, T.; Kodama, T. *J. Bacteriol.* **1987**, *169*, 4049–4054.

Development of a Phenol Biosensor Using a Genetically Modified Thermophile

*P. Oriel**, *Z. Lu, M. R. Natarajan, G. Gurujeyalakshmi*
Michigan State University

An attractive alternative to conventional contaminant analysis is the utilization of biosensors capable of rapid and specific detection of pollutants on site. Biosensors utilizing immobilized living microbial cells have proven dependable and inexpensive and are now commercially available for BOD measurement (for a recent review, see Karube 1990). Although these approaches have been successful, higher specificity for individual chemical contaminants in the environment requires the monitoring of enzyme or metabolite levels in selected biochemical pathways, rather than overall cell metabolic response. We have initiated a project to develop environmental biosensors based upon detection of unique metabolites produced in response to aromatic xenobiotics in the environment. We have chosen catechol and its halogenated derivatives as key intermediate metabolites for detection, as these are formed in the degradation of a variety of aromatic contaminants, including phenol, toluene, napthalene, halogenated benzene derivatives, 2,4-dichlorophenoxyacetic acid, and many others (Rochkind-Dubinsky *et al.* 1987). We have further selected thermophilic Bacilli as the microorganisms of

* Department of Microbiology, Michigan State University, 40 Giltner Hall, East Lansing MI 48824-1101

choice for biosensor development. Although their aromatic pathways are not as well understood as those of mesophiles, initial studies indicate higher resistance to toxicity of aromatic chemicals (Gurujeyalakshmi & Oriel 1989), possibly because thermostable enzymes are resistant to both thermal and chemical denaturation (Tombs 1985). In this communication, we report progress in the development of a biosensor for the detection of phenol, an environmental contaminant of significant concern because of its toxicity and water solubility.

Materials and Methods

Microorganisms and Growth Conditions. The isolation and characterization of *Bacillus stearothermophilus* BR219 has previously been described (Gurujeyalaskshmi & Oriel 1989). This thermophile is capable of growth in phenol to 15 mM concentration at temperatures to 70 C. The organism was usually grown in L broth. For phenol utilization, DP medium containing salts, 1.0 g/L yeast extract, and 2.0 g/L Casamino acids was used (Gurujeyalakshmi & Oriel 1989). $MgSO_4 \cdot 7H_2O$ at a concentration of 20 mg/L and phenol to the desired level was added after autoclaving and cooling. For convenience, a growth temperature of 55 C was utilized.

Genetic Modification of BR219. The genetic alteration and characterization of the catechol-accumulating mutant of BR219, designated BR321, will be described separately (Natarajan, M. R., and Oriel, P., personal communication). Briefly, electroporation of plasmid pAM120 (Gawron-Burke & Clewell 1984), which contains broad host range transposon Tn916 as an insert in the gram-negative replicon pGL101, was used to transform BR219. Catechol-accumulating insertion mutants were selected on DP plates containing 5 μgm/mL tetracyline and 5 mM phenol by the brown color of catechol oxidation products surrounding colonies with this phenotype. Confirmation of catechol accumulation by one such mutant designated BR321 was obtained following growth in liquid DP medium with 5 mM phenol, cell removal by centrifugation, and colorimetric determination of catechol by addition of an equal volume of 4-aminoantipyrine in 2 percent Na_2CO_3 and 0.02 N NaOH and measurement at 515 nm following color development (Irie *et al.* 1987).

Catechol Production Studies. For studies examining catechol formation following induction of the phenol pathway, BR321 in L broth was grown overnight at 55 C to a population density of approximately 10^8 cells/mL. The cells were centrifuged and resuspended in an equal volume of DP medium containing varied levels of phenol. Cells in a 50-mL suspension were shaken in 125-mL Delong flasks at 55 C. Catechol was determined colorimetrically from supernatants of samples taken at various times. Catechol production rates were determined from slopes of catechol plots with time. For all concentrations, maximum production occurred approximately 3 or 4 h after resuspension in phenol.

Immobilization of BR321. For immobilization, BR321 cells were grown in L broth overnight at 55 C. Cells from 35 mL of culture were resuspended in DP medium containing 0.8 g/L Gelrite polysaccharide (Scott Laboratories, Carson CA) following autoclaving, cooling to 55 C, and addition of $MgSO_4$. Samples of 10 mL of cell suspension were placed into 100-mm petri dishes and cooled at room temperature.

Catechol Formation By Immobilized BR321 Cells. Production of catechol by immobilized BR321 cells in response to the presence of phenol was determined by adding 2 mL of DP medium containing phenol in varied concentrations to BR321 cells induced for the phenol pathway and immobilized as described above. After 4 h of contact at 55 C, the phenol solution was removed and the gel solubilized by addition of 2 mL catechol colorimetric reagent. Cell debris was removed by centrifugation, and catechol was measured colorimetrically at 515 nm.

Results

Extracellular Catechol Production. *B. stearothermophilus* BR219 catabolizes phenol using the *meta* pathway (Oriel & Gurujeyalakshmi 1989) in which phenol is first oxidized to catechol, which is oxidized to 2-hydroxymuconic semialdehyde and further products. Although the site of Tn916 transposon insertion is still under investigation, the expected insertion site causing catechol accumulation is in the gene for catechol 2,3-dioxygenase, the enzyme that catalyzes the oxidation of catechol. Insertion at this site is suggested by the inability of the BR321 mutant to show the normal bright yellow intracellular color formation of 2-hydroxymuconic semialdehyde upon addition of 1 mg/mL catechol observed with BR219. The ability of BR321 cell suspensions to produce extracellular catechol in response to phenol is shown in Figure 1, where

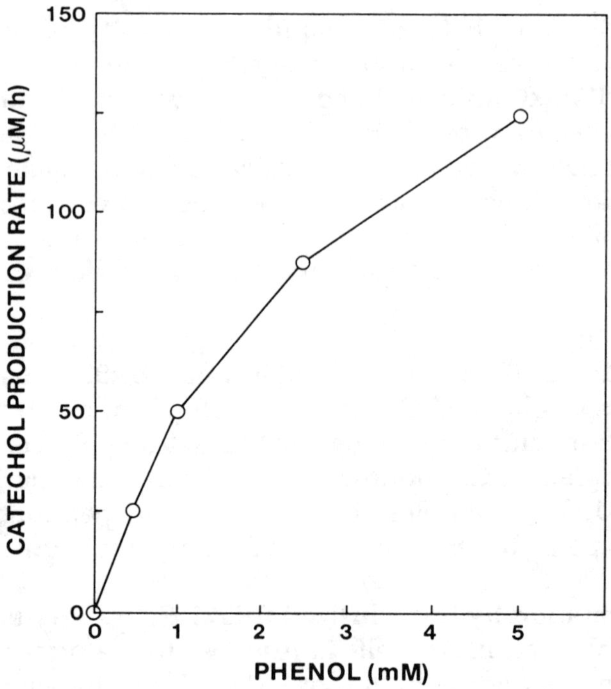

FIGURE 1. Dependence of catechol production on external phenol for suspensions of BR321.

catechol production by genetically altered *B. stearothermophilus* is seen to increase with increasing phenol at concentrations up to 5 mM.

Immobilization of BR321. Many microbial immobilization procedures including that using alginate cannot be used at elevated temperatures. We have instead investigated immobilization with the Pseudomonad polysaccharide Gelrite. This commercial agar substitute thermosets to a strong and thermostable gel of high optical clarity. We have found that the divalent cations in DP medium are sufficient for strong gel formation and that dispersion of BR321 cells in DP/Gelrite prior to cooling yields immobilized cell preparations with good viability, optical clarity, and strength. An unexpected finding was the solubilization of these gels by the alkaline catechol colorimetric reagent.

Response of Immobilized BR321 to Phenol. In preliminary studies of the response of BR321 to phenol, 2 mL of DP medium with varied amounts of phenol were placed on plates of BR321 preinduced for the phenol pathway and immobilized in Gelrite. After 4 h at 55 C, the phenol test solution was removed, and catechol formation by the immobilized BR321 cells observed colorimetrically. As seen in Figure 2, response of immobilized BR321 cells to phenol above 1 mM was readily apparent, with increased catechol produced with higher concentrations of phenol.

Discussion and Conclusions

These preliminary results offer promise for the biodetection of phenol contamination based on accumulation of the key metabolite catechol as a result of genetic modification of a phenol-catabolizing thermophile. Although significant additional work is needed to optimize the system and develop applicability for field use, we believe that the successful colorimetric detection of catechol produced by Gelrite-immobilized thermophiles offers promise for sensitive and specific detection devices for a number of aromatic contaminants of interest. Development of biodetectors for toluene and 2,4-dichlorophenoxyacetate has been initiated using this strategy.

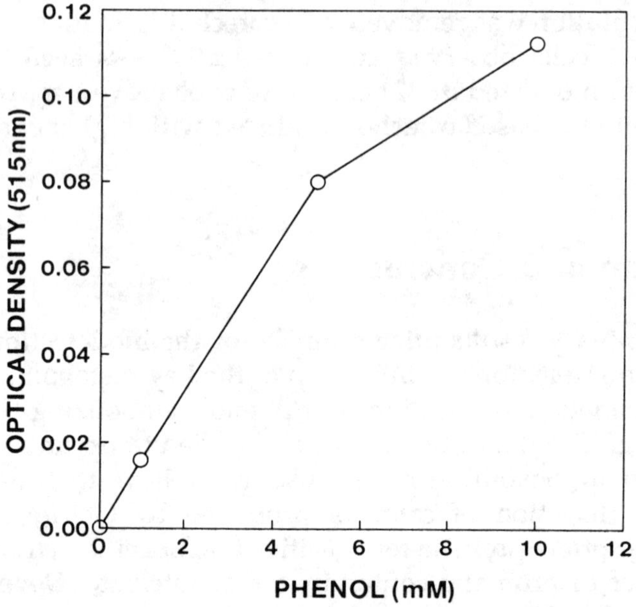

FIGURE 2. Colorimetric response of immobilized BR321 to phenol.

REFERENCES

Gawron-Burke, C.; Clewell, D. B. *J. Bacteriol.* **1984,** *159,* 214–221.
Gurujeyalakshmi, G.; Oriel, P. *Appl. Environ. Microbiol.* **1989,** *55,* 500–502.
Irie, S.; Shirai, K.; Doi, S.; Yorifuji, T. *Agric. Biol. Chem.* **1987,** *51,* 1489–1493.
Karube, I. *J. Biotechnol.* **1990,** *15,* 255–266.
Rochkind-Dubinsky, M. L.; Sayler, G. S.; Blackburn, J. W. *Microbial Decomposition of Aromatic Compounds;* Dekker: New York, 1987.
Tombs, M. P. *J. Appl. Biochem.* **1985,** *1,* 3–24.

Why Timely Bioremediation of Synthetics May Require Rational Enzyme Redesign: Preliminary Report on Redesigning Cytochrome P450cam for Trichloroethylene Dehalogenation

Rick L. Ornstein[*]
Pacific Northwest Laboratory

Cost-effective remediation in "deep" subsurface environments, where halogenated hydrocarbons (and other synthetic substances) tend to be widely dispersed, is for the most part limited to bioremediation alternatives. Since natural enzyme "lock-and-key" or "induced-fit" mechanisms have evolved to be very specific for their particular substrate(s) and product(s), microbial enzymatic degradation of synthetic substances will probably only occasionally occur at significant levels, and then usually by coincidental cometabolism. Several different enzyme systems (reviewed in Witholt *et al.* 1989), including cytochromes P450, appear to play some role in the biodehalogenation of various halogenated hydrocarbons (Castro *et al.* 1985, 1988, 1989; Lam & Vilker 1987; Miller & Guengerich 1982; Takazawa & Strobel 1986; Vilker & Khan 1989). Extant catabolic pathways in microorganisms able to degrade halogenated hydrocarbons generally have low throughput so that timely removal of synthetic compounds is not adequate from an environmental or human health perspective.

Genetic engineering strategies involving controlling selection pressures (Sangodkar *et al.* 1989), genetic transfers between existing organisms (Lehrbach *et al.* 1984), and altering regulatory genes (Ramos *et al.* 1986) offer the possibility of improving the degradative capabilities of microorganisms. Such techniques, however, presuppose that the genetic traits necessary for biodegradation of synthetic compounds either can be readily selected for under laboratory (or field) conditions

[*] Molecular Science Research Center, Pacific Northwest Laboratory, P.O. Box 999, Richland WA 99352

in relatively short time intervals, or that such functionality already exists in nature and that gene recruitment is feasible.

Genealogical analysis of different evolutionarily related proteins suggests that proteins evolve at more or less characteristic rates. The unit evolutionary period (UEP) is the time in millions of years needed to establish a 1 percent difference between the amino acid sequences of divergent lines; the UEP range for several enzymes is 2 to 20 (Dickerson 1971). For cytochromes P450, the observed range is 3 to 14 (Nebert & Gonzalez 1987). For example, an enzyme with 400 amino acids will typically take several million years to acquire four amino acid changes. Since these numbers reflect mostly surface residue changes or conservative nonsurface changes, active site changes, especially in a deeply buried pocket, are expected to take considerably longer. Thus, the evolution of an enzyme with enhanced specificity and efficiency for a non-native substrate may proceed very slowly if multiple and specific genetic changes are simultaneously required for adaptation. Since each mutational event is by itself a rare and random event, a basic premise of our research is that "designing" multiple changes into an enzyme may be the only timely alternative to waiting for essentially geological timescales for such changes to occur biologically (by some combination of random mutation, genetic recombination, and gene duplication) either *via* natural selection or under laboratory-controlled conditions.

The role of rational protein engineering in relation to developing biodegradative enzymes with enhanced remediation potential has been recently discussed (Ornstein 1991). To have a reasonable chance at rational redesign, to alter a biodegradative enzyme's specificity and/or enhance its efficiency, it is necessary to have adequate structural data about the native enzyme and native enzyme-substrate interactions, an understanding of the underlying catalytic mechanism(s), a cloned gene for site-directed mutagenesis experiments, and reliable simulation methods for studying protein dynamic fluctuations. Studying the dynamic motions of relevant enzyme-ligand complexes for adequate time periods is necessary for rational redesign since an experimental (X-ray crystal or solution nmr) structure is a static, time- and space-average structure. Considerable progress has been made in the area of validating the effectiveness of computationally efficient protein molecular dynamics simulation methods (Ornstein 1990; Ornstein *et al.* 1991) that permit useful enzyme-ligand simulations. Even so, present enzyme simulations can only be performed for about a nanosecond or less.

Cytochromes P450 are a ubiquitous superfamily of monooxygenase enzymes that have been identified in prokaryotes, plants, and eukaryotes. A wide range of alkane and aromatic compounds can serve as substrate for the P450 superfamily. Walsh (1979), in his classic book *Enzymatic Reaction Mechanisms*, notes that this enzyme class "may catalyze the widest variety of reactions of any known enzyme." The functional versatility of cytochromes P450 with respect to substrate multiplicity and catalysis has been more recently reviewed (Guengerich 1989). Sequence data from several dozen P450 sequences indicate that the P450 gene superfamily has diverged more than 2 billion years ago from a single ancestral gene (Nebert & Gonzalez 1987). The emergence of many new genes during the last 800 million years most likely reflects inter-species "warfare" (Gonzalez & Nebert 1990) resulting from selection pressures and defense mechanisms. The ecological effect of many synthetic compounds can be viewed as a continuation of this warfare.

Structurally, the most well-studied form of cytochrome P450 is the camphor hydroxylase (P450cam) from *Pseudomonas putida* (Sligar & Murray 1986). It is the only cytochrome P450 for which a 3-dimensional X-ray crystal structure has been determined. Structures have been determined for the substrate-free (Poulos *et al.* 1986), substrate-bound (Poulos *et al.* 1987; Raag & Poulos 1989a), and inhibitor-bound (Poulos & Howard 1987) forms, as well as a ternary complex with both camphor and carbon monoxide bound at the active site (Raag & Poulos 1989b). Cytochrome P450cam is unlike many other enzymes in two important and general respects. First, its active center contains few acidic or basic amino acids; therefore, successful substrate-enzyme interactions do not depend on many directionally sensitive hydrogen bonds and charge relay interactions. Instead, it is thought that P450cam could be catalytically competent for a large array of substances so long as they are sterically acceptible to the active center and spend adequate time near the iron atom of the reactive heme moiety (Atkins & Sligar 1987, 1988a, 1988b, 1989).

Cytochromes P450 have significant but modest dehalogenation capacity for various halogenated hydrocarbons. Recently, it has been shown that cytochrome P450cam is capable of dehalogenating 1,1,2-trichloroethane oxidatively and reductively under the same conditions, although the rate of dehalogenation is very modest and buildup leads to toxicity (Castro & Belser 1990). Agents that selectively inhibit P450 are known to hinder reduction of halogenated hydrocarbons (reviewed in Luke *et al.* 1987).

A major goal of our research is to redesign P450cam mutants that are stereochemically and dynamically engineered to accept specific recalcitrant and hazardous halogenated hydrocarbons as "ideal" substrates. Mutants that still look promising after visual and theoretical analysis will then be constructed, expressed, and tested for *in vitro* activity in the laboratory of Professor Stephen G. Sligar at the University of Illinois at Urbana. Mutants that exhibit adequate *in vitro* activities will then be tested under laboratory-simulated environmental conditions by our PNL collaborators in the Environmental Sciences Research Center. We here describe some preliminary results and implications of a 175-psec molecular dynamics simulation of trichloroethylene-bound (TCE) cytochrome P450cam.

Results and Discussion

Computer-aided molecular design has been used to insert TCE visually in place of camphor in the X-ray crystal structure complex; a 175-psec molecular dynamics simulation was then performed (Paulsen & Ornstein, unpub.). Preliminary analysis of the resulting trajectory and comparison with camphor-bound P450cam (Paulsen & Ornstein 1991; Paulsen *et al.* 1991) indicate that bound TCE exhibits greater rotational and translational mobility in the active site of P450cam. Our simulation results suggest that TCE can switch between orientations that appear appropriate for chlorine abstraction (reductive dehalogenation pathway) and hydrogen abstraction (oxidative dehalogenation pathway) in 10 to 20 psec. Deuterium isotope effect studies with rat liver microsomes and purified cytochrome P450 indicate that oxidative dechlorination by P450 involves hydrogen atom abstraction by a P450 intermediate oxo complex (Hales *et al.* 1987). Reductive dehalogenation under anaerobic conditions by P450 is believed to proceed by radical formation (Fujii *et al.* 1982) and then a carbanion (Ahr *et al.* 1982; Nastainczyk *et al.* 1982).

Clearly TCE is very mobile in the active site pocket. Stereoviews of representative configurations of these two mechanistic alternatives, resulting from the simulation, have been given in another recent progress report (Figure 2 in Ornstein *et al.* 1991). The substrate analogue norcamphor, which is more mobile in the active site than camphor, considerably reduces the catalytic efficiency of P450cam and diminishes the stereospecificity (Bass *et al.* 1991). Whereas 5-exo-bromocamphor is readily metabolized by P450cam to 5-ketocamphor at rates similar to that of the normal substrate, camphor (Gould *et al.* 1981), it is reasonable to anticipate the properly constructed mutants of P450cam can efficiently dehalogenate noncamphor like halogenated hydrocarbons. We are therefore designing mutants of P450cam with specific lock-and-key complementarity for TCE in an orientation suitable for either oxidative ($P450_{TCE\text{-}ox}$) or reductive ($P450_{TCE\text{-}re}$) dehalogenation. Such mutants will then be constructed and tested for *in vitro* activity in the laboratory of Professor Steve Sligar.

Since intracellular oxygen varies considerably even for aerobic organisms, our working hypothesis is that a designated TCE degrader should contain both classes of mutant genes, i.e., $P450_{TCE\text{-}ox}$ and $P450_{TCE\text{-}re}$. If mutants of P450cam can be designed to dehalogenate efficiently a single halogen atom from a synthetic substrate, the next goal is to design additional mutants of P450cam specific for dehalogenation of the most likely intermediate halogenated products. A series of such engineered genes could then be inserted on an appropriate microbial vector. Although prospects and concerns for using genetically engineered microorganisms for bioremediation purposes continue to be debated (Brown *et al.* 1984; Cairns & Pratt 1986; Loper 1989; Tiedje *et al.* 1989), the ability of naturally occurring microorganisms to degrade adequately the growing array of halogenated hydrocarbons remains uncertain. In view of the natural diversity of cytochromes P450, it seems that the intentional redesign of additional cytochromes P450 for incorporation into microorganisms for environmental bioremediation purposes could be a potentially acceptable risk, especially in cases where no other acceptable means of degrading known hazardous contaminants exist.

Acknowledgments

Pacific Northwest Laboratory is operated for the U.S. Department of Energy by Battelle Memorial Institute under Contract DE-AC06-76RLO 1830. This work is supported by Exploratory Research Projects funded within the Molecular Science Research Center and the Environmental Science Research Center.

REFERENCES

Ahr, H. J.; King, L. J.; Nastainczyk, W.; Ulrich, V. *Biochem. Pharmacol.* **1982**, *31*, 383–390.
Atkins, W. M.; Sligar, S. G. *J. Am. Chem. Soc.* **1987**, *109*, 3754–3760.
Atkins, W. M.; Sligar, S. G. *Biochemistry* **1988a**, *27*, 1610–1616.
Atkins, W. M.; Sligar, S. G. *J. Biol. Chem.* **1988b**, *263*, 18842–18849.
Atkins, W. M.; Sligar, S. G. *J. Am. Chem. Soc.* **1989**, *111*, 2715–2717.
Bass, M. B.; Paulsen, M. D.; Ornstein, R. L. Submitted for publication, **1991**.
Brown, J. H.; Colwell, R. K.; Lenski, R. E.; Levin, B. R.; Lloyd, M.; Regal, P. J.; Simberloff, D. *Ecol. Soc. Amer. Bull.* **1984**, *65*, 436–438.
Cairns, J., Jr.; Pratt, J. R. *Wat. Res. Bull.* **1986**, *22*, 171–182.
Castro, C. E.; Belser, N. O. *Environ. Toxic. Chem.* **1990**, *9*, 707–714.
Castro, C. E.; Wade, R. S.; Belser, N. O. *Biochemistry* **1985**, *24*, 204–210.

Castro, C. E.; Yokoyama, W. H.; Belser, N. O. *J. Agric. Food Chem.* **1988,** *36,* 915–919.
Castro, C. E.; Yokoyama, W.; Belser, N. O. *Environ. Toxicol. Chem.* **1989,** *8,* 13–18.
Dickerson, R. *J. Mol. Evol.* **1971,** *1,* 26–38.
Fujii, K.; Miki, N.; Kanashiro, M.; Miura, R.; Sugiyama, T.; Morio, M.; Yamano, T.; Miyake, Y. *J. Biochem. (Tokyo)* **1982,** *91,* 415–418.
Gonzalez, F. J.; Nebert, D. W. *Trends Genet.* **1990,** *6,* 182–186.
Gould, P. V.; Gelb, M. H.; Sligar, S. G. *J. Biol. Chem.* **1981,** *256,* 6686–6691.
Guengerich, F. P. In *Basis and Mechanisms of Regulation of Cytochrome P-450;* Ruckpaul, K.; Rein, H., Eds.; Taylor & Francis; London, 1989, pp 101–150.
Hales, D. B.; Ho, B.; Thompson, J. A. *Biochem. Biophys. Res. Commun.* **1987,** *149,* 319–325.
Lam, T.; Vilker, V. L. *Biotech. Bioeng.* **1987,** *24,* 151–159.
Lehrbach, P. R.; Zeyer, J.; Reineke, W.; Knackmuss, H. J.; Timmis, K. E. *J. Bacteriol.* **1984,** *158,* 1025–1032.
Loper, J. C. *Hazardous Material Control* **1989,** *2,* 24–27.
Luke, B. T.; Loew, G. H.; McLean, A. D. *J. Am. Chem. Soc.* **1987,** *109,* 1307–1317.
Miller, R. E.; Guengerich, F. P. *Biochemistry* **1982,** *21,* 1090–1097.
Nastainczyk, W.; Ahr, H.; Ulrich, V. *Adv. Exp. Med. Biol.* **1982,** *136A,* 799–808.
Nebert, D. W.; Gonzalez, F. J. *Ann. Rev. Biochem.* **1987,** *56,* 945–994.
Ornstein, R. L. *J. Biomolec. Struct. & Dyn.* **1990,** *7,* 1019–1041.
Ornstein, R. L. In *Proceedings of National Research & Development Conference on the Control of Hazardous Materials,* Hazardous Materials Control Research Institute, Greenbelt, MD, 1991, pp 314–319.
Ornstein, R. L.; Bass, M. B.; Paulsen, M. D.; Arnold, G. E. In *Proceedings of Eighth Annual HAZMACON Conference,* Hazardous Materials Control Research Institute, 1991 (in press).
Paulsen, M. D.; Ornstein, R. L. *PROTEINS: Structure, Function, and Genetics,* 1991 (in press).
Paulsen, M. D.; Bass, M. B.; Ornstein, R. L. Submitted for publication, 1991.
Poulos, T. L.; Howard, A. J. *Biochemistry* **1987,** *26,* 8165–8174.
Poulos, T. L.; Finzel, B. C.; Howard, A. J. *Biochemistry* **1986,** *25,* 5314–5322.
Poulos, T. L.; Finzel, B. C.; Howard, A. J. *J. Mol. Biol.* **1987,** *95,* 687–700.
Raag, R.; Poulos, T. L. *Biochemistry* **1989a,** *28,* 917–922.
Raag, R.; Poulos, T. L. *Biochemistry* **1989b,** *28,* 7586–7592.
Ramos, J. L., Stotz, A.; Reineke, W.; Timmis, K. N. *Proc. Natl. Acad. Sci. USA* **1986,** *83,* 8467–8471.
Sangodkar, U.M.X.; Aldrich, T. L.; Haugland, R. A; Johnson, J.; Rothmel, R. K.; Chapman, P. J.; Chakabarty, A. M. *Acta Biotechnol.* **1989,** *9,* 301–316.
Sligar, S. G.; Murray, R. L. In *Cytochrome P450: Structure, Mechanism, and Biochemistry;* Ortiz de Montellano, P., Ed.; Plenum; New York, 1986; pp 429–503.
Takazawa, R. S.; Strobel, H. W. *Biochemistry* **1986,** *25,* 4804–4809.
Tiedje, J. M.; Colwell, R. K.; Grossman, Y. L.; Hodson, R. E.; Lenski, R. E.; Mack, R. N.; Regal, P. J. *Soc. Ind. Microbiol. News* **1989,** *39,* 149–166.
Vilker, V. L.; Khan, F. *Biotech. Prog.* **1989,** *5,* 70–74.
Walsh, C. *Enzymatic Reaction Mechanisms;* W. H. Freeman; San Francisco, 1979; Chapter 15.
Witholt, B.; Janssen, D. B.; Kenning, S. In *Proceedings of the 2nd National Conference: BIOTREATMENT: The Use of Microorganisms in the Treatment of Hazardous Materials and Hazardous Wastes,* Hazardous Materials Control Research Institute, Silver Spring, MD, 1989, pp 109–117.

Hopane, a New Chemical Tool for Measuring Oil Biodegradation

Eric L. Butler, Gregory S. Douglas, William G. Steinhauer*
Battelle Ocean Sciences and Technology
Roger C. Prince, Thomas Aczel, Chang S. Hsu
Exxon Research and Engineering
Michael T. Bronson
America North Inc.
James R. Clark
U.S. Environmental Protection Agency
Jon E. Lindstrom
Alaska Department of Environmental Conservation

The ability to determine the fate of spilled oil in the environment is critically dependent on an understanding of natural oil degradation processes. Immediately after an oil spill, physical weathering such as evaporation, dispersion and solubilization alter the chemical, physical, and toxicological properties of the spilled product. After this initial weathering, microbial degradation is the principal route for the eventual removal of oil from the environment.

Unfortunately, simple gravimetric estimates of biodegradation suffer large uncertainties from the inherent heterogeneity of oiling after a spill. Previous estimates of biodegradation have relied principally on ratios of biodegradable:less degradable components such as the *n*-alkane:isoprenoid ratios. Changes in these ratios are an early indication that biodegradation has begun, but in fact isoprenoids are quite biodegradable (Kennicutt 1988; Tabak *et al.* 1991), so changes in the ratios may substantially underestimate the extent of biodegradation.

A joint Bioremediation Monitoring Program of the Alaska Department of Environmental Conservation (ADEC), the U.S. Environmental Protection Agency (EPA), and Exxon employed a novel approach in characterizing and quantifying the rate of oil

* Battelle Ocean Sciences and Technology Department, 397 Washington Street, P.O. Drawer AH, Duxbury MA 02332-0601

biodegradation on selected beaches of Prince William Sound. Hopanes, a class of polycyclic alkanes that are water-insoluble and extremely resistant to biodegradation (Kennicutt 1988; Seifert & Moldowan 1978), were used as an internal oil reference to quantify the depletion of crude oil and its individual components in oiled beach sediments. The depletions of total oil and individual oil components were calculated relative to 30 percent depleted Prudhoe Bay crude oil, a reference material that approximates the composition of the spilled oil that landed on the beaches (Figure 1). In this study, traditional weathering indicators such as C_{18}/phytane, C_{17}/pristane, and the aromatic weathering ratio were compared to the hopane oil degradation indicator.

Field Program

Three oiled beaches were selected and divided in half, and three stations on each half were marked for sampling from May through September 1990. One half of each beach received fertilizer, and nutrients, dissolved oxygen, microbiological parameters, ecological effects, and oil chemistry were monitored. We report here only on a portion of the oil chemistry data.

Analytical Methods

The analytical methods used for the analysis of aliphatic and aromatic hydrocarbons in sediments were modifications of standard EPA protocols (Sauer & Boehm 1991). Figure 1 summarizes the analytical approach. The sample GC/FID and GC/MS analyses were bracketed by a laboratory-prepared 30 percent depleted Prudhoe Bay crude (PBC) oil standard, and C_{30} 17α (H) 21β (H) hopane was measured in it and all the samples. By assuming that 17α (H) 21β (H) hopane had not undergone biodegradation and that the oil landed in a form substantually similar to the 30 percent depleted PBC, we could readily calculate the amount of oil that must have originally landed on the particular sample of sediment being analyzed, and hence the depletion since then. We calculated these percent depletions for total oil and individual oil components for all the samples.

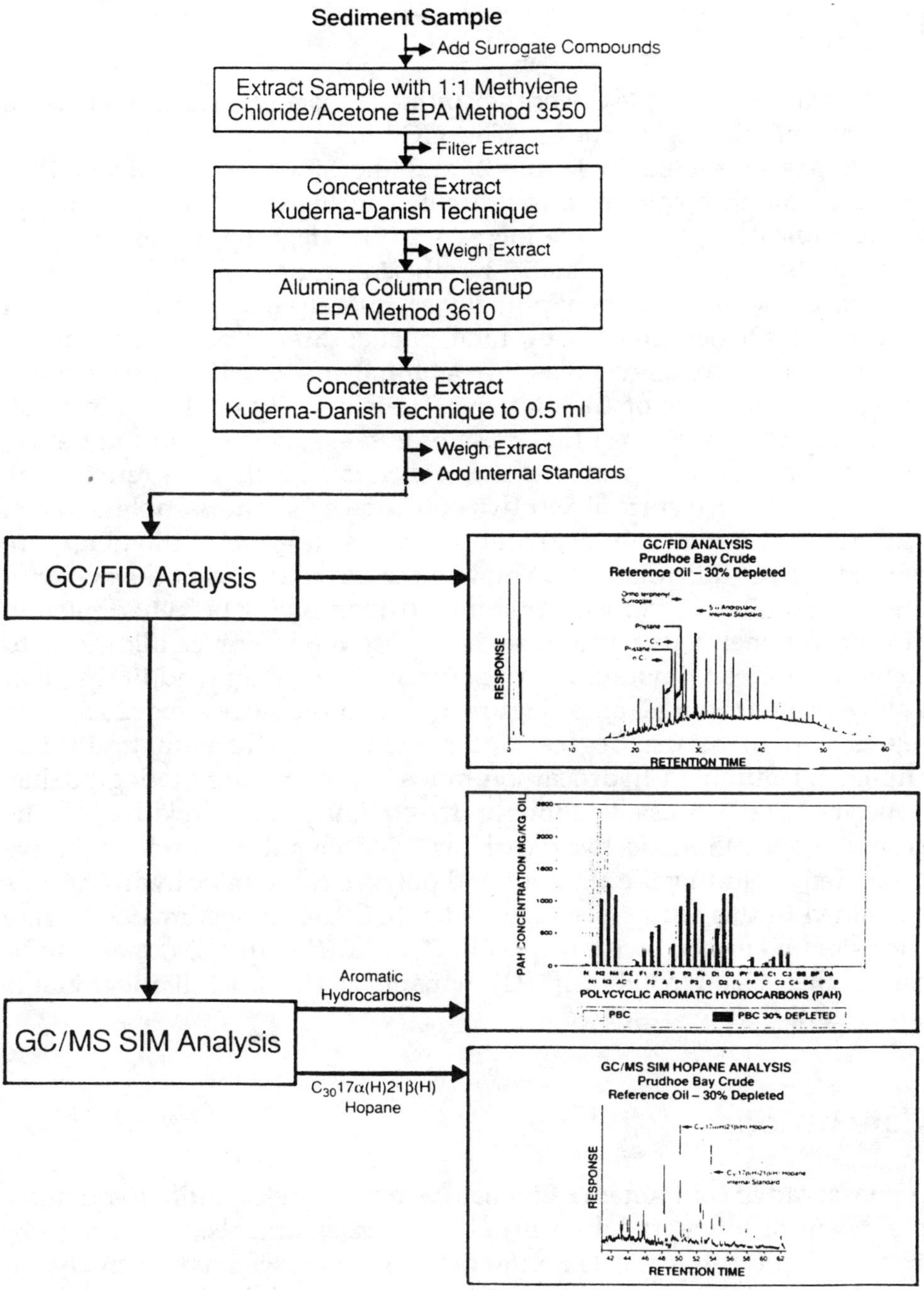

FIGURE 1. Analytical methods summary.

Discussion

Four sediment samples were examined that represent increasing amounts of oil degradation. The GC/FID chromatograms for the sample are presented in Figure 2, and the 30 percent depleted PBC reference oil chromatogram is presented in Figure 1. The chromatograms indicate a progressive increase in the degradation of the more labile *n*-alkanes. The aromatic weathering ratio (AWR; Figure 3[a]) consists of the sum of total naphthalenes, fluorenes, phenanthrenes, and dibenzothiophenes divided by total phenanthrenes and dibenzothiophenes, and approaches 1.0 as the naphthalenes and fluorenes are lost by evaporation and/or dissolution. The similarity of this ratio in all four samples may suggest that these processes have only a minor effect on the chemical composition of beached oil or that the effects are similar for the two surface and two subsurface (30-cm) samples. When traditional aliphatic biodegradation ratios (*e.g.*, C_{18}/phytane) are plotted, we see little distinction among the samples, except the one with the freshest oil. In contrast, our ability to quantify the percent depletion of oil components using hopane as an internal reference allows us to show that indeed phytane and pristane are being degraded in the oil at a slower rate than *n*-alkanes (Figure 3[b]) and provides a more accurate representation of the degree of biodegradation than do traditional aliphatic/isoprenoid hydrocarbon ratios. Hopanes are biodegradable; however, the process is thought to be fairly slow relative to the aromatic and aliphatic hydrocarbons. No chemical species that we measured, including the aliphatic and polycyclic aromatic hydrocarbons examined in this study (see analyte list in Sauer & Boehm 1991), were consistently enriched with respect to C_{30} 17α (H) 21β (H) hopane, indicating that C_{30} 17α (H) 21β (H) hopane is the least biodegradable compound that we resolved.

Conclusions

A conservative component of oil has been used successfully to quantify the extent of biodegradation in oiled beach sediments. By using C_{30} 17α (H) 21β (H) hopane as an internal reference, we have been able to calculate the amount and percent lost for both aliphatic and aromatic constituents, as well as for total oil. These measures have been found to be internally consistent, and they remain distinguishable among biodegraded samples even after traditional measures of degradation lose their utility. Application of this chemical indicator to surface and

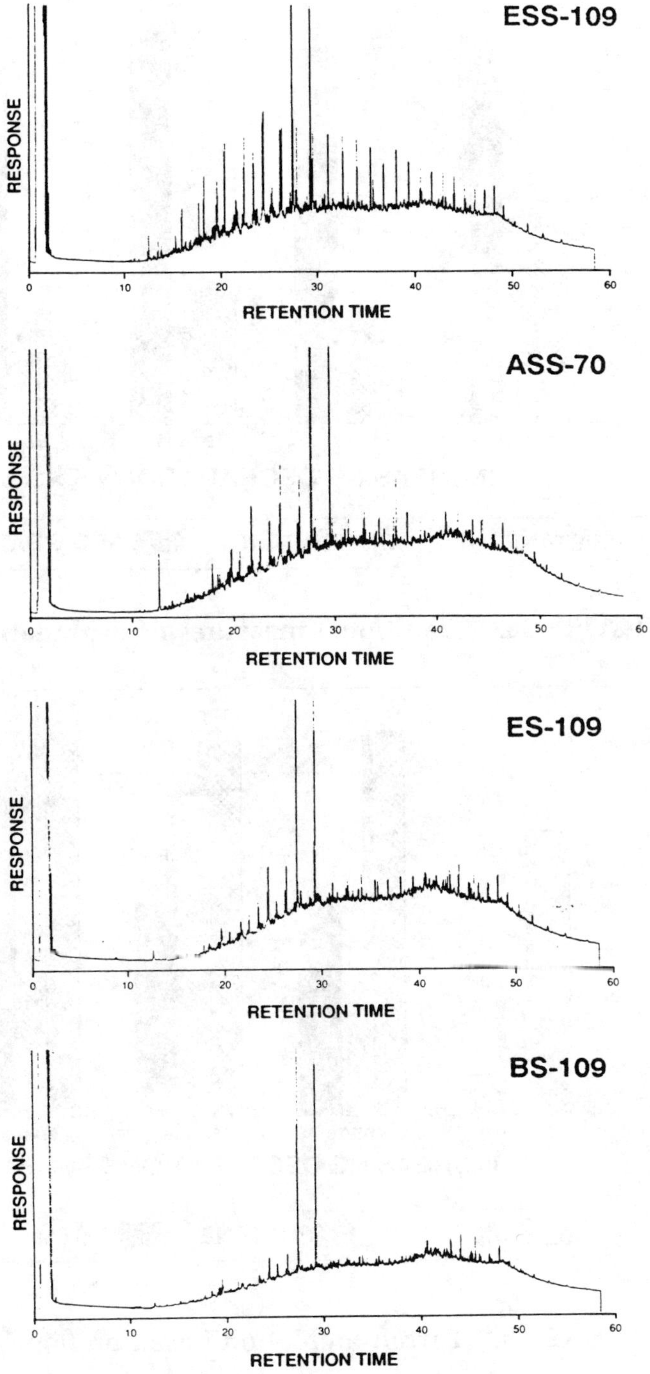

FIGURE 2. GC/FID chromatograms from selected sediment samples.

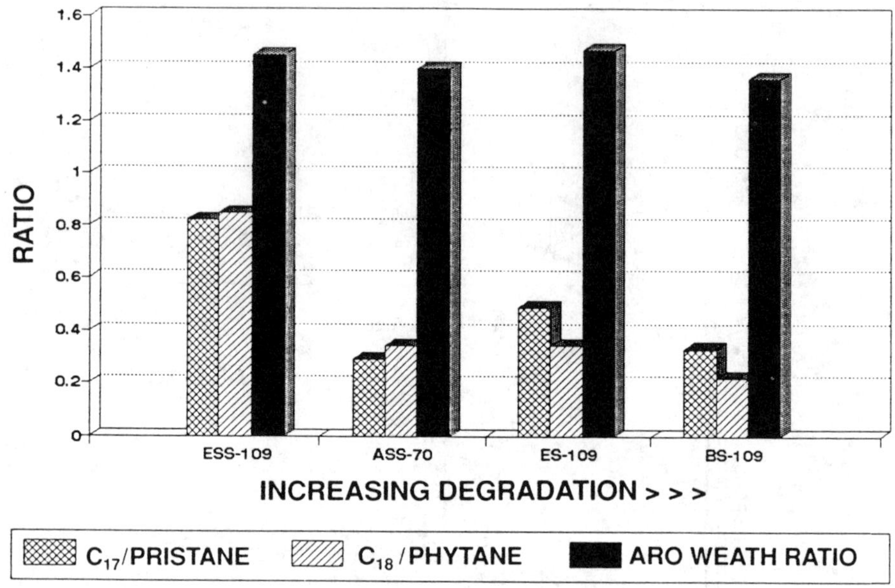

FIGURE 3a. Traditional measures of degradation.

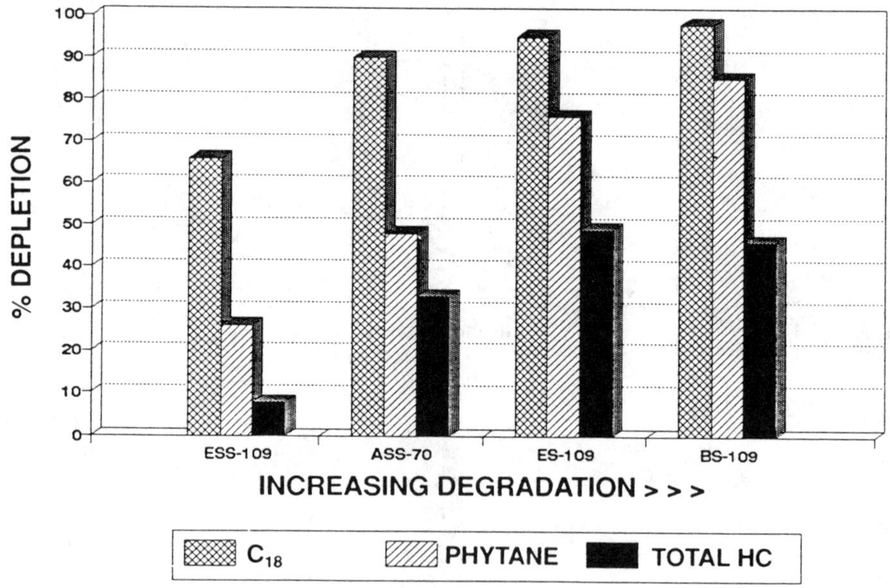

FIGURE 3b. Percent depletion based on hopane.

subsurface field sediment samples has provided valuable quantitative information that can be used to monitor the degradation of petroleum in highly variable environments and improve our understanding of the rates and mechanisms responsible for the natural cleansing of oil from beaches in Prince William Sound.

REFERENCES

Kennicutt, M. C., II. "The Effect of Biodegradation on Crude Oil Bulk and Molecular Composition." *Oil Chem. Pollut.* **1988**, *4*, 89–112.

Sauer, T.; Boehm, P. "The Use of Defensible Analytical Chemical Measurements for Oil Spill Natural Resource Damage Assessment." *Proceedings*, 1991 International Oil Spill Conference, American Petroleum Institute, Washington, DC, 1991; pp 363–369.

Seifert, W. K.; Moldowan, J. M. "The Effect of Biodegradation on Steranes and Triterpanes in Crude Oils." *Geochim. Cosmochim. Acta* **1978**, *43*, 111–126.

Tabak, H. H.; Haines, J. R.; Venosa, A. D.; Glasser, J. A.; Desai, S.; Nisamaneepong, W. "Enhancement of Biodegradation of Alaskan Weathered Crude Oil by Indigenous Microbiota with the Use of Fertilizers and Nutrients." *Proceedings*, 1991 International Oil Spill Conference, American Petroleum Institute, Washington, DC, 1991; pp 583–590.

List of Contributors

[*] Contributions of authors whose names appear in *italics* on this list can be found in the companion volume, In Situ *Bioreclamation*.

Index

* Page numbers in *italics* can be found in the companion volume, In Situ *Bioreclamation*.

B

C

F

G

H

O

P

T